IET ENERGY ENGINEERING SERIES 108

# Fault Diagnosis of Induction Motors

## Other volumes in this series:

Volume 1 **Power Circuit Breaker Theory and Design** C.H. Flurscheim (Editor)
Volume 4 **Industrial Microwave Heating** A.C. Metaxas and R.J. Meredith
Volume 7 **Insulators for High Voltages** J.S.T. Looms
Volume 8 **Variable Frequency AC Motor Drive Systems** D. Finney
Volume 10 **$SF_6$ Switchgear** H.M. Ryan and G.R. Jones
Volume 11 **Conduction and Induction Heating** E.J. Davies
Volume 13 **Statistical Techniques for High Voltage Engineering** W. Hauschild and W. Mosch
Volume 14 **Uninterruptible Power Supplies** J. Platts and J.D. St Aubyn (Editors)
Volume 15 **Digital Protection for Power Systems** A.T. Johns and S.K. Salman
Volume 16 **Electricity Economics and Planning** T.W. Berrie
Volume 18 **Vacuum Switchgear** A. Greenwood
Volume 19 **Electrical Safety: A guide to causes and prevention of hazards** J. Maxwell Adams
Volume 21 **Electricity Distribution Network Design, 2nd Edition** E. Lakervi and E.J. Holmes
Volume 22 **Artificial Intelligence Techniques in Power Systems** K. Warwick, A.O. Ekwue and R. Aggarwal (Editors)
Volume 24 **Power System Commissioning and Maintenance Practice** K. Harker
Volume 25 **Engineers' Handbook of Industrial Microwave Heating** R.J. Meredith
Volume 26 **Small Electric Motors** H. Moczala *et al.*
Volume 27 **AC–DC Power System Analysis** J. Arrillaga and B.C. Smith
Volume 29 **High Voltage Direct Current Transmission, 2nd Edition** J. Arrillaga
Volume 30 **Flexible AC Transmission Systems (FACTS)** Y.-H. Song (Editor)
Volume 31 **Embedded Generation** N. Jenkins *et al.*
Volume 32 **High Voltage Engineering and Testing, 2nd Edition** H.M. Ryan (Editor)
Volume 33 **Overvoltage Protection of Low-Voltage Systems, Revised Edition** P. Hasse
Volume 36 **Voltage Quality in Electrical Power Systems** J. Schlabbach *et al.*
Volume 37 **Electrical Steels for Rotating Machines** P. Beckley
Volume 38 **The Electric Car: Development and future of battery, hybrid and fuel-cell cars** M. Westbrook
Volume 39 **Power Systems Electromagnetic Transients Simulation** J. Arrillaga and N. Watson
Volume 40 **Advances in High Voltage Engineering** M. Haddad and D. Warne
Volume 41 **Electrical Operation of Electrostatic Precipitators** K. Parker
Volume 43 **Thermal Power Plant Simulation and Control** D. Flynn
Volume 44 **Economic Evaluation of Projects in the Electricity Supply Industry** H. Khatib
Volume 45 **Propulsion Systems for Hybrid Vehicles** J. Miller
Volume 46 **Distribution Switchgear** S. Stewart
Volume 47 **Protection of Electricity Distribution Networks, 2nd Edition** J. Gers and E. Holmes
Volume 48 **Wood Pole Overhead Lines** B. Wareing
Volume 49 **Electric Fuses, 3rd Edition** A. Wright and G. Newbery
Volume 50 **Wind Power Integration: Connection and system operational aspects** B. Fox *et al.*
Volume 51 **Short Circuit Currents** J. Schlabbach
Volume 52 **Nuclear Power** J. Wood
Volume 53 **Condition Assessment of High Voltage Insulation in Power System Equipment** R.E. James and Q. Su
Volume 55 **Local Energy: Distributed generation of heat and power** J. Wood
Volume 56 **Condition Monitoring of Rotating Electrical Machines** P. Tavner, L. Ran, J. Penman and H. Sedding
Volume 57 **The Control Techniques Drives and Controls Handbook, 2nd Edition** B. Drury
Volume 58 **Lightning Protection** V. Cooray (Editor)
Volume 59 **Ultracapacitor Applications** J.M. Miller
Volume 62 **Lightning Electromagnetics** V. Cooray
Volume 63 **Energy Storage for Power Systems, 2nd Edition** A. Ter-Gazarian
Volume 65 **Protection of Electricity Distribution Networks, 3rd Edition** J. Gers
Volume 66 **High Voltage Engineering Testing, 3rd Edition** H. Ryan (Editor)
Volume 67 **Multicore Simulation of Power System Transients** F.M. Uriate
Volume 68 **Distribution System Analysis and Automation** J. Gers
Volume 69 **The Lightening Flash, 2nd Edition** V. Cooray (Editor)
Volume 70 **Economic Evaluation of Projects in the Electricity Supply Industry, 3rd Edition** H. Khatib
Volume 72 **Control Circuits in Power Electronics: Practical issues in design and implementation** M. Castilla (Editor)
Volume 73 **Wide Area Monitoring, Protection and Control Systems: The enabler for Smarter Grids** A. Vaccaro and A. Zobaa (Editors)
Volume 74 **Power Electronic Converters and Systems: Frontiers and applications** A.M. Trzynadlowski (Editor)
Volume 75 **Power Distribution Automation** B. Das (Editor)
Volume 76 **Power System Stability: Modelling, analysis and control** B. Om and P. Malik
Volume 78 **Numerical Analysis of Power System Transients and Dynamics** A. Ametani (Editor)
Volume 79 **Vehicle-to-Grid: Linking electric vehicles to the smart grid** J. Lu and J. Hossain (Editors)
Volume 81 **Cyber-Physical-Social Systems and Constructs in Electric Power Engineering** Siddharth Suryanarayanan, Robin Roche and Timothy M. Hansen (Editors)
Volume 82 **Periodic Control of Power Electronic Converters** F. Blaabjerg, K. Zhou, D. Wang and Y. Yang
Volume 86 **Advances in Power System Modelling, Control and Stability Analysis** F. Milano (Editor)
Volume 87 **Cogeneration: Technologies, optimisation and implementation** C.A. Frangopoulos (Editor)
Volume 88 **Smarter Energy: From smart metering to the smart grid** H. Sun, N. Hatziargyriou, H.V. Poor, L. Carpanini and M.A. Sánchez Fornié (Editors)
Volume 89 **Hydrogen Production, Separation and Purification for Energy** A. Basile, F. Dalena, J. Tong, T.N. Veziroğlu (Editors)
Volume 90 **Clean Energy Microgrids** S. Obara and J. Morel (Editors)
Volume 92 **Power Quality in Future Electrical Power Systems** A.F. Zobaa and S.H.E.A. Aleem (Editors)
Volume 93 **Cogeneration and District Energy Systems: Modelling, analysis and optimization** M.A. Rosen and S. Koohi-Fayegh
Volume 94 **Introduction to the Smart Grid: Concepts, technologies and evolution** Salman K. Salman
Volume 95 **Communication, Control and Security Challenges for the Smart Grid** S.M. Muyeen and S. Rahman (Editors)
Volume 97 **Synchronized Phasor Measurements for Smart Grids** M.J.B. Reddy and D.K. Mohanta (Editors)
Volume 98 **Large Scale Grid Integration of Renewable Energy Sources** Antonio Moreno-Munoz (Editor)
Volume 100 **Modeling and Dynamic Behaviour of Hydropower Plants** N. Kishor and J. Fraile-Ardanuy (Editors)
Volume 101 **Methane and Hydrogen for Energy Storage** R. Carriveau and David S.-K. Ting
Volume 905 **Power System Protection, 4 volumes**

# Fault Diagnosis of Induction Motors

Jawad Faiz, Vahid Ghorbanian and Gojko Joksimović

The Institution of Engineering and Technology

Published by The Institution of Engineering and Technology, London, United Kingdom

The Institution of Engineering and Technology is registered as a Charity in England & Wales (no. 211014) and Scotland (no. SC038698).



First published 2017



The Institution of Engineering and Technology
Michael Faraday House
Six Hills Way, Stevenage
Herts, SG1 2AY, United Kingdom

www.theiet.org



**British Library Cataloguing in Publication Data**
A catalogue record for this product is available from the British Library

**ISBN 978-1-78561-328-9 (hardback)**
**ISBN 978-1-78561-329-6 (PDF)**

Typeset in India by MPS Limited

# Contents

# Preface

Induction machines are used more extensively than any other form of electrical machines. Low voltage machines are often used in domestic applications. However, larger induction machines with higher power and voltage ratings are widely launched in industrial environment such as pumps, mines, petrochemical and electrical power generators in order to provide motive force for major services. Nowadays, the responsibilities of induction machines are continuously growing, and the reliable operation of induction machines is strategically very important for providing essential services.

There are many faults which can prevent smooth operation of induction machines. Generally, any fault starts as a small declination from the healthy normal operation. Then, it may lead to a catastrophic failure if the fault is not detected and treated at its early stages. Therefore, a concept called "condition monitoring" of induction motors is introduced. Condition monitoring is a process of detecting and diagnosing fault-related changes of an essential motor quantity. Such action allows to schedule a proper maintenance, prevents failure, avoid significant damages, avoid economic lost, stop outages, optimizes the performance and enhances the quality control.

This book originated from the experience of the authors during researching fault diagnosis of induction machines at University of Tehran, Iran and University of Montenegro, Podgorica, Montenegro. Our research has led to many publications in induction machine fault detection, some have been referred in the book. This book is intended to help electrical engineers in industries and postgraduate/graduate students who would like to do research in this area or similar topics. Moreover, this book in an endeavor to collect recent attempts in the field of diagnosis of induction motors and sort them in a meaningful way so that potential readers can find a hierarchical discussion on different types of faults. This includes basics, developments, simulation approaches and experimental measurement of three important type of faults, namely the broken bars, eccentricity and interturn faults.

The book has been managed to cover fundamental and advanced aspects of the previously mentioned types of faults by providing step-by-step theoretical, analytical, numerical and experimental implementations. Basically, there are ten chapters as follows:

Chapter 1 provides a preliminary discussion on different types of faults and the corresponding influential factors in induction machines. Then, the necessity of the fault diagnosis procedures and their importance in maintaining a safe industrial environment are highlighted.

Chapter 2 explains the fundamentals of motor-drive operations in different supply modes including the line-start and inverter-fed applications in order to prepare readers for the next chapters.

Chapter 3 is probably the heart of this book since it provides a fundamental to advance knowledge of different types of faults, namely the broken bars, eccentricity and short-circuit faults. Moreover, a useful implementation technique of different faults and the corresponding measurement are discussed.

Chapter 4 addresses detailed analytical analysis of healthy induction motors. This chapter is considered as the basis for analyzing the time and space harmonics of induction motors, which will be used to extract fault indicators in the next parts of the book.

Chapter 5: The winding function theory is one of the most useful analytical modeling approaches of different types of faults in induction motors, which is explained in this chapter. The broken bars, eccentricity and short-circuit faults will be formulated using the winding function theory and some typical results of the corresponding simulation process are introduced.

Chapter 6: As the most reliable simulation process of electrical machines and their defects, the finite element method is targeted in Chapter 6. The mathematical implementation of the approach is provided first. Then, the way to implement different types of faults is discussed.

Chapter 7: Time, frequency and time-frequency signal processing techniques are explained and the corresponding MATLAB codes are provided in Chapter 7 so that one can easily use the programs to conduct further signal processing applied to different signals of faulty motor.

Chapter 8: The broken bars fault is the main focus of Chapter 8. In this chapter, various aspects of the corresponding diagnosis process are provided in order to get a better realization of how the broken bars fault is detected and diagnosed in different operating conditions. Different drive systems, various load levels, various speed levels and different indicators are addressed.

Chapter 9: The same process as that of Chapter 8 is followed in Chapter 9 in order to address the issues related to the eccentricity fault. Two main types of faults including the static and dynamic eccentricity faults are investigated by means of time and frequency analysis.

Chapter 10: The short-circuit fault will be studied in Chapter 10. Different time and frequency domain techniques will be used to address the challenges related to diagnosis of the short-circuit fault. Two types of machines, namely the wound and squirrel-cage induction motors, are investigated.

So many postgraduate/graduate students have helped with the preparation of the present book that we cannot properly acknowledge them all and this book would never have been possible without their excellent research over the last 15 years. We would specially like to thank University of Tehran for its previous financial supports of the projects whose outcome has been gathered here as a book.

We would like to thank Mrs. Rana Hassanpour Bradaran (M.Sc. in English language translation) for her efforts to edit and improve the language of the book, Dr. Iman Tabatabaei Ardekani, previous M.Sc. student in the University of Tehran and now senior lecturer in Unitec Institute of Technology, Auckland, New Zealand, Dr. Bashir Mahdi Ebrahimi the previous Ph.D. student in the University of Tehran and now the director in Pars Generator Company in Iran, Dr. Mansour Ojaghi the previous Ph.D. student in the University of Tehran and now Associate Professor in University of Zanjan, Iran, the previous M.Sc. students Amir Masoud Takbash, Mr. Mahmoud Ghasemi-Bijan and Mr. Mehran Keravand in the University of Tehran.

# About the authors

**Jawad Faiz** is Professor in the School of Electrical and Computer Engineering at the University of Tehran, Iran. His research interests are the design, modeling and fault diagnosis of electrical machines and transformers. He is a member of the Iran Academy of Sciences and has published more than 460 papers and a book in this field.

**Vahid Ghorbanian** received his PhD degree in Electrical Engineering from McGill University, Montreal, Canada in 2017. His research interests include analysis, design, optimization, condition monitoring and fault diagnosis of conventional and new types of line-start and inverter-fed electrical machines. He is a recipient of the McGill Engineering Doctoral Awards (MEDA) in 2013.

**Gojko Joksimović** is Professor of Electrical Power Engineering at the University of Montenegro, Montenegro. His research interests are in electrical machines, their design, analysis and modeling. He has a long research record on the modeling and fault detection of induction machines. He is a recipient of the prestigious Alexander von Humboldt research fellowship. He is a senior member of IEEE.

# *Chapter 1*

# Condition monitoring and fault diagnosis of induction motors

## 1.1 Introduction

Electrical machines are one of the most growingly manufactured devices which are widely used not only in industries but also in domestic applications. Due to their higher efficiency level, they are common sources of electromechanical energy conversion. Various applications including control, automotive and power generation applications are served by means of these devices, and no one can deny their ability in being precisely controlled in terms of position, speed and torque which are some of the essential quantities in electromechanical devices.

Among all kinds of electrical machines, induction motor has been more widely used since the beginning stage of industrial development. However, permanent magnet-assisted motors are probably going to be a good substitute for induction motors due to their higher power densities and efficiencies. Nevertheless, induction motors are still one of the most reliable machines used in both line-start and inverter-fed applications. A very good example of a latest application of this machine is doubly fed induction generator utilized in wind turbines in conjunction with inverters. High-speed induction motor for low- and medium-voltage powertrains is another example. Due to the rapidly increasing demand of energy consumption in the world, induction motors are required to operate in different environments which might apply electrical, magnetic, thermal and mechanical transients and stresses to the motor [1–11]. Therefore, the motor likely encounters undesirable operating conditions which in turn makes maintenance procedure an inseparable part of any industrial development. Equally important, maintenance could not be precisely done unless a proper monitoring and diagnosis process is devised. Induction motors are of great robustness against the mentioned stressful factors caused by any unpredictable reason. However, situation changes and sever internal or external tensions can probably affect not only the motor structure, material and smooth healthy operation [1,12–21]. Stress-causing factors interrupt the smooth and reliable operation of motors, and induction motors, although very reliable, are not an exception. In this situation, the motor becomes out of order if unhealthy conditions are not diagnosed and treated well. This undesirable condition is usually referred to as "fault," and the procedure of diagnosis is called "fault diagnosis procedure" which is the outcome of a condition monitoring system [22–35]. Fault-producing

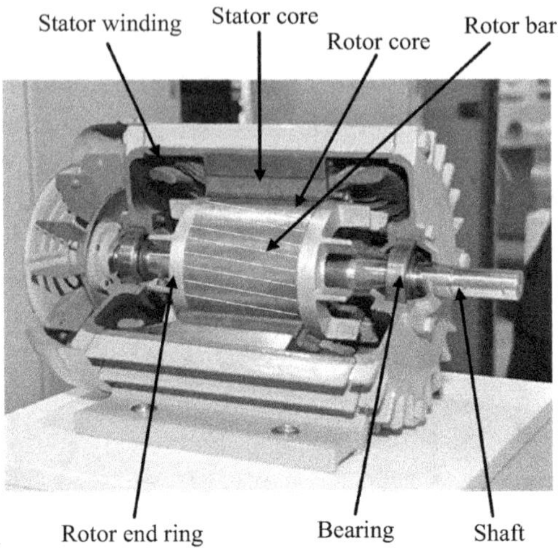


*Figure 1.1 Induction motor crosscut along its length*

factors are divided into two general categories of external or internal, depending on the motor component they impact on. A general overview of different motor components is shown in Figure 1.1 in which the main components are highlighted as follows:

- stator core,
- stator winding,
- rotor core,
- rotor bars [36],
- rotor end ring,
- bearings [37] and
- shaft.

First, let us generally describe different components and take a glance at corresponding possible stressful conditions. Figure 1.1 shows a squirrel cage induction motor in which rotor consists of the bars which are connected to each other through end rings at two ends. The end rings have the same material as the bars which are usually made of aluminum. The stator windings are made of copper and carry the stator current which flows into the motor and the network connected to the motor terminals. Depending on the number of motor phases, there should be the same number of network phases to supply the motor. On the basis of the governing rules of induction motors, the stator current produces electromotive force across the rotor bars; hence, the rotor bar currents are produced as the result of this phenomenon which is called "induction." This is the reason why these types of machines are called induction motors. The interaction between the stator and rotor magnetic fields produces the electromagnetic torque which is the main reason of the mechanical rotation of the rotor. The stator winding is located inside the stator slots which are surrounded by the stator core made of silicon–steel material. Normally, the rotor and stator cores are of the same material unless a specific application is targeted. The motor shaft is concentric with the rotor core and fixed at the two ends of the motor housing by means of the bearings. There is another type of induction motor in which the rotor consists of windings instead of bars, and this topology is called "wound-rotor" induction motor. Any winding inside the motor consists of insulation layers which separate the conductors electrically to prevent short-circuit faults.

Any magnetic, thermal, electrical and mechanical asymmetry in the stator, the rotor, the shaft or the bearing disrupts the smooth operation of the motor. The asymmetry can be inherent or fault dependent. In the former case, which is normally of a mechanical nature, the motor structure is made inherently asymmetrical during manufacturing process. This is somewhat inevitable as the cause is usually unknown or irreducible. Sometimes, the asymmetry in motor quantities is caused by the impure sinusoidal nature of induction motors. For example, it is ideally preferred to have a totally sinusoidal winding or bars distribution; however, it is impossible in practice. Therefore, some sort of undesirable harmonic components is produced. On the other hand, these harmonic components are some of the best tools for monitoring the motor behavior during faulty states; hence, they are not thoroughly undesirable or useless. Another type of inherent asymmetry of induction motors is the presence of eccentric rotor or shaft which is not concentric with the stator center. This is caused by improper manufacturing process due to the fact that assembling apparatus is not ideally accurate.

Unlike the inherent asymmetries or faulty conditions, there are some cases where an additional factor, besides the improper manufacturing process, causes some components to fail to operate correctly. One of the very well-known situations is the insulation failure of the stator or rotor winding [32]. As a result, two or more turns of the stator or rotor windings are short circuited. This is usually caused by aging or wearing of insulation due to thermal or electrical stresses applied to the insulation. The presence of the short-circuit fault is a reason for a highly unbalanced current distribution in the stator and/or the rotor, causing an asymmetrical operation of motor quantities including torque, speed, magnetic flux, current and voltage if an inverter-fed application is studied. This kind of failure necessitates a regular condition monitoring and quality investigation of motor components. The short-circuit fault is a very good example of electrical faults. As amplitude of motor current might dramatically increase depending on the location of the fault, over-saturated core and consequent hot spots are generated in motor core material as well. The harmonic distortion is another obvious result of this type of fault. As an initial tool used for the diagnostic purposes, zero sequence current might be used to detect and determine the short-circuit fault and its level. However, the short-circuit fault is still one of the hot topics in the field of condition monitoring of electrical machines. In the wound rotor motors, the rotor can also be subjected to the short-circuit fault as there are turns and coils separated from each other by means of insulations. Therefore, both the rotor and the stator should be monitored for addressing possible short-circuit faults in wound-rotor motors.

The other type of unhealthy conditions caused by mechanical stresses is the deficiency of the rotor bars, end rings and bearings [38–40]. Mechanical stresses are mostly observed in the form of disconnected joints, unaligned centers and corrosion. The broken bars, the eccentric rotor and the bearing wear are the best examples of the mentioned mechanical defects. As the broken bar fault directly affects the rotor bars current, it is sometimes considered as an electrical type of fault. However, the eccentricity fault is always of a mechanical nature. Any undesirable friction force or aging is the major reason of improper operation of bearings leading to an eccentric rotor if suitable maintenance is not performed. As a result of bearing deficiency, the eccentricity fault might occur. In the case of eccentricity fault, the air gap length becomes asymmetrical around the rotor circumference. Hence, the rotor is subjected to an unbalanced magnetic pull which in turn affects the motor current, torque, speed or any other quantity. If the air gap length varies in time at different positions around the rotor circumference, a dynamic type of the eccentricity fault exists. In fact, the rotor center rotates about the stator center. On the other hand, if the rotor center is fixed at a point close to the stator center and does not change position, it is called the static eccentricity. Both the static and dynamic types produce time- or frequency-domain variations in mechanical, electric and magnetic variables of the motor. One of the very significant variations is that of the motor torque and speed which lead to a highly unsmooth operation. This is not a desirable situation in industrial environment. Besides, it causes the other motor components such as bearings or housing to have defects if higher fault levels exist or lower levels of faults are not detected and treated properly. Therefore, many efforts are made to address the issues associated with monitoring and diagnosis of induction motors and their fault levels and types. Under some major fault levels, typically a static eccentricity of 50% or above, the rotor might rub the stator in high-speed levels due to centrifugal force, and the whole motor operation fails. This can be a catastrophe for industries. Thus, diagnosing faults in their initial levels should be an inseparable part of the diagnosis procedure.

Sometimes, due to inappropriate manufacturing and casting process, joints between bars and end rings are not tight enough to stay connected. In other words, the joint between rotor bars and end rings is broken and no current passes through broken joints. This phenomenon is called broken bars fault and is generally considered as an electrical fault while it can also be assumed a mechanical fault as well as it is initially caused by a mechanical deficiency. Although the broken bar fault is considered as the less probable fault compared to the eccentricity and short-circuit faults, it is of a great interest not only in industries but also in academia. Therefore, many researches have devoted themselves to addressing the pros and cons of this type of fault along with those related to other types of faults. The reason for holding a condition monitoring process is actually the safety and maintenance-related costs which might be unbearable if a fault takes place and is not diagnosed afterward. The motor quantities are considerably affected by the broken bars fault in both the time and frequency domains. What happens is the production or magnification of undesirable backward magnetic fields caused by asymmetrical current distribution in the rotor. It means that additional torque-producing harmonic components are generated.

Many efforts have been carried out since 30 or 40 years ago, to carefully find the main reasons and also practically possible diagnostic procedures to detect the fault. However, none of them could come up with a comprehensive approach which can be referred to in terms of diagnosing the fault in every possible situation. The reason is that the number of influential parameters affecting the motor behavior under fault conditions is very large, and almost none of the researchers could have gathered and organized existing information in a significant way by means of which one can select a proper procedure to diagnose the fault. The following are some of the main influential factors in broken bars fault diagnosis procedure:

- fault level [18],
- fault location [19],
- motor load level [10],
- motor speed which can be a function of motor load or be separately changed by means of drives [40] and
- supply mode including
  - line-start mode
  - open-loop speed control mode
  - closed-loop speed control mode.

In addition to the mentioned factors, internal adjustments of any control-loop such as the cutoff frequency of regulators, along with the bandwidth of controller of any motor–drive system, are the other significant reasons for writing the present book in which all the mentioned parameters will be carefully discussed. Moreover, the drives capability in separately controlling the motor speed and torque levels will be investigated. The increasing number of influential factors necessitates the presence of a generalized approach which enables the user to deal with different types of faults. The idea is to discriminate the concepts of detecting, determining, locating and diagnosing through analyzing various supply modes and operating conditions. The "detection" is to find an incipient fault during the motor operation. The "determination" is to find the fault severity, and "fault-location" deals with finding the fault location. The combination of detection, determination and fault-location detection processes is called "diagnosis procedure" which is a very important concept and will be defined in the next chapters.

To prepare the readers' minds and make them interested and aware of various aspects of the diagnosis procedure, let us discuss in more detail. Induction motors are electromechanical devices which generally consist of the following inputs and outputs:

- input electric variables (voltage and current) and
- output mechanical variables (torque and speed).

The aforementioned variables are measurable outside the motor by means of electrical and mechanical sensors. Depending on the sensor type and how it is installed to measure motor performance, two concepts of invasive and noninvasive sampling techniques are introduced. Generally, mechanical signals measurement requires invasive types of sensors while electrical signals measurement is usually handled in a noninvasive way. Sometimes, the motor torque is measured by means of dynamometers whose inertia is relatively larger than that of the motor shaft. This is an undesirable situation in which some of potential fault-related harmonic components might be filtered out. Motor torque might not be a good medium for the diagnosis purposes at some points. Thus, proper signal selection depending on the type of available sensors and even the supply mode should be considered. This is another major focus of this book. It is also possible to measure magnetic quantities such as the motor flux variations. The flux, if measured properly, would be the most promising signal to analyze any kind of fault as both the stator and the rotor deficiencies are reflected directly by this quantity. The flux is measured in both invasive and noninvasive ways by means of flux estimation techniques or search coils. Furthermore, measurements, either offline or online, depend on the sensor type, measured signal and signal processing technique. Of course, online approaches are mostly preferred while there are several offline methods applicable to inverter-fed motors.

As another significant factor, the motor supply mode might widely change the motor behavior, depending on the operating condition. Unlike the line-start operation in which the motor speed is a function of the motor torque, the torque and the speed are controlled independently in inverter-fed motors. On the other hand, the voltage and the current are not smooth sinusoidal signals anymore. Instead, a PWM-type voltage, along with a ripple-included current, is applied to motors in inverter-fed applications. This means that any diagnosis procedure which is based on motor torque, speed, voltage and current is indeed affected by inverters and also control strategies. Sometimes, proper ideas might be taken from conventional diagnosis approaches applied to the line-start mode, but they are not generally applicable to any kind of inverter-fed system.

On the other hand, induction motors generally operate in two modes including the transient and steady state. Analyzing different faults in these two modes of operation requires a very deep knowledge of various processing tools including time, frequency and time-frequency domains. Besides, extracting a proper fault indicator is achievable only if an inclusive behavioral study is conducted. The chapters of the book are organized in a way to provide such a knowledge based on which the final goal which is helping readers to understand and get familiar with fault diagnosis tools is achieved. It is worth noting that no one can claim a single diagnosis approach is applicable to all fault types and conditions; and the main goal of this book is to address practical diagnosis techniques, conditions and procedures. To achieve this goal, mathematical, simulation-based and experimental analysis of faulty induction motors in different operating conditions is provided by means of the following tools:

- analytical formulation of healthy and faulty induction motors,
- winding function-based modeling and analysis of healthy and faulty motors,
- finite element-based modeling and analysis of healthy and faulty motors and
- experimental implementation and analysis of healthy and faulty motors.

As this book is organized to discuss fault-related materials, it is expected that potential readers would have the primary knowledge of line-start and inverter-fed induction motors, although the book includes a useful chapter addressing the line-start and inverter-fed applications and their effects on induction motors behavior.

To deal with this, a general routine is proposed in order to gather required information and knowledge. As the starting point, a proper analysis tool should be provided. This is handled by mathematical, experimental and simulation-based materials included in the Chapters 2, 3 and 5. In practice, it is hard or sometimes impossible to measure a signal inside the motor. For example, the air gap flux density requires a highly invasive technique to be measured. In this case, the quantities will be obtained by means of the finite element or winding function package. Electrical, magnetic and mechanical quantities are investigated in different operating conditions including various faults, load and speed levels. During the experimental implementation of faults, conventional and new sensor types are discussed, and the way they are used during the process is explained. In the case of simulation processes, interesting and practical implementing techniques and details are addressed and provided for readers.

Having measured or sampled signals, signal processing techniques including time, frequency and time-frequency domains are explained in Chapter 7. The fast Fourier, wavelet and the Hilbert–Huang transforms are discussed and the corresponding codes are provided.

A comprehensive database of induction motors operating in the line-start or inverter-fed modes are gathered in the Chapters 8–10. The chapters include different aspects of a diagnosis procedure as follows:

- online or offline methods,
- line-start motors,
- inverter-fed motors including the open-loop and closed-loop schemes,
- invasive or noninvasive methods,
- different fault types,
- experimental implementation of different faults,
- simulation process of faulty motors,
- time, frequency and time-frequency analysis of faulty motors,
- different fault features,
- different fault severities,
- the impact of different load levels on fault features,
- the impact of speed variation on fault features,
- the impact of drive parameters including proportional and integral gains as well as bandwidth of controller on fault features,
- fault diagnosis in transient and steady-state modes and
- loss characterization of faulty motors.

## References

[1] Benbouzid M.E.H. A review of induction motors signature analysis as a medium for faults detection, *IEEE Transactions on Industrial Electronics*, 2000; **IE-47**: 984–993.

[2] Bellini A., Filippetti F., Tassoni C. and Capolino G.A. Advances in diagnostic techniques for induction machines, *IEEE Transactions on Industrial Electronics*, 2008; **TIA-55**: 4109–4126.

[3] Thomson W.T. and Fenger M. Current signature analysis to detect induction motor faults, *IEEE Industry Applications Magazine*, 2001; **IAM-7**: 26–34.

[4] Levi E., Bojoi R., Profumo F., Toliyat H.A. and Williamson S. Multiphase induction motor drives – a technology status review, *IET Electric Power Applications*, 2007; **EPA-1**: 489–516.

[5] Schoen R.R., Habetler T.G., Kamran F. and Bartfield R.G. Motor bearing damage detection using stator current monitoring, *IEEE Transactions on Industry Applications*, 1995; **TIA-31**: 1274–1279.

[6] Kliman G.B., Koegl R.A., Stein J., Endicott R.D. and Madden M.W. Noninvasive detection of broken rotor bars in operating induction motors, *IEEE Transactions on Energy Conversion*, 1988; **EC-3**: 873–879.

[7] Erdman J.M., Kerkman R.J., Schlegel D.W. and Skibinski G.L. Effect of PWM inverters on AC motor bearing currents and shaft voltages, *IEEE Transactions on Industry Applications*, 1996; **TIA-32**: 250–259.

[8] Bellini A., Filippetti F., Franceschini G., Tassoni C. and Kliman G.B. Quantitative evaluation of induction motor broken bars by means of electrical signature analysis, *IEEE Transactions on Industry Applications*, 2001; **TIA-37**: 1248–1255.

[9] Dorrell D.G., Thomson W.T. and Roach S. Analysis of airgap flux, current, and vibration signals as a function of the combination of static and dynamic airgap eccentricity in 3-phase induction motors, *IEEE Transactions on Industry Applications*, 1997; **TIA-33**: 24–34.

[10] Filippetti F., Franceschini G., Tassoni C. and Vas P. Recent developments of induction motor drives fault diagnosis using AI techniques, *IEEE Transactions on Industrial Electronics*, 2000; **IE-47**: 994–1004.

[11] Kastha D. and Bose B.K. Investigation of fault modes of voltage-fed inverter system for induction motor drive, *IEEE Transactions on Industry Applications*, 1994; **IA-30**: 1028–1038.

[12] Toliyat H.A. and Lipo T.A. Transient analysis of cage induction machines under stator, rotor bar and end ring faults, *IEEE Transactions on Energy Conversion*, 1995; **EC-10**: 241–247.

[13] Jung J.H., Lee J.J. and Kwon B.H. Online diagnosis of induction motors using MCSA, *IEEE Transactions on Industrial Electronics*, 2006; **IE-53**: 1842–1852.

[14] Cameron J.R., Thomson W.T. and Dow A.B. Vibration and current monitoring for detecting airgap eccentricity in large induction motors, *IEE Proceedings, Part B, Electric Power Applications*, 1986; **EPA-133**: 155–163.

[15] Benbouzid M.E.H. and Kliman G.B. What stator current processing-based technique to use for induction motor rotor faults diagnosis?, *IEEE Transactions on Energy Conversion*, 2003; **EC-18**: 238–244.

[16] Nandi S., Ahmed S. and Toliyat H.A. Detection of rotor slot and other eccentricity related harmonics in a three phase induction motor with

different rotor cages, *IEEE Transactions on Energy Conversion*, 2011; **EC-16**: 253–260.

[17] CusidÓCusido J., Romeral L., Ortega J.A., Rosero J.A. and GarcíaGarcia Espinosa A. Fault detection in induction machines using power spectral density in wavelet decomposition, *IEEE Transactions on Industrial Electronics*, 2008; **IE-55**: 633–643.

[18] Joksimovic G.M. and Penman J. The detection of inter-turn short circuits in the stator windings of operating motors, *IEEE Transactions on Industrial Electronics*, 2000; **IE-47**: 1078–1084.

[19] Benbouzid M.E.H., Vieira M. and Theys C. Induction motors' faults detection and localization using stator current advanced signal processing techniques, *IEEE Transactions on Power Electronics*, 1999; **PE-14**: 14–22.

[20] Siddique A., Yadava G.S. and Singh B. A review of stator fault monitoring techniques of induction motors, *IEEE Transactions on Energy Conversion*, 2005; **EC-20**: 106–114.

[21] Cruz S.M.A. and Cardoso A.J.M. Stator winding fault diagnosis in three-phase synchronous and asynchronous motors, by the extended Park's vector approach, *IEEE Transactions on Industry Applications*, 2001; **IA-37**: 1227–1233.

[22] Takbash A.M., Faiz J. and Ebrahimi B.M. Losses characterization in voltage-fed PWM inverter induction motor drives under rotor broken bars fault, *IEEE Transactions on Magnetics*, 2013; **MAG-49**: 1516–1525.

[23] Ojaghi M. and Faiz J. An experimental/simulation investigation to mixed eccentricity fault diagnosis of induction motors under DTC, 2014 IEEE International Conference on Industrial Technology (ICIT), Busan, 2014, 143–148.

[24] Ebrahimi B.M., Takbash A.M. and Faiz J. Losses calculation in line-start and inverter-fed induction motors under broken bar fault, *IEEE Transactions on Instrumentation and Measurement*, 2013; **IM-62**: 140–152.

[25] Faiz J., Ebrahimi B.M., Akin B. and Toliyat H.A. Comprehensive eccentricity fault diagnosis in induction motors using finite element method, *IEEE Transactions on Magnetics*, 2009; **MAG-45**: 1764–1767.

[26] Faiz J. and Ojaghi M. Instantaneous-power harmonics as indexes for mixed eccentricity fault in mains-fed and open/closed-loop drive-connected squirrel-cage induction motors, *IEEE Transactions on Industrial Electronics*, 2009; **IE-56**: 4718–4726.

[27] Faiz J., Ebrahimi B.M. and Toliyat H.A. Effect of magnetic saturation on static and mixed eccentricity fault diagnosis in induction motor, *IEEE Transactions on Magnetics*, 2009; **MAG-45**: 3137–3144.

[28] Faiz J., Ebrahimi B.M., Asaie B., Rajabioun R. and Toliyat H.A. A criterion function for broken bar fault diagnosis in induction motor under load variation using wavelet transform, *International Conference on Electrical Machines and Systems (ICEMS)*, 2007; 1249–1254.

[29] Abu-Elhaija W.S., Ghorbanian V., Faiz J. and Ebrahimi B.M. Impact of closed-loop control on behavior of inverter-fed induction motors with rotor

broken-bars fault, 2012 *IEEE International Conference on Power Electronics, Drives and Energy Systems (PEDES)*, 2017; 1-4.

[30] Faiz J., Ebrahimi B.M., Toliyat H.A. and Akin B. Diagnosis of a mixed eccentricity fault in a squirrel-cage three-phase induction motor using time stepping finite element technique, *IEEE International Electric Machines & Drives Conference*, 2007; 1446–1450.

[31] Faiz J., Ghorbanian V. and Ebrahimi B.M. A survey on condition monitoring and fault diagnosis in line-start and inverter-fed broken bar induction motors, IEEE *International Conference on Power Electronics, Drives and Energy Systems (PEDES)*, 2012; 1–5.

[32] Faiz J., Keravand M. and Ghasemi-Bijan, M. Impact of rotor inter-turn short circuit fault upon performance of a wound rotor induction motor, *2015 Intl Aegean Conference on Electrical Machines, Power Electronics (ACEMP), and Intl Conference on Optimization of Electrical & Electronic Equipment (OPTIM) and 2015 Intl Symposium on Advanced Electromechanical Motion Systems (ELECTROMOTION)*, Side, 681–686.

[33] Faiz J., Ebrahimi B.M., Akin B. and Toliyat H.A. Finite-element transient analysis of induction motors under mixed eccentricity fault, *IEEE Transactions on Magnetics*, 2088; **MAG-44**: 66–74.

[34] Faiz J. and Ojaghi M. Stator inductance fluctuation of induction motor as an eccentricity fault index, *IEEE Transactions on Magnetics*, 2011; **MAG-47**: 1775–1785.

[35] Sabouri M., Ojaghi M., Faiz J. and Cardoso A.J.M. New technique for identifying bearing faults in three-phase induction motors, *IECON 2016 – 42nd Annual Conference of the IEEE Industrial Electronics Society*, Florence, Italy, **2016**; 1530–1535.

[36] Abu-Elhaija W.S. and Faiz J. Review of series connected wound rotor three phase induction motors, *International Symposium on Power Electronics, Electrical Drives, Automation and Motion*, 2014; 613–618.

[37] Faiz J., Ebrahimi B.M., Akin B. and Toliyat H.A. Dynamic analysis of mixed eccentricity signatures at various operating points and scrutiny of related indices for induction motors, *IET Electric Power Applications*, 2010; **EPA-4**: 1–16.

[38] Faiz J. and Tabatabaei I. Extension of winding function theory for non-uniform air gap in electric machinery, *IEEE Transactions on Magnetics*, 2002; **MAG-38**: 3654–3657.

[39] Faiz J., Ghorbanian V. and Ebrahimi B.M. EMD-based analysis of industrial induction motors with broken rotor bars for identification of operating point at different supply modes, *IEEE Transactions on Industrial Informatics*, 2014; **TII-10**: 957–966.

[40] Faiz J., Ghorbanian V. and Ebrahimi B.M. A new criterion for rotor broken bar fault diagnosis in line-start and inverter-fed induction motors using Hilbert-Huang transform, *IEEE International Conference on Power Electronics, Drives and Energy Systems (PEDES)*, 2012; 1–6.

# Chapter 2
# Theory of line-start and inverter-fed induction motors

## 2.1 Introduction

Induction motors, which hire fundamental rules of electromagnetism in order to produce rotational or translational mechanical movement when supplied by an input electric power, are energy conversion apparatus [1–6]. These electromechanical devices are very well known for their smoothly run capability if a balanced input power supply, along with a symmetric motor structure, is used simultaneously. This is the reason why a balanced and symmetric three-phase motor supply system is usually preferred in industry due to its gentle and low noise. However, a higher number of phases, despite its rather higher topological complexity, are also possible. Notably, the underlying idea of the operation of this kind of motor is basically not related to different possible structures used in various applications. It quite depends on the nature of the induction phenomenon which couples the stator and rotor magnetic fields. Regardless of the stator geometry, there are generally two types of induction motors as the following:

- wound rotor [3]
- squirrel cage rotor [6].

The first type usually presents a less reliable structure in terms of mechanical stresses as well as the electrical deterioration of the brushes connecting rotor windings to the output terminals. The wound-rotor structure is relatively fragile compared to the second type, i.e. the squirrel cage rotor, while revealing a more complex topology that requires a relatively greater challenge for placing windings than casting rotor bars into the slots. On the other hand, the cage rotor presents an easy-to-build topology which merely requires a casting process to build complete bar and end-ring connections playing exactly the same role as brushes in a wound-rotor induction motor. Due to the robustness of a cage rotor, it is mostly utilized as a medium not only in industrial but also in domestic environment since the corresponding maintenance process is somehow less demanding regardless of the corresponding types of faults introduced by the presence of the cage. In addition, absence of brushes is a very important factor in choosing a cage rotor against a wound rotor.

The stator can be the same for both types of the rotors. It is generally a symmetric winding topology that encloses the supply current applied by the supply voltage with the goal of generating a rotating or translational magnetic field that induces currents into rotor windings or bars. On the other hand, thanks to the advances in power electronics devices, supply voltage can be precisely controlled in terms of both amplitude and frequency. As a result, many distinct operations of induction motors are observed and different electrical, magnetic and mechanical characteristics and motor profiles are obtained by means of controlling the frequency and amplitude of the supply. Moreover, any motor signature is subjected to change due to a fault occurrence. Thus, understanding the signatures of a healthy motor should be the very beginning step of any fault diagnosis process since faults introduce their own signatures which might be totally different. Three approaches including mathematical developments, simulation-based analysis and experimental measurements are used to comprehensively address induction motors behavior in terms of electrical, magnetic and mechanical quantities in the healthy case, but before moving forward, a structural analysis of a real induction motor is provided in the next section of this chapter to help beginners of the field to understand and find out how different parts are assembled. Then, we will step toward developing fundamental mathematical formulations. It should be noted that this chapter only focuses on the healthy motor operations, and further discussions in terms of different kinds of fault and the corresponding consequences are discussed in the next chapters. Therefore, this chapter focuses on the analysis of the operating principals of healthy induction motors and addressing electrical and mechanical aspects of the motor by means of a preliminary linear analytical model. Then, the model is improved with the goal of incorporating a more comprehensive magnetic aspect by considering the winding distributions. Finally, a detailed harmonic analysis of the healthy induction motor which is a turning point of the study from a conventional one to an advanced investigation of the motor behavior is proposed.

Generally speaking, the healthy motor behavior in the "line-start" supply mode is targeted. Therefore, readers could potentially follow up future details in terms of more complex operations such as inverter-fed applications.

Otherwise, the advanced topics might be of a little bit vague if one does not have the proper background knowledge. All the motor-drive principles required for a better realization of motor behavior are investigated as well. As the starting point, the physical structure of the motor is first illustrated and analyzed in the following section.

## 2.2 Induction motor structure

This section provides a brief explanation of the main parts in Figure 2.1 which illustrates a complete induction motor structure with two different wound and cage rotor.

The stator, as the combination of the nonmoving (stationary) parts supporting the motor, basically includes the magnetic circuit parts, and it carries the windings located inside the slots in a symmetric manner with the goal of generating the initial magneto-motive force (MMF) of the motor at its air gap level [7]. The stator is usually made of insulated steel laminations sorted side by side to navigate the

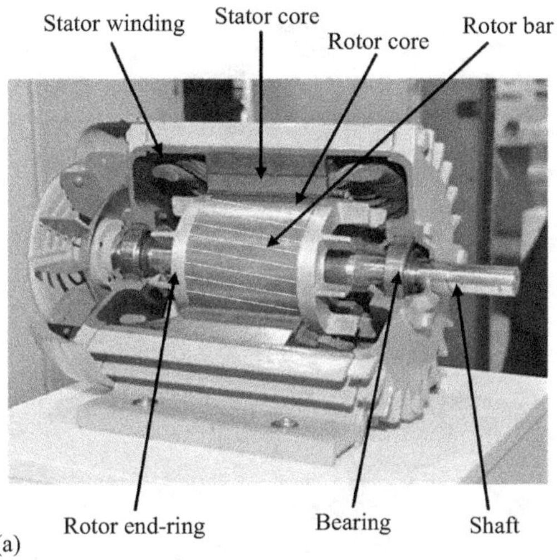


(a)

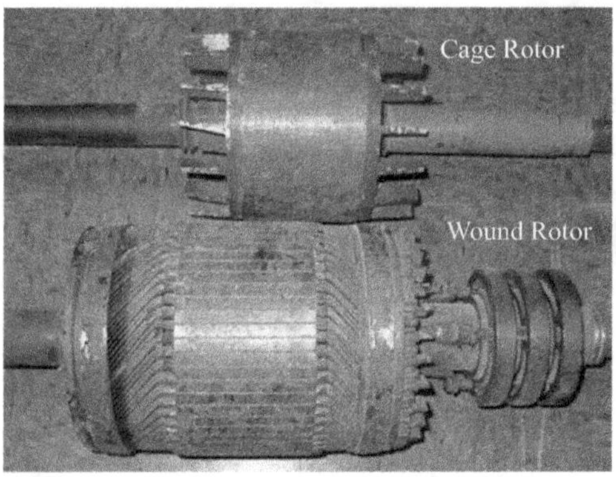


(b)

*Figure 2.1 (a) Cut-view of an induction motor and (b) different rotor types*

magnetic flux density produced by the windings through proper paths while reducing the magnetic losses inside the core material. Magnetic losses are generally divided into two distinct categories: the hysteresis and eddy losses which are related to the nonlinear magnetization phenomenon of laminations and the magnetically induced electrical currents in separate laminations, respectively. The three-phase stator windings are inserted into the slots of the stator and cover the inner circumference of the stator by a quasisinusoidal spatial distribution that guarantees a less spatial harmonic distortion of the motor quantities such as current, flux and electromagnetic torque (see Figure 2.2(a)). Each winding consists of several series-connected coils, and each coil can be a solid wire or parallel stranded wires. Increasing the number of coils, as well as the number of turns, started in slot $i$ and ended in slot $j$ (see Figure 2.2(b)), improves the torque of the motor while

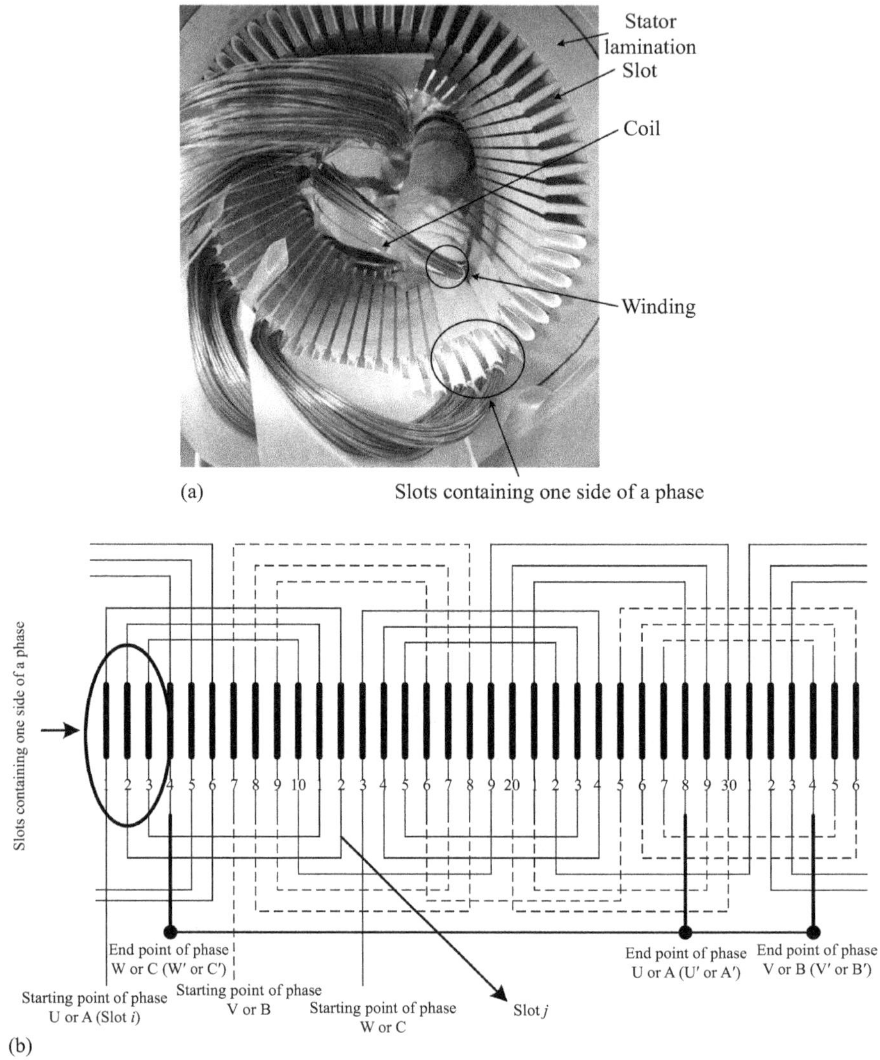


*Figure 2.2 (a) A 48-slot stator*[1] *and (b) stator-winding layout, a three-phase Δ-connected stator with 36 slots*

there should be a trade-off between the amount of the output torque and the increase in the winding resistance which, in turn, reduces the motor efficiency. It is noted that the standard routines ask to assign the letters (U, V and W) or (A, B and C) for the starting points of the three phases/windings and the letters (U′, V′ and W′) or (A′, B′ and C′) for the end point of the corresponding windings. The terms "phase" and "winding" might be used interchangeably. The windings present a similar distribution

[1]http://www.edisontechcenter.org (access date: October 2016)

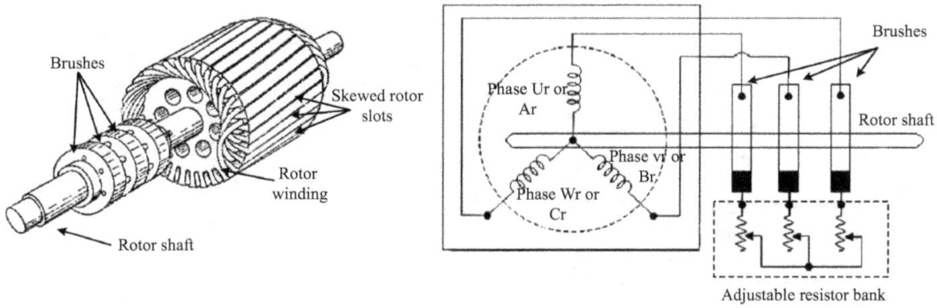


*Figure 2.3 Structure of a wound rotor and its connections with the brushes*

but with a spatial shift in the neighboring slots. Each winding has two ends called "terminal" to which the supply voltage which produces the magnetic field is applied. The three-phase stator windings can be *Y* or delta connected. The type of connection can be changed by changing the configuration of the connection of the six available terminals shown in Figure 2.2. The amount of voltage which can be applied to the windings highly depends on the torque production capability of the motor as well as the quality of insulations. 230VΔ–400VY or 400VΔ–690VY are some of the most common voltage standards in industrial zones.

As shown in Figure 2.1(b), the main difference between two types of the rotors, the wound and cage rotors, is related to the apparatus which carries the electrical current induced from the stator to the rotor. What plays this role in a wound rotor is a three-phase winding which is inserted symmetrically into the slots of the rotor while it is handled by the bars, normally made of aluminum, in a cage-rotor motor. Figure 2.3 illustrates the structure of a wound rotor along with the electrical connections of its terminals with the brushes. In fact, there would be no current flowing inside the rotor windings unless the rotor consists of an electrically short-circuited system. If the current is entirely zero in the rotor, the motor operation fails. Generally, the rotor windings possess a *Y* (star) connection, and only one terminal of each winding is available for purpose of resistance adjustment. This is a major task which is held to control the start-up torque by means of an adjustable resistor bank. The resistor bank is connected to the rotor windings and changes the total resistance of the rotor. By increasing the resistance, the start-up torque increases as well. The symbols $U_r$, $V_r$ and $W_r$ represent the rotor phases/windings. Due to the weakness of the brushes and also presence of an additional electrical connection between the rotor and the equipment outside the motor, the wound rotor is not usually preferred in ordinary application. Instead, a squirrel cage rotor is mostly selected due to its solid structure and connection-free nature (Figure 2.4). In fact, the wound rotor is very fragile while the cage rotor reveals interesting robustness in a real-world application. However, the new technologies are beneficial of specific types of wound-rotor motors that are called doubly fed induction motor with the goal of controlling both the stator and rotor circuit currents by means of inverters. The only shortcoming of a cage rotor is its inability to provide an adjustable terminal resistance required for a strong start-up of the motor. Once a cage rotor is casted and

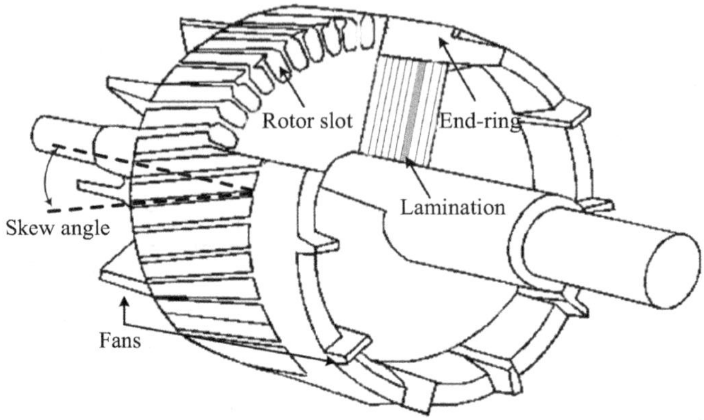


*Figure 2.4 Structure of a squirrel cage rotor*

built, its structure is totally enclosed inside the motor, and there is no access to it. However, it makes the structure more robust against stresses.

The common feature of two types of rotors shown in Figures 2.3 and 2.4 is the skewed rotor slots which are intended to reduce the total harmonic distortion (THD) of the motor. Although the skew reduces also the fundamental motor torque component, it is always proposed to keep a specific level of skewing in order to have a smooth operation of the motor. Otherwise, the torque ripple increases. The torque ripple highly depends on the configuration of the windings, the number of slots and the number of poles. The number of poles is indeed an influential factor in determining the motor characteristics, particularly the supply frequency and also the output speed. This topic will be further discussed in the next sections of the current chapter. Like the stator, the rotor also consists of laminations (see Figure 2.4) to reduce the magnetic (iron) losses. The end-rings shown in Figure 2.4 play the same role as the brushes in a wound rotor. During the manufacturing process, both the bars and the end-rings which are located at the two ends of the rotor are casted together simultaneously to have an integrated unit of rotor electric circuit. There are also some other components, shown in Figure 2.1, which deal with holding the entire motor and cooling it. These components are as follows:

- motor housing
- chassis
- bearings
- fans
- cooling fins.

The motor housing is the outer part which is electrically grounded by means of an electrical connection in order to prevent injuries. Mostly, the housing consists of the cooling blades, so-called cooling fins, through which the air generated by the cooling fan flows. Bearings are located at the two ends of the housing which holds the rotor. They consist of inner and outer races in the middle of which the balls rotate while the

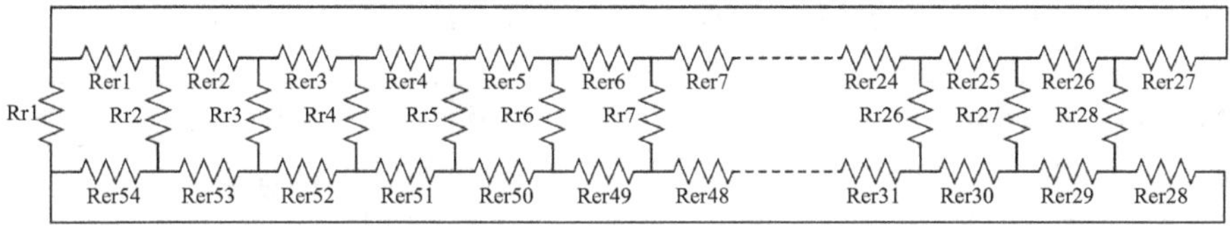


*Figure 2.5 Equivalent resistive network of the rotor*

rotor rotates. The bearings are the main sources of mechanical types of faults in induction motors. The eccentricity fault is a very good example of this kind of fault which is the result of an improper placement of the bearings or defect of balls. So the bearings must be monitored all the time to avoid future mechanical faults.

In order to provide a better demonstration of a rotor, an equivalent electric circuit that looks like a resistance network, as shown in Figure 2.5, is used for a rotor with 28 slots/bars. The horizontal and vertical resistors represent the equivalent circuit of the end-ring and the bars, respectively. There are also inductances connected in series to each resistor. Therefore, the resistor must be replaced by resistive-inductive impedances. However, for the sake of conciseness, the resistors are used as the candidates to represent the impedances. The number of the vertical elements is the same as the number of bars and that of the horizontal elements should be equal to (number of bars − 1). The horizontal elements take care of short-circuiting the rotor bars so that the rotor current can be built-up. Any electrical disconnection of joints of the elements shown in Figure 2.5 results in a rotor fault. For example, if the joint of one bar and one end-ring is broken, the bar current is zero. Consequently, the balanced operation of the motor is violated.

The discussed short-circuited equivalent network rotates around the inner surface of the stator with the same speed as that of the rotor. The movement across the stator windings induces electro-motive forces (EMF) in every bar or winding inserted inside the rotor slots. Consequently, an electric current is induced into the rotor conductors, so the electromagnetic torque is developed as a result of the interaction between the stator and rotor MMFs. This is the fundamental principle of the operation of induction motors and will be fully described mathematically in this chapter. The most critical part of the rotor circuit is the joint between the rotor bars and the end-rings which is produced during the casting process. This is exactly where the broken bar fault occurs. As there is no such connection in the wound-rotor motor, the broken bar fault has no meaning in wound rotor case. Table 2.1 characterizes the differences between the two types of induction motors.

Getting familiar with the fundamentals of the operation of induction motors is indeed a must-do step since any fault-related analysis will be useless unless readers have the idea of how motor quantities are in the healthy and normal condition. Therefore, to conduct the analysis, a single-harmonic motor model is investigated here in this chapter. Then, a more advanced study on the principal slot harmonics of the motor is introduced. These two steps prepare the reader's mind for our future efforts in terms of how motor quantities might be affected by various faults. The only required background is the concept of the electromagnetism.

*Table 2.1 Characteristics of wound- and cage-induction motors*

| Wound-rotor-induction motor | Cage-rotor-induction motor |
|---|---|
| Slip ring and brushes make the construction much more complicated | Possess a very simple construction |
| Windings are the same in the stator and rotor sides | Instead of windings, solid bars are replaced inside the rotor slots |
| Rotor resistance is adjustable | Rotor resistance is fixed |
| Starting torque is improvable | Starting torque is fixed |
| It is rarely used | Due to its simple structure and less maintenance requirements, it is mostly preferred |
| Lower efficiency due to the larger rotor copper losses | Higher efficiency due to smaller rotor copper losses |

## 2.3 Line-start induction motor: linear and single harmonic model of a healthy motor

Analytical representation of induction motors is fundamentally initiated by assuming that the motor consists of the spatial harmonic components that are the result of an entirely sinusoidal winding distribution across the stator and rotor circumferences. Therefore, for a three-phase motor, one can assume that the rotor conductors, whether the wound rotor or the cage rotor, act as a balanced three-phase configuration demonstrating three similar field vectors (see Figure 2.6) [8–15]. The stator and the rotor are each identified by three identical windings (phases) which must be practically located 120° mechanically away from each other. This situation is shown in Figure 2.6. With this intention in mind, one concentrated-at-one-point winding is ideally assigned to each stator and rotor winding. In the case of a cage rotor, it is for now assumed that the behavior can be equivalently modeled by the three-phase windings of the rotor (see Figure 2.6). The stator phase axes are fixed with respect to the origin of the rotation which is mostly assigned to the direction of the stator winding/phase A. However, the rotor windings rotate during the motor operations. Therefore, its corresponding angle is measured by means of the rotor mechanical angle ($\theta_m$) which is equal to $(1/p)(\theta_e)$, where $p$ is the number of motor pole pairs and $\theta_e$ is the electrical angle of the rotor.

The next step is to make a couple of assumptions to eliminate the dependency on nonlinearity of the core material, thermal variations as well as the geometry.

- Motor is healthy.
- Nonlinearity of the stator and rotor core materials is neglected. Moreover, the core losses including the hysteresis and eddy losses are ignored.
- Every side of the windings and bars of each phase can be modeled as a concentrated-at-one-point winding. Therefore, the slotting effect is neglected. As a result, the air gap is smooth and uniform.
- Thermal variation is neglected.

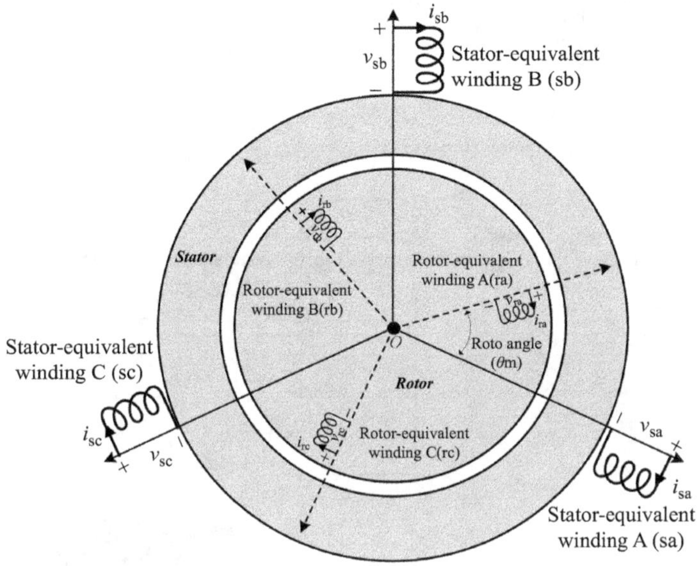


*Figure 2.6 Three-phase equivalent circuit of an induction motor*

- End-windings are the two ends of the coils outside the stator or rotor cores and their effects are neglected.
- Skewing effect is neglected. Otherwise, a 3D model must be developed. Nevertheless, developing an analytical 3D model is a very advanced topic that cannot managed in a simple way in this chapter.

These are the essential in producing a perfectly formulated analytical model of an induction motor which has been greatly used to analyze the healthy motor behavior in time domain. This also provides us with a promising tool for describing how an inverter-fed motor is and should be controlled.

The line-start motor model is the starting point of developing the analytical solutions to the analysis of motor operation. Therefore, a common line-start configuration is presented in Figure 2.7. The line-start mode is the conventional approach of starting up and supplying an induction motor in industry. This network supplies a three-phase voltage system, shown in Figure 2.8, which is formulated as follows.

$$[V_s(V)] = \begin{bmatrix} v_{sa} \\ v_{sb} \\ v_{sc} \end{bmatrix} = \begin{bmatrix} V_m \sin(2\pi f_s t) \\ V_m \sin(2\pi f_s t - 2\pi/3) \\ V_m \sin(2\pi f_s t - 4\pi/3) \end{bmatrix},$$

$$\text{where} \left\{ \begin{array}{l} V_s \rightarrow \text{Stator voltage vector} \\ V_m \rightarrow \text{Amplitude of the stator voltage} \\ f_s \rightarrow \text{Electrical supply frequency (Synchronous frequency)} \\ t \rightarrow \text{Time in seconds} \\ V \rightarrow \text{Voltage unit in Volts} \end{array} \right\} \qquad (2.1)$$

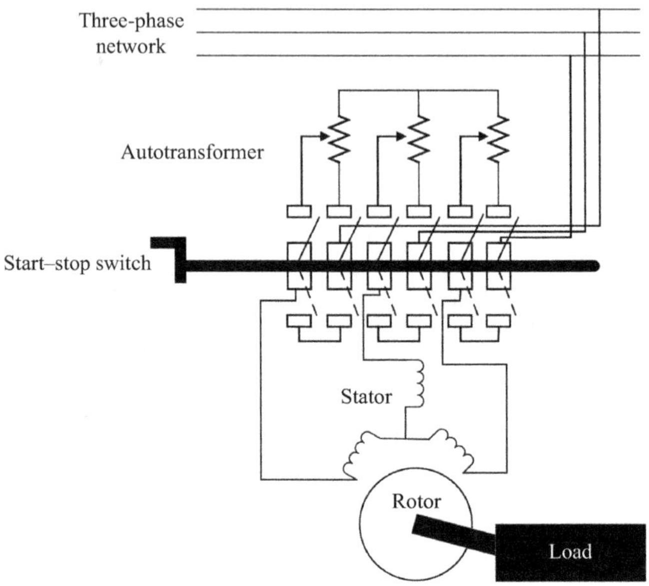


*Figure 2.7 Configuration of a line-start supply mode of an induction motor*

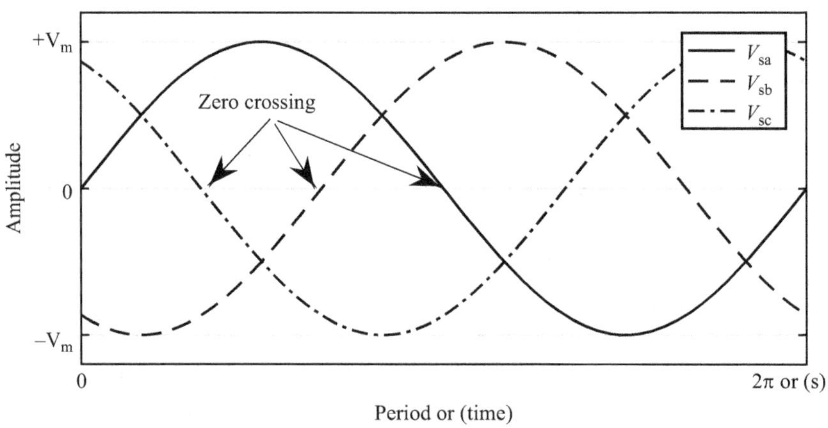


*Figure 2.8 Balanced three-phase voltage system*

$v_{sa}$, $v_{sb}$ and $v_{sc}$ are the stator phase voltages as shown in Figure 2.6. In a symmetric and balanced voltage system, all the phases exhibit the same amplitude but different zero-crossing points. The zero crossings determine the electrical phase shift of the three phases. $2\pi f_s$ is the electrical angular frequency of the supply voltage, the so-called $\omega_s$. $f_s$ is fixed for any line-start application while the amplitude of the voltages might be easily adjusted by means of an autotransformer which applies minor regulations to the voltages. An autotransformer, seen in Figure 2.7, is the only interfering device that might be used to stabilize the supply voltage in order to make sure that all the phases have the same amplitudes (magnitude). It has nothing to do with the frequency of the supply network and only changes the voltage magnitude. Most of the time, transformers consist of an adjusting bottom by means of which all the three-phase voltages are changed simultaneously. According to this discussion, it is quite clear that why this kind of motor supply is called a "line-start" supply mode. In fact, the motor is directly connected to a three-phase network (line) without any interconnection such as drives. Together with (2.1), the other stator quantities including the current and flux vectors are formulated as the following:

$$[I_s(A)] = \begin{bmatrix} i_{\text{sa}} \\ i_{\text{sb}} \\ i_{\text{sc}} \end{bmatrix}, \quad [\Lambda_s(\text{Wb})] = \begin{bmatrix} \lambda_{\text{sa}} \\ \lambda_{\text{sb}} \\ \lambda_{\text{sc}} \end{bmatrix}$$

$$\text{where} \left\{ \begin{array}{l} I_s \rightarrow \text{Stator current vector} \\ i_{\text{sa}},\ i_{\text{sb}} \text{ and } i_{\text{sc}} \rightarrow \text{Stator phase currents} \\ \Lambda_s \rightarrow \text{Stator flux vector} \\ \lambda_{\text{sa}},\ \lambda_{\text{sb}} \text{ and } \lambda_{\text{sc}} \rightarrow \text{Stator phase fluxes} \\ A \rightarrow \text{Current unit in Amperes} \\ \text{Wb} \rightarrow \text{Flux unit in Webers} \end{array} \right\} \tag{2.2}$$

Likewise, a balanced three-phase system is assigned to the rotor windings (phases) (see (2.3)). The rotor currents and voltages are illustrated in Figure 2.6. If a cage-rotor motor is investigated, the rotor voltages should be set to zero since the rotor circuit is always short-circuited by means of the end-rings. In the case of a wound rotor, the rotor voltage vector is equal to the voltage drop across the external resistance connected through the brushes to the rotor windings (see Figure 2.3).

$$[V_r] = \begin{bmatrix} v_{\text{ra}} \\ v_{\text{rb}} \\ v_{\text{rc}} \end{bmatrix}, \quad [I_r] = \begin{bmatrix} i_{\text{ra}} \\ i_{\text{rb}} \\ i_{\text{rc}} \end{bmatrix}, \quad [\Lambda_r] = \begin{bmatrix} \lambda_{\text{ra}} \\ \lambda_{\text{rb}} \\ \lambda_{\text{rc}} \end{bmatrix}$$

$$\text{where} \left\{ \begin{array}{l} V_r \rightarrow \text{Rotor voltage vector} \\ v_{\text{ra}},\ v_{\text{rb}} \text{ and } v_{\text{rb}} \rightarrow \text{Rotor phase voltages} \\ I_r \rightarrow \text{Rotor current vector} \\ i_{\text{ra}},\ i_{\text{rb}} \text{ and } i_{\text{rc}} \rightarrow \text{Rotor phase currents} \\ \Lambda_r \rightarrow \text{Rotor flux vector} \\ \lambda_{\text{ra}},\ \lambda_{\text{rb}} \text{ and } \lambda_{\text{rc}} \rightarrow \text{Rotorphasefluxes} \end{array} \right\} \tag{2.3}$$

### 2.3.1 Flux equation

This section intends to deal with the linear electrical equations of an induction motor considering some assumptions related to the nonlinearity and thermal variation of the motor's materials. Focusing on the two main parts of the motor, the stator and the rotor, the input energy variation provided by the input three-phase network is split as follows:

$$\mathrm{dW_{ei}}(J) = \mathrm{dW_{es}}(J) + \mathrm{dW_{er}}(J)$$

$$\text{where}\left\{\begin{array}{l} \mathrm{dW_{ei}} \rightarrow \text{Input electrical energy change} \\ \mathrm{dW_{es}} \rightarrow \text{Stator electrical energy change} \\ \mathrm{dW_{er}} \rightarrow \text{Rotor electrical energy change} \\ d \rightarrow \text{Differential operator} \\ J \rightarrow \text{Energy unit in Joules} \end{array}\right\} \tag{2.4}$$

In this section, the focus is on the electrical energy since the magnetic energy losses of the motor components, the stator and the rotor, are assumed to be zero. This is an acceptable assumption in the event that one does not look for a precise loss characterization of an electromagnetic system such as induction motors. On the other hand, the electrical qualities, the voltage and the current, are the mediums to develop any electrical energy in an induction motor. In view of the electrical energy, the time-dependent variation of the product of the voltage and current signals should be integrated over time.

$$\begin{aligned} \mathrm{dW_{es}} &= [I_s]^T[V_s]dt \\ \mathrm{dW_{er}} &= [I_r]^T[V_r]dt \end{aligned} \tag{2.5}$$

Then the matrix form of the energy equation is used to express the total energy variation caused by all the windings of the stator and the rotor. Therefore, all the phases play a role in developing the energy of the system. On the other hand, the electrical energy losses, considering that the magnetic losses are equal to zero, are formulated as follows:

$$\begin{gathered} \mathrm{dW_{e_{loss}}}(J) = [I_s]^T[R_s][I_s]dt + [I_r]^T[R_r][I_r]dt \\ [R_s(\Omega)] = \begin{bmatrix} R_s & 0 & 0 \\ 0 & R_s & 0 \\ 0 & 0 & R_s \end{bmatrix}, \quad [R_r(\Omega)] = \begin{bmatrix} R_r & 0 & 0 \\ 0 & R_r & 0 \\ 0 & 0 & R_r \end{bmatrix} \\ \text{where}\left\{\begin{array}{l} R_s \rightarrow \text{Stator winding resistance} \\ R_r \rightarrow \text{Rotor winding resistance} \\ \Omega \rightarrow \text{Resistance unit in Ohms} \end{array}\right\} \end{gathered} \tag{2.6}$$

In the case of any rotor bar breakage or open-circuit fault of the rotor windings, the corresponding equivalent resistance increases leading to an unbalanced resistance matrix which, in turn, produces undesirable harmonic components. For now, assume that all the phases possess a similar resistance value. Having ignored the magnetic

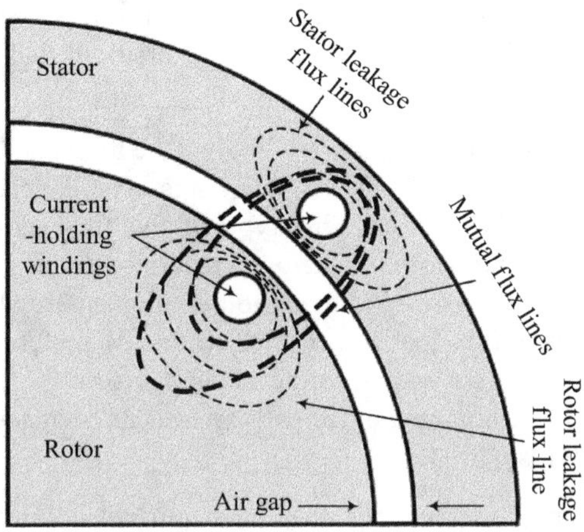


*Figure 2.9 Representation of mutual and leakage fluxes*

losses, the electrical energy converted into the active electromagnetic energy, which contains both the required magnetizing and mechanical energy of the motor, is obtained as (2.7).

$$\begin{aligned} &\mathrm{dW}_{\text{electrical}}(J) = \mathrm{dW}_{\text{ei}} - \mathrm{dW}_{\mathrm{e}_{\text{loss}}} \\ &\text{where } \{\mathrm{dW}_{\text{electrical}} \rightarrow \text{Loss} - \text{less electromagnetic energy change}\} \end{aligned} \tag{2.7}$$

By integrating (2.4–2.7), one can easily end up with the following equation representing the differential form of the Faraday's law:

$$\begin{aligned} \mathrm{dW}_{\text{electrical}} &= [I_s]^T d[\Lambda_s] + [I_r]^T d[\Lambda_r] \\ \frac{d[\Lambda_s]}{dt} &= [V_s] - [R_s][I_s]|_{\text{flux linkage variation of stator}} \\ \frac{d[\Lambda_r]}{dt} &= [V_r] - [R_r][I_r]|_{\text{flux linkage variation of rotor}} \end{aligned} \tag{2.8}$$

By integrating the differential forms of the flux linkages expressed by (2.8), the total fluxes which cover the stator or rotor windings all over their circumferences are calculated. As a well-known fact, the total stator or rotor fluxes are made of two distinct types of flux lines, i.e. the mutual and leakage fluxes shown in Figure 2.9. The mutual flux is the part enclosing both the rotor and stator windings, and the leakage flux encloses only one side, whether the stator or the rotor. The underlying idea of the aforementioned definition is the way that the flux lines contribute to the stator–rotor magnetic coupling leading to the electromagnetic torque production of the motor. In other words, the mutual flux is the unique part of the flux that guarantees the coupling of stator and rotor components while the leakage flux serves as additional magnetic and electrical losses that lead to a higher temperature rise of the motor components. Therefore, it should be pointed out that

- The mutual flux: A flux that transforms power or energy from the primary winding (the stator) to the secondary winding (the rotor) and is produced by the coupled MMF of the stator and rotor windings. The concept of MMF will be discussed later in this chapter. This type of the flux should definitely pass across the motor air gap from the stator to reach the rotor and vice versa. In an aligned position of the two windings, the maximum mutual flux is achieved.
- The leakage flux: A flux that is confined to one of the current-holding windings. It might also partly pass into the air gap. However, it is inclined to the corresponding core to make a close path around its own source of production, one of the windings.

With this in mind, one realizes that each current-holding winding possess two flux components, namely the leakage flux and the mutual flux. The mutual fluxes are generally divided into two categories: the self-magnetizing and mutual-magnetizing fluxes. The former one corresponds to the flux produced by one of the windings and encapsulated by the same winding. The latter corresponds to the encapsulated by two different windings.

A magnetic flux is defined as a combination of a current passing through a coil and a geometry-dependent quantity so-called inductance. The combination is normally in a form a product of the mentioned quantities, the current and the inductance, if a linear magnetic system is investigated. Moreover, the flux that crosses a winding might be composed of components built by different windings. Therefore, depending on the number of windings, contributing to the flux linkage of a specific winding, different combinations of current-inductance are introduced. For example, for an induction motor with three stator and three rotor windings, the flux linkage of one of the stator windings consist of six terms (see (2.10)) including the one related to the self-magnetizing and self-leakage inductance, and the remaining five terms that are related to the mutual magnetizing fluxes produced by the other windings. Flashing back to the main assumptions made at the beginning of this subsection, a linear model of the motor, as a result of which the magnetic flux can be written as the superposition of its governing terms (the leakage and mutual parts), is intended. If a nonlinear system is analyzed, the principle of superposition is not valid anymore. For a linear case, we can simply formulate the self-inductances as follows:

$$
\begin{aligned}
l_s &= l_{s\sigma} + l_{\mathrm{sm}} \\
l_r &= l_{r\sigma} + l_{\mathrm{rm}}
\end{aligned}
$$

$$
\text{where} \left\{
\begin{array}{l}
l_s \rightarrow \text{Stator total inductance} \\
l_{s\sigma} \rightarrow \text{Stator leakage inductance} \\
l_{\mathrm{sm}} \rightarrow \text{Stator mutual inductance magnitude} \\
l_{r\sigma} \rightarrow \text{Rotor leakage inductance} \\
l_{\mathrm{rm}} \rightarrow \text{Rotor mutual inductance magnitude}
\end{array}
\right\} \tag{2.9}
$$

Considering a linear characteristic for the silicon steel material, the motor inductances would only depend on the physical geometry of the motor. In (2.9), the term "magnitude" has been used to denote that the real values of the mutual inductance vary as a function of the position of the two windings whose mutual inductances are calculated. Considering all these, the magnetic flux linkage of the stator winding "A" or "a" is formulated as follows:

$$\lambda_{\text{sa}} = l_{s\sigma} i_{\text{sa}} + l_{\text{sm}} i_{\text{sa}} + M_s \cos\left(\frac{2\pi}{3}\right) i_{\text{sb}} + M_s \cos\left(\frac{4\pi}{3}\right) i_{\text{sc}} + M_{\text{sr}}\left(i_{\text{ra}} \cos(\theta_e) + i_{\text{rb}} \cos\left(\theta_e - \frac{2\pi}{3}\right) + i_{\text{rc}} \cos\left(\theta_e - \frac{4\pi}{3}\right)\right) \quad (2.10)$$

where $M_s = l_{\text{sm}}$ and $M_{\text{sr}}$ is a function of the configuration of the stator–rotor windings. It is noted that considering the inductance variations of the motor and its different possible compositions is a key point to understand the behavior of an eccentric motor since the inductance change is a kind of signature for this type of fault. Therefore, it is important to clearly understand the initial step in introducing the motor fundamentals. All the details provided in this chapter are sorted in a way that leads to a meaningful step-by-step realization of the characteristics of an induction motor which are highly significant in view of transition from healthy to faulty motors.

Going back to our main discussion, the mutual inductance of the winding (phase) A with itself is called "self-magnetizing inductance" and is equal to "$l_{\text{sm}}$" while the magnitude of the mutual inductance of the phase A and other stator windings is a function of the position of the windings and equal to

$$M_s = l_{\text{sm}} \rightarrow M_s \cos\left(\frac{2\pi}{3}\right) = l_{\text{sm}} \cos\left(\frac{2\pi}{3}\right) = -\frac{1}{2} l_{\text{sm}} \quad (2.11)$$

If the slotting effects and nonsinusoidal distribution of the windings come into play, the (2.11) is not valid anymore. Similarly, the terms which relate the stator windings to the rotor windings will be invalidated. All we have had so far is the consequence of the sinusoidal distribution of the concentrated-at-one-point windings. The rotation of the rotor is the compelling reason for the presence of the $\theta_{\text{e}}$ dependency of the stator–rotor mutual inductances. In this case, if the rotor rotates, the stator–rotor mutual inductances follow a sinusoidal alternation. Moreover, it is the electrical rotor angle which is incorporated into the formulations, not the mechanical one. However, wherever mechanical quantities are dealt with, the mechanical angle should be used correspondingly. Given the three-phase stator and rotor winding system, the matrix representation of the stator fluxes, i.e. (2.2), can be further extended as (2.12). Noticeably, the mutual and self-inductance matrices relate the stator and rotor current vectors to the fluxes such that the stator flux vector is a linear function of the stator currents while the dependency on the rotor current vectors exhibit a $\theta_{\text{e}}$-associated function. Given a zero-stator current vector, the maximum flux enclosing the stator flux takes place when the stator and rotor windings are completely aligned. Additionally, the coupling is zero when they are located 90° mechanical or equivalently 180° electrical degrees away from each other. When a faulty condition such as the eccentricity or misalignment fault takes place, the M matrix (see (2.12)) which is the representation of the mutual inductance of the stator and the rotor should be modified according to the change in the length of motor air gap which is not a smooth and fixed quantity anymore.

$$[\Lambda_s] = \begin{bmatrix} \lambda_{sa} \\ \lambda_{sb} \\ \lambda_{sc} \end{bmatrix} = \underbrace{\begin{bmatrix} l_{s\sigma} + l_{sm} & -\frac{1}{2} l_{sm} & -\frac{1}{2} l_{sm} \\ -\frac{1}{2} l_{sm} & l_{s\sigma} + l_{sm} & -\frac{1}{2} l_{sm} \\ -\frac{1}{2} l_{sm} & -\frac{1}{2} l_{sm} & l_{s\sigma} + l_{sm} \end{bmatrix}}_{[L_s]} \begin{bmatrix} i_{sa} \\ i_{sb} \\ i_{sc} \end{bmatrix}$$

$$+ M_{sr} \underbrace{\begin{bmatrix} \cos(\theta_e) & \cos\left(\theta_e - \frac{2\pi}{3}\right) & \cos\left(\theta_e - \frac{4\pi}{3}\right) \\ \cos\left(\theta_e - \frac{4\pi}{3}\right) & \cos(\theta_e) & \cos\left(\theta_e - \frac{2\pi}{3}\right) \\ \cos\left(\theta_e - \frac{2\pi}{3}\right) & \cos\left(\theta_e - \frac{4\pi}{3}\right) & \cos(\theta_e) \end{bmatrix}}_{[M]} \begin{bmatrix} i_{ra} \\ i_{rb} \\ i_{rc} \end{bmatrix} \quad (2.12)$$

As a result of mechanical deformations, the motor-inductance matrices, either $L_s$ or *M,* will contain multi-harmonic sinusoidal components. Thus, a nonsmooth variation of the motor fluxes should be expected. The saturation profile represented and modeled by nonlinear B–H curves of silicon steel materials is another factor that changes the inductance matrices. Thus, any factor that affect the motor saturation level brings up a nonlinear behavior of not only the inductance matrices but also all the resultant quantities such as the motor flux or current vectors. As another example, although broken rotor bars do not change the air gap length, they introduce a highly saturated region close to the breakage. This leads to an unbalanced magnetic flux distribution that creates a nonsinusoidal inductance matrix. The concepts discussed above are only some of the hints to relate the motor fundamental parameters such as inductances to the fault diagnosis applications, and the corresponding detailed investigations will be discussed in the next chapters. Correspondingly, the rotor flux equations are obtained as follows:

$$[\Lambda_r] = \begin{bmatrix} \lambda_{ra} \\ \lambda_{rb} \\ \lambda_{rc} \end{bmatrix} = \underbrace{\begin{bmatrix} l_{r\sigma} + l_{rm} & -\frac{1}{2} l_{rm} & -\frac{1}{2} l_{rm} \\ -\frac{1}{2} l_{rm} & l_{r\sigma} + l_{rm} & -\frac{1}{2} l_{rm} \\ -\frac{1}{2} l_{rm} & -\frac{1}{2} l_{rm} & l_{r\sigma} + l_{rm} \end{bmatrix}}_{[L_r]} \begin{bmatrix} i_{ra} \\ i_{rb} \\ i_{rc} \end{bmatrix}$$

$$+ M_{rs}\begin{bmatrix} \cos(\theta_e) & \cos\left(\theta_e - \frac{2\pi}{3}\right) & \cos\left(\theta_e - \frac{4\pi}{3}\right) \\ \cos\left(\theta_e - \frac{4\pi}{3}\right) & \cos(\theta_e) & \cos\left(\theta_e - \frac{2\pi}{3}\right) \\ \cos\left(\theta_e - \frac{2\pi}{3}\right) & \cos\left(\theta_e - \frac{4\pi}{3}\right) & \cos(\theta_e) \end{bmatrix}^T \begin{bmatrix} i_{sa} \\ i_{sb} \\ i_{sc} \end{bmatrix}$$

$$\text{where } \{ M_{rs} = M_{sr} \} \tag{2.13}$$

What is significant is the importance of the mutual inductances in studying the eccentricity fault while the self- and leakage inductances are not usually an appropriate index to realize the effect of the eccentric rotor. It is also interesting that depending on the topology of the winding connections, delta ($\Delta$) or star ($Y$) connections, the stator–stator and rotor–rotor mutual inductances might be or might not be present in the motor flux. As a matter of fact, if a star ($Y$) connected and ground node-free stator and rotor-winding system are considered, the stator–stator mutual inductances will be zero. In the case where the three-phase currents of the stator and the rotor do not allow to have nonzero summation in a $Y$-connected system, one of the phase currents can be written as the negative of the sum of the other two currents, either in the stator or the rotor. Applying this situation to (2.10) and (2.13) causes the stator–stator and rotor–rotor matrices to have zero non-diagonal components. In contrast, the stator–rotor inductance matrix remains intact. This conveys the idea of using the mutual inductances as the tools for a consistent and connection-type free approach to analyze the motor behavior. Therefore, not only is it proposed to thoroughly understand the materials of this chapter, but it is also beneficial to become familiar with the advanced harmonic-included motor-inductance analysis.

So far, the flux-inductance relevance has been demonstrated by means of a single harmonic approximation of the motor geometry and quantities. Now, it is time to propose a closed-form equation for the electromagnetic torque which takes care of rotating the motor shaft. This is exactly where several mechanical defects are imposed to the motor, especially to the moving parts such as the rotor shaft and bearings which hold the entire weight of the rotor. Bearing wear and shaft mis-alignment which are all caused by mechanical deficiencies in the rotor are examples of the mentioned defects. It is also important to know that the movement can obviously worsen the situation. On the other hand, motor transients highly depend on the way that the electromagnetic torque is developed. In the first place, it is the motor geometry that affects the developed torque. Then, it is the input values and subsequently the resultant electrical and magnetic quantities, which are functions of the material characteristics and also the motor geometry that affects the electromagnetic torque. Therefore, it seems so much information can be extracted from the motor torque signal as well as the flux and the current. As a matter of fact, in Chapter 3, it is going to be mathematically proved that the electromagnetic torque contains unbelievably helpful information on the faults such as the broken bar and end-ring faults. Taking this brief discussion on the gravity of the torque signal into account, we try to present the single harmonic representation of the motor torque.

### 2.3.2 Electromagnetic torque equation

The torque is developed from the position-dependent variation of the mechanical energy and is applied to the rotor through the air gap. Therefore, the hint to calculate the torque is to find the variations of the mechanical energy in time. To this end, recall and combine (2.8), (2.12) and (2.13):

$$\begin{aligned}
\mathrm{dW}_{\text{electrical}} &= [I_s]^T[L_s]d[I_s] + [I_r]^T[L_r]d[I_r] + [I_s]^T M_{\text{sr}} d([M][I_r]) + [I_r]^T M_{\text{sr}} d([M]^T[I_s]) \\
&= [I_s]^T[L_s]d[I_s] + [I_r]^T[L_r]d[I_r] + [I_s]^T M_{\text{sr}} d([M])[I_r] + [I_s]^T M_{\text{sr}}[M]d([I_r]) \\
&\quad + [I_r]^T M_{\text{sr}} d([M])[I_s] + [I_r]^T M_{\text{sr}}[M]d([I_s]) \\
&= [I_s]^T[L_s]d[I_s] + [I_r]^T[L_r]d[I_r] + 2[I_s]^T M_{\text{sr}} d([M])[I_r] + [I_s]^T M_{\text{sr}}[M]d([I_r]) \\
&\quad + [I_r]^T M_{\text{sr}}[M]d([I_s]).
\end{aligned} \tag{2.14}$$

On the other hand, in a magnetically loss-less system, the stored magnetic energy variation is expressed as (2.15).

$$\begin{aligned}
\mathrm{dW}_{\text{magnetic}} &= [I_s]^T[L_s]d[I_s] + [I_r]^T[L_r]d[I_r] + [I_s]^T M_{\text{sr}} d([M])[I_r] \\
&\quad + [I_s]^T M_{\text{sr}}[M]d([I_r]) + [I_r]^T M_{\text{sr}}[M]d([I_s])
\end{aligned} \tag{2.15}$$

Coupled with

$$\mathrm{dW}_{\text{mechanical}} = \mathrm{dW}_{\text{electrical}} - \mathrm{dW}_{\text{magnetic}} \tag{2.16}$$

The mechanical energy variation, $\mathrm{dW}_{\text{mechanical}}$, is equal to

$$\mathrm{dW}_{\text{mechanical}} = [I_s]^T M_{\text{sr}} d([M])[I_r] \tag{2.17}$$

In addition, the electromagnetic torque is equal to the variations of the mechanical energy with respect to the angular position of the rotor.

$$\begin{aligned}
T_{\text{em}}(N.m) &= p\frac{\mathrm{dW}_{\text{mechanical}}}{d\theta_m} = p\left([I_s]^T M_{\text{sr}} \frac{d([M])}{d\theta_m}[I_r]\right) = p\left([I_r]^T M_{\text{sr}} \frac{d([M])}{d\theta_m}[I_s]\right) \\
&= \frac{p}{2}\left([I_s]^T M_{\text{sr}} \frac{d([M])}{d\theta_m}[I_r] + [I_r]^T M_{\text{sr}} \frac{d([M])}{d\theta_m}[I_s]\right)
\end{aligned}$$

$$\text{where } \{T_{\text{em}} \rightarrow \text{Electromagnetic torque in a loss} - \text{less system}\} \tag{2.18}$$

There should also be a way to relate the developed electromagnetic torque to the motor speed, and the first-order mechanical differential equation of the motor handles this task.

$$T_{\text{electromagnetic}} - T_{\text{load}} = J\frac{d\omega_m}{d\theta_m} - B_v\omega_m$$

$$\text{where}\begin{Bmatrix} \omega_m \rightarrow \text{Mechanical angular frequency} \\ J \rightarrow \text{Rotor Inertia (kgm}^2) \\ B_v \rightarrow \text{Viscosity coefficient (kg/(sm))} \end{Bmatrix} \tag{2.19}$$

$T_{\text{load}}$ is the load torque that can be a constant or an oscillating load expressed in N m. Oscillating load torques introduce a new challenge to the field of fault diagnosis of induction motors because the motor-harmonic components amplified by oscillating loads sometimes overlap the fault-related components. This leads to a misunderstanding in terms of the severity of faults. The best example of the kind is the broken bar motor connected to an oscillating load with the oscillation frequency of $2sf_s$ where $s$ is the motor "slip." In particular, broken bars or broken end-rings amplify the magnetic backward fields which have the same frequency pattern as that of the oscillating load presenting the same fashion of harmonic change in the motor quantities, specifically the motor current signal. Therefore, devising a strategy to discriminate between the fault and oscillating loads should be an inseparable part of the diagnosis process. However, a better understanding of the healthy motor behavior is still the best suggestion to tackle more complicated fault-related problems. Therefore, a helpful temporal study of the healthy motor quantities is provided herein. Through the following investigation, the requirements and concepts, such as the synchronous frequency or speed, the slip, the transient operation and the steady-state operation, etc., are fully discussed. To do so, a line-start supply mode of an induction motor with the data listed in Table 2.2 is simulated, and the results are illustrated in Figure 2.10.

Three motor signals are illustrated in detail in Figure 2.10. The first signal is the motor current at three different load levels. It is important to note that whenever

*Table 2.2 Simulated motor data*

| | |
|---|---|
| Power | 11 kW |
| Line voltage | 460 V |
| Frequency | 50 Hz |
| Line current | 25 A |
| Number of poles | 4 |
| $R_s$ | 0.087 |
| $R_r$ | 0.187 |
| $M_{sr}$ | 26.7 mH |
| $l_s$ | 15.8 mH |
| $l_r$ | 16.1 mH |
| $J$ | 1.662 kg m$^2$ |
| $B_v$ | 0.01 kg/(s m) |

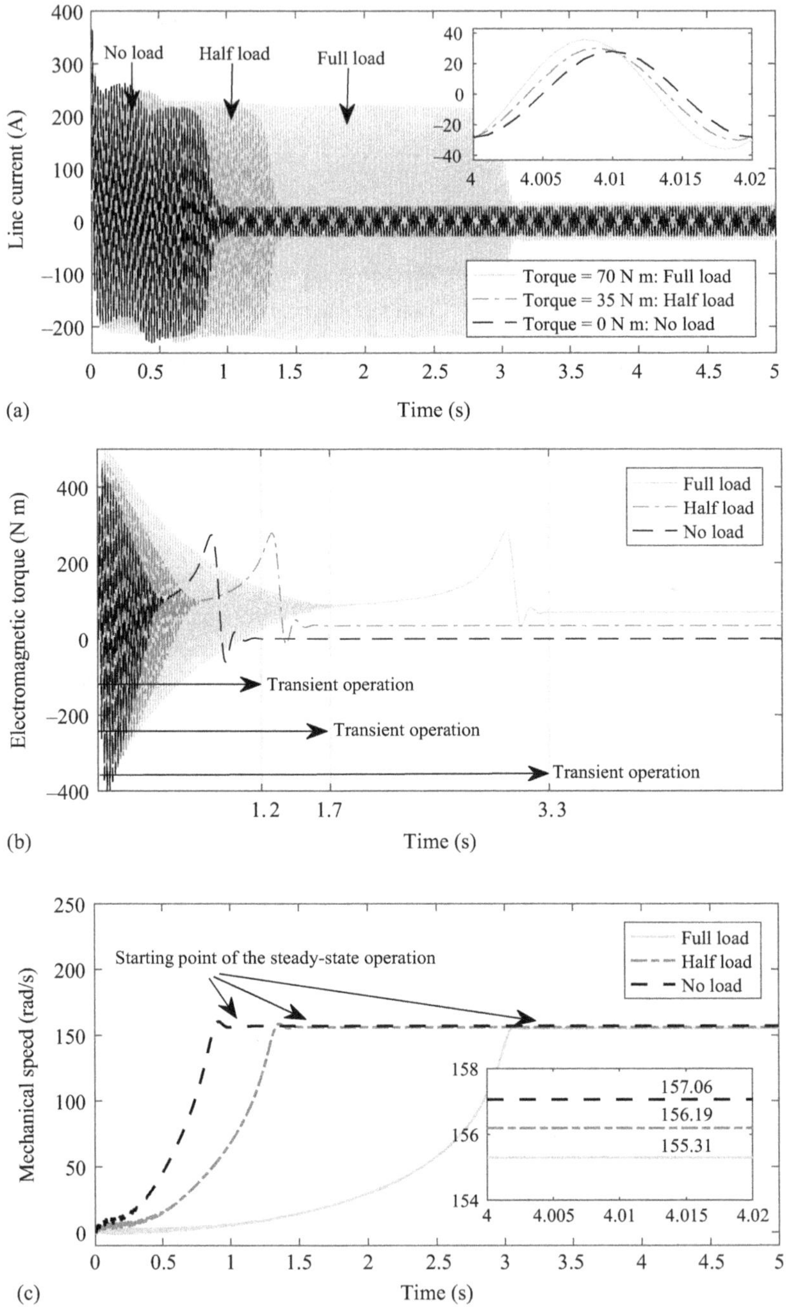


*Figure 2.10 Impact of load variation in a line-start motor (a) current, (b) electromagnetic torque and (c) speed*

we talk about the motor load in this book, we mean the motor torque, expressed in N m, connected to the motor shaft unless it is specified otherwise. If the mechanical losses are zero, this value should be equal to the electromagnetic torque. According to Figure 2.10(a), the time required for the motor current to reach a steady-state point, beyond which the current magnitude does not change, greatly increases with the increase in the load level. Moreover, the amplitude of the motor steady-state current also increases by the load. The terms "steady-state operation" or sometimes called "stationary operation" is one of the key terms of this book.

The fact that most of the fault indices are defined, formulated and extracted for the steady-state operating regime of the motor signals. Contrary to the steady-state operation, there is the "transient operation," mostly defined as the start-up period of the motor from a zero speed to the steady-state point (see Figure 2.10(b) and (c)). The transient period is exactly the same period in which the motor current possesses a larger speed value along with a swinging amplitude while the current amplitude becomes constant for a fixed load level (see Figure 2.10(a)).

Since a single-harmonic model of the induction motor has been used so far, the fundamental frequency of the motor current is equal to that of the source, 50 Hz in this study. Accordingly, there should be a balanced three-phase current system with an electrical phase shift of 120° with respect to each other. Of course, the phase angles are different from that of the voltages because induction motors serve as highly inductive loads that produce a lagging phase angle. Given a healthy motor and that the higher order harmonic components are absent, the speed and the electromagnetic torque always follow a very smooth and nonoscillating trend which is simply expected from a linear and single harmonic model (see Figure 2.10 and Figure 2.10c). However, the transient operation is totally different. That is usually why most of the recent efforts in the field of diagnosis merely focus on the steady-state operation in which a well-behaved behavior is followed. The transient operation is somehow challenging specifically in the case of "inverter-fed" motors with a wholly nonsinusoidal supply voltage. Likewise, there is another strong excuse for not incorporating the transient which is the short-time period of the motor signal presence. In fact, the period in which the transient signals are present greatly affects the processing quality of the extracted signals such as the motor current. Besides, the diminishing nature of the fault-related components during the transient operation makes the process of extracting fault information a complex procedure. This is the compelling reason for insufficiency of the existing literature in terms of the diagnosis tools in the transient operations. Companies and researchers are markedly focused on the steady-state operation unless the under-investigation motor is to perform in a highly stressful start–stop application. In high-power industrial applications, motors usually operate in a nonstop way unless a maintenance procedure is undertaken. Therefore, it is proposed to diagnose faults during start-up and prior to any aggressively growing fault level that normally happens in a lone-term steady-state operation.

Having focused on the discussion provided so far, the following guidelines are extracted.

- Any motor signal including the current, the flux, the electromagnetic torque and the speed consists of two main regimes, the transient and the steady-state operations.
- The transient operation reveals relatively large amplitude but a short-time presence of the motor signals.
- The motor signals possess a highly nonlinear time-varying behavior in the transient operation.
- In the steady-state operation, a smooth and favorably nonoscillating motor signal is observed in the case of a healthy motor. If faults occur, the situation becomes totally different, and a highly oscillating and nonsmooth motor quantity should be detected depending on the fault type, the location and the severity.
- The smoothness of the healthy signals depends on the number of slots, the number of bars, the skewing angle, the saturation profile, etc.
- Increasing the load level increases the motor rise-time, the time that motor speed comes to a stable point.
- Increasing the load increases the developed electromagnetic torque. This makes sense because the equilibrium of (2.19) should be always valid.
- The transient torque and current levels are considerably larger than that of the steady-state one. This is why the motor starts up very fast. Sometimes, the start-up period is typically less than a second, depending on the motor size, but it is not larger than a couple of seconds in general.
- The steady-state current amplitude increases by increasing the load level.
- There are undershoots associated with the electromagnetic torque right before the rise-time of the motor speed.
- There are overshoots associated with the motor speed right before the rise-time.
- In line-start applications, the motor speed increases almost exponentially while it is linearly controlled in the inverter-fed applications.
- Undershoots and overshoots are easily controlled and eliminated if a well-tuned drive is connected to the motor. If so, the amplitude of the transient parts of the motor current is constant in time. As a result, an almost constant electromagnetic torque is developed in the motor. This reduces the motor stresses, but it challenges the diagnosis process.
- Overload applications, larger than 100% of the full load are also possible.
- Note that the no-load operation does not mean a zero-current level. Normally, the no-load current possesses one-third, more or less, of the full-load current. This part of the motor current is called magnetizing current. Usually, it is very difficult, sometimes impossible, to diagnose the broken bar fault in no-load operation.
- Sometimes, the motor speed or load level changes from a value other than zero. Actually, the operation is not a zero-starting one, and the speed has a predetermined value, below or over the rated speed. The rated speed is the level at which the motor load is equal to the rated value. The term "rated" means that all the motor quantities including the voltage, the current, the

torque and the speed are set at their nominal values. Therefore, the output power is equal to the rated value on the motor nomenclature plate. In this case, the voltage is definitely equal to the rated value.

It is important to note that the steady-state speed level decreases by increasing the load/torque level (see Figure 2.10(c)). This arises from the fact that the rotor speed is **NOT** equal to the synchronous speed. This concept which is called the "motor slip" is thoroughly discussed below (see (2.20)–(2.27)).

$$\text{Electrical synchronous angular frequency or } \omega_s = 2\pi f_s = \frac{15pn_s}{\pi}\left(\frac{\text{rad}}{s}\right) \tag{2.20}$$

$p$: Number of motor poles, $f_s$: Electrical synchronous frequency (Hz) (2.21)

$$\text{Mechanical synchronous speed or } n_s = \frac{120f_s}{p}(\text{rpm}) \tag{2.22}$$

$$\text{Rotor mechanical speed or } n_m = (1-s)n_s(\text{rpm}) \tag{2.23}$$

$$\text{Rotor mechanical angular speed or } \omega_m = (1-s)\omega_s(\text{rad}/s) \tag{2.24}$$

$$\text{Slip or } s = \frac{\omega_s - \omega_m}{\omega_s} = \frac{f_s - f_m}{f_s} = \frac{n_s - n_m}{n_s} \tag{2.25}$$

$$\text{Rotor electrical frequency or } f_r = sf_s(\text{Hz}) \tag{2.26}$$

$$f_s = f_r + f_m \tag{2.27}$$

The motor slip is, in fact, the most noticeable factor in induction motor operation such that in zero slip, there is no induced voltage in the rotor bars, and the corresponding currents are zero. As a result, there will be no field produced by the rotor for developing the electromagnetic torque. This is indeed why these kinds of machines are called "induction motors" since the main operating principle is based on the induced currents. Increasing the slip means decreasing the rotor mechanical speed and therefore increasing the induced voltage and current quantities in the rotor side. On the other hand, increasing the motor load increases the motor current which in turn causes an increase in the rotor bar currents. Hence, it is expected to have a larger slip value upon increasing the motor load and this is crucial for developing a higher developed electromagnetic torque. As a result of increasing the slip, according to (2.23) and assuming that the motor synchronous frequency is fixed, the rotor speed, $n_m$, should decrease, and the same trend is observable in Figure 2.10(c). A better realization of the mentioned points could be achieved by monitoring the steady-state operation of the motor quantities. The transient operation is a little bit challenging due to the fact that the frequency components of the motor quantities expect those which are related to the source increase gradually during the motor start-up from a zero speed to the steady-state mode. If the motor is subjected to a de-acceleration mode, the frequency components, specifically those of the rotor would represent a totally opposite behavior compared to the acceleration mode. It should be noted that as long as the supplied synchronous frequency is fixed (the case in any line-start application), the only load-dependent quantities would be those of the rotor only if an ideal single harmonic motor is tested or analyzed. Such kind of motors does not exist and all types of existing induction machines, whether operating in the motoring or generating modes, reveal some sort of undesirable harmonic components producing load- and so slip-dependent frequency components even in the stator side. All the claims provided in this chapter will be shortly addressed quantitatively or by means of the analytical solutions. However, since the main focus of this chapter is to provide the essential healthy motor information, it is preferred to skip the details of the harmonic components which might be used in fault diagnosis applications and switch to the main discussion on the principles of the motor operation. Hence, looking back at Figure 2.6 in which the stator and rotor coils have been demonstrated as a set of concentrated-at-one-point windings, and assuming that the supply or synchronous electrical frequency is $f_s$, and the rotor bars or windings electrical frequency is $f_r$, the back-EMF equations in a single harmonic system can be expressed as follows:

$$\begin{aligned} E_{sa} &= 4.44 N_s \lambda_{\text{sam}} f_s \\ E_{ra} &= 4.44 N_r \lambda_{\text{ram}} f_r = 4.44 N_r \lambda_{\text{rbm}} s f_s = s E_{ra0} \end{aligned}$$

$$\text{where} \left\{ \begin{array}{l} E_{\text{sa}} \rightarrow \text{rmsvalueofstatorback} - \text{emf} \\ E_{\text{ra}} \rightarrow \text{rms value of rotor back} - \text{emf} \\ E_{\text{ra0}} \rightarrow \text{rms value of rotor slip} - \text{free back} - \text{emf} \\ \lambda_{\text{sam}} \rightarrow \text{Amplitude of stator flux linkage} \\ \lambda_{\text{ram}} \rightarrow \text{Amplitude of rotor flux linkage} \end{array} \right\} \tag{2.28}$$

According to the theory of rotating fields, the three-phase stator currents with the electrical and spatial phase angle of 120° away from each other generate a rotating magnetic field with a fixed magnitude, but a changing position in space. The electrical frequency of rotation is equal to $f_s$. Based on the Faraday's law, the rotating field passes across the rotor coils or bars and induces an EMF which is the main source of generating currents of the rotor conductors. Subsequently, the magnetic field of the rotor is generated. If the motor starts up from a zero speed, $f_r$ gradually decreases, so the corresponding EMF decreases also decreases. The flux linkage shown by $\lambda$ in (2.28) is the replacement for the pole cross section area multiplied by the air gap flux density. The magnetic quantities including the flux and the flux density will be further analyzed since these are actually the main driving forces of induction motors. For now, we assume that all the mentioned quantities including the voltage, the current, the flux-linkage, the back-EMF and the magnetic flux density are pure sinusoidal function of time and position of the rotor. The only difference is between the magnitudes as well as the phase angles compared to a common measure which is the voltage phase angle. Bearing this in mind and considering Figure 2.6, every stator and rotor windings are modeled as series-connected inductances and resistances which represent the inductivity and

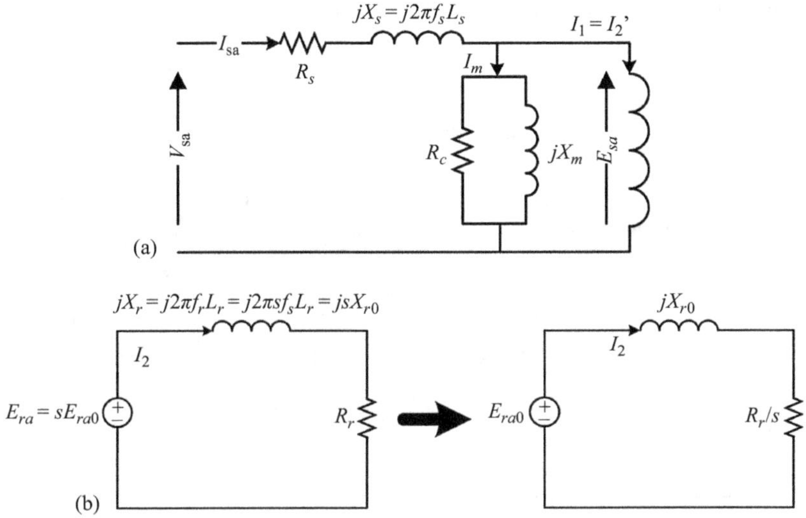


*Figure 2.11 Electrical equivalent circuit of (a) stator and (b) rotor*

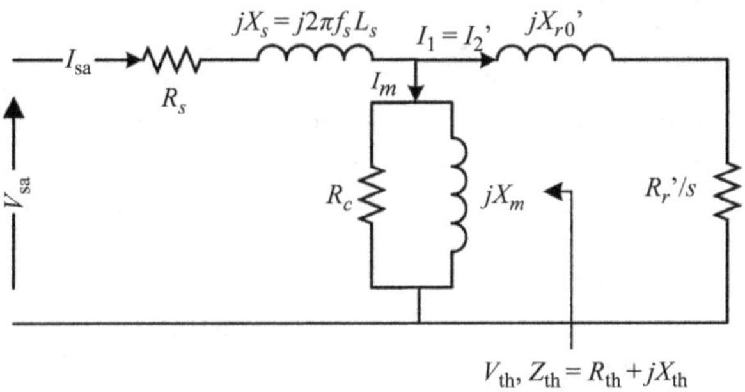


*Figure 2.12 Electrical equivalent circuit for an induction motor*

the resistivity of the windings, respectively (see Figure 2.11). For as much as the inherent operating principle of induction motor is based on induced magnetic fields, it is usually modeled as a rotating transformer in which the rotor side parameters depend on the motor slip.

In Figure 2.11, $R_c$ and $X_m$ are the equivalent resistance of the core losses and the magnetizing inductance, respectively. $L_r$ and $L_s$ are the leakage part of the rotor and stator winding inductances. If the ratio of the effective number of turns $n_{\mathrm{eff}} = \frac{N_s}{N_r}$ is applied to the rotor circuit with the goal of referring the elements to the stator side, we will end up with the electrical circuit shown in Figure 2.12 in which $R_r' = n_{\mathrm{eff}}^2 R_r$, $X_{r0}' = n_{\mathrm{eff}}^2 X_{r0}$.

In Figure 2.12, $V_{\mathrm{th}}$ and $Z_{\mathrm{th}}$ are the Thevenin equivalents of the voltage source and the impedances on the stator side, respectively. By incorporating the Thevenin equivalents and solving the circuit for $I_1$, (2.29) is obtained:

$$I_1 = \frac{V_{\text{th}}}{\sqrt{\left(R_{\text{th}} + \frac{R_r'}{s}\right)^2 + \left(X_{r0}' + X_{\text{th}}\right)^2}} \tag{2.29}$$

So the absolute value of the active power transferred to the rotating mechanical parts of the motor can be calculated by the following equation:

$$\begin{cases} P_m = \left(\frac{1}{s} - 1\right) R_r' I_1^2 = \dfrac{(1-s) R_r' n V_{\text{th}}^2}{s\left(\left(R_{\text{th}} + \dfrac{R_r'}{s}\right)^2 + \left(X_{r0}' + X_{\text{th}}\right)^2\right)} \\ P_m = T_e \omega_m \end{cases} \xrightarrow{\text{Equality of the two } P_m \text{ equations}}$$

$$T_e = \frac{R_r' n V_{\text{th}}^2}{s\omega_s\left(\left(R_{\text{th}} + \dfrac{R_r'}{s}\right)^2 + \left(X_{r0}' + X_{\text{th}}\right)^2\right)} \tag{2.30}$$

The number of phases, $n$, also contributes to the development of electromagnetic torque (see (2.30)). The only rotor component consuming the active power is $\frac{R_r'}{s}$. However, not all of its portion is transferable to the motion parts, and a part of it disappears through the copper losses consumed in $R_r'$. Therefore, the active part of the rotor resistance would be equal to $\left(\frac{1}{s} - 1\right)R_r'$.

When it comes to the motor torque–speed profile analysis, (2.30) is exactly what is usually referred to. As seen, the above equation relates the torque to the voltage and the slip which together determine the operating point of the machine, of course in the steady-state operation, in both generating and motoring mode. Figure 2.13 illustrates the torque–speed profile of a line-start induction motor in the

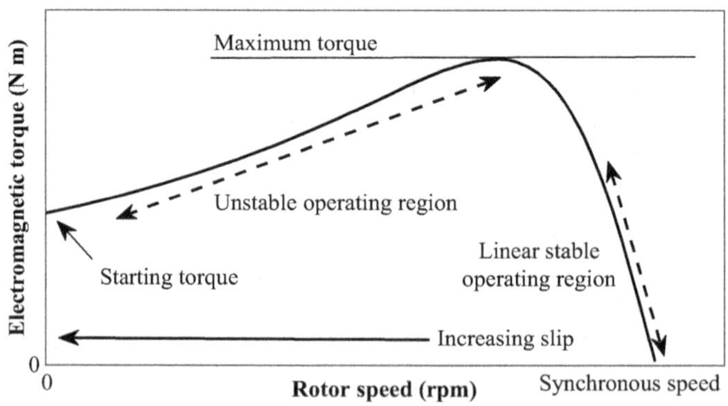


*Figure 2.13 Steady-state torque–speed profile of an induction motor*

steady-state operation. The maximum torque and the points behind the maximum torque, the speed values ranging from zero to the corresponding point of the maximum torque, are called "Unstable operating regions" in which the motor does not usually operate unless transient operations are dealt with. On the other hand, the "linear stable operating region" is the most probable operating point of an induction motor. There is no shortage of agreement that an increasing motor load leads to a decrease in the motor speed and consequently the slip value. If the motor slip is equal to 1, the starting torque is obtained at the speed of zero. Moreover, the developed torque is zero at the synchronous speed.

Different from the demonstrated steady-state torque–speed profile in which the motor comes to a stable operating condition for every single pair of torque and speed, the transient behavior is somehow distinct particularly for the lower speed ranges (see Figure 2.14). Although Figures 2.13 and 2.14 are similar in many ways, they are totally different representations of the motor behaviors. In Figure 2.13, the effect of time has not been included while the second figure is obtained by sampling the torque and the speed at different time steps during a constant load of 50 N m. However, the load is changed and the corresponding steady-state speeds are measured to obtain Figure 2.13. A highly fluctuating torque component is observed at the very beginning instances of Figure 2.14, i.e. a wide range of [−500,500] N m is detected. This part of the torque profile, along with the following torque curve, forms the transient mode of the motor. Finally, the motor torque–speed profile comes to a stable point at around 50 N m (Figure 2.14) which is equal to the motor average torque in the steady-state operation. The corresponding speed is a little bit away from the synchronous speed and demonstrates the slip phenomenon. During the steady-state operation, the torque–speed profile is completely focused on one single point. There would only be a minor discrepancy caused by the inherent oscillations of the motor torque and speed components. The situation gets worse in case of the fault occurrence in which a bunch of undesirable fluctuations are added not only to the end-point of Figure 2.14, shown as the "Final steady-state torque"

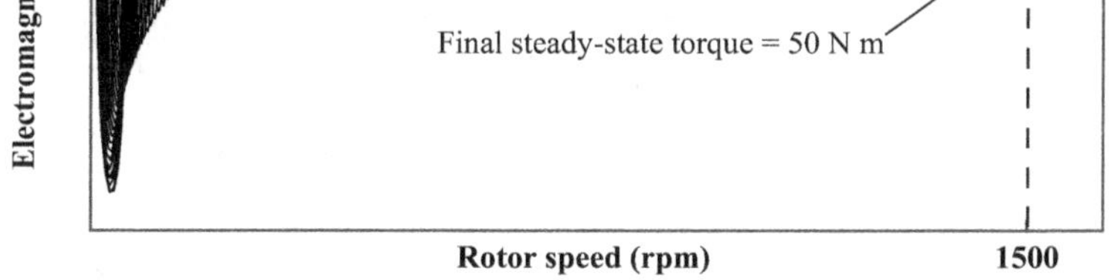


*Figure 2.14 Transient torque–speed profile of an induction motor*

but also to the transient operation. This issue will be further discussed when the fault diagnosis concepts are initiated in the next chapters. The transient profile illustrated in Figure 2.14 is obtained by simulating the single-harmonic motor model described in (2.1)–(2.19).

A very small part of the motor speed goes over the synchronous speed, 1,500 rpm, in Figure 2.14. This is related to the speed overshoot, also visible in Figure 2.10, during the transition through the steady-state operation. The highly fluctuating transient regime makes it difficult, sometimes impossible, to extract proper fault indicators and that is why the fault diagnosis during the start-ups/stops is a rather new and intact field of study. However, this regime only matters when a large-scale machine with a considerably large start-up period, for example a couple of seconds, is under investigation. In the case of low- or medium-power applications or the applications in which the motor mostly operates in the steady-state mode, the transient analysis in not worth being analyzed. It is also interesting to know that incorporating the drive circuits into the fault diagnosis process brings up new challenges, specifically in the transient mode, since the motor behavior is not merely a function of the load any more. It is besides affected by other factors such as the variation of the voltage and frequency which in turn introduce new burdens to the problem. For instance, changing the motor frequency changes the magnitude, and the speed of the magnetic rotating field of the motor, and consequently all the quantities are affected. As a result, an extremely different motor behavior should be expected for inverter-fed applications. Nowadays, all the change in the motor voltage and frequency is handled by a control device which is called "drive." This so-called drive device generally contains two main parts, namely the control unit and the inverter shown in Figure 2.15. The control unit is somehow straightforward

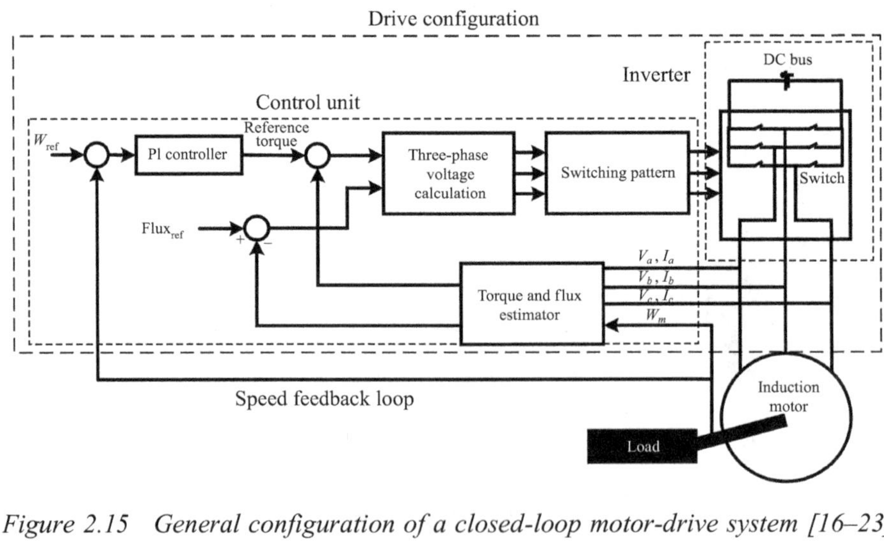


*Figure 2.15 General configuration of a closed-loop motor-drive system [16–23]*

and takes care of translating the input commands in terms of the motor speed, torque or flux into the electrically applicable waveforms. The major responsibility of the control unit is to find an appropriate switching pattern applied to the inverter switches which modulate the DC bus voltage into the motor terminal voltages with a specific fundamental frequency component. Generally speaking, the whole drive circuit operates as a translator from a reference or command value to real supply voltages applied to the motor terminals. By means of this configuration, both the supply voltage and the frequency can be finely controlled and adjusted. There can be a speed feedback loop as well. If so, the corresponding configuration is called "closed-loop control strategy." Otherwise, the strategy would be an "open-loop" one. Both the methods are shown and discussed in the next subsection. It is noteworthy that the torque, the speed, the voltage, the current and also the flux could be controlled by drives. A closed-loop configuration operates better than an open-loop one since there is a direct control on the speed or even the torque. In Figure 2.15, the terms $\omega_{\text{ref}}$ and $\text{flux}_{\text{ref}}$ are the reference values of the speed and the flux, respectively.

Now, let us take a brief look at different drive types. Then, we will move toward the detailed analysis of their operating principles. The following strategies are essentially the most traceable techniques existing in the literature:

- Scalar Constant Voltage per Frequency Control (CV/F)
- Field-Oriented Control (FOC)
- Direct Torque Control (DTC).

Although the closed-loop implementation is also available, the first type is usually preferred for open-loop applications in which several motors should operate in parallel at the same time. The other two options are used in a closed-loop form in which a single motor is connected to the drive circuit. The CV/F strategy is more often used as a part of the other two general types in order to handle the acceleration or de-acceleration of the motor. The underlying idea of all the methods is based on controlling the torque and flux commands applied to the motor separately like what is done in DC motor control. Since quantities of induction machines are highly inter-related, unlike the DC machines, it is tried to first isolate the flux and torque components by any means, for example a mathematical transformation from a time-dependent to a time-independent space, and then control the quantities separately. After generating proper torque and flux commands, the values should be translated into the voltage signals and this is exactly where the switching pattern plays its role in the motor-drive system operation. In practice, there are many existing patterns each of which has its own advantages and disadvantages while the most frequent one is the space vector modulation (SVM) technique regardless of its different extensions. The switching pattern outputs the signals applied to the switches connecting the DC bus voltage to the motor terminal voltages. By means of the briefly discussed configuration, the speed and the torque can be independently controlled and adjusted depending on operating point characterized by application requirements.

There is a wide range of operating conditions in inverter-fed induction motors compared with the line-start one and this is the main difference between the line-start and inverter-fed applications. In a line-start mode, the only controllable quantity is the amplitude of the motor voltage while the supply frequency, the speed and the motor torque can be easily controlled in the inverter-fed mode as well. This, in fact, makes the diagnosis process more troublesome since the number of factors influencing the fault indicators definitely increases. This is indeed the state of the art nowadays, and many researchers are trying to first realize how faulty motor quantities are affected if drives are incorporated and then find a novel solution or at least a complement for previous line-start-related diagnosis technique in order to outline the fault diagnosis of inverter-fed machines. Although attempts have been initiated couple of years ago, the idea is still new and worth dealing with.

Having discussed the major idea behind using drives and their applications in electrical motor industries, now it is time to go in depth of its principles, not very technical rules in terms of the structures or the design process of the drives, but a more comprehensive behavioral study of its effect on the motor quantities and how they should be translated into the drive circuit signals. In view of fault diagnosis procedures of electrical machines, what really matters is to process the motor signals, either the original or transformed signals, and extract the fault information out of the analysis. So we do not aim at working around different type of switches, various inverter topologies or switching techniques which might not be directly related to the condition monitoring of electrical machines. However, no one can deny that any of the mentioned factors might also have an impact on the quality of the process, but this is an absolutely uninvestigated problem, and there is no information on it.

## 2.4 Inverter-fed induction motors

### *2.4.1 Constant voltage per frequency strategy (CV/F)*

The CV/F is one of the popular scalar methods of control in induction motors. There are generally both the open-loop and closed-loop modes of this control strategy. It is preferred to start the discussion by targeting the open-loop regime. The corresponding closed-loop strategy differs in only one feedback loop which is used to stabilize the motor speed at a reference value. The underlying idea comes from a very basic relationship as $\text{Motorflux}(\Lambda) \propto \text{Voltage}/\text{frequency}$ (see (2.28)). If the number of turns and also the motor geometry is fixed, the flux is proportional to the supply voltage and inverse-proportional to the supply frequency. Figure 2.16 shows the real variation trend of the motor voltage versus its frequency. There is a nonlinear initial curve located in $[0\text{–}f_{\text{cutt-off}}]$ Hz related to the part of the voltage which should be added to the minimum voltage applied to the motor to compensate for the voltage drop across the stator winding resistance. The typical value should be around 10% of the rated voltage where the reference frequency or speed is zero.

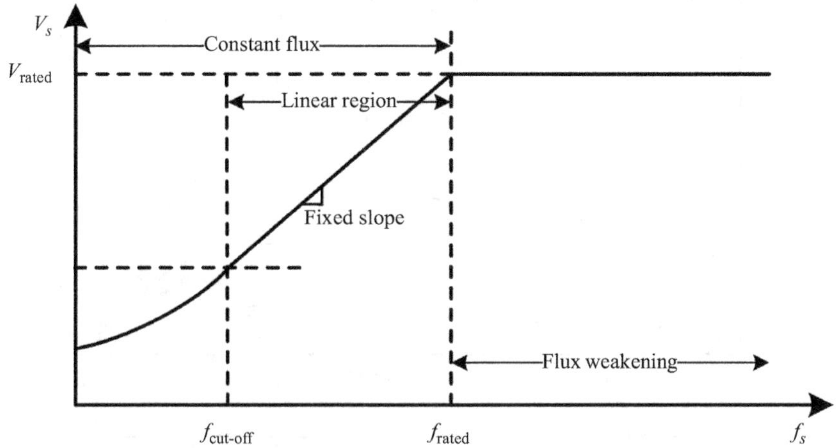


*Figure 2.16 Stator voltage versus supply frequency*

Then the voltage or rather equivalently and more technically the EMF should follow a linear direction toward the rated values with a constant voltage/frequency which represents the line slope. Practically speaking, when the voltage comes across the rated value, the rated frequency should also be satisfied. This is an extremely important requirement ensuring a rated flux at this point. Along the linear region, there is a perfectly tuned rated flux. Otherwise, the motor power density falls below its maximum capability if the slope is less than the rated. On the other hand, if the slope is larger than enough, the magnetic flux and consequently the magnetic losses increase dramatically. This is the reason that all the electrical machines are designed in a way to operate around the knee point of the magnetic saturation profile.

Beyond the rated frequency, a very well-known operation, the so-called Flux weakening is present. Although the frequency can be freely increased up to a certain high value, there is a limitation on the motor voltage which should be restricted to the tolerable insulation limit. Otherwise, insulations would fail. Therefore, the voltage/frequency decreases by increasing the frequency beyond the rated value. In this operating mode, the motor definitely possesses a speed higher than that of a normal line-start mode. However, the speed or frequency cannot be raised infinitely due to the mechanical and magnetic losses limits. So there should always be trade-off between the speed and the efficiency of the motor.

This simple but practical curve (see Figure 2.16) determines the possible range of variation of the torque–speed of the induction motors controlled by inverters. This curve includes the constant flux and flux weakening regions which are basically brought up by the primary torque–speed profile shown in Figure 2.13 by assigning various voltage and frequency to (2.30) concerning the voltage limits of the motor. The same is done by changing the frequency and the voltage from very

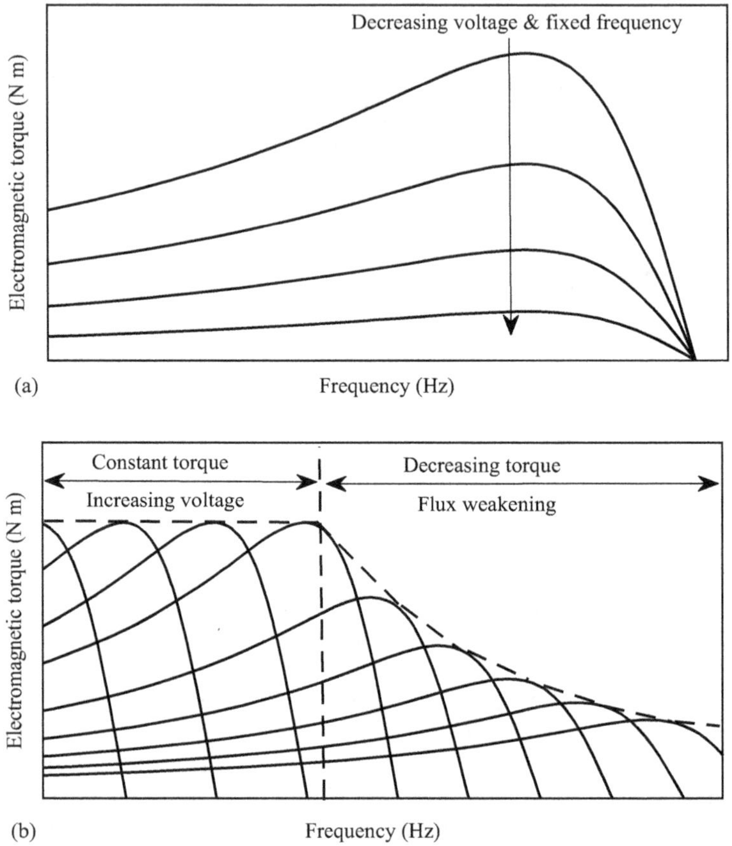


*Figure 2.17 (a) Torque–speed profile for different voltages in line-start mode and (b) torque–speed profile for different voltages and frequency in inverter-fed mode*

small values up to their rated and even larger values in the case of supply frequency. Then the torque–speed curves are plotted in Figure 2.17(b). In addition, the effect of a variable voltage and fixed frequency in line–start mode is observed Figure 2.17(a). As long as the synchronous frequency is fixed, the curves are not shifted toward the positive frequencies, and the change is only applied to the torque-axis which is proportional to voltage squared. This means that if the voltage becomes half of the rated value, the motor torque productivity is reduced by a factor of 4. This considerably affects the motor start-up quality due to the fact that the starting torque decreases as well. Decreasing the motor voltage in the line-start mode is also a reason for increasing the slip losses and this can undeniably be understood from the decreasing slope of the stable linear region of the profile (see Figure 2.17(a)).

In the same manner, a comprehensive judgment in terms of the inverter-fed induction motor could be held herein as follows:

- The vertical line of the rated frequency defines two distinct operating modes, namely the constant flux (constant torque) and flux weakening (constant power) regions.
- Below the rated frequency, the maximum torque (breakdown torque) is fixed upon keeping the voltage/frequency ratio fixed. In fact, the voltage is kept increasing linearly as a function of the frequency up to the rated value which is required in the rated frequency point. In this case, the curves look like each other but with a specific frequency shift.
- The slope of the linear region of the curves is the same; hence, it is expected to have a close to similar slip losses in the constant torque region.
- Increasing the supply frequency beyond the rated values demands a decrease in the terminal voltage in order to follow the insulation safety. Accordingly, the motor flux and the torque productivity of the motor are reduced. An almost exponential reduction rate is normally observed for regular motors.
- Not only is there a frequency shift of the curves, but also the break-down torques gradually decrease by increasing the frequency.
- The variation observed in Figure 2.17(b) explicitly explains the claim that torque and the speed can be adjusted independently if an inverter-fed application is targeted. Specifically, the constant flux region clearly reveals this fact.

As seen, the reference torque and speed can also play a role in determining the motor behavior based on which any fault diagnosis procedure is performed. Therefore, it is not only the motor load affecting the detection process, but several other factors including the control strategy, the reference speed, the torque and the DC bus voltage might also have an impact on the process. This is the reason why we insist on getting familiar with the behavior of the motor quantities in both the line-start and inverter-fed cases and discriminating between them if needed. Sometimes, instead of the motor original signals such as the current, their transformations used in drive circuits are utilized for the fault diagnosis purposes. On the other hand, these transformations are simply available in the most industrial drives such as ABB ACS800.

The CV/F control strategy is implemented based on intentionally following the inherent curves of the motor in terms of the voltage, the frequency, the torque and the flux with the goal of pertaining the safe and also possible motor limits. This is handled by means of the control strategy shown in Figure 2.18. This is a demonstration of an open-loop strategy while a close-loop one is also possible and will be discussed soon.

The references could be the speed in RPM or the frequency in Hz. No matter which one is selected, the frequency control unit takes care of transforming the input into the electrical frequency, using (2.20)–(2.22). The voltage regulator keeps following the $V/F$ curve illustrated in Figure 2.16. What it does is to define the maximum terminal voltage, $V_m$, based on the voltage limits forced by the $V/F$ curve.

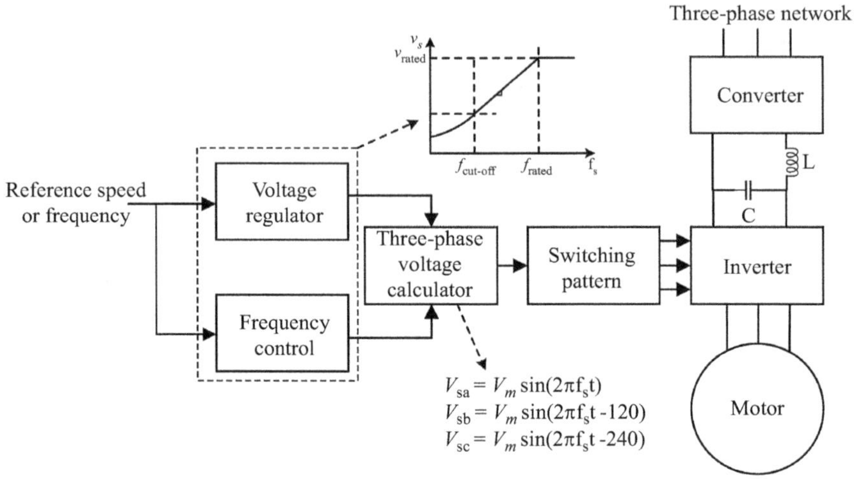


*Figure 2.18 Open-loop constant voltage per frequency drive*

To do so, the rated values should be definitely assigned to the drive. This is the first step that an operator needs to do when dealing with a drive. In the next step, the reference voltage commands are generated based on the reference frequency/speed and are passed through the switching topology to generate the switching pulses. Finally, the terminal voltages of the inverter connected to a DC bus the motor runs. Since there is no feedback loop applied to the electrical or mechanical motor signals, this topology is inherently an open-loop control mode. It is not guaranteed to have a very small current or torque ripples in practice. Moreover, the motor speed is always smaller than the applied reference due to the fact that induction motor operates based on the slip phenomenon which causes a speed reduction in the rotor compared to the synchronous speed.

There is also a combination of an inductance and a capacitor between the converter and the inverter (see Figure 2.18). This combination acts as a perfect filter as well as the DC bus maintaining an almost constant DC voltage on the bus. The DC current flowing into this filter is sometimes referred to as a proper index of the broken bars fault in inverter-fed applications. Therefore, any part of the drive ranging from the control circuit to the DC bus might be affected by faults.

Like what has been mentioned before, there are several topologies to be hired at this point. However, just in the case of introducing the fundamentals and also providing industrial points of view, we will try to go through the SVM technique which is dominantly used in industry.

### 2.4.2 Space vector modulation

SVM technique is the most popular switching topology in industry. The most interesting advantage of this technique is its capability to simply control the amplitude

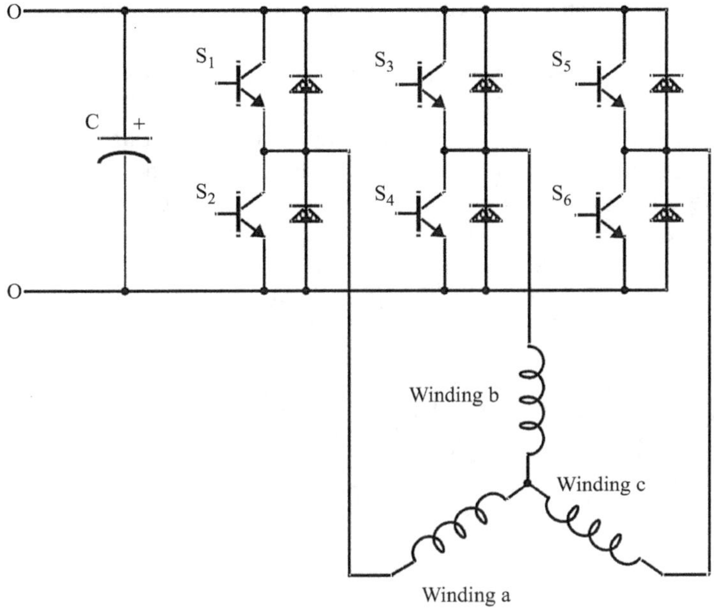


*Figure 2.19 A two-level three-phase inverter*

*Table 2.3 SVM switching states*

| Vector | $S_1$ | $S_3$ | $S_5$ | $S_2$ | $S_4$ | $S_6$ | $V_{AB}$ | $V_{BC}$ | $V_{CA}$ | |
|---|---|---|---|---|---|---|---|---|---|---|
| $V_0 = \{000\}$ | OFF | OFF | OFF | ON | ON | ON | 0 | 0 | 0 | Zero vector |
| $V_1 = \{100\}$ | ON | OFF | OFF | OFF | ON | ON | $+V_{dc}$ | 0 | $-V_{dc}$ | Active vector |
| $V_2 = \{110\}$ | ON | ON | OFF | OFF | OFF | ON | 0 | $+V_{dc}$ | $-V_{dc}$ | Active vector |
| $V_3 = \{010\}$ | OFF | ON | OFF | ON | OFF | ON | $-V_{dc}$ | $+V_{dc}$ | 0 | Active vector |
| $V_4 = \{011\}$ | OFF | ON | ON | ON | OFF | OFF | $-V_{dc}$ | 0 | $+V_{dc}$ | Active vector |
| $V_5 = \{001\}$ | OFF | OFF | ON | ON | ON | OFF | 0 | $-V_{dc}$ | $+V_{dc}$ | Active vector |
| $V_6 = \{101\}$ | ON | OFF | ON | OFF | ON | OFF | $+V_{dc}$ | $-V_{dc}$ | 0 | Active vector |
| $V_7 = \{111\}$ | ON | ON | ON | OFF | OFF | OFF | 0 | 0 | 0 | Zero vector |

of higher-order harmonic components of the motor voltage, so the THD is somehow controllable. It is normally based on determining the exact period in which switches should be on or off. Before going through analyzing SVM, we will have a look at a typical two-level three-phase inverter shown in Figure 2.19.

Assuming that not both switches in one leg become on or off at the same time, there can be eight possible switching states for this circuit (see Table 2.3). The DC bus voltage is often divided into two parts between the capacitors each containing half the total voltage, and the middle point of the capacitors is assigned as the neutral point. Based on this convention, the maximum and minimum phase-neutral voltages are $+V_{dc}/2$ and $-V_{dc}/2$, respectively.

According to the definitions, when all upper or lower switches of all the legs are off or on, the state is called "zero vector." Otherwise, it is called "active vector." The underlying idea is brought up by a vector representation of the voltage in the stationary reference frame. Given a group of switch states and phase voltages (see Figure 2.19), the vector components of the vector representation are derived as follows:

$$\vec{V} = V_\alpha + jV_\beta = \frac{2}{3}\left(V_{A0}\vec{a}^0 + V_{B0}\vec{a}^1 + V_{C0}\vec{a}^2\right) \quad \text{where} \quad \vec{a} = e^{j\frac{2\pi}{3}} \tag{2.31}$$

$$\begin{bmatrix} V_\alpha \\ V_\beta \end{bmatrix} = \frac{2}{3}\begin{bmatrix} 1 & \frac{-1}{2} & \frac{-1}{2} \\ 0 & \frac{\sqrt{3}}{2} & -\frac{\sqrt{3}}{2} \end{bmatrix}\begin{bmatrix} V_{A0} \\ V_{B0} \\ V_{C0} \end{bmatrix} \tag{2.32}$$

$$|\vec{V}| = \sqrt{V_\alpha^2 + V_\beta^2}, \qquad \alpha = \tan^{-1}\left(\frac{V_\beta}{V_\alpha}\right) = \omega_s t = 2\pi f_s t \tag{2.33}$$

As a well-known fact, there is a space vector of constant magnitude but rotating in real-imaginary plane if a set balanced three-phase network voltages is considered. The angular frequency of the rotation is equal to $\omega_s = 2\pi f_s$. The vector representation of (2.31) turns into zero if the zero vector states described in Table 2.3 are met. Generally, the zero along with the nonzero vectors are demonstrated in a $\alpha$–$\beta$ plane illustrated in Figure 2.20.

Assuming a voltage vector located in the first slice of Figure 2.20, between $V_1$ and $V_2$, the voltage vector V can be modulated into the $V_1$ and $V_2$. This means that

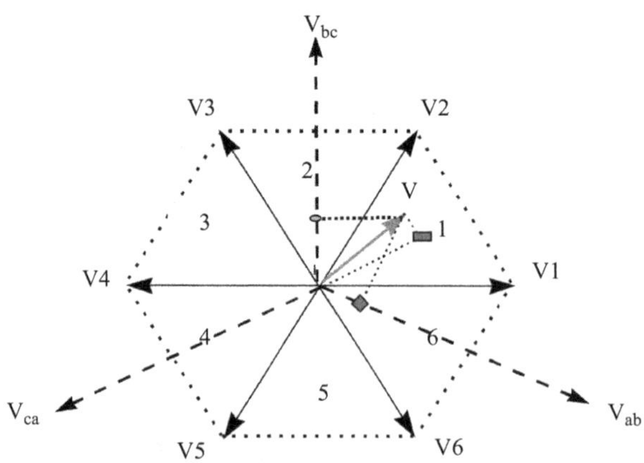


*Figure 2.20 Vector representation of the stator reference voltages*

the vector $V$, the reference voltage, can be obtained by sum of two adjacent vectors, $V_1$ and $V_2$ each presenting a specific period of time, $T_1$ and $T_2$, within the switching period, $T_s$. "T" defines the duty cycles of the switches. If $T_1 + T_2$ is less than half the switching cycle, the remaining period is handled by one of the zero vectors close to the vector $V$. The choice depends on the fact that transition from one state to the other one should be performed by changing only one leg of the inverter. In the next half cycle, the switching order is reversed.

$$T_s\vec{V} = T_1V_1 + T_2V_2 \tag{2.34}$$

$$T_1 + T_2 + T_0 = \frac{T_s}{2} \tag{2.35}$$

As a result of this modulation of the voltage, the terminal voltage would be constant for very short-time periods which in turn defines the concept of the pulse width modulation (PWM). Hence, the voltage consists of positive, negative and zero pulses and not a pure sinusoidal form any more. It is also possible to produce the pulses by adding nonadjacent voltage vectors which often produce a higher THD level, so it is not proposed in high-power applications.

SVMs are centered on calculating the correct duty cycles of the switches and this can be handled by taking the average value of the reference voltage in one cycle and setting it equal to the sum of the average of the vectorized voltages (see (2.36)). The calculations are done for the first sector including $V_1$ and $V_2$ and then extended to a general case. Symmetrically averaging during the switching period is the simplest continuous form of the SVM technique while other options also exist.

$$\int_0^{T_s} \vec{V}\,dt = \int_0^{T_1} V_1dt + \int_{T_1}^{T_1+T_2} V_2dt + \int_{T_1+T_2}^{T_s} V_0dt$$

$$\xrightarrow{\text{if the voltages are constant in one duty cycle}} T_s\vec{V} = T_1V_1 = T_2V_2 \rightarrow T_s|\vec{V}|\begin{bmatrix}\cos(\alpha)\\ \sin(\alpha)\end{bmatrix}$$

$$= \frac{2}{3}T_1V_{dc}\begin{bmatrix}1\\0\end{bmatrix} + \frac{2}{3}T_2V_{\text{dc}}\begin{bmatrix}\cos\left(\frac{\pi}{3}\right)\\ \sin\left(\frac{\pi}{3}\right)\end{bmatrix} \xrightarrow{\text{yields}} \begin{cases} T_1 = T_s\,.\,a\,.\dfrac{\sin(\pi/3-\alpha)}{\sin(\pi/3)} \\ T_2 = T_s\,.\,a\,.\dfrac{\sin(\alpha)}{\sin(\pi/3)} \\ T_0 = T_s - (T_1 + T_s) \end{cases}$$

$$\text{where } T_s = \frac{1}{f_{\text{switching}}} \quad \text{and} \quad a = \frac{|\vec{V}|}{\frac{2}{3}V_{\text{dc}}} \tag{2.36}$$

Switching time during any sector

$$T_1 = \frac{\sqrt{3}T_s|\vec{V}|}{V_{dc}}\left(\sin\left(\frac{\pi}{3} - \alpha + \frac{n-1}{3}\pi\right)\right)$$

$$T_2 = \frac{\sqrt{3}T_s|\vec{V}|}{V_{\mathrm{dc}}}\left(\sin\left(\alpha - \frac{n-1}{3}\pi\right)\right) \tag{2.37}$$

$$T_0 = T_s - (T_1 + T_s)$$

*where* $n = 1:6$ *defining the sector numbers*

Ultimately, the gate timings are shown in Figure 2.21. All the curves are the same except in the phase shifts which obviously show different on/off periods of the switches. The $x$-axis represents the period of the reference voltage which is equal to $1/f_{\mathrm{ref}}$ or $s$.

## 2.4.3 Analysis of motor behavior in open-loop CV/F mode

Having acquired the gate timings and applied them to the transistors (the switches), the line voltage is obtained. It is a PWM signal whose fundamental harmonic component is the same as the reference voltage calculated by the frequency control module implemented in the CV/F diagram (see Figure 2.18). Compare Figure 2.22 to one of the phase voltages in Figure 2.8 which corresponds to a line-start application. Assuming that the periods are the same, the two signals are totally different in shape and harmony. The aim of any PWM technique is to satisfy the possibility of different reference frequencies and fundamental voltages.

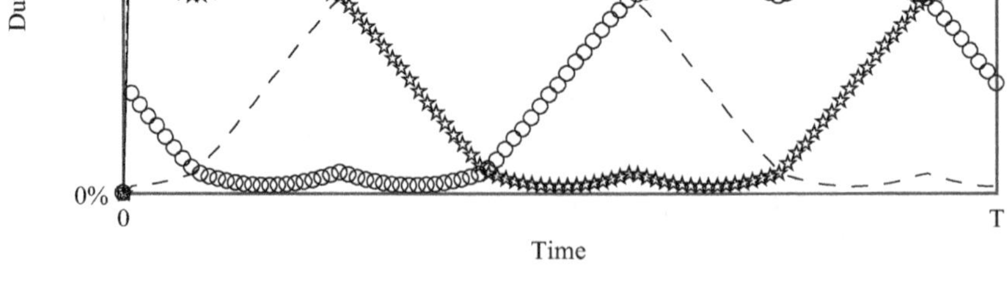


*Figure 2.21 Gate timings (duty cycles) of six switches*

However, it ruins the quality of the signal in terms of the THD which is certainly larger in the case of an inverter-fed machine. In fact, higher-order harmonic components, which produce the spike-like signals and are present in the supply voltage introduce considerably higher saturation and loss levels. Furthermore, they are mostly a source of elimination of the fault components, especially those located in higher frequency bands, for example higher than twice or triple the fundamental component. This is one of the trickiest but sometimes useful parts of the inverter-fed motor which facilitates the diagnosis process if utilized correctly. The discussion on harmonic components necessitates another important argument in terms of the Fourier Transform (FT) of the signals. It grounds some of the significant aspects of any diagnosis procedures based on which the harmonic components of the signals are analyzed and explored with the goal of finding proper fault indicators. A guideline on implementation and analysis of a Fourier Transform will be

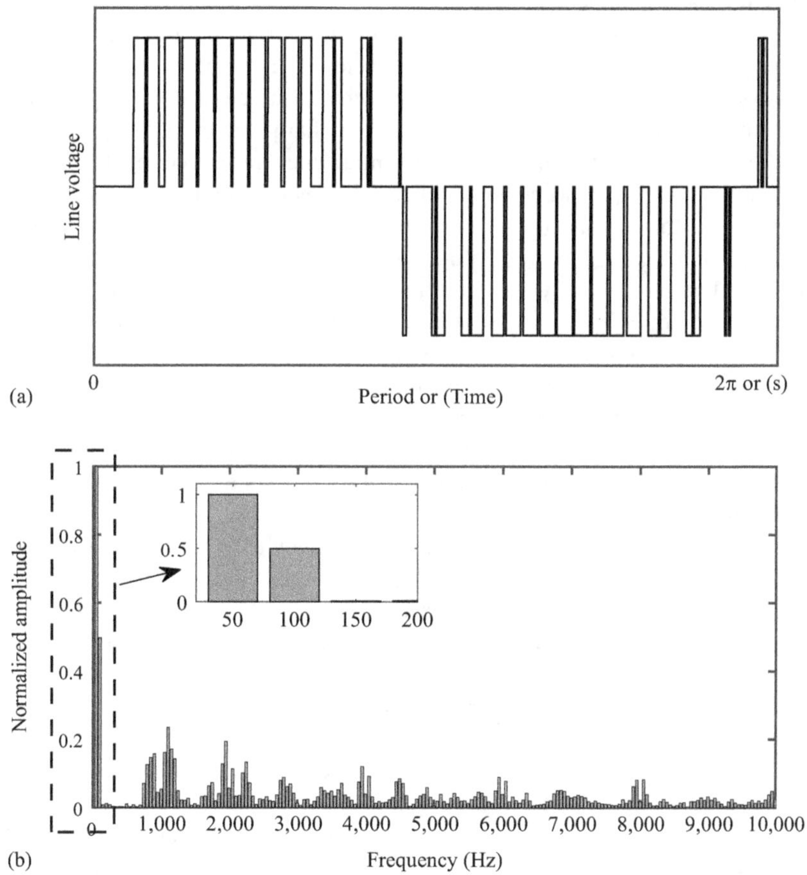


*Figure 2.22 (a) Time-domain line voltage in a symmetric SVM technique and (b) frequency-domain line voltage in a symmetric SVM technique*

provided in chapter 7. For now, assuming an existing implemented algorithm, the FT of the line voltage is taken and provided in Figure 2.22(b) within the range [0–10,000] Hz.

Accordingly, the following specifications are observed:

- There is an almost decreasing trend of amplitudes of the harmonic components upon increasing the frequency. This claims that the original time-domain signal contains a wide range of various-frequency sub-signals added to each other. However, the higher the component order toward the higher frequencies is, the smaller the amplitude will be.
- The largest amplitude corresponds to the fundamental component equal to 50 Hz in this study. This means that the frequency of the reference voltage applied as the input of the CV/F approach is equal to 50 Hz.
- The normalized amplitude of the largest component is equal to one, and the rest are less than 1. The spectrum has been normalized with respect to the fundamental frequency, 50 Hz.
- Although higher-order components exit in the supply voltage, they are not able to pass through the electromagnetic cycle of the motor because induction motors normally act as a low-pass filter which filters out the components with the order of larger than 20 to 25 times the fundamental component. This is due to the presence of a relatively large inductance produced by the windings. Depending on the winding inductance, different harmonic orders might pass into the motor operation and can be observed by means of monitoring the signals other than the motor voltage. Let us focus on the fault-related components. They are usually low-frequency components detectable around the fundamental one. Therefore, it is expected that a group of strong signals exist in the motor magnetic, electrical and may be mechanical components. This conveys the idea of exploring and monitoring the motor frequency signature and check which components are sensitive or insensitive to the fault, the load and the speed, etc. Almost all the fault diagnosis approaches utilize the harmonic components and their variations as a tool to detect, determine and locate different faults.
- Figure 2.22 includes a carrier/switching frequency ($f_{\text{carrier}}$) equal to 1 kHz. The corresponding period, $T_s$, used in (2.36), is equal to $1/f_{\text{carrier}}$. The larger the $f_{\text{carrier}}$ is, the larger the number of pulses in Figure 2.22(a) is. A higher-frequency resolution would also be achieved by increasing the carrier frequency.
- The amplitude of the voltage signal and the corresponding harmonic components directly depend on the DC bus voltage. Some researchers have claimed that changing the supply voltage amplitude or even the current does not change the normalized values; hence, they always rely on this normalized diagram in the diagnosis applications. We are going to show that this is a totally incorrect claim since the motor saturation level is definitely changed if the voltage or current are changed. Therefore, the fault components which are highly saturation-dependent undoubtedly change. This is a very demanding and at the same time challenging aspects being dealt with in this book in detail.

So far, we have explained the route to a proper voltage extraction out of the applied reference value which is regularly the motor synchronous speed or frequency. Following the CV/F diagram, the line voltages are generated out of the DC bus voltage and then applied to the motor terminals. The synchronous frequency is equal to the reference frequency, so the rotor should rotate with a frequency smaller than the synchronous one in order to satisfy the motor slip. The motor quantities including the current, the torque and the speed are shown in Figure 2.23. An acceleration unit can also be implemented prior to the reference value to control the motor transient start-up and make it a smoothly increasing regime. So the observed rattle behavior at the beginning of the curves shown in Figure 2.23 will become smoother. This is actually the case in industrial drives, although herein a harsh start-up such as what happens in the line-start mode is practiced just in the case of preparing fair comparisons. Moreover, the speed overshoot can be completely eliminated if a proper start-up curve, usually a linear one, is applied to the system. It is also important to note that a motor with different characteristics listed in Table 2.2 is used here. This is a 11 kW#, 4-pole, 400 V induction motor which has 36 stator slots and 28 rotor bars. The complete list of specifications will be further provided in the Chapter 6 related to the finite element-based modeling of this motor. The simulated model used in this case is not a single harmonic model, but a more comprehensive one in which the higher-order spatial harmonics are incorporated as well. The model is based on the winding function theory (WTA) and will be discussed in the chapter corresponding to the simulation process of faulty motors. However, introducing new components to the model, a more nonsmooth and/or nonsinusoidal signal is observed in Figure 2.23. The electromagnetic torque and the speed do not reveal a smooth steady-state performance and some obvious periodic oscillations are present which coupled with the previous discussion prove the existence of higher-order harmonic components in the motor-drive system. The components related to the switching phenomenon could be detected in the motor current, i.e. Figure 2.23(a), in which the ripples are carried by the main signal. Not to mention that the positive slopes of the ripples are relevant to the positive voltage pulses and vice versa. The switching-related ripples are barely detectable in the speed or the torque since the high-frequency components are usually filtered out of the mechanical signals due to the low-pass nature of the mechanical loop of the induction motor. This concept will be further proved mathematically. If the aforementioned acceleration block is added to the control strategy, the motor rise time is even controllable. As a technical issue, the amplitude of the ripples and so their improper effects on the motor operation can be reduced by increasing the switching frequency. Nevertheless, there should always be a trade-off between the switching frequency and the motor losses.

In the case where a single harmonic model is used for simulating the CV/F mode, the torque and the speed would be smooth enough even if the switching frequency is very small. This statement brings up the significant effect of the additional spatial harmonic components used in the winding function theory by means of which the components produced by the nonsinusoidal winding distributions and also

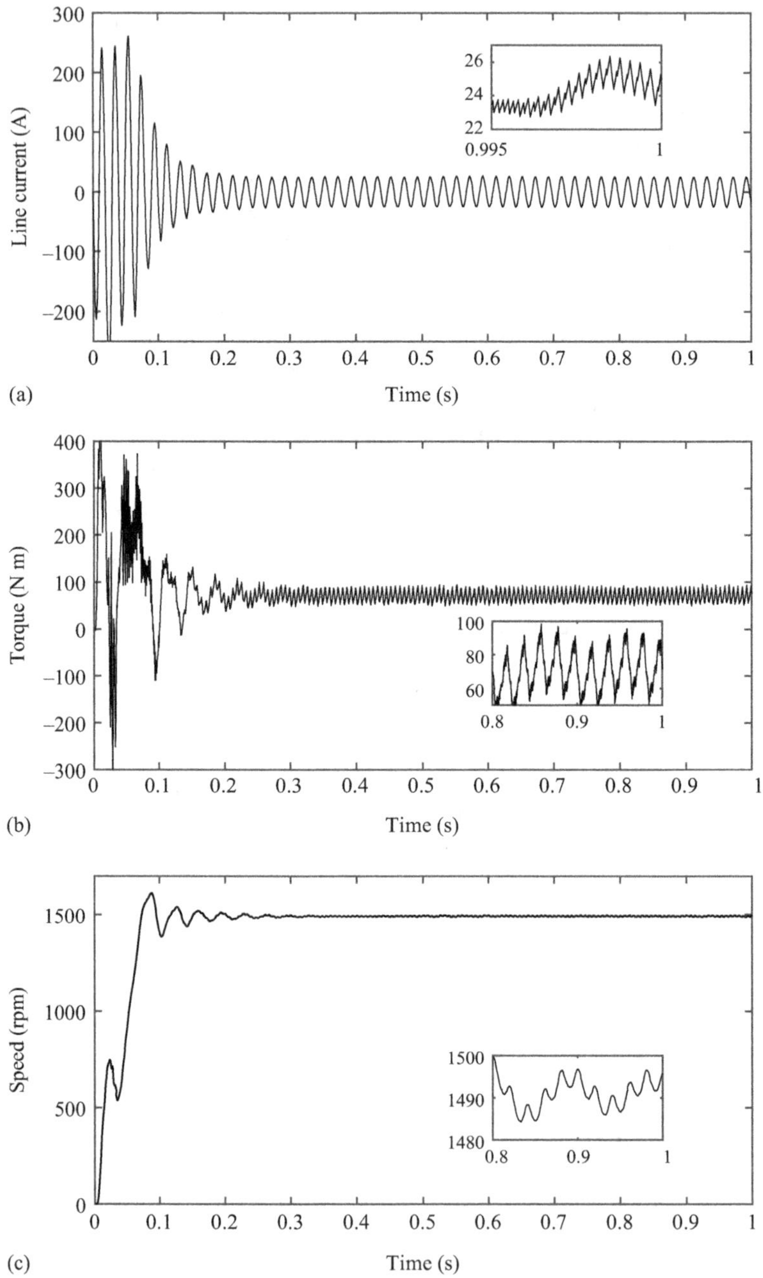


*Figure 2.23 (a) Current, (b) electromagnetic torque and (c) speed in CV/F mode without acceleration unit*

the slotting effects are taken into account. Therefore, it is claimed that the low-frequency torque and speed periodic oscillations are not because of the switching phenomenon. The reason is the use of the accurate motor model. These low-frequency components are so highly fault-sensitive that even a small level of the fault such as the bar breakage can change the amplitude of the oscillations. Surprisingly, the fault-related components already exist in the motor natural frequencies and the fault occurrence merely changes their amplitude. However, some of the researchers claim, of course by mistake, that the fault occurrence is a reason to introduce new harmonic components. Maybe, in some particular and rare cases, they are right.

Previously, it was noted that the fault-related components possess a variable trend in any transient mode, and hence any factor influencing the transient operation certainly affects the fault-related components. Therefore, a new challenge is developed and that is the infinite possibility of the transient profiles if a user-defined acceleration or de-acceleration adjustment is allowed. This matter has not so far been investigated deeply since the number of situations is not identifiable. So we highly recommend switching this case if a new research work is to be done. The analysis of steady-state analysis has been turned into a technology today and only minor improvements are suggested.

A closed-loop implementation of the CV/F control strategy is also possible by adding a closed speed loop and comparing it with the reference value. Doing so, the rotor speed approaches the reference value, and the synchronous speed goes beyond that due to the motor slip. This is handled by the very well-known slip compensation technique usually applied by a simple sum-loop to the system. It might also be noted that since there are more referable closed-loop strategies like the FOC and DTC modes, it is preferred to switch the discussion to analyzing them instead of working on the CV/F mode. The CV/F mode can also be a part of the strategies discussed later.

### *2.4.4 Reference frame theory of induction motors*

In an FOC method, some transformations are always applied to the three-phase motor quantities including the voltage, the current and the flux in order to eliminate the time-dependency and simply to control them as separate values like what is done in DC motors in terms of the torque and the flux [24–28]. Actually, an abstract concept called the "reference frame theory" is introduced. In this theory, a three-phase system is mapped onto a two-phase time-dependent or independent system. Conventionally, the two phases are highlighted by $D$ and $Q$ notations. A general $DQ$ representation of an induction motor is provided in Figure 2.24. Instead of $W_a$, $W_b$ and $W_c$ which represent the three-phase system windings, $W_{\text{qs}}$, $W_{\text{qr}}$, $W_{\text{ds}}$ and $W_{\text{dr}}$ are used to demonstrate the $DQ$-related windings/coils through the analysis. $D$ and $Q$ windings are fixed with respect to each other and the relative speed between the axes is zero. However, the $DQ$ axis might or might not rotate depending on the

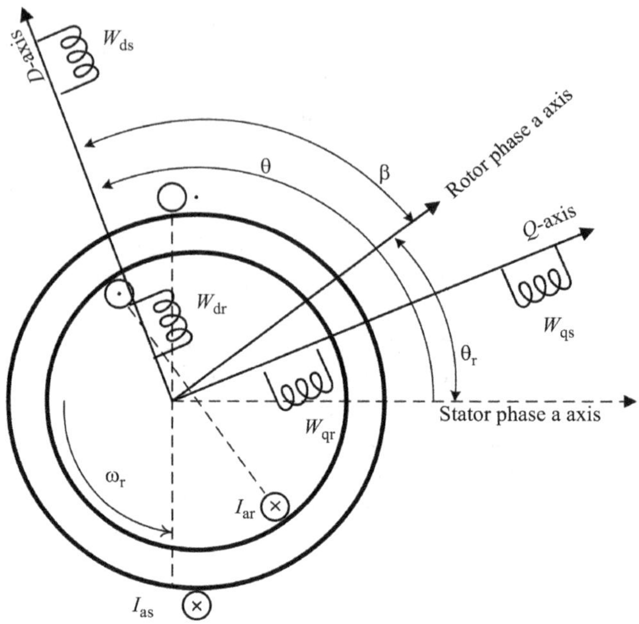


*Figure 2.24 DQ reference frame of an induction motor*

user's choice. The speed of rotation can be chosen as zero, the synchronous speed, the rotor speed as well as a user-defined speed. Likewise, different reference frames called the stationary, synchronously rotating and rotor frames are achieved, respectively. By definition, the *D*-axis leads the *Q*-axis in a counter clockwise direction.

All motor variables must be projected to the DQ reference frame, using a mathematical transformation matrix, the so-called "the Park's transformation matrix" formulated as (2.38). The idea can also be extended to a system with larger number of phases.

Park's Transformation Matrix *K*

$$= \sqrt{\frac{2}{3}} \begin{bmatrix} \cos(\theta) & \cos\left(\theta - \frac{2\pi}{3}\right) & \cos\left(\theta + \frac{2\pi}{3}\right) \\ -\sin(\theta) & -\sin\left(\theta - \frac{2\pi}{3}\right) & -\sin\left(\theta + \frac{2\pi}{3}\right) \\ \frac{\sqrt{2}}{2} & \frac{\sqrt{2}}{2} & \frac{\sqrt{2}}{2} \end{bmatrix} \tag{2.38}$$

Transformed motor variables: $x_{dq0} = Kx_{abc}$ (2.39)

Applying the transformation and replacing the results into (2.1)–(2.19) leads to the following voltage balance equations with an arbitrary rotation speed of the DQ frame.

$$v_{ds} = R_s i_{ds} + p\Lambda_{ds} + \Lambda_{qs} p\theta$$
$$v_{qs} = R_s i_{qs} + p\Lambda_{qs} - \Lambda_{ds} p\theta$$
$$v_{dr} = R_r i_{dr} + p\Lambda_{dr} + \Lambda_{qr} p\beta$$
$$v_{qr} = R_r i_{qr} + p\Lambda_{qr} - \Lambda_{dr} p\beta$$

$$\text{where} \begin{bmatrix} \Lambda_{ds} \\ \Lambda_{qs} \\ \Lambda_{dr} \\ \Lambda_{qr} \end{bmatrix} = \begin{bmatrix} l_{s\sigma} + l_{sm} & -0.5l_{sm} & 0 & 0 \\ -0.5l_{sm} & l_{s\sigma} + l_{sm} & 0 & 0 \\ 0 & 0 & l_{r\sigma} + l_{rm} & -0.5l_{rm} \\ 0 & 0 & -0.5l_{rm} & l_{r\sigma} + l_{rm} \end{bmatrix} \begin{bmatrix} i_{ds} \\ i_{qs} \\ i_{dr} \\ i_{qr} \end{bmatrix} \tag{2.40}$$

Certain simplifications in (2.40) are achieved if the term "$\theta$" is predefined. Here, the stationary reference frame for which $p\theta = 0$, is discussed. The other two important cases including the rotor frame and also the synchronous frame are obtained by simply replacing the $p\theta = \omega_r$ and $= \omega_s$, respectively.

$$\omega = p\theta = 0 \xrightarrow{\Delta} \beta = \theta - \theta_r = -\theta_r, \quad p\beta = -\omega_r$$

Results in

$$v_{ds} = R_s i_{ds} + p\Lambda_{ds}$$
$$v_{qs} = R_s i_{qs} + p\Lambda_{qs}$$
$$v_{dr} = R_r i_{dr} + p\Lambda_{dr} - \omega_r \Lambda_{qr}$$
$$v_{qr} = R_r i_{qr} + p\Lambda_{qr} + \omega_r \Lambda_{dr}$$

$$\xrightarrow{\text{yields}} [v] = [R][i] + [L]p[i] + \omega_r[G][i] \text{ where } [\,] \text{ is a matrix representation of variables} \tag{2.41}$$

The following features are valid in terms of the reference frames:

- In the stationary frame, all the DQ variables are time-dependent in both the transient and the steady-state modes. Therefore, the complexity of the analysis is somehow maintained.
- In the synchronous frame, any stator quantities such as voltage or current possess a constant value for the steady-state operation while the transient operation still reveals a fluctuating trend. The rotor quantities are also varied since they are projected to a frame with different rotating speed. The frequency of the electrical variables of the rotor should ideally be equal to that of the rotor in real world, $sf_s$, of course in steady-state operation.

- In the rotor reference frame, the stator variables are also of an oscillating nature in both the transient and steady-state operations since they are projected to a frame whose rotating speed is equal to $\omega_r$ which is definitely different from the synchronous speed. In other words, the complexity of the analysis and also the level of time-dependency increases in the following order: the stationary frame, the rotor frame and the synchronous frame. Due to the variability of the existing alternatives, various control strategies are also available. However, a competent industrial drive should be able to control the motor with any possible topology.

Together with the FOC strategy, the reference frame theory helps control the motor flux and the speed separately. This is what really required in terms of a wide speed range control of induction motors. The flux must be independent of the motor speed in the constant flux region while it should be controlled as an exponential function of the speed for the flux weakening region. On the other hand, the variables of a three-phase motor are totally interrelated. So changing one parameter would change the others. The reference frame theory aims at decoupling the flux and speed control loops of the motor in a way that they are controlled somehow independently. The decoupling is not 100% accurate. However, it works well at least in terms of ordinary applications. The idea comes from the fact that D and Q components are literally independent proved by the inductance matrix shown in (2.40). However, there is always a cross saturation effect between the D and Q axes caused by the saturation of the silicon steel material. Thus, a saturation-based decoupling approach is sometimes applied to the control circuit to eliminate this effect as much as possible.

### *2.4.5 Field-oriented control of induction motors*

The FOC of induction motors is fundamentally based on measuring the speed and the position of the motor by means of the equipment such as the encoder or by estimating analytically (see Figure 2.25). This is the critical part of any closed loop application. If the speed and position estimation is handled analytically as describing the motor state space, the method would be called "sensor-less closed loop control." Otherwise, it is a with-sensor solution. Normally, the sensor-less applications are less accurate in terms of control goals. Regardless of the type of the speed estimation, having measured the motor speed, it is compared to the reference speed value which is actually the input of the control process. This is the value which should be met by the control strategy. In fact, the rotor speed should precisely follow the reference value. Consequently, the synchronous speed should be a little bit larger than the reference speed in compensation for the slip. The other input is the reference value of $i_d$ in the stator side. So it is likewise noted as $i_{ds}$ in some papers. This parameter can be fixed or adjusted depending on the required flux level forced by the flux control block shown in Figure 2.18. To do so, another block shown in Figure 2.26, is usually added to the very beginning of the FOC circuit. The acceleration block takes care of smoothly increasing or decreasing the speed reference so that a very smooth start-up is obtained. This is

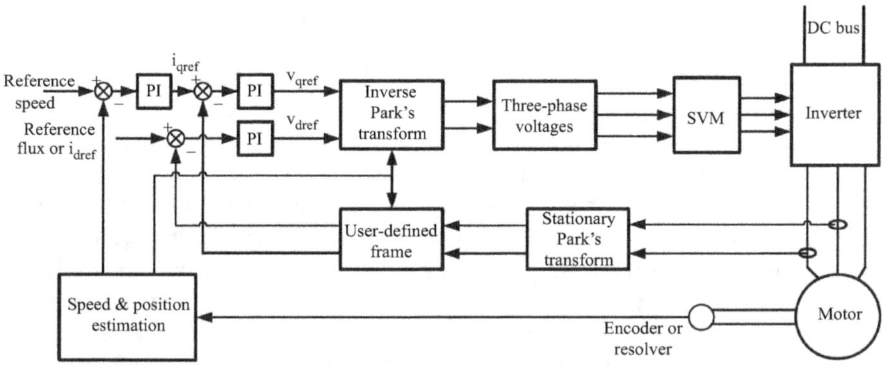


*Figure 2.25 Field-oriented control of induction motors*

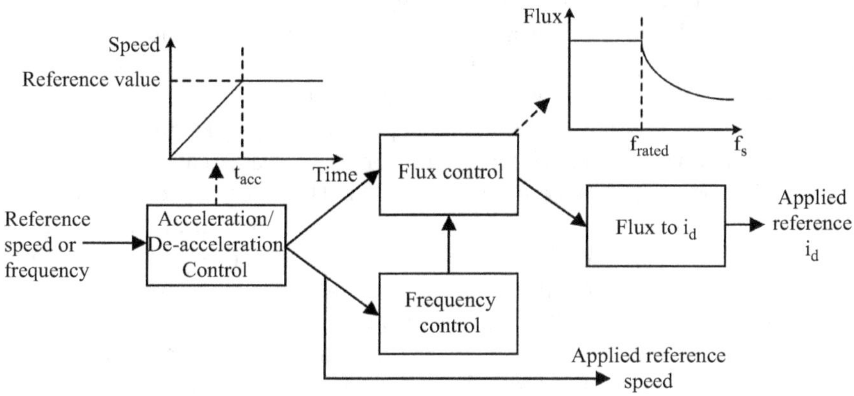


*Figure 2.26 Reference values of speed and $i_d$*

finalized within the time period 0 to $t_{acc}$. Then, a constant reference value equal to the desired one is applied to the motor. During the transient operation, the CV/F concept is utilized to maintain a constant rated flux if the motor speed should be kept below the rated value. Otherwise, by means of the exponential part of the flux controller, a flux value proportional to the motor speed is calculated and applied to the block which determines the value of $i_d$. It is obvious that the D-axis component of the motor current and consequently its voltage is related to the flux level. Indeed, it is the $i_d$ which controls the flux level required for various operating speeds. This is where the CV/F concept contributes to the close-loop applications.

On the other hand, there is a speed control loop separate from the flux control loop in any FOC control strategy. The input value is compared to the real rotor speed estimated or measured by the feedback and then passed through a linear PI controller to generate the reference $i_q$ value which is in relation with the motor torque.

Actually, $i_q$ is called the torque component of the DQ frame. In this level of control, the reference $i_d$ and $i_q$ pass through a set of PI controller to generate the $v_d$ and $v_q$ (2.41) simply verify the applicability of the PI controllers. Basically, since the relation between ($i_d$, $i_q$) and ($v_d$, $v_q$) is linear during the steady-state operation, a linear controller could be used to relate the voltages to the currents. The only challenging part is to find the appropriate proportional and integral coefficients (gains). The typical strategy is to choose the proportional gain 10 times larger than that of the integral one. The difference between the real and reference values of the currents producing the reference voltages, not the reference currents themselves.

$v_{\text{dref}}$ and $v_{\text{qref}}$ are the D and Q components of the required voltages for the requested speed and flux values. There is always a third voltage component, i.e. $v_0$, as one of the outputs of the Park's transform. Meanwhile, this value should be ideally set to zero if a balanced three-phase system is desired. Then, according to the three-phase voltages, appropriate switching signals are forced to the inverter, and the motor terminal voltages are ready.

Now it is time to provide some clues in terms of the contribution of the FOC strategy in fault diagnosis procedures. As mentioned before, one of the biggest objections in the time-domain analysis of the faulty motor signals is the closeness of some of the fault indicators to the fundamental harmonic component of the motor variables such as the current. While applying the projection from the three-phase system to a DQ frame, the dependency on the fundamental frequency will be eliminated if a proper reference frame is selected. Therefore, a useful alternative is to analyze the DQ components instead of the components in an ABC reference frame. This really matters in the case of a broken bar motor. Sometimes, the effect of the proportional and integral gains on the components of the ABC frame is not tractable. Nevertheless, the DQ components reveal a very well-behaved variation with respect to a change in the gain levels. The change in the gain levels is another significantly influential factor in any diagnosis process. Unfortunately, a few works has been reported in this field, so it is still an open area of research.

The other objection is the negative effect of the closed-loop control strategies on the fault indicators. For example, if an eccentric motor under an FOC drive is investigated, the torque or speed fluctuations caused by the fault will not be visible anymore if a very well-tuned drive is utilized. In case the fault goes beyond a specific level, the diagnosis process will be able to detect a mal-operation. This phenomenon makes the detection, determination and finally the diagnosis process way difficult and probably impossible at least in the case of a lower fault level.

The third and probably the most important challenge is the presence of two control loops each acting like a low-pass filter eliminating some of the fault-related components. This fact will be analyzed deeply by mathematical and experimental methods in the next chapters. Moreover, different drive types and technologies reveal different band passes and also various effects on the variation rates of the fault indicators. This is an unaddressed issue in the literature, and there should be a proper investigation of the commonly used inverters and their impact on the diagnosis process. This is suggested as a new field of research for potential researchers in academia and even research and development centers of industries.

At this point, it is preferred to move toward investigating the next well-known closed-loop strategy called "direct torque control (DTC)" prior to providing the behavioral study of the motor quantities. The reason is that the corresponding behavior of both FOC and DTC approaches really look similar. Of course, there are differences which might not be considered as problems in fault diagnosis procedures. The main difference is the type of variables used to control the motor torque and the flux. Otherwise, the whole concept is the same. Depending on the control strategy, the type of variables which must be chosen for the diagnosis purposes should also be different because not all variables contributed in a FOC drive are simply available in a DTC drive. Therefore, in the fault analysis, the variable selection process might be the most demanding aspect instead of the major difference in how these two types of drives might operate on the motor.

Acknowledging the aforementioned points, the DTC strategy is studied; then a detailed behavioral analysis of the motor quantities in both ABC and DQ frames is provided.

### *2.4.6 Direct torque control of induction motors*

Figure 2.27 shows a generalized DTC strategy along with its fundamental elements including a PI controller, two hysteresis controllers, the torque/flux estimator, etc. In DTC approach, there is no transformation of the motor variables from the ABC to DQ frame. It is basically grounded on the comparison of the real torque and flux values with the corresponding reference values. The reference values of the speed and the flux are obtained in the same way as that of the FOC drive. Then, the values are assigned to the inputs of the drive system. The inputs are compared to the real values estimated by the "torque/flux estimator block" in which the Park's transform is implemented.

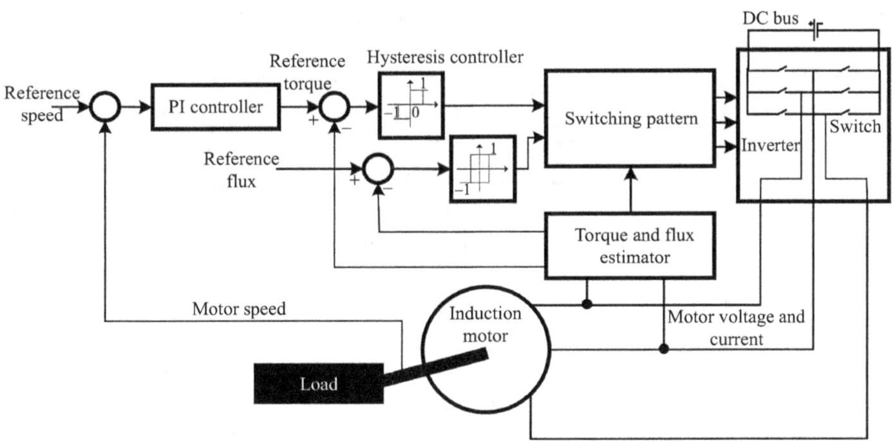


*Figure 2.27 Direct torque control of induction motors*

The estimated torque and flux functions are formulated as follows:

$$\begin{cases} \text{Stator flux magnitude } \Lambda_s = \sqrt{\Lambda_{ds}^2 + \Lambda_{qs}^2} \\ \text{Stator flux angle } < \Lambda_s = \tan^{-1}\left(\dfrac{\Lambda_{qs}}{\Lambda_{ds}}\right) \end{cases} \tag{2.42}$$

$$\text{Electromagnetic torque } T_e = \frac{2}{3}P\left(\Lambda_{ds}i_{qs} - \Lambda_{qs}i_{ds}\right) \tag{2.43}$$

The differences between the real and reference values, i.e. $\Delta T_e$ and $\Delta\Lambda_s$, enters the hysteresis controllers which directly manage the torque and flux magnitudes. The controller corresponding to the flux and torque loops is a two-level and a three-level controller, respectively represented as follows:

$$d_\Lambda = \begin{cases} -1; \quad \text{if } \Delta\Lambda_s \leq -\dfrac{\Delta\Lambda_{s\max}}{2} \\ 1; \quad \text{if } \Delta\Lambda_s \geq \dfrac{\Delta\Lambda_{s\max}}{2} \\ 1 \text{ or } -1; \quad \text{if } -\dfrac{\Delta\Lambda_{s\max}}{2} \leq \Delta\Lambda_s \leq \dfrac{\Delta\Lambda_{s\max}}{2} \end{cases} \tag{2.44}$$

$$d_{\text{Te}} = \begin{cases} -1; \quad \text{if } \Delta T_e \leq -\Delta T_{\max} \\ 1; \quad \text{if } \Delta T_e \geq \Delta T_{e\max} \\ 0; \quad \text{if } \Delta T_e = 0 \\ 0 \quad \text{or} \quad -1; \quad \text{if } -\Delta T_{e\max} \leq \Delta T_e < 0 \\ 0 \quad \text{or} \quad 1; \quad \text{if } 0 < \Delta T_e \leq \Delta T_{e\max} \end{cases} \tag{2.45}$$

$d_\Lambda$ and $d_{\text{Te}}$ are the maximum allowed range of variations of the flux and the torque, respectively. These two parameters are the maximum and minimum limits of the hysteresis controllers. The value of 1 in the torque controller means that the motor torque should increase. Similarly, the value of $-1$ corresponds to the torque reduction region, meaning that the real torque is larger than that of the reference value. If the real and the reference values are the same, there is no need to change the torque command and the corresponding voltage vector, so the output must be zero. Likewise, the flux controller consists of two states, 1 and $-1$. The state 1 is related to the case where the stator real flux is smaller than the reference value and vice versa. The combinations of two states of the flux controller and three states of the torque controller generate six different switching arrangements listed in Table 2.4. "N" is the number of voltage sectors similar to that discussed in the SVM technique. The states are shown by a three-digit number "xxx." If the first x from the left is equal to 1, then it means the upper switch of the first leg of the inverter must be turned on. If it is zero, the lower switch should be turned on. The second and third digits correspond to the second and third legs, respectively.

*Table 2.4 States of the switches in DTC strategy*

| λ | *T* | *N* = 1 | *N* = 2 | *N* = 3 | *N* = 4 | *N* = 5 | *N* = 6 |
|---|---|---|---|---|---|---|---|
| | | 110 | 010 | 011 | 001 | 101 | 100 |
| | | 111 | 000 | 111 | 000 | 111 | 000 |
| | −1 | 101 | 100 | 110 | 010 | 011 | 001 |
| 1 | | 010 | 011 | 001 | 101 | 100 | 110 |
| | | 000 | 111 | 000 | 111 | 000 | 111 |
| | −1 | 001 | 101 | 100 | 110 | 010 | 011 |

The switch states directly depend on the torque, the flux and the switching frequency cannot be controlled by the DTC method. The switching pattern used in this technology is not based on the SVM or even the sinusoidal PWM techniques. These are the values in Table 2.4 which define the switching pattern, and switching frequency is somehow random. This is one of the drawbacks of the DTC method. Hence, it is sometimes combined by a FOC-based technique to satisfy the fixed switching frequency requirement. Generally, the FOC and DTC strategies are different in the ways listed in Table 2.5.

Following the description of the FOC and DTC strategies, a behavioral study of the motor quantities is offered herein to see how a closed-loop approach affects the motor variables. The simulated motor specifications are listed in Table 2.6. A single harmonic model of the motor like what was previously discussed in this chapter is utilized. Here a very large motor of 150 kW is considered. The rated frequency is 60 Hz, and the number of poles is 2. Therefore, the rated synchronous speed is 3,600 rpm. However, the attempt is made to simulate the motor in a speed other than the synchronous speed in order to show the capability of the drive in controlling the motor.

Here are the features and characteristics of the motor variables for a given speed and torque command applied to a FOC drive. The speed command is set at 2,000 rpm while the required torque at 0.

- For all closed-loop applications, maximum current and torque are applied to the motor in order to achieve as fast-speed response as possible. So even though the torque command is set at 0, the electromagnetic torque is a considerably large during the start-up (see Figure 2.28(b)). The acceleration rate has been set at 2,000 rpm/s. It is usually an adjustable term in any industrial drive. It should not be essentially a linear curve like what is shown in Figure 2.28(c). It can take any user-defined curve, of course if the inverter has enough potential to produce the required voltage and current. Actually, this is always the inverter limits which restrict the integrated motor-drive system operation. Therefore, beyond the inverter limits, no operation is available.
- The motor current and torque at start-up is easily adjustable by means of the current limiters implemented inside the PI controllers. The larger the limits are, the faster the response can be.

*Table 2.5 Comparison of DTC and FOC control strategies*[2]

| Comparison property | DTC | FOC |
|---|---|---|
| Dynamic response to torque | Very fast | Fast |
| Coordinates reference frame | alpha, beta (stator) | d, q (rotor) |
| Low speed (<5% of nominal) behavior | Requires speed sensor for continuous braking | Good with position or speed sensor |
| Controlled variables | torque and stator flux | rotor flux, torque current iq and rotor flux current id vector components |
| Steady-state torque/current/flux ripple and distortion | Low (requires high quality current sensors) | Low |
| Parameter sensitivity, sensorless | Stator resistance | d, q inductances, rotor resistance |
| Parameter sensitivity, closed-loop | d, q inductances, flux (near zero speed only) | d, q inductances, rotor resistance |
| Rotor position measurement | Not required | Required (either sensor or estimation) |
| Current control | Not required | Required |
| PWM modulator | Not required | Required |
| Coordinate transformations | Not required | Required |
| Switching frequency | Varies widely around average frequency | Constant |
| Switching losses | Lower (requires high-quality current sensors) | Low |
| Audible noise | Spread spectrum sizzling noise | Constant frequency whistling noise |
| Control tuning loops | Speed (PID control) | Speed (PID control), rotor flux control (PI), id and iq current controls (PI) |
| Complexity/processing requirements | Lower | Higher |
| Typical control cycle time | 10–30 μs | 100–500 μs |

- The ripples caused by the switching are obviously seen in all the motor signals. As far as the motor model is a single harmonic one, there should not be any ripple in the motor quantities. However, it is not the case here, and there exist ripples. Generally, the amplitude of the ripples can be reduced by increasing the switching frequency, but it cannot be reduced below a certain point due to the limits on the switching frequency. For normal applications, a maximum value of 20 kHz is assigned.
- Unlike the line-start mode, any other mode has an open-loop nature, the current and torque magnitudes are fixed during the transient operation of the motor. This is the astounding feature of any closed-loop operation which allows the motor to start up as quickly as possible.

[2]https://en.wikipedia.org/wiki/Direct_torque_control (access date: October 2016)

*Table 2.6 Simulated motor data*

| **Power** | **150 kW** |
|---|---|
| Line voltage | 460 V |
| Frequency | 60 Hz |
| Number of poles | 2 |
| $R_s$ | 14.85 mΩ |
| $R_r$ | 9.29 mΩ |
| $M_{\mathrm{sr}}$ | 10.46 mH |
| $l_s$ | 0.3027 mH |
| $l_r$ | 0.3027 mH |
| $J$ | 3.1 kg m$^2$ |
| $B_v$ | 0.08 kg/(s m) |

- The right subplot of Figure 2.28(a) is a DQ representation of the stator current. There are two semi-circle shapes demonstrating the transient and steady-state operations. The circle with the larger radius is of a transient nature, and the smaller one corresponds to the steady-state operation. Always, the DQ components of the motor variables produce circle-like locus assuming a balanced and healthy three-phase system. If faults occur, the locus will reshape depending on the fault level and type. So this is one of the great potentials of drive circuits in providing fault-friendly signals. It is noted that the system has been simulated in a rotor reference frame.
- In Figure 2.28(c), the overshoot is also very small compared to the line-start mode. This value could be zero if a very well-tuned drive is used.
- During the transient operation, the variables, those of either the stator or the rotor, have sinusoidal nature. However, the rotor variables become constant during the steady-state operation since the motor was modeled in the rotor frame. The stator variables are still sinusoidal.
- The rotor voltages are set to zero deliberately (see Figure 2.28(e)). In practice, there is a close to zero induced voltage on the bars.
- When the broken bar fault occurs, the voltage across the broken bar is not zero anymore and this can be considered as a fault indicator. However, measuring the bar voltage is not a generally applicable task in the industry.
- Depending on the drive and inverter types, different motor variables are directly controlled. For example, in simple DTC drive, there is closed loop applied to the motor speed and consequently the torque. Therefore, the torque and speed will finally come to a very smooth operation even in case a fault occurs. This is why the proper signal selection should be a part of the fault diagnosis process. Likewise, if a current source control inverter is used, the motor current would be stabilized regardless of the fault presence. In this case, the motor torque and speed might be fluctuating because of the fault. In the line-start mode, none of the mentioned challenges exists and all the motor quantities including the magnetic, electrical and mechanical ones are simply affected even with a very small fault level. So if an inverter-fed motor is

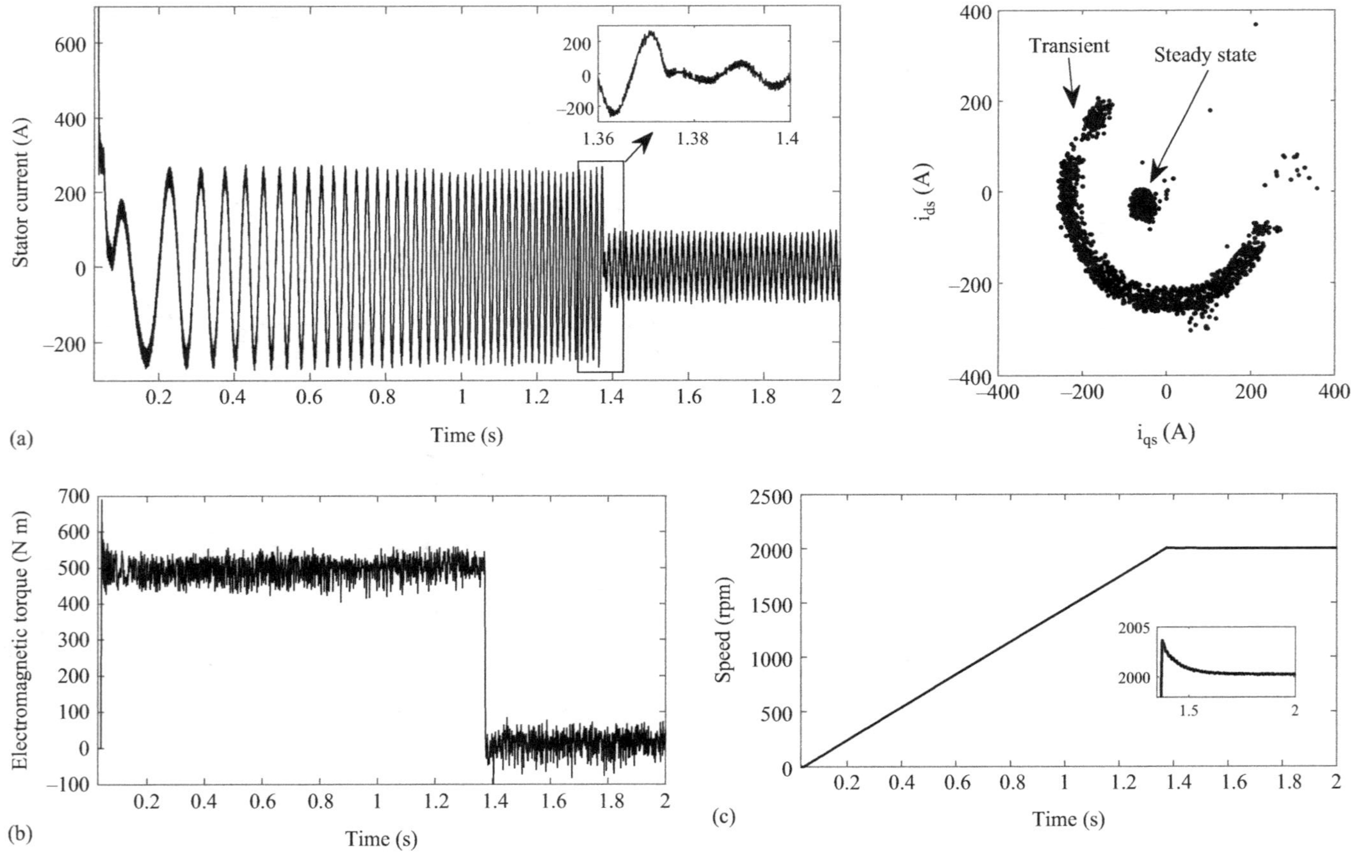


*Figure 2.28* *FOC controlled induction motor behavior (a) stator current, (b) torque, (c) speed, (d) stator DQ fluxes, (e) rotor DQ voltages and (f) rotor DQ currents*

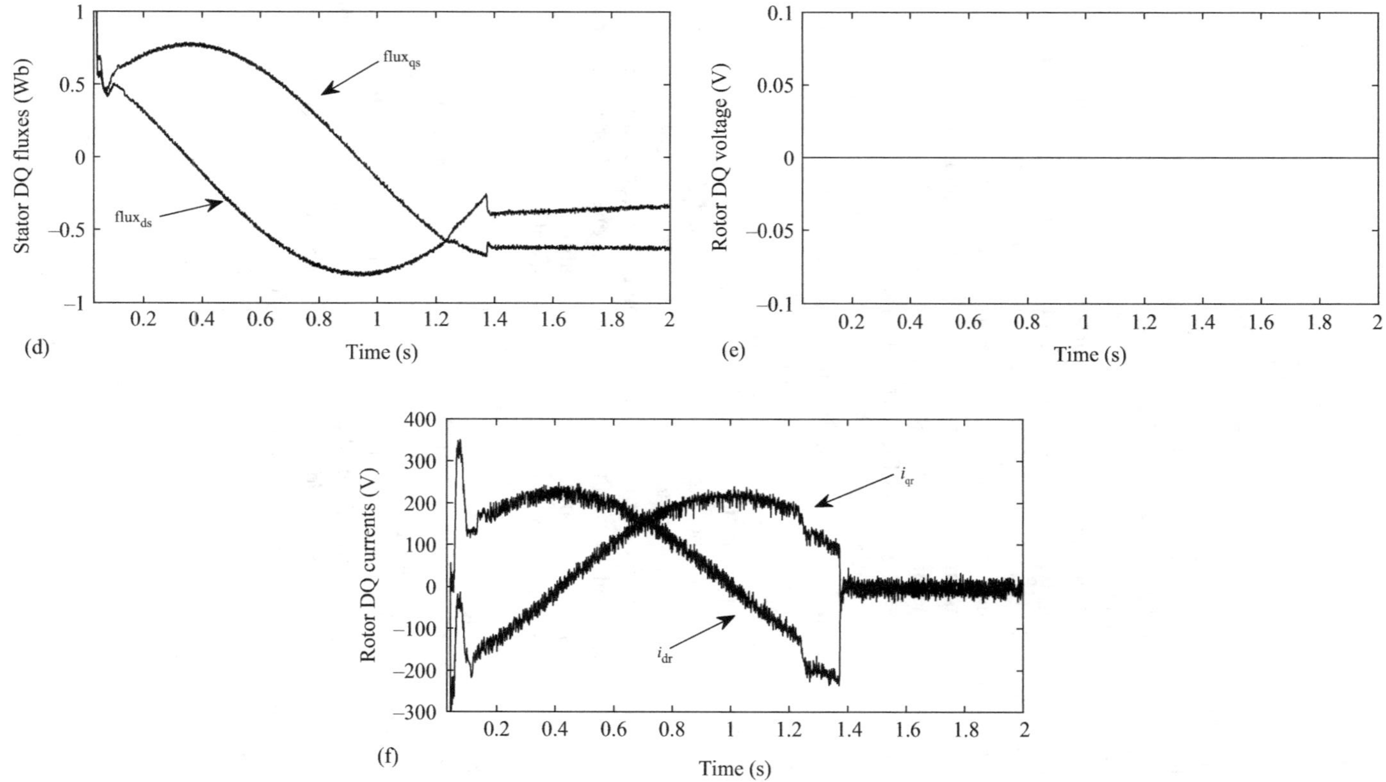


*Figure 2.28 (Continued)*

investigated, more fault-sensitive processors should be hired. Otherwise, it is almost impossible to diagnose the fault.

- Due to the importance of the topic, it is noted again that the transient mode of faulty inverter-fed motors has not been addressed yet and more efforts are needed in the field. The point is that the transients matter only if a considerably high-power motor with the corresponding start-up of around a couple of seconds is under investigation.

So far in this chapter, an essentially required insight of the induction motor and its operation has been provided. Then, through mathematical and simulation-based developments, the challenges in terms of the behavioral study of a healthy and sometime the faulty motor were addressed. According to the discussion, the following are highlights of this chapter and the guidelines for the next chapters as well, in terms of the fault diagnosis procedures.

- Distinct parts of an induction motor are subjected to different types of fault. For example, the short-circuit fault occurs in the motor windings consisting of the insulation while the broken bar fault occurs in the rotor bars. Therefore, a comprehensive study on the types of fault and their target places should be done in the next chapters. This is a must-do before going through the advanced topics.
- Equally important, the motor signals including the voltage, the current, the flux, the flux density, the torque, the speed and also the vibration might be affected in different ways upon the fault occurrence. Furthermore, the type of effect differs with the type of fault as well as the operating mode. Therefore, the process of selecting proper signals should be discussed deeply in order to get familiar with various practical situations and their requirements.
- So far, the type of the processor used to extract the fault information has been discussed. Depending on the fault features in time, frequency or even time-frequency domains, various processors each revealing specifically useful aspects must be utilized. For example, for a given faulty steady-state operation, the FT and sometimes the time-domain processors are the most popular. However, they lose accuracy when it comes to the transient analysis.
- Extracting proper indices, the fault indicators, is for sure an important step to take. So far, a lot of indices each dealing with a specific aspect of fault has been proposed. None of the existing indices is completely able to detect, determine and diagnose the fault. For example, assuming that the motor operates in a light- or no-load condition, the conventional FT processors might not be able to diagnose the fault unless a super high resolution of the spectrum is set. Instead, the Hilbert–Huang transform (HHT) which is of a time-frequency type easily detects the fault even in light loads. So being aware of different fault indicators and their pros and cons will be useful in introducing a competent fault diagnosis procedure.
- Unlike the researcher's claim based on which one single index or process is able to cover all the faulty detection conditions, it will be shown that an accurate diagnosis procedure is a matter of operating conditions, and the process should be updated for any major change in the conditions. For instance, if the number of the rotor bars changes, one broken bar fault would have a

different effect in terms of magnitude and variation rate of the corresponding fault indicators. Therefore, it is not a straightforward task as simply mentioned in the literature. In addition, any change in the shape of slots affects the process. Even any change in the manufacturer of the drive might change the fault indicator which makes the detection a trickier problem. This aspect of the diagnosis task will be further addressed.

- If a research and development center intends to practice the fault diagnosis process, the experiments related to different types of faults come to mind first. Thus, a correct implementation of the faults should be considered. This is the reason why we will devote a part of a chapter to this topic. Meanwhile, proper sensor selection, sampling process and finally index extraction are proposed.
- Sometimes, it is impossible to examine all the fault types and conditions, specifically in academic environments. Therefore, accurate motor and fault models become important. In this case, the finite element method (FEM), as well as the WFA, will be deeply challenged and analyzed. So a very useful basis will be provided for the academic researchers who suffer from lack of test rigs.

These are some significant aspects of this book which differentiate it from the existing references. Actually, the goal is to merely focus on one single type of motors, i.e. the induction motor, and develop really useful and engineering ideas for the diagnosis process, instead of wandering around general topics of all types of motors without going through the details of materials.

## References


[1] Ceraolo M. and Poli D. Induction machines and drives, *Fundamentals of Electric Power Engineering: From Electromagnetics to Power Systems*, 2014; Hoboken, NJ: Wiley-IEEE Press.

[2] Krause P.C., Wasynczuk O., Sudhoff S.D. and Pekarek S. Induction motor drives, *Analysis of Electric Machinery and Drive Systems*, 2013; Hoboken, NJ: Wiley-IEEE Press.

[3] Bose B.K. Electrical machines for drives, *Power Electronics and Variable Frequency Drives: Technology and Applications*, 1997; Piscataway, NJ: Wiley-IEEE Press.

[4] Das J.C. Effects of Harmonics, *Power System Harmonics and Passive Filter Designs*, 2015; Piscataway, NJ: Wiley-IEEE Press.

[5] Chikwanda H.S. and Bolton H.R. Analysis of the naturally commutated, convertor-fed induction motor drive system using a flux vector method, *IEE Proceedings, Part B, Electric Power Applications*, 1993; **EPA-140**: 401–415.

[6] Gandhi A., Corrigan T. and Parsa L. Recent Advances in modeling and online detection of stator interturn faults in electrical motors, *IEEE Transactions on Industrial Electronics*, 2011; **IE-5**: 1564–1575.

[7] Huangfu Y., Wang S., Qiu J., Zhang H., Wang G. and Zhu J., Transient performance analysis of induction motor using field-circuit coupled finite-element method, *IEEE Transactions on Magnetics*, 2014; **MAG-50**: 873–876.

[8] Ito M.N., Okuda F.H., Takahashi N. and Miyata T. Analytical model for magnetic field analysis of induction motor performance, *IEEE Transactions on Power Apparatus and Systems*, 1981; **PAS-100**: 4582–4590.

[9] Davey K.R. Analytic analysis of single- and three-phase induction motors, *IEEE Transactions on Magnetics*, 1998; **MAG-34**: 3721–3727.

[10] Chan T.F. and Shi K. Philosophy of induction motor control, *Applied Intelligent Control of Induction Motor Drives*, 2011; Singapore: Wiley-IEEE Press: 9–30.

[11] Depenbrock M. Direct self-control (DSC) of inverter-fed induction machine, *IEEE Transactions on Power Electronics*, 1998; **PE-3**: 420–429.

[12] Kubota H., Matsuse K. and Nakano T. DSP-based speed adaptive flux observer of induction motor, *IEEE Transactions on Industry Applications*, 1993; **IA-29**: 344–348.

[13] Huber L. and Borojevic D. Space vector modulated three-phase to three-phase matrix converter with input power factor correction, *IEEE Transactions on Industry Applications*, 1995; **IA-31**: 1234–1246.

[14] Levi E.R., Profumo B.F., Toliyat H.A. and Williamson S. Multiphase induction motor drives – a technology status review, *IET Electric Power Applications*, 2007; **EPA-1**: 489–516.

[15] Zhao Y. and Lipo T.A. Space vector PWM control of dual three-phase induction machine using vector space decomposition, *IEEE Transactions on Industry Applications*, 1995; **IA-31**: 1100–1109.

[16] Holtz J. Sensorless control of induction machines—with or without signal injection, *IEEE Transactions on Industrial Electronics*, 2006; **IE-53**: 7–30.

[17] Lascu C., Boldea I., and Blaabjerg F. A modified direct torque control for induction motor sensorless drive, *IEEE Transactions on Industry Applications*, 2000, **IA-36**: 122–130.

[18] Matsuo T. and Lipo T.A. A Rotor parameter identification scheme for vector-controlled induction motor drives, *IEEE Transactions on Industry Applications*, 1985; **IA -21**: 624–632.

[19] Casadei D., Serra G. and Tani K. Implementation of a direct control algorithm for induction motors based on discrete space vector modulation, *IEEE Transactions on Power Electronics*, 2000; **PE-15**: 769–777.

[20] Leonhard R.W. and Nordby C.J. Field-oriented control of a standard AC motor using microprocessors, *IEEE Transactions on Industry Applications*, 1980; **IA-16**: 186–192.

[21] Kowalska T.O. and Dybkowski M. Stator-current-based MRAS estimator for a wide range speed-sensorless induction-motor drive, *IEEE Transactions on Industrial Electronics*, 2010; **IE-57**: 1296–1308.

[22] Nash J.N. Direct torque control, induction motor vector control without an encoder, *IEEE Transactions on Industry Applications*, 1997; **IA-33**: 333–341.

[23] Lascu C., Boldea I. and Blaabjerg F. Direct torque control of sensorless induction motor drives: a sliding-mode approach, *IEEE Transactions on Industry Applications*, 2004; **IA-40**: 582–590.

[24] Kim S.H. and Sul S.K. Maximum torque control of an induction machine in the field weakening region, *IEEE Transactions on Industry Applications*, 1995; **IA-31**: 787–794.

[25] Correa P., Pacas M. and Rodriguez J. Predictive torque control for inverter-fed induction machines, *IEEE Transactions on Industrial Electronics*, 2007: **IE-54**: 1073–1079.

[26] Idris N.R.N. and Yatim A.H.M. Direct torque control of induction machines with constant switching frequency and reduced torque ripple, *IEEE Transactions on Industrial Electronics*, 2004; **IE-51**: 758–767.

[27] Dorrell D.G., Shek J.K.H., Mueller M.A. and Hsieh M.F. Damper windings in induction machines for reduction of unbalanced magnetic pull and bearing wear, *IEEE Transactions on Industry Applications*, 2013; **IA-49**: 2206–2216.

[28] Casadei D., Profumo F., Serra G. and Tani A. FOC and DTC: two viable schemes for induction motors torque control, *IEEE Transactions on Power Electronics*, 2002; **PE-17**: 779–787.

# *Chapter 3*
# Induction motor faults: basics, developments and laboratory-scale implementation

## 3.1 Introduction

In the previous chapters, we devoted ourselves to the analysis of behavior of the healthy induction motor in different possible operating modes including transient and rated operations each having their own effects on the electrical, magnetic and mechanical quantities of the motor. In this chapter, we deal with an easy and comprehensible way of formulating and describing motor quantities by means of very useful mathematical equations. The goal is initially to see how motor quantities might vary in time and partly in frequency domains. However, the frequency domain is only mentioned as a black box while there would be a sufficient discussion in the next chapter corresponding to various time, frequency and time-frequency processors. The materials provided earlier in the previous chapters were interesting and, of course an essential step through understanding the induction motor faults and the corresponding fundamentals. By "fundamentals," the causes, the effects and also any development related to the faults are targeted. Moreover, various types of the supply modes including the line-start, open-loop and closed-loop applications were discussed in the previous chapter to wrap up what happens in industry in terms of the motor control strategies. So, a potential reader of this book will get used to the terminologies and possible motor–drive interactions [1–6]. A general guideline based on which the motor operations are justified in different modes were provided as well. As the goal of this book is not to discuss every single aspect of drive circuits and also the control strategies, they were explained in a way to cover a great portion of existing operations.

On the other hand, drive circuits are not directly related to the fault diagnosis procedure of induction motors. They mostly apply a kind of impact on the quantities of the faulty motor. It means that the drive and the inverter are not usually involved in the diagnosis procedure. However, they definitely affect the process in comparison to a line-start application. Having discussed and investigated the behavior of the healthy induction motor, now it is time to go through the theoretical and practical aspects of the faulty induction motor. This is a step-by-step process which illustrates and formulates various faults and their effects on the magnetic, electrical, mechanical and also thermal, of course if possible, quantities. Unfortunately, the fault diagnosis of electrical machines really suffers from lack of knowledge in terms of thermal analysis of induction motors. Therefore, the main focus will be on the analysis of other types of physics. In the meanwhile, you will sometimes find a subsection related to some specific aspects of thermal modeling or analysis of a faulty motor.

According to the early materials of Chapter 2, any induction motor, regardless of the type and the configuration, consists of two main parts, the stator and the rotor each having their own particular components. Depending on the location and the type of the component, various faults might occur. The magnetic, thermal, electrical and also mechanical stresses are the major causes of faults in induction motors. Generally, there are two types of faults as follows:

- **Electrical faults:** Any type of fault which has an electrical source is called an "electrical fault." The most important and well-known kind of this type of fault is the "short-circuit" or "interturn" fault. This fault is common for all types of machines which include windings and coils. The main cause of the interturn fault is the deterioration or damage of winding insulations. Overvoltages or over-currents are the main causes. In fact, if the voltage applied to the turns of windings goes over the tolerable thresholds of insulations, insulations fail. Over-current, which is the main cause of heat-up in the machine core and winding materials, is another factor in damaging insulations. As a result of this phenomenon, two or more coils in a winding or even between two windings are short-circuited, and an uncontrollable large current circulates through the windings without contributing to the main torque component of the motor. Therefore, it is only considered as a source of losses, unbalanced operation and failure in the symmetrical operation of the motor. This is why, this kind of unbalanced/asymmetrical operation is called "fault."

  There are also other causes which lead to an unbalanced electrical operation of the motor quantities, mainly the magnetic flux density. The best examples are an unbalanced supply voltage, reversed-phase voltages as well as the gearbox fault. However, they have nothing to do with the internal structure of the motor. Therefore, they should not be considered as a type of fault assigned to the motor because the source is not located inside the motor. Unfortunately, some of research works take them as a kind of motor fault by mistake. In this book, the term "fault" refers to a phenomenon which directly affects an interior motor component so that a change in the shape, the material or the motor operation takes place. Basically, if an external factor leads to a harsh motor operation, it does not necessarily damage the motor components. consequently, if the factor is removed, the motor will be back to its normal, symmetrical and balanced operation. Nonetheless, if the harsh conditions go on for a while, the external factor might lead to a destruction of the building blocks of the motor. This is exactly what we call the "fault." Therefore, all the materials of this book are valid as internal faults take place. This discussion necessitates the presence of an early fault diagnosis procedure usually called an "incipient fault diagnosis" which should be an essential part of any diagnosis task. Incipient diagnosis is an extremely important condition monitoring process existing in high-power applications. In fact, diagnosing the transition from a healthy state to a faulty one is always appreciate in industry.

- **Mechanical faults:** In this case, the main cause of the fault has a mechanical nature. The "broken bar," "broken end-ring" and "eccentricity" faults are the best examples of the kind. There is no shortage of agreement that all the mechanical faults are caused by a corrosion, improper casting or unsuitable placement of the components. Sometimes, an internal defect of the motor materials also produces a mechanical fault. For instance, the bearing fault sometimes is a result of a partial breakage of the interior balls. The broken bar fault is generally assigned to the event of dis-joining the end-rings and the bars in the rotor. As the bars are connected to the rings through a molded or casted area, they are likely subjected to a breakage if an overwhelming tension is applied to the joints. Unbalanced magnetic pulls (UMP) are mostly the main reasons. The UMP itself can be an output of a misaligned or eccentric rotor. When a rotor center is not aligned with that of the stator, the motor air-gap experiences an UMP around the outer circumference of the rotor. As a result, the bars and the rings are impacted. When a broken bar motor comes across an eccentric rotor, it is expected to observe a higher level of the broken bar indicators. It is worth noting that any mechanical or even electrical fault eventually contributes to an improper magnetic and/or thermal tension and imbalance. Undoubtedly, the UMP is the best example of the magnetic imbalance. In terms of the thermal stress, the short-circuited turns or windings reveal a hot spot adjacent to the fault area and introduce the possibility of a molded silicon steel material due to the high temperatures around the faulty region. Hence, some specific types of fault faults might create another type of fault. Thus, any methodology leading to a fast, reliable and most importantly early detection of the fault is much respected.

In general, three major types of defects, namely the broken bars/end-rings, the eccentricity/misalignment and the short-circuit faults are going to be investigated in this book in details. The materials of this chapter are sort in the way described below:

- the broken bars/end-ring fault
- the eccentricity/misalignment fault
- the short-circuit fault.

Each type will be analytically and experimentally dealt with. The broken bars fault is practically the most tackled and analytically addressed one, and the other two types somehow contain fewer databases in terms of analytical extensions and also the drive-connected problems. Nevertheless, the authors try to gather up every required detail for the purpose of providing the fundamentals of advanced knowledge of the field. To this end, it is preferred to first discuss the basics of all types of fault and then move toward some of the analytical demonstrations of the faulty motor behavior. Then, the experimental approaches corresponding to the laboratory-scale implementation of various faults, along with the required sampling and processing tools and equipment, are studied. Hence, this chapter seems to be interesting and, for sure, indispensable for almost all of the readers who are amateurs. If the readers already have the idea of how faults are examined, the current chapter might be skipped. Here are the main outlines of this chapter:

- to investigate the fundamental concepts of fault occurrence
- to study the time-domain effects of the faults on the motor quantities
- to provide some important analytical descriptions of the faulty motor behavior
- to provide the guidelines of the experimental implementation of various faults in a laboratory
- to explain how the faulty motor signals are measured and then processed experimentally
- to discuss various type of sensors such as the wireless apparatus which could be used during the sampling and measurement task
- to go through the advantages and disadvantages of the measurement techniques provided in this chapter and then introduce the most competent one.

As the starting point of the discussion, the broken bar fault is targeted as follows.

## 3.2 Broken bar/end-ring fault in induction motors

Bars breakage usually takes place in applications which specifically require high-power motors, and a highly stressful environment normally exists [7–10]. In such conditions, oscillating loads, improper motor assembly, mechanical stresses, as well as the hysteresis stresses, might weaken the joints connecting the bars and the end-rings. Consequently, weakly connected bars might be disconnected from end-rings in one or both of the motor ends. This phenomenon, which partially or fully eliminates the current in the broken bar or bars, is known as the "broken bars/end-rings" (see Figure 3.1). Sometimes, there can also be a crack or breakage in the end-ring, not in the location of the bar/end-ring joint, but somewhere between two bars. This type is usually caused by improper casting process during the motor construction. Equally important, the air bubbles located inside the joints or the materials could be a decisive factor in increasing the breakage likelihood. As it is conveyed by the name of this type of fault, it only affects the motors with a squirrel-cage rotor, and because the squirrel-cage motors are widely and prominently used in industry, the broken bars fault is very common.

The term "partially" which was assigned to the broken bar fault in the previous statements reinforces the idea that there might be a partial or full broken bar or bars. What happens in the case of partial broken bar is that the connection between the bar and the end-ring still exists, although the breakage has already happened and caused to disconnect a part of the joint. As a result, the partial broken bars fault is introduced. Likewise, when there is a fully disconnected joint from

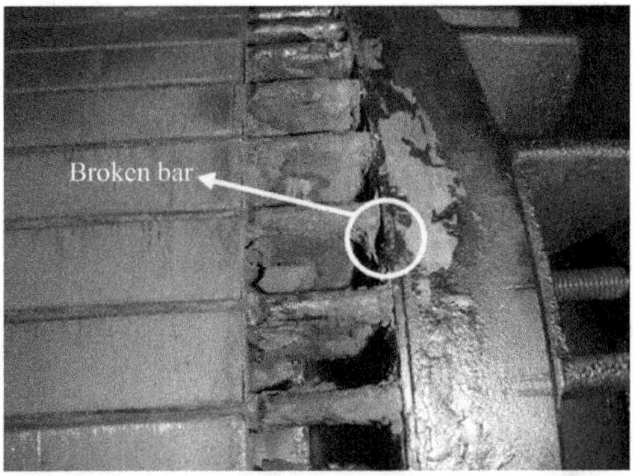


*Figure 3.1 Broken bar fault in a squirrel-cage induction motor*

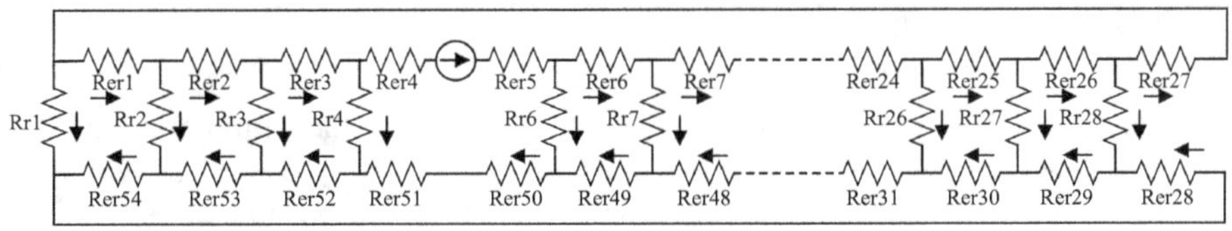


*Figure 3.2 Equivalent network for a broken bar rotor*

which the bar current is not able to flow, the full broken bars fault is introduced. Depending on the severity of the breakage, different levels of the partial broken bar are addressed and investigated in the literature. Moreover, more than one breakage might also be present at the same time. However, increasing the number of broken bars, whether partial or full breakage, does not necessarily mean that the effect of the fault on the motor quantities increases as well. It sometimes depends on another factor which is the location of the bars in the rotor across the rotor circumference. So, a new influential factor is introduced as well; that is the fault location. It should be noted that increasing the severity of the breakage in the case of partial broken bar always amplifies the fault effect, while it is not always the case where the number of broken bars increases.

In Chapter 2, the rotor circuit was modeled and represented by an impedance network containing one impedance per bar or one end-ring (see Figure 2.5). In the same fashion, the broken bar rotor is modeled in a way that the bar subjected to a full breakage is totally removed from the network. Instead, the current previously flowing in the broken bar finds a pathway through the adjacent bars leading to an increase in the bar currents close to the breakage, and the removed current is modeled by a current source in Figure 3.2. The amount of the current source should be ideally equal to the bar current removed from the circuit. The main reason for

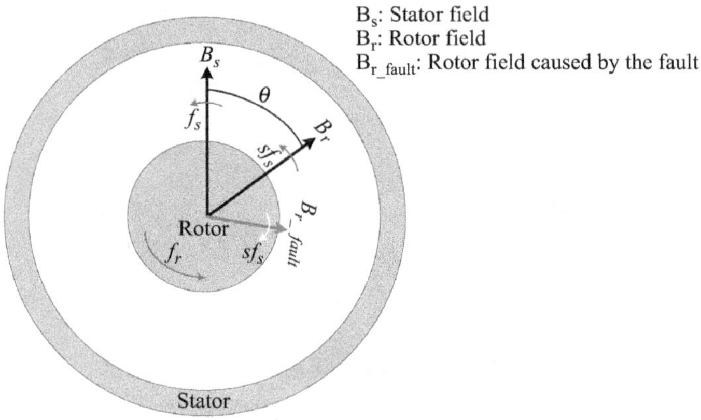


*Figure 3.3 Schematic field distribution in an induction motor with rotor broken bar*

the increase in the current level of the adjacent bars is the elimination of the armature reaction generated by the broken bar in a healthy case. The voltages applied to the adjacent bars increase; hence, their total current increases as well. The amount of divergence forming a symmetric and healthy motor depends on the severity of the breakage. The more sever the joint is broken, the larger the asymmetry of the motor variables will be.

The first variable affected by the broken bar is the current space distribution. Then, the corresponding magnetic field pattern is distorted. The development process of the broken bars fault is as follows:

**Mechanical asymmetry → Electrical distortion→ Magnetic distortion→ Thermal stress**

At the final stage of this fault, due to the dramatically increased magnetic saturation caused by the over-current phenomenon, the local losses and subsequently the local temperature increase as well. This is the general qualitative flow observed in any broken bar event.

Now, it is time to take a bit distance from the philosophical aspects of the broken bars fault and move along with the real-world happenings in terms of formulating this kind of fault. Without exception, if a symmetric, healthy and single harmonic motor is under investigation, the ideal magnitude of the backward field should be zero. The term "backward" is assigned to a field which rotates in the opposite direction to that of the synchronous speed or frequency. When an inherent asymmetry such as the nonsinusoidal spatial distribution of the bars exists, a very well-known backward field whose frequency is equal to $sf_s$ is produced (see Figure 3.3, the clock-wise rotating $sf_s$ components). The other $sf_s$ component always exists in a motor.

According to the fundamentals of induction motor, the following relations hold [11–15]:

$$\begin{aligned}&\text{Stator electrical frequency: } f_s\\&\text{Rotor rotating electrical frequency: } f_r = (1-s)f_s\\&\text{Motor slip: } s = \frac{f_s - f_r}{f_s}\\&\text{Rotor bar electrical frequency: } sf_s\end{aligned} \tag{3.1}$$

Two magnetic fields, $B_s$ and $B_r$, which are spatially placed with a specific angle, $\theta$, with respect to each other produce the electromagnetic torque. In fact, in electrical motors, the electromagnetic torque is produced by cross product of two fields generated by the stator and the rotor. Like $B_s$, the forward magnetic field produced by the rotor rotates with the electrical frequency of

$$f_s = f_r + sf_s \tag{3.2}$$

There is another backward field, $B_{r\_\text{fault}}$, which rotates in the direction opposite to the main rotor and stator fields. Therefore, the resultant electrical frequency in the stator reference frame is equal to $f_r - sf_s = (1-2s)f_s$ which is an indication of the armature reaction of the clock-wise rotating backward field on the stator variables including its current and back-EMF. This is actually the basis of investigating an asymmetry in the rotor. Not only the fault, but also the rotor inherent asymmetries produce a backward field causing additional harmonic components of the motor. Thus, it is expected to observe the backward field in the case of a healthy motor as manufacturing process does not guarantee a hundred percent ideal rotor design.

Given the fact that even healthy motors consist of the backward fields, the fault occurrence only amplifies the magnitude of the corresponding field and has nothing to do with creating the frequency component. Most of the time, researchers make this big mistake that the broken bar/end-ring fault generates the backward $sf_s$ component, while this is a totally wrong notion. In reality, the broken bars fault only impacts the magnitudes of some existing harmonic components which are normally the outcomes of the motor structure.

By means of the overviewed concept and also the one related to the single harmonic motor model investigated in Chapter 2, one can develop the fundamental formulas of the induction motor with broken bars assuming a linear silicon steel material with a constant permeability of both the stator and the rotor. This makes the process of relating the magnetic field to their producing currents a simple one. The fundamental component of the model reveals a synchronous frequency of $f_s$ in the stator. Therefore, the fundamental harmonic component of the air-gap flux density produced by the stator current is

$$B_{\text{fundamental}} = B_m \sin(\omega_s t) \tag{3.3}$$

where $B_m$ is the magnitude of the fundamental component of the flux density at the air gap. On the other hand, there is a flux density component with the frequency of $(1-2s)f_s$ produced by the rotor backward field [16]. Hence, the corresponding flux density is formulated as the following:

$$B_{\text{backward}} = B_{mb}\sin((1-2s)\omega_s t - \alpha_{B_{\text{backward}}}) \tag{3.4}$$

where $\alpha_{B\_\text{backward}}$ is the electrical phase angle of the backward field. $B_{mb}$ is the magnitude of backward magnetic flux density caused by rotor asymmetry. The assumption of a linear magnetic material leads to

$$NI = kB = \psi R \tag{3.5}$$

where $I$, $N$, $k$, $\psi$ and $R$ are the current, number of turns, constant coefficient and reluctance of the flux path, respectively. So the motor current is related to the magnetic flux density as follows:

$$I = \frac{k}{N}B = k^{'}B \tag{3.6}$$

As far as the material is linear, the superposition is applicable. Taking advantage of the superposition rule, each component of the magnetic flux density, the synchronous and fault-related components, could be separately substituted in (3.4). If it is applied to the fault component, the following equation is achieved.

$$\begin{aligned} I_{\text{sideband}} &= k'B_{\text{backward}} = k'B_{mb}\sin((1-2s)\omega t - \alpha_{B_{\text{backward}}}) \\ &= I_{m_{\text{sideband}}}\sin((1-2s)\omega t - \alpha_{I_{\text{backward}}}) \end{aligned} \tag{3.7}$$

where $I_{\text{sideband}}$ is a current component with the frequency and magnitude of $(1-2s)$ $f_s$ and $I_{m\text{sideband}}$, respectively. This component is the outcome of the backward rotating field. The term "Sideband" is assigned to this component because it is located within a specific frequency range around the fundamental component if the current spectrum is analyzed. For the moment, we do not intend to talk about the frequency-domain analysis and will only stick to the time-domain variations for justifications.

Suppose, there was an asymmetry in the rotor, whether a fault or inherent casting problem. Then, we ended up with the fact that describes the reason for the existence of a backward field and subsequently the corresponding $(1-2s)f_s$ components of the flux density and the current. This clearly proves that the sideband components are always present in a motor with asymmetries, regardless of the rotor fault occurrence. Nevertheless, the broken bar/end-ring fault, a kind of asymmetry in the rotor, helps the component get stronger compared to a healthy motor. Again, it should be noted that a healthy motor is different from an ideal motor in which the forward rotating fields are only observable.

Let us now flash back to the main discussion. Phase angle of faulty current component should be different from that of the flux density as nonlinear relationship always holds in reality instead of a linear correspondence. This claim can be further discussed by the FFT analysis of both the current and the flux density and will be done in Chapter 4. For now, we do not miss the chance to continue with the analytical formulations which have been provided so far. Considering the fact that the electromagnetic torque is the main motive force to produce rotation, it is used here. The fundamental and fault-related components of the electromagnetic torque are expressed as follows:

$$\begin{cases} \Lambda_{\text{fundamental}} = \Lambda_m \sin(\omega_s t - \alpha_{\Lambda_{\text{fundamental}}}) \\ I = I_{\text{fundamental}} + I_{\text{sideband}} \\ I_{\text{fundamental}} = I_m \sin(\omega_s t - \alpha_{I_{\text{fundamental}}}) \\ I_{\text{sideband}} = I_{m_{\text{sideband}}} \sin((1-2s)\omega_s t - \alpha_{I_{\text{backward}}}) \\ T = \dfrac{3p}{2} \Lambda_{\text{fundamental}} I \end{cases} \xrightarrow{\text{yields}}$$

$$\begin{aligned} T &= \frac{3p}{2}[\Lambda_m \sin(\omega_s t - \alpha_{\Lambda_{\text{fundamental}}})[I_m \sin(\omega_s t - \alpha_{I_{\text{fundamental}}}) \\ &\quad + I_{m_{\text{sideband}}} \sin((1-2s)\omega_s t - \alpha_{I_{\text{backward}}})]] \\ &= \frac{3p}{4}[\Lambda_m I_m [\cos(\alpha_{I_{\text{fundamental}}} - \alpha_{\Lambda_{\text{fundamental}}}) - \cos(2\omega_s t - \alpha_{\Lambda_{\text{fundamental}}} - \alpha_{I_{\text{fundamental}}})] \\ &\quad + \Lambda_m I_{m_{\text{sideband}}} [\cos(2s\omega_s t + \alpha_{I_{\text{backward}}} - \alpha_{\Lambda_{\text{fundamental}}}) \\ &\quad - \cos((1-s)2\omega_s t - \alpha_{\Lambda_{\text{fundamental}}} - \alpha_{I_{\text{backward}}})]] \end{aligned} \quad (3.8)$$

where $\Lambda_{\text{fundamental}}$, $\Lambda_m$, $\alpha_{\Lambda_{\text{fundamental}}}$, $I_{\text{fundamental}}$, $I_m$, $\alpha_{I_{\text{fundamental}}}$, $T$ and $p$ are the fundamental components of flux linkage, the amplitude of flux linkage, the phase angle of flux linkage, the fundamental current component, the amplitude of fundamental current component, the phase angle of fundamental current component, the developed electromagnetic torque and the number of poles, respectively. Here, the coupling between the fundamental flux linkage component and the stator current variables including the fundamental and sideband components are dealt with. The resultant torque consists of DC, $2\omega_s t$, $2s\omega_s t$ and $(1-s)2\omega_s t$ components. The DC component produces the average torque required for the main rotor operation. The terms $2\omega_s t$, $(1-s)2\omega_s t$ are higher frequency components which are located around twice the fundamental synchronous frequency and are mostly subjected to being filtered out from the motor operating cycle due to the fact that the rotor plus load inertia is so high that the only the low-frequency components with the order of a couple of Hertz would remain in the torque. As a result, the remaining part which comes into considerations and might produce ripples in the motor torque is $2s\omega_s t$. This is exactly the component which is the consequence of a rotor asymmetry such as the broken bar fault. As it is a linear function of the motor slip, it is located at very small frequencies, typically below 10 Hz, depending on the motor supply frequency and the slip level.

As a very interesting point, the motor torque and its harmonic components are controlled directly by means of a drive circuit. Therefore, any low-frequency fault-related component which is located inside the pass band of the PI regulators is normally affected by the closed loop of the drive circuit. As a result, it is not likely to have a lower torque oscillation even in case of higher fault levels. In Chapter 8 which is related to the fault diagnosis of the broken bar motors, this issue will be discussed in details.

Now, consider the torque fluctuations caused by the $2s\omega_s t$ component as follows:

$$\Delta T = \frac{3p}{4}\Lambda_m I_{m_{\text{sideband}}} \cos(2s\omega_s t + \alpha_{I_{\text{backward}}} - \alpha_{\Lambda_{\text{fundamental}}}) \tag{3.9}$$

On the other hand, the motor torque and speed are related to each other by the following equation given that the viscosity of the shaft is zero. Then, one comes up with the following equation which is demonstration of speed fluctuations caused by rotor fault.

$$\begin{aligned}\Delta T &= J\frac{d\omega_r}{dt} \xrightarrow{\text{yields}} \Delta\omega_r = \frac{1}{J}\int \Delta T dt \\ &= \frac{3p\Lambda_m I_{m_{\text{sideband}}}}{4J}\int \cos(2s\omega_s t + \alpha_{I_{\text{backward}}} - \alpha_{\Lambda_{\text{fundamental}}})dt \\ &= \frac{3p\Lambda_m I_{m_{\text{sideband}}}}{8Js\omega_s}\sin(2s\omega_s t + \alpha_{I_{\text{backward}}} - \alpha_{\Lambda_{\text{fundamental}}}) = \Delta\omega_r\end{aligned} \tag{3.10}$$

$J$ is the motor inertia and $\Delta\omega_r$ is the illustration of the motor speed fluctuations caused by the broken bars fault. It is obviously seen that the asymmetry in the rotor, for example the broken bar fault, brings the speed to oscillate with the frequency of $2sf_s$. This fluctuation happens over the average value of the speed. As the next step of the analysis, electromotive force (EMF) produced by the speed fluctuations is calculated and added to the fundamental component as follows:

$$\begin{aligned}\text{EMF} &= \omega_r\Lambda_{\text{fundamental}} = (\omega_{r_{\text{average}}} + \Delta\omega_r)\psi_m \sin(\omega_s t - \alpha_{\Lambda_{\text{fundamental}}}) \\ &= \left[\omega_{r_{\text{average}}} + \frac{3p\Lambda_m I_{m_{\text{sideband}}}}{8Js\omega_s}\sin(2s\omega_s t + \alpha_{I_{\text{backward}}} - \alpha_{\Lambda_{\text{fundamental}}})\right] \\ \Lambda_m \sin(\omega_s t - \alpha_{\Lambda_{\text{fundamental}}}) &= \omega_{r_{\text{average}}}\Lambda_m \sin(\omega_s t - \alpha_{\Lambda_{\text{fundamental}}}) \\ &\quad + \frac{3p\Lambda_m^2 I_{m_{\text{sideband}}}}{16Js\omega_s}[\cos((1-2s)\omega_s t + \alpha_{I_{\text{backward}}}) \\ &\quad - \cos((1+2s)\omega_s t + \alpha_{I_{\text{backward}}} - 2\alpha_{\Lambda_{\text{fundamental}}})]\end{aligned} \tag{3.11}$$

where $\omega_{r_{\text{average}}}$ is the average speed of the motor. Equation (3.11) reveals a very interesting and significant aspect of the broken bars fault which is actually the production of a sideband component with the frequency of $(1+2s)f_s$. It means that not only does a left sideband component of the fundamental current harmonic exist, but a right sideband component is also produced as a result of the motor-speed fluctuations. Both are the functions of the motor slip. These harmonic components are generally called the **sideband components of the motor current** shown by the pattern $(1 \pm 2s)f_s$. Accordingly, if the speed is fixed by any means, for example, if it is connected to a closed-loop drive with a very well-refined speed control loop, there is no right sideband component, $(1+2s)f_s$ or it exists with a considerably small magnitude compared to the left sideband. In practice, eliminating speed fluctuations is impossible. Hence, the right sideband component always exists while its amplitudes would be so much sensitive.

Each sideband component appears as a current component in the stator windings with the same frequency of the mentioned pattern. In turn, the corresponding magnetic flux densities are produced at the air gap (see (3.12)). The corresponding magnetic flux density terms are named the fault component of the flux density as the source could be related to an asymmetry such as the broken bar phenomenon. Taking this along with the Biot Savart law into account, two terms defining the right and left sideband components of the magnetic flux density are formulated as (3.12). In fact, hereafter, the terms "right" and "left" sideband components are used to address $(1+2s)f_s$ and $(1-2s)f_s$ elements in any motor variable, respectively. Accordingly, the left and right sideband components of the magnetic flux density are as follows:

$$B_{\text{fault}} = B_{m_{\text{left}}}\sin((1-2s)\omega_s t - \alpha_{B_{\text{left}}}) + B_{m_{\text{right}}}\sin\left((1+2s)\omega_s t - \alpha_{B_{\text{right}}}\right) \tag{3.12}$$

where $B_{\text{fault}}$, $B_{m_{\text{left}}}$, $B_{m_{\text{right}}}$, $\alpha_{B_{\text{left}}}$ and $\alpha_{B_{\text{right}}}$ are the fault component of flux density, the amplitude of left sideband component, the amplitude of right sideband component, the phase angle of left sideband component and the phase angle of right sideband component, respectively. In a similar fashion the current components are formulated as follows:

$$I_{\text{fault}} = I_{m_{\text{left}}}\sin((1-2s)\omega_s t - \alpha_{I_{\text{left}}}) + I_{m_{\text{right}}}\sin\left((1+2s)\omega_s t - \alpha_{I_{\text{right}}}\right) \tag{3.13}$$

where $I_{\text{fault}}$, $I_{m_{\text{left}}}$, $I_{m_{\text{right}}}$, $\alpha_{I_{\text{left}}}$ and $\alpha_{I_{\text{right}}}$ are the fault components of motor current, the amplitude of left sideband component, the amplitude of right sideband component, the phase angle of left sideband component and the phase angle of right sideband component, respectively. So the total motor current containing the fundamental and faulty components is as follows:

$$\begin{aligned} I_{\text{motor}} = {} & I_m \sin(\omega_s t - \alpha_{I_{\text{fundamental}}}) + I_{m_{\text{left}}}\sin((1-2s)\omega_s t - \alpha_{I_{\text{left}}}) \\ & + I_{m_{\text{right}}}\sin\left((1+2s)\omega_s t - \alpha_{I_{\text{right}}}\right) \end{aligned} \tag{3.14}$$

On the basis of the fact that the motor flux has been affected by the sideband components, (3.15) is conducted. It is the output of a fundamental electromagnetic rule based on which the product of a flux and flux-producing current results in an electromagnetic torque if the corresponding flux vectors make a specific angle. Therefore, as both the flux and the current are affected by the fault, the left and right sideband components contribute to the motor torque production as follows:

$$
\begin{aligned}
T &= \frac{3p}{2}\Lambda_{\text{motor}} I_{\text{motor}} \\
&= \frac{3p}{2}\Big[\Lambda_m \sin\left(\omega_s t - \alpha_{\Lambda_{\text{fundamental}}}\right) \\
&\quad + \Lambda_{m_{\text{left}}} \sin\left((1-2s)\omega_s t - \alpha_{\psi_{\text{left}}}\right) \\
&\quad + \Lambda_{m_{\text{right}}} \sin\left((1+2s)\omega_s t - \alpha_{\psi_{\text{right}}}\right)\Big]\Big[I_m \sin\left(\omega_s t - \alpha_{I_{\text{fundamental}}}\right) \\
&\quad + I_{m_{\text{left}}} \sin\left((1-2s)\omega_s t - \alpha_{I_{\text{left}}}\right) + I_{m_{\text{right}}} \sin\left((1+2s)\omega_s t - \alpha_{I_{\text{right}}}\right)\Big] \\
&\xrightarrow{\text{yields}} \frac{3p}{2}
\begin{bmatrix}
+\Lambda_m I_m \sin\left(\omega_s t - \alpha_{\Lambda_{\text{fundamental}}}\right)\sin\left(\omega_s t - \alpha_{I_{\text{fundamental}}}\right) \\
+\Lambda_m I_{m_{\text{left}}} \sin\left(\omega_s t - \alpha_{\Lambda_{\text{fundamental}}}\right)\sin\left((1-2s)\omega_s t - \alpha_{I_{\text{left}}}\right) \\
+\Lambda_m I_{m_{\text{right}}} \sin\left(\omega_s t - \alpha_{\Lambda_{\text{fundamental}}}\right)\sin\left((1+2s)\omega_s t - \alpha_{I_{\text{right}}}\right) \\
+\Lambda_{m_{\text{left}}} I_m \sin\left((1-2s)\omega_s t - \alpha_{\Lambda_{\text{left}}}\right)\sin\left(\omega_s t - \alpha_{I_{\text{fundamental}}}\right) \\
+\Lambda_{m_{\text{left}}} I_{m_{\text{left}}} \sin\left((1-2s)\omega_s t - \alpha_{\Lambda_{\text{left}}}\right)\sin\left((1-2s)\omega_s t - \alpha_{I_{\text{left}}}\right) \\
+\Lambda_{m_{\text{left}}} I_{m_{\text{right}}} \sin\left((1-2s)\omega_s t - \alpha_{\Lambda_{\text{left}}}\right)\sin\left((1+2s)\omega_s t - \alpha_{I_{\text{right}}}\right) \\
+\Lambda_{m_{\text{right}}} I_m \sin\left((1+2s)\omega_s t - \alpha_{\Lambda_{\text{right}}}\right)\sin\left(\omega_s t - \alpha_{I_{\text{fundamental}}}\right) \\
+\Lambda_{m_{\text{right}}} I_{m_{\text{left}}} \sin\left((1+2s)\omega_s t - \alpha_{\Lambda_{\text{right}}}\right)\sin\left((1-2s)\omega_s t - \alpha_{I_{\text{left}}}\right) \\
+\Lambda_{m_{\text{right}}} I_{m_{\text{right}}} \sin\left((1+2s)\omega_s t - \alpha_{\Lambda_{\text{right}}}\right)\sin\left((1+2s)\omega_s t - \alpha_{I_{\text{right}}}\right)
\end{bmatrix}
\end{aligned}
\tag{3.15}
$$

The above demonstration of the electromagnetic torque is the more comprehensive case compared to the simple representation in which all the flux and current signals were ideal and healthy. In the last illustration of the electromagnetic torque, (3.15), there are nine terms added up. By means of (3.16), the mentioned terms are separated into their fundamental terms expressing the corresponding frequencies.

$$
\begin{aligned}
&\text{Term 1: } \Lambda_m I_m \sin(\omega_s t - \alpha_{\Lambda_{\text{fundamental}}})\sin(\omega_s t - \alpha_{I_{\text{fundamental}}}) \\
&= \frac{\Lambda_m I_m}{2}[\cos(\alpha_{I_{\text{fundamental}}} - \alpha_{\Lambda_{\text{fundamental}}}) - \cos(2\omega_s t - \alpha_{\Lambda_{\text{fundamental}}} - \alpha_{I_{\text{fundamental}}})] \\
&\text{Term 2: } \Lambda_m I_{m_{\text{left}}} \sin(\omega_s t - \alpha_{\Lambda_{\text{fundamental}}})\sin((1-2s)\omega_s t - \alpha_{I_{\text{left}}}) \\
&= \frac{\Lambda_m I_{m_{\text{left}}}}{2}[\cos(2s\omega_s t + \alpha_{I_{\text{left}}} - \alpha_{\Lambda_{\text{fundamental}}}) - \cos((1-s)2\omega_s t - \alpha_{\Lambda_{\text{fundamental}}} - \alpha_{I_{\text{left}}})] \\
&\text{Term 3: } \Lambda_m I_{m_{\text{right}}} \sin(\omega_s t - \alpha_{\Lambda_{\text{fundamental}}})\sin\left((1+2s)\omega_s t - \alpha_{I_{\text{right}}}\right) \\
&= \frac{\Lambda_m I_{m_{\text{right}}}}{2}\left[\cos\left(-2s\omega_s t + \alpha_{I_{\text{right}}} - \alpha_{\Lambda_{\text{fundamental}}}\right) - \cos\left((1+s)2\omega_s t - \alpha_{\Lambda_{\text{fundamental}}} - \alpha_{I_{\text{right}}}\right)\right]
\end{aligned}
$$

$$
\begin{aligned}
&\text{Term 4: } \Lambda_{m_\text{left}} I_m \sin((1-2s)\omega_s t - \alpha_{\Lambda_\text{left}}) \sin(\omega_s t - \alpha_{I_\text{fundamental}}) \\
&= \frac{\Lambda_{m_\text{left}} I_m}{2} [\cos(-2s\omega_s t + \alpha_{I_\text{fundamental}} - \alpha_{\Lambda_\text{left}}) - \cos((1-s)2\omega_s t - \alpha_{\Lambda_\text{left}} - \alpha_{I_\text{fundamental}})] \\
&\text{Term 5: } \Lambda_{m_\text{left}} I_{m_\text{left}} \sin((1-2s)\omega_s t - \alpha_{\Lambda_\text{left}}) \sin((1-2s)\omega_s t - \alpha_{I_\text{left}}) \\
&= \frac{\Lambda_{m_\text{left}} I_{m_\text{left}}}{2} [\cos(\alpha_{I_\text{left}} - \alpha_{\Lambda_\text{left}}) - \cos((1-2s)2\omega_s t - \alpha_{\Lambda_\text{left}} - \alpha_{I_\text{left}})] \\
&\text{Term 6: } \Lambda_{m_\text{left}} I_{m_\text{right}} \sin((1-2s)\omega_s t - \alpha_{\Lambda_\text{left}}) \sin\left((1+2s)\omega_s t - \alpha_{I_\text{right}}\right) \\
&= \frac{\Lambda_{m_\text{left}} I_{m_\text{right}}}{2} \left[\cos(-4s\omega_s t + \alpha_{I_\text{right}} - \alpha_{\Lambda_\text{left}}) - \cos(2\omega_s t - \alpha_{\Lambda_\text{left}} - \alpha_{I_\text{right}})\right] \\
&\text{Term 7: } \Lambda_{m_\text{right}} I_m \sin((1+2s)\omega_s t - \alpha_{\Lambda_\text{right}}) \sin(\omega_s t - \alpha_{I_\text{fundamental}}) \\
&= \frac{\Lambda_{m_\text{right}} I_m}{2} \left[\cos(2s\omega_s t + \alpha_{I_\text{fundamental}} - \alpha_{\Lambda_\text{right}}) - \cos((1+s)2\omega_s t - \alpha_{\Lambda_\text{right}} - \alpha_{I_\text{fundamental}})\right] \\
&\text{Term 8: } \Lambda_{m_\text{right}} I_{m_\text{left}} \sin((1+2s)\omega_s t - \alpha_{\Lambda_\text{right}}) \sin((1-2s)\omega_s t - \alpha_{I_\text{left}}) \\
&= \frac{\Lambda_{m_\text{right}} I_{m_\text{left}}}{2} \left[\cos(4s\omega_s t + \alpha_{I_\text{left}} - \alpha_{\Lambda_\text{right}}) - \cos(2\omega_s t - \alpha_{\Lambda_\text{right}} - \alpha_{I_\text{left}})\right] \\
&\text{Term 9: } \Lambda_{m_\text{right}} I_{m_\text{right}} \sin((1+2s)\omega_s t - \alpha_{\Lambda_\text{right}}) \sin((1+2s)\omega_s t - \alpha_{I_\text{right}}) \\
&= \frac{\Lambda_{m_\text{right}} I_{m_\text{right}}}{2} \left[\cos(\alpha_{I_\text{right}} - \alpha_{\Lambda_\text{right}}) - \cos((1+2s)2\omega_s t - \alpha_{\Lambda_\text{right}} - \alpha_{I_\text{right}})\right]
\end{aligned}
\tag{3.16}
$$

Table 3.1 lists the included frequency components in the electromagnetic torque equations.

DC component corresponds to the average motor torque. The higher order harmonic components present themselves in the shape of fluctuations carried by the average value. The amplitude of fluctuations increases with the increase in fault level. As mentioned before and what will be proved in Chapter 4, the high-order harmonic components of the motor torque including $2f_s$, $(1 \pm s)2f_s$ and $(1 \pm 2s)2f_s$ are usually filtered by the motor transfer function which is in practice a low-pass filter. So the only remaining components are those which are the product of the motor slip. If one takes electromagnetic torque spectrum, s/he will notice the higher

*Table 3.1 Frequency components of the motor torque caused by the faulty current and flux signals*

| | |
|---|---|
| DC | Defines the average value of the torque |
| $2f_s$ | High-frequency component |
| $2sf_s$ | Low-frequency component |
| $(1 \pm 2s)2f_s$ | High-frequency component |
| $(1 \pm s)2f_s$ | High-frequency component |
| $4sf_s$ | Low-frequency component |

harmonic components as well. The reason is that usually the motor mechanical torque is measured in which high-order harmonic components eliminate, depending on rotor and coupling inertias. The calculated electromagnetic torque will, of course, contain the high-order components.

Finally, the active power of the motor could be calculated mathematically by multiplying the torque and speed equations. There should certainly be an average value providing the motor with an active power to run. The active or sometimes the reactive power of the motor gives us the opportunity to discriminate between the fault-related $2sf_s$ component and the one produced by an oscillating load. In fact, any load oscillating with the frequency of $2ksf_s$ introduces some levels of change of the sideband components. Thus, they might be mistaken as the real broken bar fault if simply the motor current signature analysis (MCSA) is used. In this case, the motor power spectrum is usually preferred, of course in the steady-state regime, to find the principle source of sideband components change.

As shown in (3.11), the backward field whose frequency is equal to $(1-2s)f_s$ generates the right sideband component with the frequency of $(1+2s)f_s$. This was revealed by substituting speed fluctuations caused by $2sf_s$ component with the EMF equation (see (3.11)). If such a trend is repeated for the $4sf_s$ component existing in motor torque equation, the corresponding torque and EMF will contain new components with the frequency of $4sf_s$ and $(1 \pm 4s)f_s$, respectively. On closer inspection, if one continues the process of mutation of the sideband components repeatedly, the following pattern is introduced and could be extracted by the frequency analysis of motor current and EMF.

$$\text{Frequency pattern of the sideband components: } (1 \pm 2ks)f_s \tag{3.17}$$

where $k$ is an integer starting from 1. This pattern exists in the vicinity of the fundamental component of the current and the flux. However, it might be seen around the products of the fundamental components including the third, fifth, seventh product and so forth [17–28]. Equation (3.17) prepares a very significant engineering basis for the starting point of the fault diagnosis procedure in terms of the broken bars fault. Actually, the amplitudes of these components are the key feature in the fault diagnosis procedure. The point is that the amplitudes are functions of so many influential factors including the fault severity, the fault location, the load level, the speed level, the supply mode, the motor voltage and the motor structure. Therefore, so many efforts have been made so far to find a meaningful trend of variations between the influential factors and the corresponding amplitude of indicator. The studies include the time, frequency and time-frequency processors which deal with stationary and nonstationary operations.

### *3.2.1 Time-domain behavior of induction motors with broken bar/end-ring faults*

A very useful analytical approach has been already provided for understanding what theoretically happens to the motor quantities due to the broken bar/end-ring fault occurrence. The statements should be completed by additional schematic illustration of the motor quantities such as current. Figure 3.4 presents the current, torque, speed and flux signals for an induction motor with one broken bar.

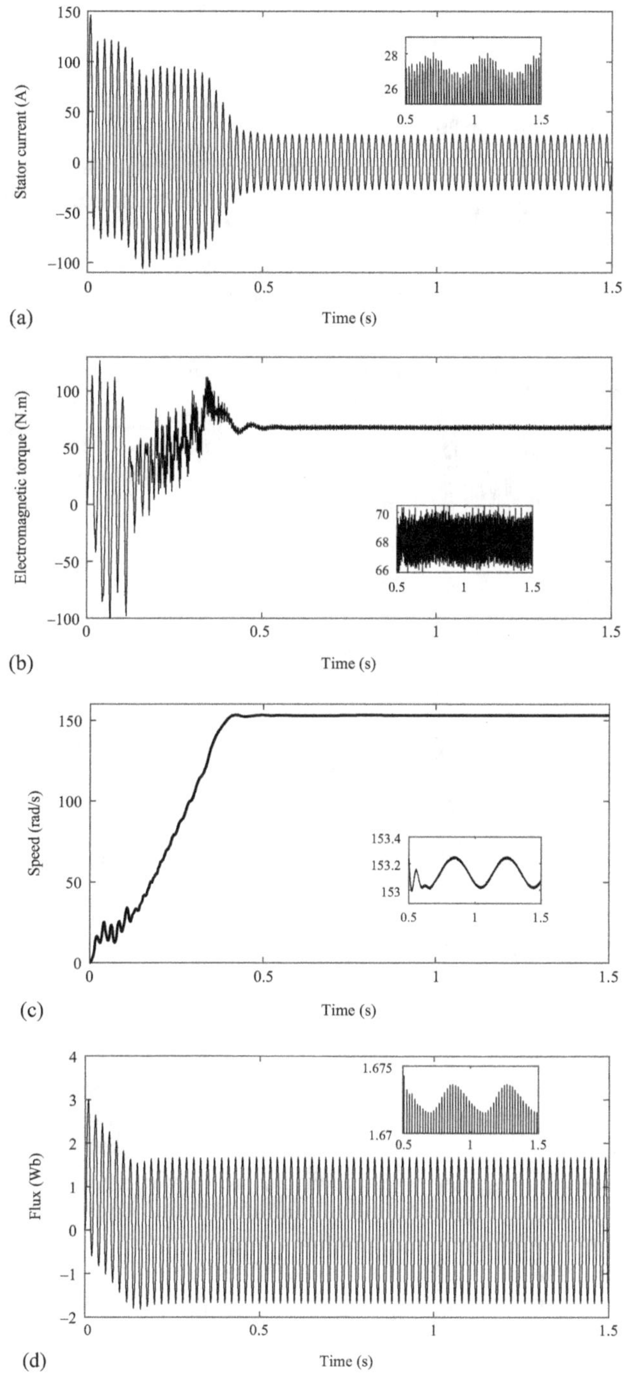


*Figure 3.4 Line-start motor variables (a) current, (b) torque, (c) speed and (d) flux in full-load condition, one broken bar case*

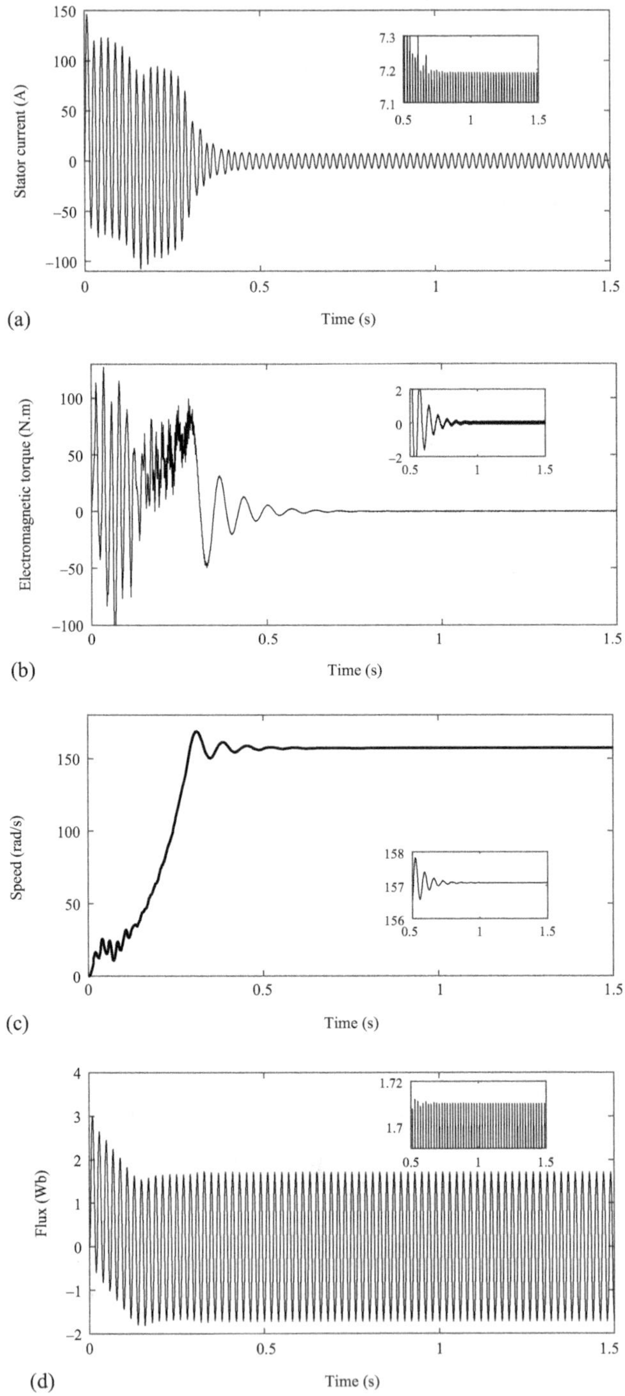


*Figure 3.5 Line-start motor variables (a) current, (b) torque, (c) speed and (d) flux in no-load condition, one broken bar case*

The investigated motor is the same as the one with the power of 11 kW used in Chapter 2. This motor consists of 28 rotor bars along with 36 stator slots. The broken bar characteristics reflected in the motor variables are listed below.

Transient operation is always a tricky and challenging mode of diagnosis in every type of fault as the corresponding faulty signals are only present for a very short period of time. As a result, it is often impossible to track the frequency components dealing with the fault. Besides, the fault components have no fixed amplitude and frequency wise, and they reveal a dramatically changing trend. Thus, diagnosing the faults in the transient regime is still one of the trending aspects of the field. This is not merely restricted to the broken bars fault and is a matter of all the types.

On the other hand, although no/light-load operating condition might be very helpful in diagnosing the faults such as the eccentricity, it diminishes or at least reduces the possibility of detecting the broken bars fault. It is generally because of the masking effects of the fundamental harmonic component which is very close to the fault-related patterns, specifically if the motor current is the signal which is used for the diagnosis purposes. In the no/light-load condition, the slip value is so small that the sideband components almost stick to the fundamental component and they are hopelessly undetectable unless a frequency-domain analysis is hired (Figure 3.5).

The hints are provided here in terms of the theoretical and schematic behavioral study of an induction motor with one full broken bar. There are several other aspects associated with this problem such as:

- the location of multiple broken bars
- the partial broken bar fault
- the effect of various supply modes including the open- or closed-loop modes
- the effect of the drive reference parameters on the diagnosis procedure.

Having provided the fashion based on which the motor faults could be studied, let us move on to the next type of faults which corresponds to the rotor shaft and its displacement from the stator center. This is called the "misalignment fault" or generally the "eccentricity fault." First, a set of mathematical developments are addressed; then a couple of illustrations are provided in terms of the behavioral study of the faulty motor.

## 3.3 Eccentric/misaligned and bearing faults in induction motors

In induction motors, there are three physical dimensions $x$, $y$ and $z$ indicating the Cartesian axes in 3D space (see Figure 3.6(a)). In addition, the 2D demonstration is also provided in Figure 3.6(b) which illustrates the rotor symmetry center ($A_r$), the stator symmetry center ($A_s$) and the rotor rotation/whirling center ($A_w$). Besides, the Cartesian axes are highlighted in the following fashion:

- The $x$-axis is aligned with the spatial axis of the phase "a" of the stator.
- The $y$-axis is located exactly 90 mechanical degrees away, in a counter clockwise direction, from the $x$-axis.
- The $z$-axis is perpendicular to both of the $x$ and $y$ axes in the direction of the motor shaft.

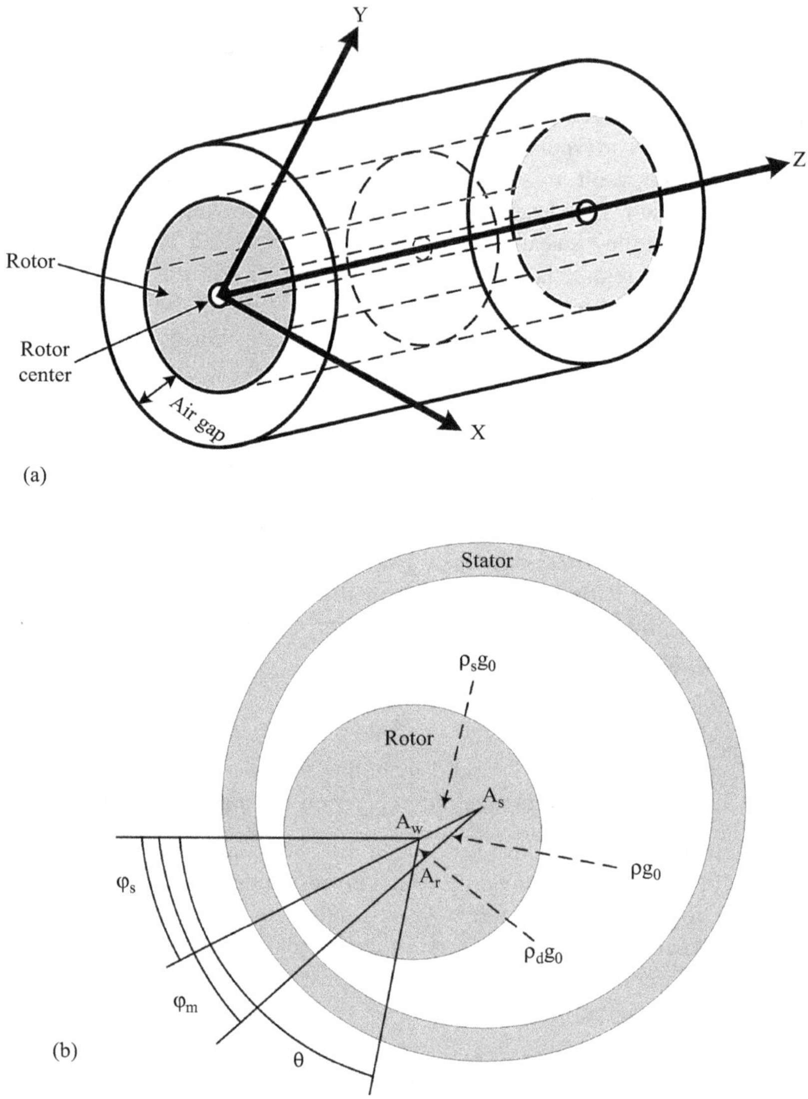


*Figure 3.6 Induction motor schematic (a) 3D space and (b) 2D considering possible rotor displacements*

In healthy motors all the three centers perfectly coincide with the spatial center of the stator, i.e., $A_s$, which is fixed in one point. This point is the origin of any mechanical displacement analysis. So the air-gap length, highlighted in Figure 3.6(a), would be the same all over across the stator circumference and equals to $g_0$. The rotational movement of the rotor does not disturb this uniformity while a hundred percent match between the mentioned centers is impractical due to several imperfections such as the improper assembly or placement of the bearings. This is the reason that the bearing fault could also be a potential source of eccentricity fault. This phenomenon could be easily detected even in the case of healthy and new brand motors. This type of eccentricity is called the inherent eccentricity. Other factors including the misalignment of the load and the shaft axes, as well as the mechanical stresses and imbalances, fortify the situation. Consequently, the eccentricity fault takes place, and the uniformity of the flux distribution in the air gap is distorted. This literally means that the air-gap length is not uniform across the stator circumference. Using a very acceptable approximation of the air-gap length variation, the following representation should always hold correctly [29–35]:

$$g(\varphi, \varphi_m, \rho) = g_0[1 - \rho\cos(\varphi - \varphi_m)] \tag{3.18}$$

where $g_0$, $\rho$, $\varphi$ and $\varphi_m$ are the uniform air-gap length in healthy condition, the ratio of the distance of the rotor and stator centers over $g_0$ (the severity of the eccentricity fault), the mechanical angle of the rotating and nonrotating point is the stator reference frame and the angle separating the stator and rotor centers in the direction of the rotation, respectively (see Figure 3.6(b)). It is obvious that the distance between the stator and rotor center could not exceed $g_0$. As a result, $\rho$ should geometrically remain within the range [0–1] if the rotor is locked. However, while the rotor rotates, the UMP, along with the centrifugal force, leads to the increase in the fault severity, so $\rho$ might not go beyond 0.72 (proved by the experiments). In such situation, the rotor touches the inner surface of the stator at both ends while the motor operates.

The center of rotation of the rotor has not been taken into account yet and it has only been assumed that $A_w$ is fixed. In fact, it is the position of $A_w$ which defines the type of the eccentricity fault. If the center of rotation remains concentric with $A_s$, it is called the "static eccentricity." The term static conveys the idea that although the air-gap length is nonuniform across the stator circumference, it is fixed and does not change with time. However, if the center of rotation is aligned with the stator center during any eccentric condition, it is called "dynamic eccentricity" fault. In this case, the rotor center rotates about the stator center with the same frequency as that of the rotor. As a result, the nonuniform air-gap length rotates by rotating the rotor, and the air-gap length is not fixed anymore. For example, $\varphi_m$ which is the position of the minimum air-gap length, increases by the rotor speed as follows:

$$\varphi_m = \alpha_d + \theta \tag{3.19}$$

where $\alpha_d$ is the angle in the rotor reference frame along which $A_r$ and $A_s$ diverge and $\theta$ is the angular position of the rotor with respect to the stator. As $\alpha_d$ does not have any considerable change in the quantities expressed in the stator reference frame, it is neglected due to the sake of simplicity. In general, when the eccentricity fault occurs, both static and dynamic eccentricities could be detected, and a new type called "mixed eccentricity fault" is introduced. This means that the air-gap status changes regarding two aspects:

- the level of the eccentricity fault ($\rho$)
- the spatial distribution of the nonuniform air gap.

Considering this point, $\rho$ and $\varphi_m$ are generally formulated as follows [36–38]:

$$\rho = \sqrt{\rho_s^2 + \rho_d^2 + 2\rho_s\rho_d\cos(\theta - \varphi_s)} \tag{3.20}$$

$$\varphi_m = \varphi_s + \tan^{-1}\left(\frac{\rho_d\sin(\theta - \varphi_s)}{\rho_s + \rho_d\cos(\theta - \varphi_s)}\right) \tag{3.21}$$

where $\rho_s$ is the ratio of the distances of $A_w$ and $A_s$ from $g_0$. $\rho_d$ and $\varphi_s$ are the ratio of the distances of $A_w$ and $A_r$ from $g_0$ and the angle along which $A_w$ and $A_s$ diverge, respectively. The static and dynamic eccentricities are only some simpler forms of the mixed eccentricity fault where $\rho_d$ or $\rho_s$ are equal to zero. Therefore, $\rho_s$ defines the severity of the static eccentricity fault. Likewise, $\rho_d$ defines the severity of the dynamic eccentricity fault. If (3.20) and (3.21) are substituted with (3.18), the real value of the air-gap length is obtained as a function of the rotor position, the static eccentricity severity and also the dynamic eccentricity severity. The inverse function of (3.18) is called the inverse air-gap function or the permeance function which plays a vital role in analyzing the eccentricity fault as well as the misalignment fault which is a rather complex version of the eccentricity fault in which the motor shaft bends toward the $z$-axis shown in Figure 3.6(a). This issue will be further investigated.

Depending on the values of $\rho_s$ and $\rho_d$, one of the following three cases is possible:

(a) $\rho_d > \rho_s$
In this case, the dynamic eccentricity dominates the static one. Using the equations represented so far, the air-gap function, along with its position, can be calculated upon changing the rotor position (see Figure 3.7). This figure illustrates such calculation according to which the minimum air-gap length oscillates between two extreme points including $g_0(1 - \rho_d - \rho_s)$ and $g_0$ $(1 - \rho_d + \rho_s)$ with respect to the rotor rotation. The frequency of $\varphi_m$ is the same as that of the rotor rotation.

(b) $\rho_d = \rho_s$
This situation is usually called a balanced mixed eccentricity fault. The corresponding variations of $g_{\min}$ and $\varphi_m$ are shown in Figure 3.8. As seen, $g_{\min}$ oscillates within the range $g_0(1 - \rho_d - \rho_s)$, and $g_0$ and $\varphi_m$ are limited to the $[-\pi/2, \pi/2]$ period. Therefore, the minimum air-gap length is eliminated where the rotor position equals one of the odd factors of $\pi$. Again, the frequency of $\varphi_m$ is the same as that of the rotor rotation.

(c) $\rho_d < \rho_s$
In the above situations, the static component of the eccentricity dominates the dynamic one, and the corresponding air-gap length looks like Figure 3.9. Accordingly, the minimum air-gap length oscillates between two extreme points including $g_0(1+ \rho_d - \rho_s)$ and $g_0(1 - \rho_d - \rho_s)$ with respect to the rotor rotation. The range of variations of $\varphi_m$ is relatively small and it oscillates very close to $\varphi_s$. The movement looks like a pendulous oscillation, and the corresponding frequency is again equal to the rotor frequency.

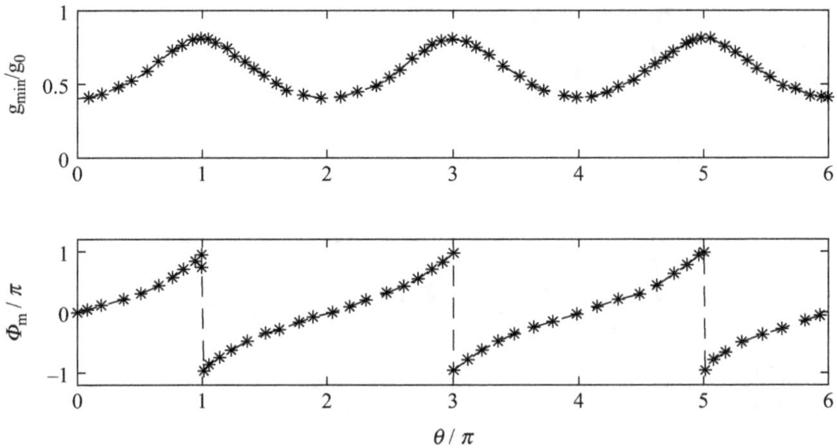


*Figure 3.7 Variation of the minimum air-gap length (up) and its position (down) with respect to the rotor position in mixed eccentricity fault ($\rho_d = 0.4$, $\rho_s = 0.2$ and $\varphi_s = 0$)*

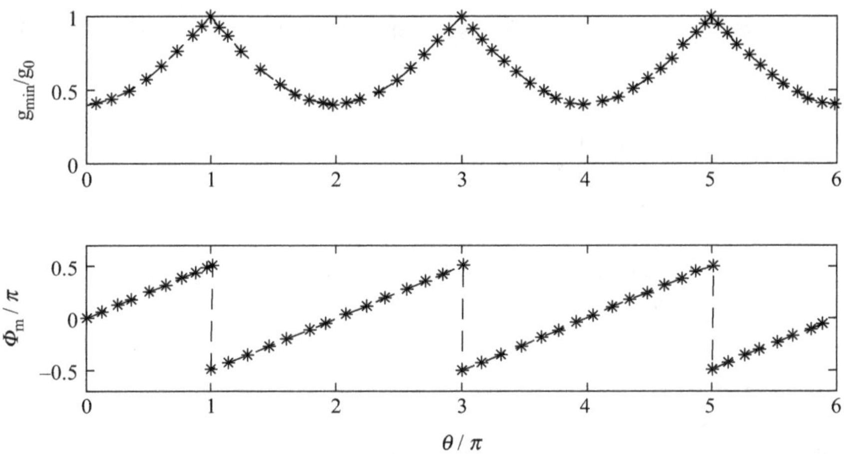


*Figure 3.8 Variation of the minimum air-gap length (up) and its position (down) with respect to the rotor position in balanced mixed eccentricity fault ($\rho_d = 0.3$, $\rho_s = 0.3$ and $\varphi_s = 0$)*

Three general cases, namely a dominant dynamic, a dominant static, as well as a balanced mixed eccentricity fault, were discussed. Considering the discussed cases, the air-gap length affected by the eccentricity fault could have several different shapes each having its own effect on the motor quantities.

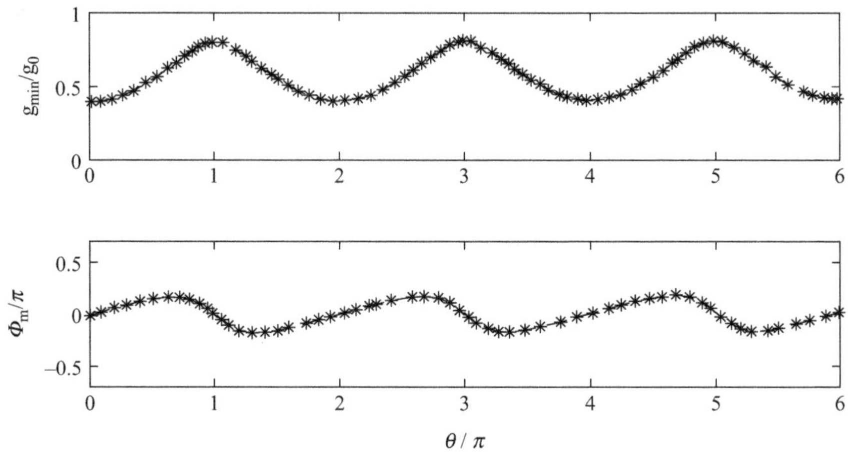


*Figure 3.9 Variation of the minimum air-gap length (up) and its position (down) with respect to the rotor position in mixed eccentricity fault ($\rho_d = 0.2$, $\rho_s = 0.4$ and $\varphi_s = 0$)*

It should be mentioned that many assumptions are associated with the analysis such as:

- ignoring saturation effect
- ignoring slotting effect
- ignoring end winding effect
- the static or dynamic eccentricity level is applied to the rotor along the *z*-axis, and the rotor should be displaced from the origin. This means there is still symmetry in *z*-direction, and all the 2D analyses are still valid in 3D if the thermal issues are not of interest.

### 3.3.1 Misalignment inclined rotor

The main focus has been already on a symmetric *z*-directional eccentricity fault which guarantees the symmetry in the third dimension of the motor. However, in a specific situation, an asymmetric distribution of the motor air-gap length along the *z*-axis might also be dealt with (see Figure 3.10). Although Figure 3.10 is a kind of exaggerated representation of the mentioned defect, it easily and clearly conveys the issue according to which the rotor axis $A_r$, inclines and takes a distance from $A_s$. $\beta$ is the amount of divergence from the concentric point, but it is totally different from a simple static eccentricity as the rotor ends move in the opposite directions, i.e., one goes up and the other falls down. So along the *z*-axis, the eccentricity severity of the fault is not constant anymore. In this case, we only refer to a static misaligned/inclined rotor to make sure that the analytical descriptions are easily extractable while a mixed misalignment fault should always take place in reality.

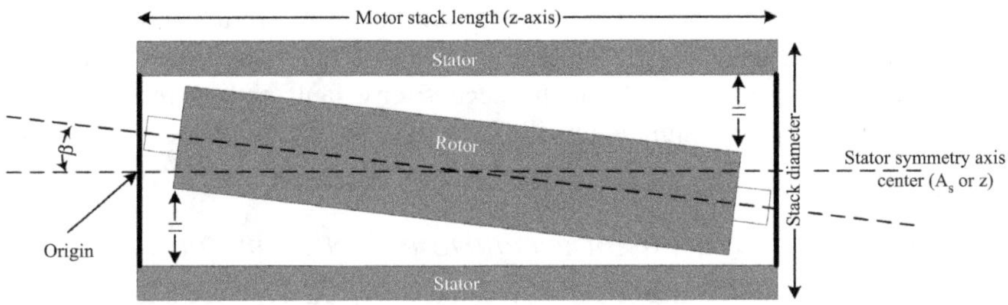


*Figure 3.10 Statically misaligned/ inclined rotor*

Some of the most important causes are as follows:

- bearings wear caused by aging
- bent shafts
- swinging mechanical loads
- inherent assembly and manufacturing defects.

After a long motor operation, these factors result in a mechanical failure of the motor if a proper diagnostic and maintenance process is not held. The misaligned rotor not only harms the motor itself, but also produces an extensively fluctuating movement of loads leading to an improper operation outside the motor as well. This might be the most compelling evidence which necessitates the fault diagnosis procedures. Now, we would like to formulate the air-gap length in a misaligned motor, using the assumption that only the static form of the fault exists and there is no additional swing caused by a dynamic movement of the rotor center. In this case, there is an air-gap length distribution similar to (3.18) but with a variable $\rho$ along the third dimension ($z$) of the motor. The air-gap length and consequently the corresponding permeance are the functions of $\beta$.

$$g(\varphi, \varphi_m, \rho) = g_0[1 - \rho(z)\cos(\varphi - \varphi_m)] \tag{3.22}$$

$$\rho(z) = \frac{(2\rho_{s0} - L_s \sin(\beta)/g_0)}{2} + z\frac{\tan(\beta)}{g_0} \tag{3.23}$$

where $\rho_{s0}$ is the static eccentricity fault severity right in the middle of the rotor. $L_s$ is the motor stack length, and $z$ is the distance from the origin shown in Figure 3.10. According to the above equations, the air-gap length is not fixed in different motor crosssections. For the future analysis, if the permeance function is required, (3.22) should be used but in an inverse form. Incorporating the dynamic eccentricity into the misalignment formulations should be so sufficient that you might not be able to find a suitable theoretical resource in this field. Therefore, it is preferred to skip the philosophical discussion of the eccentricity/misalignment fault at this point and switch to the mathematical and physical descriptions of the phenomenon by working on the impacted motor quantities including the magnetic flux density and the current. Then, we will end up with a pretty closed-form formula associated with the UMP existing in an eccentric induction motor. What is provided there is an intensive analytical representation of the eccentricity fault and its resultant components; then a few motor signals are illustrated to justify the effects in the time domain.

### *3.3.2 Theoretical analysis of eccentric induction motor*

The basis for providing the materials is introduced by the Ampere's law [38–43]. The main assumption here is the presence of a single-harmonic motor model based on which all the healthy motor signals only contain a single sinusoidal distribution. So the stator current density should also follow the same rule. It is noted that the statement is true if the discussion is held in the steady-state mode. Considering this, the current density of the stator windings can be formulated as (3.24).

$$J_s(\varphi, \varphi_m) = J_{sm}\sin(\varphi - \varphi_m) \tag{3.24}$$

where $J_{sm}$ and $\varphi$ are the stator current density (in A/mm$^2$), the synchronous field position (in $\omega_s t$, $t$ is time). $\varphi_m$ is equal to $pky$ where $k$ is the inverse air-gap function, and $y$ is the linear distance around the stator circumference. Using the Ampere's circuital law and also neglecting the angular component of the air-gap flux density which is practically a correct assumption, one will end up with the following equation.

$$B_s(y,t) = \wp(y,t)\int \mu_0 j_s(y,t)\cdot dy \tag{3. 25}$$

where $\wp(y, t)$ is the permeance function described previously in this chapter. As seen, the magnetic flux density is related to the current density by an integral operator over the stator circumference. Reforming (3.18)–(3.21) returns, the following representation of the air-gap length demonstrating the components corresponding to the static and dynamic eccentricities:

$$g(y,t) = g_0[1 - \rho_s\cos(ky) - \rho_d\cos(\omega_r t - ky)] \tag{3.26}$$

It is assumed that the air-gap variation is a sinusoidal function if the stator and rotor slotting effects are not taken into account. Otherwise, the air-gap length must be equal to sum of many sinusoidal terms expressing the Fourier transform of the air-gap function. This matter will be discussed later. It might also be noted that experiments and investigations show that the above assumption is correct in the case of a nonsalient pole machine. $\omega_r$, $\rho_s$, $\rho_d$ and $g_0$ are rotor rotational speed, static eccentricity severity, dynamic eccentricity severity and average air-gap length in healthy condition, respectively. To avoid stator-rotor rub term $(\rho_s + \rho_d) < 1$ must hold. Under small values of eccentricity fault, inverse air-gap function is approximated as follows:

$$\wp(y,t) = \frac{1}{g_0}[1 + \rho_s\cos(ky) + \rho_d\cos(\omega_r t - ky)] \tag{3.27}$$

In practice, if the eccentricity fault severity goes beyond 30%, rotor may rub the stator and probably fail to operate appropriately. By combining (3.24)–(3.27), one obtains:

$$B_s(y,t) = \frac{\mu_0 J_{sm}}{kpg_0}\cos(\omega_s t - pky)(1 + \rho_s\cos(ky) + \rho_d\cos(\omega_r t - ky)) \tag{3.28}$$

$B_s$ is the stator magnetic field observed at the air-gap level considering the presence of a mixed dynamic eccentricity fault. Multiplying the left cosine term by the terms residing inside the righter parentheses leads to an almost straightforward formulation of the magnetic flux density produced by the stator as follows:

$$\begin{aligned} B_s(y,t) = {} & B^{p}_{sm}\cos(\omega_s t - pky) + B^{p-1}_{sm(s)}\cos(\omega_s t - (p-1)ky) \\ & + B^{p+1}_{sm(s)}\cos(\omega_s t - (p+1)ky) \\ & + B^{p-1}_{sm(d)}\cos((\omega_s - \omega_r)t - (p-1)ky) \\ & + B^{p+1}_{sm(d)}\cos((\omega_s + \omega_r)t - (p+1)ky) \end{aligned} \tag{3.29}$$

where

$$\begin{aligned} B^{p}_{sm} &= \frac{\mu_0 J_{sm}}{kpg_0} \\ B^{p\pm1}_{sm(s)} &= \frac{\mu_0 J_{sm}}{2kpg_0}\rho_s \\ B^{p\pm1}_{sm(d)} &= \frac{\mu_0 J_{sm}}{2kpg_0}\rho_d \end{aligned} \tag{3.30}$$

There are five components associated with the calculated eccentricity-related magnetic flux density. From left to right of the left-hand side term, they are accordingly produced by the fundamental component, the static eccentricity and the dynamic eccentricity fault, respectively. It is clearly observable that the static eccentricity components, the ones with the magnitude of $B^{p\pm1}_{sm(s)} = \frac{\mu_0 J_{sm}}{2kpg_0}\rho_s$, merely depend on the functions of the synchronous frequency and have nothing to do with the rotor speed or position. On the other hand, the dynamic eccentricity fault-related components, the ones with the magnitude of $B^{p\pm1}_{sm(d)} = \frac{\mu_0 J_{sm}}{2kpg_0}\rho_d$, are imposed to be the functions of the rotor position as well as the synchronous speed of the stator. This is another proof of the dependency of the motor quantities on the rotor rotation in the case of dynamic eccentricity fault. The developed and discussed formulation is a pure analytical practice with a lot of imprecise assumptions made just to allow us to extract a closed-form relationship. Therefore, although the magnitudes (see (3.30)) provide a useful common basis to compare the healthy and fault-related components, are not accurate in terms of the absolute values. However, the $B^{p\pm1}_{sm(s)}$ and $B^{p\pm1}_{sm(d)}$ are smaller than $B^{p}_{sm} = \frac{\mu_0 J_{sm}}{kpg_0}$, proving the fact that the fundamental component still possesses the largest magnitude among the motor frequency components. As another perceivable fact, the fault-related components are indeed number-of-poles dependent. In fact, two faulty terms are separately assigned to each of the static and dynamic eccentricities. In the case of dynamic eccentricity, these terms are located within a specific frequency range, scaled by $(\omega_s \pm \omega_r)$, of the fundamental component.

So far, the air-gap field produced by the stator current, which in turn applies a specific level of the EMF to the rotor bars, has been evaluated. Inspecting a short-circuited rotor circuit, the voltages induced by the mentioned EMF produce the bar currents flowing into the rotor circuits. The same principle is valid for a wound rotor. The point is that the calculations provided so far are based on a stator reference frame while the rotor-related calculations require to transfer the quantities from the stator to the rotor side. As a result, (3.29) should be mapped on the rotor reference frame using a wise substitution of $ky = \omega_r t + ky'$ and $\omega_r = \frac{(1-s)}{p}\omega_s$ with (3.29). The stator magnetic flux density formulated in the stator reference frame is converted to (3.31).

$$\begin{aligned} B_s(y',t) = {} & B_{sm}^{p}\cos(s\omega_s t - p(ky' - \alpha_1)) \\ & + B_{sm(s)}^{p-1}\cos\left(\frac{1+(p-1)s}{p}\omega_s t - (p-1)ky' - p\alpha_2\right) \\ & + B_{sm(s)}^{p+1}\cos\left(\frac{-1+(p+1)s}{p}\omega_s t - (p+1)ky' - p\alpha_3\right) \\ & + B_{sm(d)}^{p-1}\cos(s\omega_s t - (p-1)ky' - p\alpha_4) \\ & + B_{sm(d)}^{p+1}\cos(s\omega_s t - (p+1)ky' - p\alpha_5) \end{aligned} \tag{3.31}$$

A very clear interpretation of (3.31) is that the rotating frequency corresponding to the magnetic flux density is of course slip-dependent. This arises from the previous transformation from the stator to rotor reference frame. The induced EMF is derived by taking derivative of (3.31) with respect to time.

$$\begin{aligned} \mathrm{EMF}_s(y',t) = {} & \mathrm{EMF}_{sm}^{p}\cos(s\omega_s t - p(ky' - \alpha_1)) \\ & + \mathrm{EMF}_{sm(s)}^{p-1}\cos\left(\frac{1+(p-1)s}{p}\omega_s t - (p-1)ky' - p\alpha_2\right) \\ & + \mathrm{EMF}_{sm(s)}^{p+1}\cos\left(\frac{-1+(p+1)s}{p}\omega_s t - (p+1)ky' - p\alpha_3\right) \\ & + \mathrm{EMF}_{sm(d)}^{p-1}\cos(s\omega_s t - (p-1)ky' - p\alpha_4) \\ & + \mathrm{EMF}_{sm(d)}^{p+1}\cos(s\omega_s t - (p+1)ky' - p\alpha_5) \end{aligned} \tag{3.32}$$

where

$$\begin{aligned}\mathrm{EMF}_{sm}^{p} &= \frac{-s\omega_s B_{sm}^{p}}{pk}\\ \mathrm{EMF}_{sm(s)}^{p\pm 1} &= \frac{-[(p\pm 1)s \mp 1]\omega_s B_{sm(s)}^{p\pm 1}}{(p\pm 1)k}\\ \mathrm{EMF}_{sm(d)}^{p\pm 1} &= \frac{-s\omega_s B_{sm(d)}^{p\pm 1}}{(p\pm 1)k}\end{aligned} \tag{3.33}$$

It is really important to realize that unlike the magnetic flux density, the amplitude of the induced EMF is the function of the motor slip, so increasing the motor slip should practically lead to an increase in the magnitude of the induced EMFs regardless of the fault type [44–46]. Thus, the rotor current must increase as well. This is consistent with the fact that increasing the motor load requires a larger current flowing into the windings and bars as a larger slip is needed. Following the explanations, the rotor bar or winding currents are simply calculated by dividing the EMFs with a factor of rotor bar or winding resistance while neglecting the phase shift between the EMF and the current caused by the inductive nature of the rotor. Then, rotor bar current densities would be obtained as follows:

$$\begin{aligned}J_r(y',t) = {}& J_{rm}^{p}\sin(s\omega_s t - p(ky' - \alpha_1))\\ &+ J_{rm(s)}^{p-1}\sin\left(\frac{1+(p-1)s}{p}\omega_s t - (p-1)ky' - p\alpha_2\right)\\ &+ J_{rm(s)}^{p+1}\sin\left(\frac{-1+(p+1)s}{p}\omega_s t - (p+1)ky' - p\alpha_3\right)\\ &+ J_{rm(d)}^{p-1}\sin(s\omega_s t - (p-1)ky' - p\alpha_4)\\ &+ J_{rm(d)}^{p+1}\sin(s\omega_s t - (p+1)ky' - p\alpha_5)\end{aligned} \tag{3.34}$$

Equation (3.34) has been developed in rotor reference frame based on the stator supply frequency. $\alpha_1$, $\alpha_2$, $\alpha_3$, $\alpha_4$ and $\alpha_5$ are the phase angle of the rotor bar currents with respect to the induced EMF from the stator side. If it is assumed that the rotor inductance is very smaller than its resistance as rotor electric frequency is a fraction of rated frequency, the amplitude of harmonic components of bar current, i.e., ${J_{rm}}^{p}$, ${J_{rm(s)}}^{p-1}$, ${J_{rm(s)}}^{p+1}$, ${J_{rm(d)}}^{p-1}$, ${J_{rm(d)}}^{p+1}$, could be simply obtained by dividing the induced EMFs by rotor bar resistance as follows:

$$\begin{aligned}J_{rm}^{p} &= \frac{-s\omega_s B_{sm}^{p}}{pkR_{bar}}\\ J_{rm(s)}^{p\pm 1} &= \frac{-[(p\pm 1)s \mp 1]\omega_s B_{sm(s)}^{p\pm 1}}{(p\pm 1)kR_{bar}}\\ J_{rm(d)}^{p\pm 1} &= \frac{-s\omega_s B_{sm(d)}^{p\pm 1}}{(p\pm 1)kR_{bar}}\end{aligned} \tag{3.35}$$

where $R_{\text{bar}}$ is the rotor bar resistance in the rotor reference frame. Generally, there are three current components rotating with the frequency of $s\omega_s$, $\frac{1+(p-1)s}{p}\omega_s$ and $\frac{-1+(p+1)s}{p}\omega_s$. Although this is a very rough approximation, it could be insightful. The last two frequencies correspond to the static eccentricity fault while the dynamic eccentricity is of the same nature of the fundamental frequency. Rotor bar currents produce their own magnetic fields in the motor air gap and then induce the corresponding MMFs into the stator winding through being modulated by the air-gap permeance similar to what has been previously done in (3.28). Working on the formulations and extracting the terms related to the nonsupply-frequency components and also the $p$ pole-pair components, we will end up with the following equations:

$$\begin{aligned} B_r(y,t)\big|_{(\neq p\pm 1)}^{(\neq\omega)} = & \left[\frac{J_{rm(s)}^{p-1}}{k(p-1)g_0}\cos(\omega_s t-(p-1)ky-p\alpha_2)\right. \\ & \left.+\frac{J_{rm(s)}^{p+1}}{k(p+1)g_0}\cos(\omega_s t-(p+1)ky-p\alpha_3)\right]\cos(\omega_r t-ky) \\ & +\left[\frac{J_{rm(d)}^{p-1}}{k(p-1)g_0}\cos((\omega_s-\omega_r)-(p-1)ky-p\alpha_4)\right. \\ & \left.+\frac{J_{rm(d)}^{p+1}}{k(p+1)g_0}\cos((\omega_s+\omega_r)-(p+1)ky-p\alpha_5)\right]\cos(ky) \end{aligned} \tag{3.36}$$

$$\begin{aligned} B_r(y,t)\big|_{p}^{(\neq\omega)} = & \left[\frac{J_{\text{rm}(s)}^{p-1}}{2k(p-1)g_0}\cos((\omega_s+\omega_r)t-\text{pky}-p\alpha_2)\right. \\ & \left.+\frac{J_{\text{rm}(s)}^{p+1}}{2k(p+1)g_0}\cos((\omega_s-\omega_r)t-\text{pky}-p\alpha_3)\right] \\ & +\left[\frac{J_{\text{rm}(d)}^{p-1}}{2k(p-1)g_0}\cos((\omega_s-\omega_r)-\text{pky}-p\alpha_4)\right. \\ & \left.+\frac{J_{\text{rm}(d)}^{p+1}}{2k(p+1)g_0}\cos((\omega_s+\omega_r)-pky-p\alpha_5)\right] \end{aligned} \tag{3.37}$$

From (3.36), it is seen that the rotor magnetic flux density waves caused by the static eccentricity are regulated with the dynamic component of the eccentricity and produce the sideband current components with the frequency pattern of $(f_s \pm f_r)$.

The point is that the components listed in (3.36) do not contribute to the torque-developing process as the number of poles is not equal to that of the stator, whereas (3.37) illustrates the torque-generating fault components containing the same frequency pattern as that of (3.36).

The $(f_s \pm f_r)$ pattern is the very well-known index to diagnose the eccentricity fault; of course, in the case of a mixed eccentricity fault, both the static and dynamic types have some level of correlation with the mentioned pattern. This pattern is also a function of the motor slip revealing a nonconstant frequency position if the motor slip is changed by any means. Moreover, the corresponding magnitudes are load-dependent deduced from the rotor current densities in (3.37).

It should also be remembered that an UMP which results in additional noise or vibration is an output of any eccentricity fault. Nowadays, due to the lack of engineering importance, very little is known about the subject of noise in electrical machines. All motor designers treated noise mitigation as an art-like cookery. For example, they had a list of low-noise empirical rules and some forbidden combinations of stator and rotor slots. All that was perfectly clear. However, these are some general laws that do not aim at explicitly describing the faulty motor behavior which is the main focus of the book. Therefore, due to the significance of the predrawn topic, we are seeking to provide a satisfactory on the vibration and probably noise analysis throughout this book. The starting point is to calculate the magnetic force between rotor and stator by means of the following commonly used approximation of the magnetic force.

$$\text{Force}_{\text{radial}} = \frac{B_{\text{radial}}^2}{2\mu_0} \tag{3.38}$$

where $B_{\text{radial}}$ is the radial component of the air-gap flux density all over the stator inner circumference. Considering a mixed eccentric rotor, 4 main flux components other than the fundamental one are present as the spatial harmonics of the flux density according to (3.29). It is worth noting that it is the interaction between the rotor and the stator flux densities which finally produces the net flux density at the air gap. So on the basis of the existing components shown in (3.29), the net flux density should also contain the same harmonics, but with different values of the magnitude and the phase angles. The general formulation should be as follows:

$$\begin{aligned} B_n(y,t) = {} & B_{nm}^{p}\cos(\omega_s t - pky - \theta_1) + B_{nm(s)}^{p-1}\cos(\omega_s t - (p-1)ky - \theta_2) \\ & + B_{nm(s)}^{p+1}\cos(\omega_s t - (p+1)ky - \theta_3) \\ & + B_{nm(d)}^{p-1}\cos((\omega_s - \omega_r)t - (p-1)ky - \theta_4) \\ & + B_{nm(d)}^{p+1}\cos((\omega_s + \omega_r)t - (p+1)ky - \theta_5) \end{aligned} \tag{3.39}$$

By replacing (3.39) with (3.38) and simplifying the equation, a closed-form expression of the radial forces at the motor air gap is obtained where the fault is present.

$$\begin{aligned}
\text{Force}_{\text{radial}} = \frac{1}{2\mu_0}\Bigg[ & \frac{\left(B^p_{nm}\right)^2}{2}\cos(2\omega_s t - 2pky - 2\theta_1) \\
& + B^p_{nm}\left[B^{p-1}_{nm(s)}\cos(2\omega_s t - (2p-1)ky - \theta_1 - \theta_2)\right] \\
& + B^{p+1}_{nm(s)}\cos(2\omega_s t - (2p+1)ky - \theta_1 - \theta_3) \\
& + B^{p-1}_{nm(d)}(\cos((2\omega_s - \omega_r)t - (2p-1)ky - \theta_1 - \theta_4) \\
& + \cos(\omega_r t - ky - \theta_1 - \theta_4)) \\
& + B^{p+1}_{nm(d)}(\cos((2\omega_s + \omega_r)t - (2p+1)ky - \theta_1 - \theta_5) \\
& + \cos(\omega_r t - ky + \theta_1 + \theta_4)) \\
& + B^{p-1}_{nm(d)}B^{p+1}_{nm(d)}\cos(2\omega_r t - 2ky + \theta_4 - \theta_5) \\
& + \frac{\left(B^{p-1}_{nm(d)}\right)^2}{2}\cos(2(\omega_s - \omega_r)t - 2(p-1)ky - 2\theta_4) \\
& + \frac{\left(B^{p+1}_{nm(d)}\right)^2}{2}\cos(2(\omega_s + \omega_r)t - 2(p+1)ky - 2\theta_5)\Bigg]
\end{aligned} \tag{3.40}$$

The most significant aspects in terms of the above expression are as follows:

- The minimum frequency included in the magnetic force is the same as that of the rotor mechanical quantities, i.e., $f_r$.
- The maximum available frequency is equal to twice the synchronous frequency, for example if there is a supply frequency of 50 Hz, the largest observable frequency will be 100 Hz.
- There are also some medium frequency components including $(2f_s + f_r)$, $(2f_s - f_r)$, $2(f_s + f_r)$, $2(f_s - f_r)$ and $2f_r$.
- Not all the frequency components would probably be detected in the vibration signal analysis of induction motors, and it actually depends on the motor structure, the mechanical damping factors, etc.
- Practically, relying on $2f_s$ component to detect the fault is somehow a wrong technique as it cannot be clearly diagnosed most of the time.

According to the literature, the vibration signal is usually considered as one of the most interesting signals available throughout the sensors mounted in the motor body.

This is not only the case in induction motors; and vibration and subsequently noise analysis have been a matter of various investigations in terms of different types of machines. However, we do not aim at discussing the diagnosis procedures in this chapter; they will be dealt with in the next chapters. For the time being, it seems terrific to have an insight into the real motor quantities including the current, the torque, the speed and even the motor flux when the eccentricity fault exists. Therefore, we aim at providing finite element (FE) results of 28-bar, 36-slot and delta-connected induction motor to which 30% dynamic eccentricity fault is applied. A 2D FE analysis is conducted as a complete 3D one will be so time-consuming that it might not be handled by means of low-power computational devices. A quasi-2D model in which several slices of one single motor in $z$-direction are modeled, paralleled and simulated at the same time is also an appropriate alternative of a 3D operation. However, there is still a high computational demand compared to 2D simulations. Considering these points, the motor is modeled in 2D. This is an acceptable approximation as long as the stack length is not small compared to the stator diameter. If so, the $z$-axis component of the motor quantities would be symmetric to that of the 2D simulations to a great extent. Otherwise, the $z$-axis should be definitely modeled due to the fact that $z$-axis components play a vital role in motor operation. With this in mind, a group of results illustrated in Figure 3.11 in terms of an eccentric motor are provided.

Figure 3.11 shows the motor quantities in the no-load and full-load conditions. The simulated motor has already been discussed in Figure 3.4. The difference is that the skewing effect resulting in a smoother response as well as a larger motor rise time has not been considered in the FE-based represented plots in Figure 3.11. Therefore, it is expected to observe more fluctuations in terms of the variables illustrated in Figure 3.11 compared to Figures 3.4 and 3.5. Significantly, skew effect is always used to reduce the magnitude of undesirable higher order spatial harmonic components caused by the slotting and nonsinusoidal distribution of winding effects while the magnitude of the fundamental component is reduced accordingly as well. So it is obvious that the motor torque production capability degrades by decreasing the fundamental magnetic field which leads to an increase in the motor rise time. These comments are made here to prepare minds for the next chapters in which a comprehensive discussion on the analytical and FE-based modeling of the healthy and faulty induction motors is going to be included. So different aspects including the winding topology, the skewing effect, the material modeling, and the fault modeling will be further discussed. Now, take a look at the following features extracted from Figure 3.11.

- Then again, the current decreases considerably by reducing the load. It gets more and more small up to the no-load current in which the steady-state current signal diverges from a pure sinusoidal curve and turns into a flat top signal. The situation gets worse if the phase current of the motor is investigated. If a 3D model is used, the shape will be certainly more sinusoidal due to the fact that the skewing effect is incorporated. Otherwise, the uncomplicated model should always reveal higher signal ripple. However, a flat top no-load

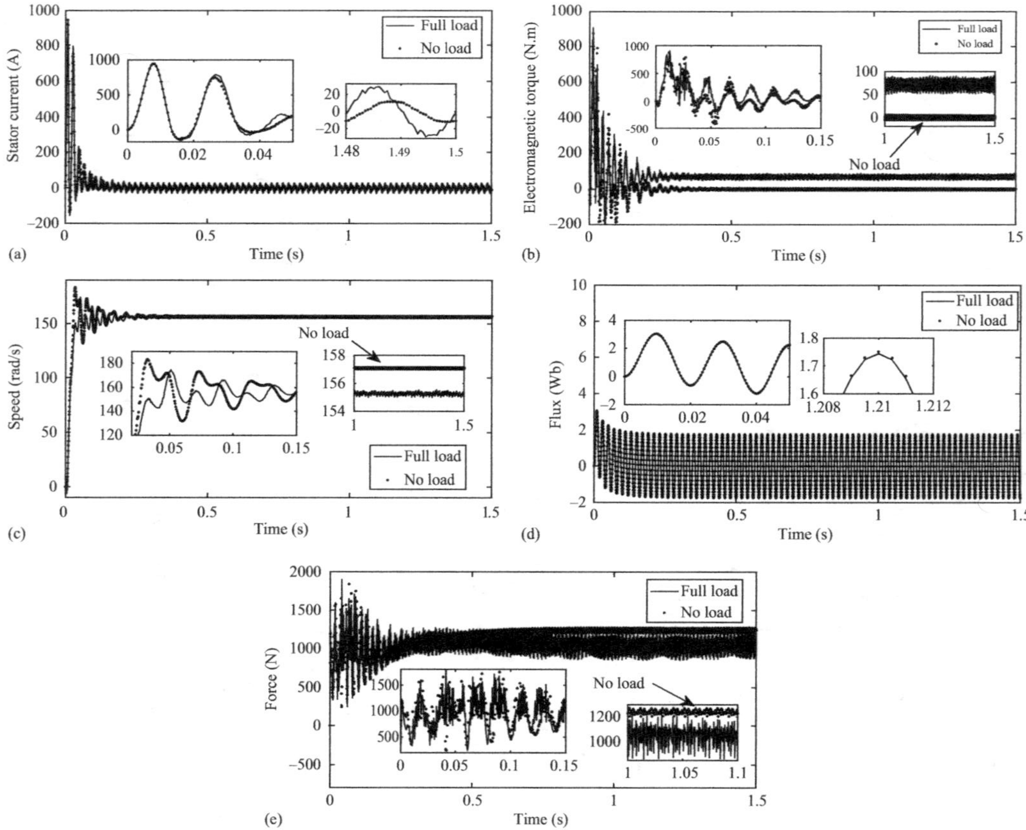


*Figure 3.11 Line-start motor variables (a) current, (b) torque, (c) speed, (d) flux and (e) force magnitude in full- and no-load condition, under a dynamic eccentricity of 30%*

signal is expected even in practice. This is caused due to the presence of the harmonic components, which are powerful enough to change the shape of the signal other than the fundamental one. In view of the fault-related oscillations, the current envelope is not a good medium for reflecting the eccentricity fault as the effect is not clearly distinguishable. It, of course, leads to some sort of fluctuations, not very useful to follow up the fault.

- The motor rise-time is considerably larger than that of Figures 3.4 and 3.5. This evidently proves the claim based on which the fundamental component of the magnetic flux is weakened if the skewing effect is applied. In Figure 3.11, the skewing effect has been completely ignored.
- The time-domain representation of the motor signals, except that of the flux, clearly conducts the distinction of the no- and full-load condition in terms of the eccentricity fault and its effect on the motor. In contrast to the broken bar fault, connecting a larger load to the motor shaft lowers the impact of the eccentricity fault on the time-domain signals. In other words, increasing the load damps the oscillations caused by the eccentricity fault.
- The effect of the sideband components produced by the eccentric rotor and discussed throughout (3.24)–(3.39) is not properly distinguishable in time-domain signals. This might be because of the smaller magnitude of the eccentricity fault compared to that of the broken bar fault. Notably, the side-band components associated with the eccentricity fault are not generated or produced by the fault. They are already there in the motor structure while their magnitude is changed if a fault takes place.
- The transient mode is apparently the best candidate for demonstrating the differences between the no- and full-load conditions. The differences are less observable in the steady-state operation. According to the transient parts of the signals, the amplitude of the fluctuations in the no-load condition is more than that of the full-load condition. This conveys the fact that the motor load acts as a damping factor in terms of the mechanical faults. This is also true in terms of the steady-state analysis. However, the simulated FE model, due to the absence of the skewing effect, does not represent the idea.
- A very interesting plot is the one related to the magnetic force applied to the inner surface of the stator as shown in Figure 3.11(e). The applied force of the no-load motor is obviously more fluctuating than that of the stator in the transient mode of operation. This is the net value of the applied force to the stator surface.

Generally, time-domain signals do not provide a reliable fault detection approach in terms of the dynamic eccentricity fault. They can only be considered as tools to detect a defection or improper operation. Nevertheless, some esoteric fault indicators such as the Gyration Radius (GR) address some unknown time-domain aspects of the eccentricity fault. However, these kinds of indicators are not popular anymore nowadays. A more reliable frequency-based diagnosis is usually preferred.

Having discussed and explained the eccentricity fault's commitment to the unsafe and unreliable performance of the motor, we now prefer to concisely talk

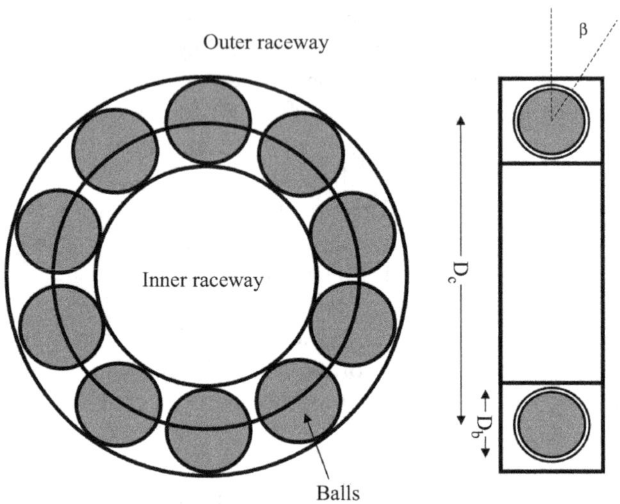


*Figure 3.12 Bearing structure*

about one of the major causes of eccentricity which is called "bearing fault." Bearing fault itself produces additional noise and consequently vibration in the motor body. Depending on the location of fault, inner or outer race, various frequency patterns are provoked and used to detect the fault, but what is interesting is the contribution of this fault to the future eccentricity fault, and this is exactly what has been greatly focused in the literature.

### *3.3.3 Bearing faults in induction motor*

Bearings are indeed one of the most important parts of an induction motor in maintaining the reliability and safety of the machine performance. They should be checked constantly to make sure if the motor operation will not fail. What makes them highly significant is their mechanical operation and being subjected to higher levels of friction leading to wear and tear of their internal and external part. In fact, no rotation exits unless something holds the rotor concentric with respect to the stator and this crucial task is handled by bearings which not only take care of maintenance of a symmetrical operation but also tolerate the rotor weight which might run over hundreds of kilograms. The studies show a considerably high percentage, almost 40%, of contribution of bearing fault to induction motor failures.

Figure 3.12 shows a general bearing structure which consists of the following main parts:

- the outer raceway
- the inner raceway
- the balls
- the cage.

More often, three types of faults are associated with bearings as follows:

- outer raceway defect
- inner raceway defect
- ball defect.

Noticeably, any bearing defect is of a mechanical nature which makes the motor vibrate due to the presence of a radially unbalanced force caused by the defect. This statement reminds the term UMP which was previously discussed in terms of an eccentric motor. So the final result of any bearing defect is somehow converted to an eccentricity type of fault based on which some principle harmonic components are used to detect the fault. Resulting from an eccentric-like fault, it is expected to observe a general frequency pattern of

$$f_{\text{bearing}} = \left| f_s \pm m \cdot f_{i,o} \right| \tag{3.41}$$

where $m = 1, 2, 3, \ldots$ and $f_{i,o}$ is a frequency characterizing and modulating bearing dimension into the motor variables. As a very well-known practice, the corresponding value is obtained as follows:

$$f_{i,o} = \frac{N_b}{2} f_r \left( 1 \pm \frac{D_b}{D_c} \cos(\beta) \right) \tag{3.42}$$

where $N_b$, $f_r$, $D_b$, $D_c$ and $\beta$ are the number of bearing balls, the mechanical rotor speed in Hz, the ball diameter, the bearing pitch diameter and the contact angle of the balls, respectively. The mentioned frequencies are the so-called "characteristic frequencies." Although different bearings produce different characteristic frequencies, depending on the number of ball and their dimensions, as a rule of thumb, the corresponding values can be calculated as follows if the number of balls is between six and twelve.

$$\begin{aligned} f_o &= 0.4 N_b f_r \\ f_i &= 0.6 N_b f_r \end{aligned} \tag{3.43}$$

Although the simplification is brief, it is certainly convincing and no more discussion seems to be necessary at this point. Just as a hint, vibration and noise analysis is one of the best approaches to detect this kind of fault.

## 3.4 Short-circuit fault in induction motors

One of the phenomenon which has been carefully addressed so far is the problem of short-circuit fault in induction motors [47–50]. Due to the dramatically wide application of induction motors in industry and also bring up the safety features among which the short-circuit fault relays or indicators are the most significant ones, we also aim at including this topic in the book. This type of fault is interesting enough due to the fact that it might similarly happen in both rotor and stator if a wound rotor is targeted. Otherwise, it is restricted to stator side if a cage induction motor is analyzed. Although the wound-rotor motors are less requested, they are promising apparatuses to provide a higher starting torque respecting to a minimum starting current. However, in addition to the stator, the short-circuit fault can also take place in the rotor. The literature shows that the effect of the short-circuit fault in the stator is very higher than that of the rotor.

According to the investigations, almost 37% of the induction motor faults are related to its insulation failure leading to a short-circuit fault [51–54]. As a result of this kind of failure, the current density of coils/windings runs over the rated values dangerously causing a hot spot area around and through the faulty coil/winding. This is one of the major sources of degradation and also aging of motor coils/windings.

In general, there are specific types of this fault including:

- turn-to-turn
- turn-to-ground
- phase-to-phase
- phase-to-ground.

Regardless of the type, the nature of the abovementioned faults is the same. Actually, a considerably high current level circulates inside the windings without contributing to the torque production capability of the motor. So, it is considered as a kind of loss. The only difference is the severity of the fault which increases with the increase in the number of short-circuited turns as well as the decrease in the short-circuiting path resistance. Ignoring the fault and letting it progress might also cause an irreversible damage to the motor tank and bearings. Therefore, diagnosis of the short-circuit fault is of great importance.

The following are some of the main reasons of generating the fault [55–60].

- **Thermal stresses:** These are produced by thermal aging or over-loading. Respecting to a 10°C of increase in temperature, insulation lifetime reduced by a factor of two. A better class insulation is recommended if the motor is going to be used in a highly stressful environment.
- **Electrical stresses:** This type of stress is usually classified into two general categories, namely the insulation breakdown and the partial discharge. When a relatively large voltage variation rate, i.e., $dv/dt$, is applied, insulations are subjected to a breakdown and destruction. This is usually the case where the voltage goes beyond 5 kV.
- **Mechanical stresses:** This kind of stress is predominantly caused by several start–stop operations of the motor causing frequent warm-up and cool-down of insulations. As a result, cracks are produced and become larger if start–stop operations are not avoided.
- **Environmental stresses:** External and polluting substances can also disturb the motor operation. For example, pollution might cause an improper thermal exchange between motor and surrounding environment and consequently it leads to motor temperature rise. This in turn increases the risk of electrical failure of coils/windings.

Many efforts have been already undertaken just in case of analyzing different aspects of short-circuit fault. We are going to focus more on the original fault-related harmonic components which are the results of the short-circuit fault. like all the other types of faults, the natural frequencies which are used for detecting the short-circuit fault are present in the motor structure and are the functions of the number of poles, the number of slots as well as the saturation profile of the machine. The presence of the short-circuit fault only changes the amplitude of the harmonic components and it has nothing to do with creating new components. This was also the case in previous types of faults including the broken bar and eccentricity faults. So, it is time to address a part of principle harmonics generated by geometrical placement of the rotor and stator slots and windings. These harmonic components are called principle slot harmonics (PSH) which are sometimes used to detect a specific type of fault. Both stator and rotor introduce their own PSHs.

It is very well known that induction machines, of course in healthy case, consists of a sort of MMF components formulated as (3.44):

$$F_s(t,\theta) = \sum_{\mu} F_{\mu}\cos(\omega_s t - \mu p\theta) \tag{3.44}$$

where $p$ is the number of pole pairs, $\omega_s$ is the fundamental angular frequency and $\mu = 6g + 1$, $g = 0, \pm 1, \pm 2, \ldots$. Assuming that the number of rotor loops is n, the MMF component corresponding to the first rotor loop is obtained in the rotor reference frame as follows:

$$F_{\text{loop1}}(t,\theta_r) = \sum_{\upsilon=1}^{\infty} [k_{\upsilon}\cos(\upsilon\theta_r + s\omega_s t) + k_{\vartheta}\cos(\upsilon\theta_r - s\omega_s t)]$$

where

$$k_{\upsilon} = \frac{2}{\upsilon\pi}\left(1 - \frac{1}{n}\right)\sin\left(\upsilon\frac{\pi}{n}\right)I_{r\max} \tag{3.45}$$

where $I_{r\max}$ is the magnitude of the rotor bar current. Following the same fashion, the MMF component of the neighboring rotor loop is analytically derived as follows:

$$\begin{aligned} F_{\text{loop2}}(t,\theta_r) = \sum_{\upsilon=1}^{\infty} \bigg[ & k_{\upsilon}\cos\left(\upsilon\theta_r + s\omega_s t - (\upsilon + p)\frac{2\pi}{n}\right) \\ & + k_{\upsilon}\cos\left(\upsilon\theta_r - s\omega_s t - (\upsilon - p)\frac{2\pi}{n}\right)\bigg] \end{aligned} \tag{3.46}$$

It is worth noting that the upper index of the summation is set at infinity due to the fact that the rotor loop MMFs looks like a spike which can only be modeled accurately by incorporating infinite terms of its Fourier transform. The total MMF produced by rotor is calculated by adding up the loop MMF.

$$F_r(t,\theta_r) = \sum_{i=0}^{n-1}\sum_{\upsilon=1}^{\infty}\left[k_\upsilon \cos\left(\upsilon\theta_r + s\omega_s t - (\upsilon+p)\frac{2\pi}{n}i\right) + k_\upsilon \cos\left(\upsilon\theta_r - s\omega_s t - (\upsilon-p)\frac{2\pi}{n}i\right)\right] \tag{3.47}$$

On the basis of the derived closed-form representation of the rotor MMF waves, it is clearly observed that MMF waves are only present for the cases $\upsilon = p, \upsilon + p = \pm\lambda n$ and $\upsilon - p = \pm\lambda n, \lambda = 1, 2, 3, \ldots$. As $\upsilon$ can only take positive integers, it follows that only for $\upsilon = p$ and $\vartheta = \lambda n \pm p$ the MMF waves exist. Therefore, in addition to the fundamental rotor harmonic component for $\upsilon = p$ which deals with the armature reaction of the fundamental component of the stator, the rest obtained by $\upsilon = \lambda n \pm p$ demonstrate the rotor slot harmonics (RSH). Regarding the stator reference frame, the RSH is expressed as follows:

$$F_r(t,\theta) = F_{r1}\cos\left(\left(1 - \lambda\frac{n}{p}(1-s)\right)\omega_s t + (\lambda n - p)\theta\right) + F_{r2}\cos\left(\left(1 + \lambda\frac{n}{p}(1-s)\right)\omega_s t - (\lambda n + p)\theta\right) \tag{3.48}$$

Coupled with previous formulation, the higher frequency MMF waves produced by higher order components are given by:

$$F_{r\mu}(t,\theta) = F_{r\mu 1}\cos\left(\left(1 - \lambda\frac{n}{p}(1-s)\right)\omega_s t + (\lambda n - \mu p)\theta\right) + F_{r\mu 2}\cos\left(\left(1 + \lambda\frac{n}{p}(1-s)\right)\omega_s t - (\lambda n + \mu p)\theta\right) \tag{3.49}$$

The corresponding frequency is the same, but the number of effective pole pairs differ. Multiplying the MMF components by the air-gap permeance, which is a content term in a healthy idea motor, returns the magnetic flux density wave which produces the EMFs and subsequently the currents in the stator. From (3.48) and (3.49), it is comprehensible that the stator EMFs and currents will only contain the additional slot frequencies $(1 \pm \lambda\frac{n}{p}(1-s))f_s$ which are now the time harmonic components extractable using a Fourier transform. As seen, the spatial components introduced by the slotting effect are eventually reflected into the time-domain signals.

Under short-circuited turns or phases, a new set of waves will apply changes to the stator MMFs described as:

$$F_{add}(t,\theta) = \sum_{\substack{k=-\infty \\ k\neq 0}}^{\infty} F_{addk}\cos(\omega_s t - k\theta) \tag{3.50}$$

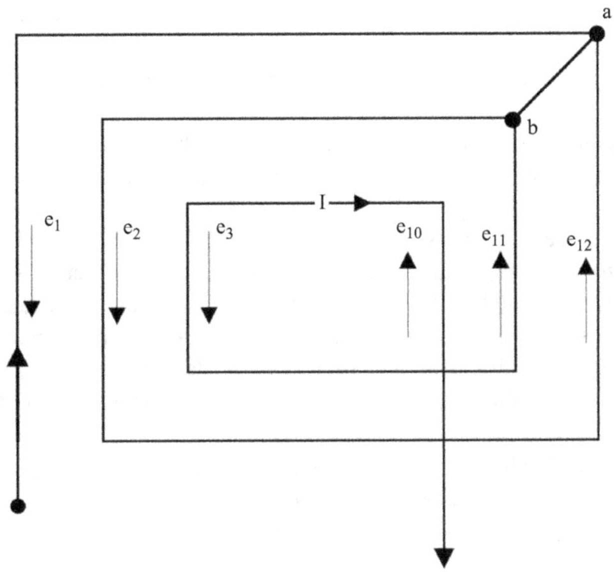


*Figure 3.13 Short circuit between turns a and b*

Accordingly, there exist the MMF waves at all numbers of pole pairs and direction of rotations. One of the waves rotates with the same frequency of the fundamental component but in an opposite direction. If a normalized Fourier transform is used, no change will be observed in the current spectrum as it is always normalized to zero. However, there should be a change in the magnitude of the RSH. No new frequency should be added to the current spectrum if the short-circuit fault takes place in the stator. The fault only contributes to an increase in the RSHs.

In the first instance, consideration is given to the simplest case where the coil has only one turn from which it is possible to draw some important conclusions. Figure 3.13, however, shows one-phase group of three coils. It is assumed that interturn short-circuit arises between points a and b, as illustrated. It is clear that the circulating current has a closed path. From simple theory, it is clear that the path A–X can be expanded to two independent circuits. From Figure 3.13, we can say that the phase current and the current which flows through the short-circuited coil, produce opposite MMF. Therefore, interturn short-circuits have a cumulative effect in decreasing the MMF in the vicinity of the short-circuited turn(s). First, when a short-circuit occurs, the phase winding has less turns and less MMF. Second, the MMF of short-circuited part is opposite to the MMF of the phase winding. Clearly, interturn shorts with more turns can be analyzed in a similar manner.

In most commercially available induction motors, coils are insulated from one another in slots as well as in the end winding region. Therefore, the highest probability for the occurrence of interturn fault is between turns in the same coil.

Here, it is assumed that the interturn short-circuit is between two turns in the same coil, and that one-half of the coil is short-circuited; this means that approximately 8% of turns of one phase are short-circuited. Simulation was carried out for this condition.

As a consequence of the interturn short-circuit, the MMF of the phase winding in which interturn short circuit exists changes, as does the mutual inductance between that phase and all other circuits in the machine. In addition, a new "phase," which we call the short-circuited phase D is introduced. It should be assumed that for modeling, this phase has no conductive contact with other phases, but it is mutually coupled with all other circuits on both the stator and rotor sides. The currents in stator circuits and rotor loops are assumed independent.

The machine with the following specifications is analyzed using the numerical model based on multiple coupled circuit approach and winding function analysis.

$$\begin{aligned} &3\text{kW}, 415\text{V}, \Delta, 50\text{Hz}, \text{p} = 3(\text{six poles machine}) \\ &\qquad S = 36 \text{ stator slots} \\ &\qquad R = 32 \text{ rotor bars} \end{aligned}$$

Stator phase winding consists of 6 coils, 1 coil per pole with 77 turns in one coil, i.e., $N = 462$ series turns per pole per phase. Stator phase A winding scheme is:

$$A - 1 - 6' - 12 - 7' - 13 - 18' - 24 - 19' - 25 - 30' - 36 - 31'$$

The connection diagram of the stator windings of the experimental machine is additionally shown in Figure 3.14.

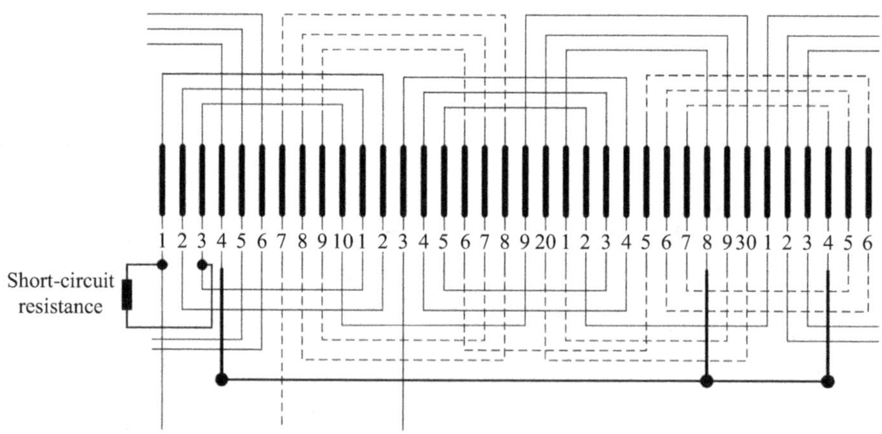


*Figure 3.14 Stator winding scheme of experimental motor. Interturn short-circuit occurs in coil placed in slots 1 and 6*

The machine parameter:

| | |
|---|---|
| Rating power | $P_r = 3$ kW |
| Rated voltage | $U_r = 415$ V |
| Rated frequency | $f = 50$ Hz |
| Number of pole pairs | $p = 3$ |
| Number of phases | $q = 3$ |
| Rated speed | $n_r = 950$ rev/min |
| Full-load current | $I_r = 6.4$ A |
| Stator winding connection | Delta |
| Turns per coil | $N = 77$ |
| Turns per slot | $N_{slot} = 77$ |
| Number of stator slots | $S = 36$ |
| Number of rotor bars | $R = 32$ |
| Rotor inertia | $J = 0.0258$ kg m$^2$ |
| Average radius of air gap | $r = 0.0681$ m |
| Length of stator stack | $l = 0.115$ m |
| Air-gap length | $g_0 = 0.58$ mm |
| Phase resistance | $R_{ph} = 5.7\ \Omega$ |
| Rotor bar resistance | $R_b = 72.8\ \mu\Omega$ |
| Rotor end-ring resistance | $R_e = 438$ n$\Omega$ |
| Rotor bar leakage inductance | $L_b = 30$ nH |
| Rotor end-ring leakage inductance | $L_e = 6$ nH |
| Resistance of short-circuit "phase" | $R_d = (38/462) \cdot R_{ph} = 0.469\Omega$ |

Motor is loaded with 30 N m under the steady-state condition. Fault is made on such manner that 38 out of 77 turns is short-circuited in one stator phase coil, under one pole. We provide an interesting illustration of the healthy and faulty motors for different load (see Figure 3.15)

By referring to Figure 3.15, it is observed that

- Increasing the load level leads to an increase in the motor current regardless of the short-circuit fault.
- The current reveals an increasing trend in both the transient and steady-state region upon the fault occurrence.
- The average developed electromagnetic torque is almost the same for the faulty and healthy conditions. However, the faulty motor contains some sort of low frequency harmonic components carried by the average value. This clear illustration of harmonic components in time domain is only achievable by the analytical models such as the winding function theory or the magnetic equivalent circuit. If an FE model shown in Figure 3.11 is used for simulations, the torque ripple caused by slotting and saturation usually dominates the fault component unless the fault severity becomes very high.
- Similar oscillations are observed in the speed. Surprisingly, the average steady-state speed decreases by the fault. This increases the power losses caused by the slip.

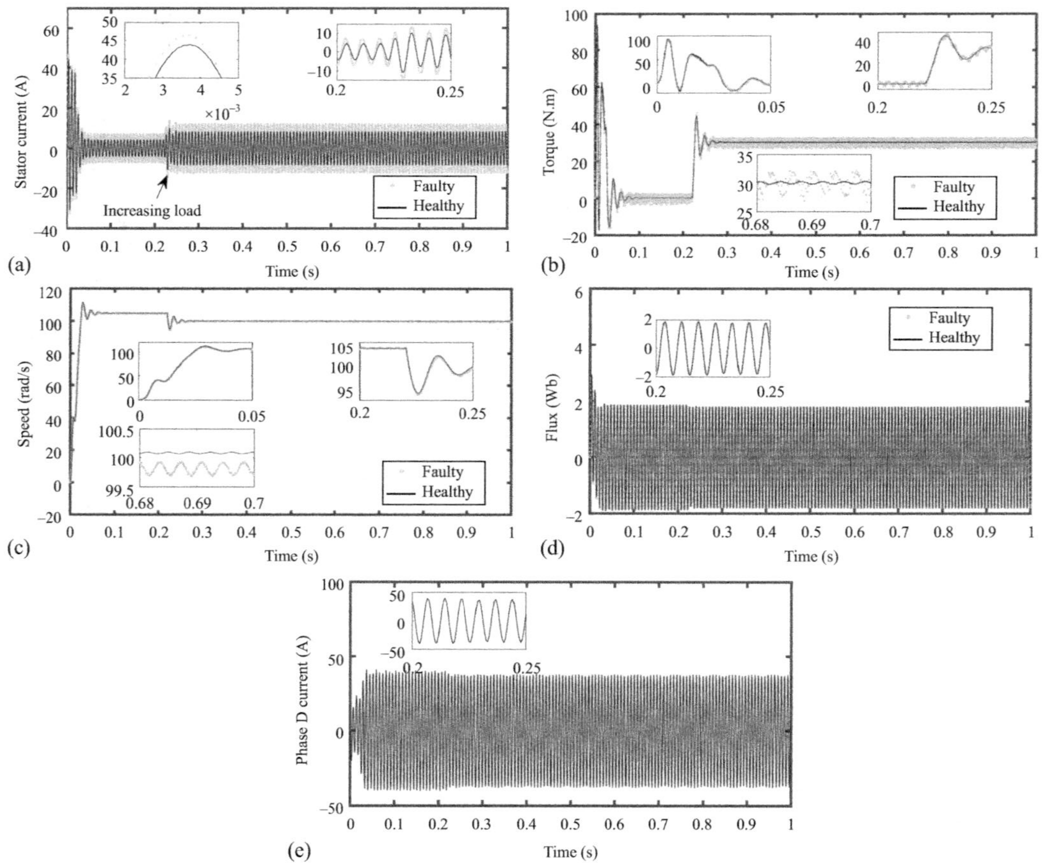


*Figure 3.15* *Line-start motor variables (a) current, (b) torque, (c) speed, (d) flux and (e) phase D current in full- and no-load condition, healthy and faulty motor signals*

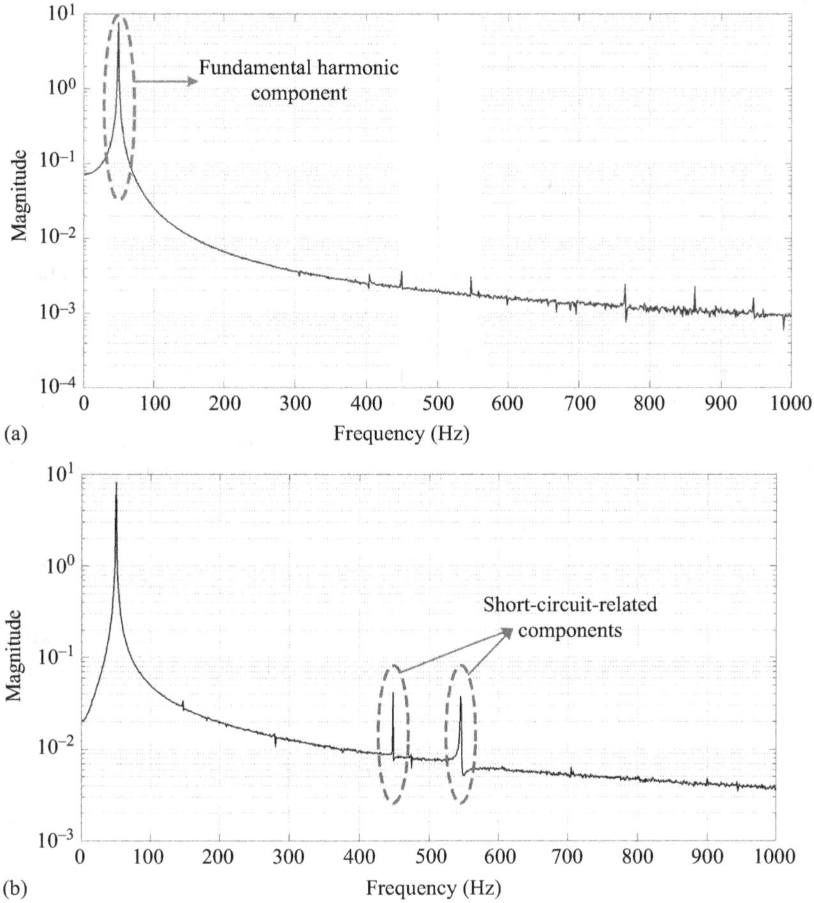


*Figure 3.16 The line current spectrum (a) healthy and (b) faulty motor*

- The phase D (short-circuited phase) current is considerably more than that of the rest of the windings. The current circulates and makes the thermal stress so worse that the other healthy insulations also fail to operate well.
- Not all the motor signals get larger by the fault. The phase D current marks down probably due to the increase in the armature reaction in the rotor side.

So far, a lot of contexts have been mentioned in terms of the motor harmonic components, the PSH, the fault-related harmonic components etc., without directly addressing how these frequency-domain signals might be observed or studied. A couple of spectra regarding the motor current signal are put forward without going in depth of how they are calculated to explain first what a spectrum is and second how different frequency analyses are performed by means of a frequency spectrum. To this end, refer to Figure 3.16.

The output of a Fourier transform is usually demonstrated by frequency and magnitude planes. Normally, the horizontal axis is an indication of frequency contents of the processed signal, for example the current and the vertical axis shows the corresponding power or the magnitude. Therefore,

- The synchronous frequency, i.e., 50 Hz for the tested motor, possesses the largest magnitude in the spectrum. That is why it is called the fundamental component (see Figure 3.16(a)).
- The rest of the spectrum drops down the mentioned fundamental component in terms of the magnitude. This distinctly explains the lower interest of higher order components in forming the shape of the current which is the processed signal.
- The entire spectrum is an almost featureless curve representing negligible information in terms of the higher order components. This arises from the fact that the number of rotor bars and pole pairs are equal to 32 and 3, respectively. So none of the principal RSH could be expected. In order that lower rotor slot harmonic exists in stator current spectrum, the following condition should be fulfilled:

$$R_{L\_RSH} = (6n + 2)p, \quad n = 0, 1, 2, \ldots \tag{3.51}$$

- For $p = 3$ and $n = 1, 2, \ldots$, the lower RSH ($R_{L\_RSH}$) is equal to 24, 42, .... On the other hand, the condition for the existence of the upper RSH is satisfied for rotors with the following number of bars.

$$R_{U\_RSH} = (6n - 2)p, \quad n = 0, 1, 2, \ldots \tag{3.52}$$

- This leads to $R_{U\_PSH} = 12, 30, 48, \ldots$ for a six-pole machine.

However, in a faulty motor, as a consequence of interturn short circuit, there are all harmonics of "phase" D MMF. In other words, in the following magnetic flux density waves,

$$\begin{aligned} B_2 &= B_{2m}\cos\left(\left(1 - \frac{\lambda R}{p}(1-s)\right)\omega_s t + \left(\frac{\lambda R}{p} - v\right)p\theta_s\right) \\ B_3 &= B_{3m}\cos\left(\left(1 + \frac{\lambda R}{p}(1-s)\right)\omega_s t + \left(\frac{\lambda R}{p} + v\right)p\theta_s\right) \end{aligned} \tag{3.53}$$

The order of space harmonic $v$ could be $v = \pm 1/3$, $\pm 2/3$, $\pm 1$, $\pm 4/3$, ... So in general, the RSH could arise in current spectrum (Figure 3.17).

For $v = -1/3$ and $\lambda = 1$ lower rotor slot harmonic exists:

$$\frac{R}{p} - v = \frac{32}{3} + \frac{1}{3} = \frac{33}{3} = 11 \tag{3.54}$$

while for $v = 1/3$ and $\lambda = 1$ it is true for upper slot harmonic:

$$\frac{R}{p} + v = \frac{32}{3} + \frac{1}{3} = \frac{33}{3} = 11 \tag{3.55}$$

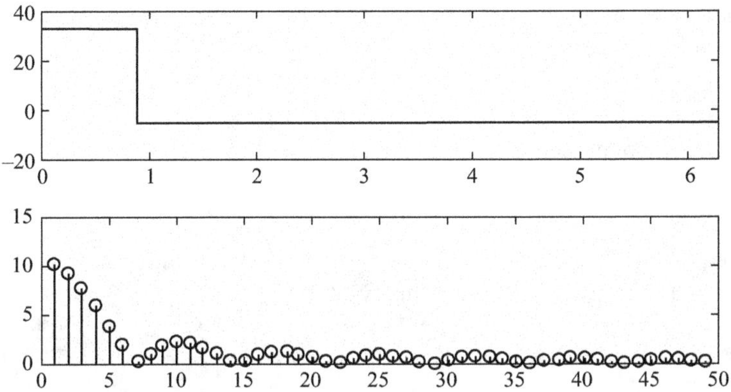


*Figure 3.17 MMF of "phase" D – as analyzed motor is motor with p = 3, fundamental harmonic ν = 1 means six poles. In case of short circuit in one coil, it means that phase D also produces subharmonics*

In addition, for $v = 11/3$

$$\frac{R}{p} - v = \frac{32}{3} - \frac{11}{3} = \frac{21}{3} = 7 \tag{3.56}$$

while for $v = -11/3$ and $\lambda = 1$, it is true for upper slot harmonic:

$$\frac{R}{p} + v = \frac{32}{3} - \frac{11}{3} = \frac{21}{3} = 7 \tag{3.57}$$

Therefore, in healthy machine with $S = 36$, $R = 32$ and $p = 3$, none of RSHs exists, as it could be seen in Figure 3.16(a). However, in case of interturn short-circuit, both of RSHs exist in the line current spectrum, i.e., Figure 3.16(b), and they are very prominent. This effect could be additionally amplified by permeance harmonics waves:

$$\begin{aligned} B_{sv3}(t,\theta_s) &= B_{sv3max}\cos\left(\left(1-\frac{R}{p}(1-s)\right)\omega_s t + \left(\frac{S-R}{p}+v\right)p\theta_s\right) \\ B_{sv4}(t,\theta_s) &= B_{sv4max}\cos\left(\left(1+\frac{R}{p}(1-s)\right)\omega_s t + \left(\frac{S-R}{p}-v\right)p\theta_s\right) \end{aligned} \tag{3.58}$$

The lower slot harmonic will appear for $v = 11/3$,

$$\frac{S-R}{p} + v = \frac{36-32}{3} + \frac{11}{3} = 5 \tag{3.59}$$

while the upper exists for $v = 1/3$:

$$\frac{S-R}{p} - v = \frac{36-32}{3} - \frac{1}{3} = 1 \tag{3.60}$$

In addition, new harmonic component in current spectrum appears due to the saturation phenomenon.

$$B_{sat} = B_{satm} \cos(3\omega_s t - (\upsilon + 2)p\theta_s) \tag{3.61}$$

now could attain 1 for $\upsilon = -1$,

$$\upsilon + 2 = -1 + 2 = 1 \tag{3.62}$$

which means that new harmonic component at 150 Hz could appear as a result of the fault due to the following flux density wave:

$$B_{sat} = B_{satm} \cos(3\omega_s t - p\theta_s) \tag{3.63}$$

However, this component does not exist in models in which the saturation is neglected. To validate the results from the dynamic model, an experimental investigation was conducted. The experiment was performed in the following manner. A standard commercially available motor was dismantled and isolation of the few turns from the same coil (in the end region) was mechanically injured, i.e., scratched. These spots were soldered conductors which were taken out from the motor. Short circuit was made between these conductors. Therefore, turns were shorted externally. By measuring EMF between conductors and having known winding details, we were able to conclude how many turns in one coil were shorted. In the case of inrush current which was experienced particularly when two turns which are in the neighborhood (in the electrical sense) were shorted, the short-circuit current was reduced by means of an externally placed resistor. In these cases, current was limited to the value of double-rated current for a short time during the experiment.

Figure 3.18 shows the spectra of line current for a loaded machine for a healthy and a faulty condition, respectively. Figure 3.18(a) shows that in a healthy machine, frequency components, the result of the saturation of magnetic material (150, 250, 350 Hz, etc.), exist. In the healthy condition, only the upper rotor slot harmonic is visible at 568 Hz ($s = 2.8\%$). From Figure 3.18(b), it is clear that as it is predicted in the simulation model, the most significant changes arise at harmonic components of $(1 \pm \lambda R(1 - s)/p)f_s$. Now, the lower rotor slot harmonic (at 469 Hz) is prominent and the upper harmonic has risen. Moreover, the 150-Hz harmonic component is considerably higher under the fault condition.

What has been already proposed in terms of PSHs existing in induction motors is actually the basis for future analysis not only in the case of short-circuit fault but also in the case of other types of faults. Furthermore, the appreciable practice of the frequency-domain analysis of the motor current was extracted and analyzed. What is not really deductible about faults in a time-domain analysis can be somehow easily detected, followed and analyzed by means of a frequency domain. The PSHs related to the short-circuit fault are the best examples of this kind and in the same fashion, the broken bar and eccentricity fault will be addressed shortly in the next chapters. Considering the mentioned points, three apparently distinct but inherently correlated processing domains including time, frequency and

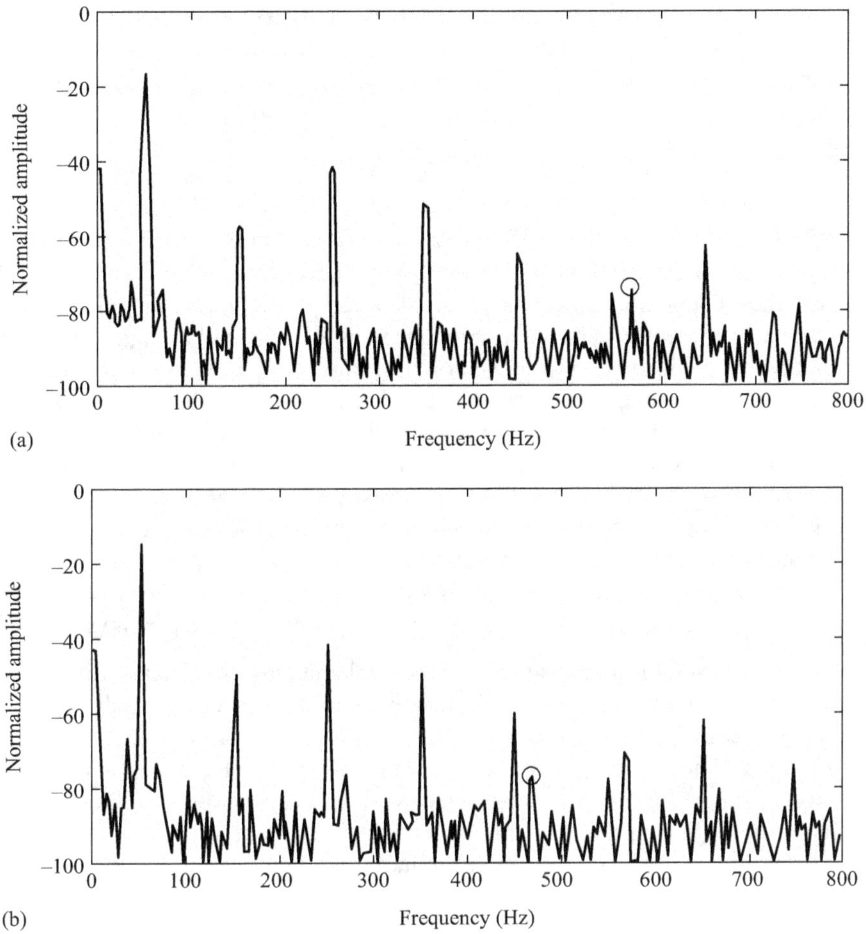


*Figure 3.18 Current spectrum of a line start motor at slip = 2.8% (a) healthy and (b) short-circuit fault*

time-frequency ones are going to be discussed in Chapters 8–10. In the current chapter, the time-domain variations are emphasized more while the frequency analysis will be further studied and combined with the time-domain information to introduce a generalized solution to fault diagnosis challenges.

Measuring the magnetic, electrical, mechanical and partly thermal quantities of the motor, we will also go directly through the property of time and frequency components of every single motor variable to see how different faults apply changes to motor variables, how the fault information could probably be extracted and how the extracted information is used for a precise detection, determination and diagnosis procedure.

## 3.5 Laboratory-sale implementation of induction motor faults

Any fault diagnosis procedure will be nonsense if experimental investigations and validations are not included in the procedure. In fact, after all the theoretical and simulation-based analysis, what really matters in industry is measurements reflecting the real motor behavior in faulty conditions. Moreover, although developing the basic ideas mathematically or even by means of simulations, whether accurate FE or analytical models, helps researchers and industries investigate practically the impossible number of situations, models still suffer from some sort of inaccuracy depending on the number of assumptions made to build models. Therefore, experimental study of what happens in reality should be an inseparable part of any diagnosis procedure of electrical machines. More importantly, some of the real world influential factors such as the thermal stresses, the nonhomogenous magnetic materials and also every single multiphysics-related problem cannot be analyzed accurately unless a real motor and drive system is incorporated. This discussion necessitates presence of a straightforward implementation of the motor–drive systems along with the corresponding faulty conditions including all three major types of fault discussed in this chapter.

There are some prerequisites which should be considered prior to any setup preparation for the experimental test rig. The prerequisites are as follows:

- The motor should not be so small that implementing the fault is difficult or practically impossible. This really makes sense as motors in which the faults take place are of a large power, for instance hundreds of kilowatts.
- The motor should not be very large because of two main reasons. First, as the focus is on preparing a logical laboratory-scale setup, dealing with a relatively large motor during the fault implementation will be a very tricky and overwhelming task. Second, the larger the motor is, the higher the investment will be. Most of the time, academic research budgets are limited, and having a solid investment plan is essential.
- Peak a motor for which an industrial drive could be easily found. Do not go through specifically dedicated motors built for special applications such as ultrahigh speed ones. Catching a drive for this kind of applications is not easy; moreover, research output will not probably be general enough to be used in other applications.
- Sometimes, more than one rotor is required if the goal is to study the effect of various broken bar positions. Therefore, include the corresponding prices in the investment plan.
- Keep in mind to apply reversible faults such as eccentricity prior to applying an irreversible defect like bar breakage if more than one type of fault should be investigated.
- Take the safe side and start from the low fault levels. Even if a reversible fault such as eccentricity of the rotor is studied, the reassembled healthy motor might not be as the same as it was initially right after the factory production. If a higher fault level is applied, the motor might be subjected to unrepairable damage and not be able to be used later.

- Again, choosing a medium or not very small power motor is usually preferred as it is also robust enough against the environment noise which might affect the faulty motor signals and consequently the diagnosis procedure.
- Try to prepare a general application drive such ABB ACS800 which can be applied to all types of motors regardless of their specifications. The only restricting factor in choosing general application drives is their power limits which should be matched with the motor power.
- Try to prepare a drive consisting of both the open- and closed-loop control strategies. Unfortunately, there is only a DTC or FOC strategy implemented in one single drive. If one looks for a wider range of useful strategies, s/he should go through providing more than one drive. It is not recommended to build the drive on your own. If you wish so, make sure to implement every influential factor which might be included in an industrial drive. Otherwise, the results might not be convincing.
- In the case of a line-start application, make sure to have balanced three-phase network. Otherwise, the diagnosis results achieved from one phase might be different from the one obtained from another phase. To deal with an unbalanced network, using an autotransformer is a good solution. In the case of inverter-fed applications, it is not a requirement as the drive itself outputs balanced three-phase signals unless a switch fault or problem occurs.
- Prepare safe data acquisition instruments.
- Prepare a processor such as a computer or digital signal processors (DSP) to process the measured data to extract the fault indicators. A very useful and comprehensive discussion on this topic will be shortly provided in this chapter.

### *3.5.1 Three-phase induction motor*

The key element of any test rig built for the diagnosis purposes targeted in this book is an induction motor. The discussion is continued by considering a three-phase induction motor with the following specifications (see Table 3.2):

The real motor is shown in Figure 3.19. This is exactly the same motor as what was previously simulated using the FE approach. So it consists of 28 rotor bars and 36 stator slots. The stator and rotor are both made of silicon steel. The rotor bars are made of aluminum. The rotor consists of two cages facilitating the motor start-up

*Table 3.2 Induction motor specifications*

| | |
|---|---|
| Rated power | 11 kW |
| Rated frequency | 50 Hz |
| Number of poles | 4 |
| Rated line voltage (rms) (Δ-connected) | 380 V |
| Rated line current | 24 A |
| Rated power factor | 0.83 |
| Rated torque | 70 N · m |
| Rated slip | 3% |
| Number of turns | 27 |

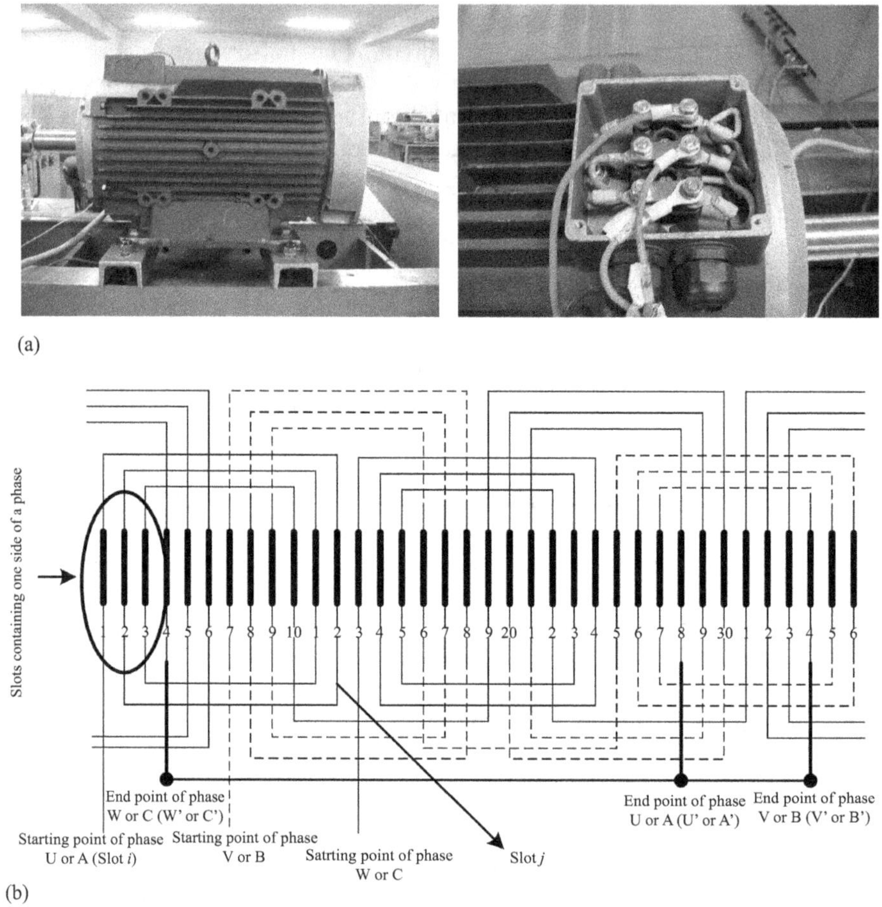


*Figure 3.19 (a) Tested induction motor and (b) corresponding winding layout*

by increasing the start-up resistance of the rotor. The number of slots per pole per phase is equal to 3, so the winding layout should be the same as Figure 3.19(b) as it consists of a concentric single layer winding topology. In high-power applications, double layer winding topology is more popular. By "layer," we mean the number of separate layers of phases placed inside one single stator slot. The available network to supply the motor contains three phases with an effective value of 380 V. Therefore, the peak value of the supply voltage is equal to 538 V. The network should be connected to the motor terminals with a delta connection (see Figure 3.19(a)). For a star connection, the motor operation will be degraded due to the lack of supply voltage. Coupled with the mentioned points, presence of a rigid bench to which the motor is hanged firmly is another crucial requirement which reduces the unwanted vibrations caused by improper or loose motor placement. The more the degree of freedom of the bench is, the more flexible the motor assembling and disassembling would be.

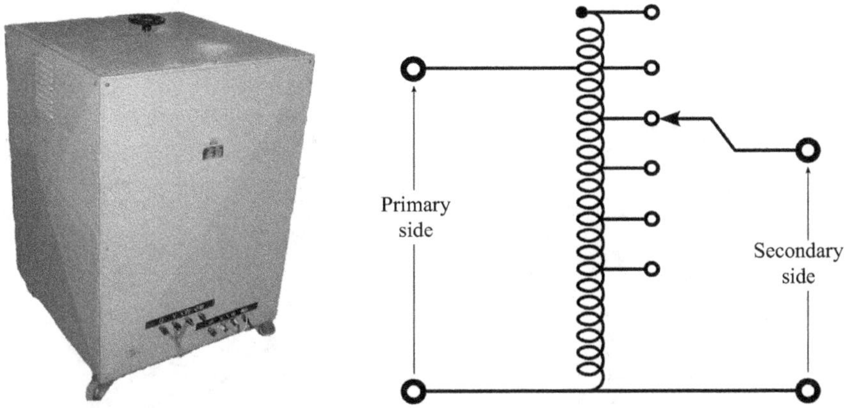


*Figure 3.20 Autotransformer*

### *3.5.2 Autotransformer*

Figure 3.20 shows a typical autotransformer used in this study. There is one input and one output set containing four terminals three of which is related to phases, and the remaining one is the ground terminal if required. The adjusting wheel is used to control the voltage level. An autotransformer is an electrical device with only one single coil per phase and one or more terminals at the secondary to provide various voltage levels for the user. The utilized autotransformer consists of the adjusting wheel instead of the output terminals to make it possible to obtain smoothly variable voltage range. The network voltage is applied to the primary winding(s) and the secondary winding(s) returns the required voltage(s). The scale of the voltage change applied by the transformer is not considerably large in fault diagnosis procedures. The main responsibility of this device is to stabilize the voltage amplitude partly different from the rated one. This apparatus has nothing to do with the supply frequency and is simply used to control the voltage level.

**SAFETY NOTE**: Before turning on the setup, ascertain that the adjusting wheel is located in a fair position in terms of the ratio of turns. Autotransformers are normally capable of increasing the voltage up to three times the rated one and this might cause life-threatening risk if it is not taken into account.

### *3.5.3 Drive*

Another key point to be considered is how the motor is managed to be used in inverter-fed applications. To do so, a very famous drive, i.e., ABB ACS800 is used in this study (see Figure 3.21).

ACS800 is an industrial drive including the following features:

- controlling the time-domain characteristics of the motor speed and torque
- regenerative Braking mode
- DC magnetization to obtain the maximum torque capabilities of the motor
- flux weakening mode
- parallel motor control mode.

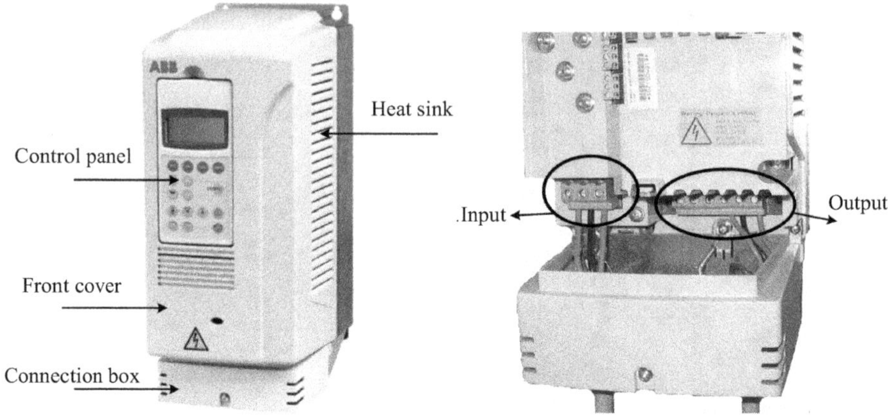


*Figure 3.21 ABB ACS800*

One of the most incredible advantages of this inverter is its ability to control motor speed in sensor-less method. In fact, there is no resolver or encoder connected to the motor shaft to measure its mechanical speed, although it is also applicable using this inverter. Mechanical speed is estimated by measuring electrical variables with a very high precision. The operating mode can also be selected among SCALAR and DTC. The main circuit of the drive is shown in Figure 3.22. The significant parts are listed as follows:

- the control panel
- the start-up prevention switch (X41)
- the I/O board
- the input, output, DC bus and external braking resistor connections
- the six-step inverter.

The starting point of any motor–drive operation is to define the motor parameters including the rated values as well as its electrical quantities, namely the resistance and inductance of the drive, using the control panel. This is a must-do step prior to any other progress. The entered values facile the torque and speed estimation performed internally by the drive if a sensor-less control strategy is used. Otherwise, the speed and torque signals can be passed to the drive through the I/O board using sensors which are going to be discussed shortly. Moreover, the estimated torque and speed signals are likewise available in the I/O board. So the I/O board is really helpful in case if an advanced data acquisition hardware is not accessible. Different drives provide different numbers and types of I/Os. Therefore, it is proposed to refer to the user manual to be informed of details. Furthermore, input and output power cables should be connected to the embedded ports shown in Figure 3.21. There are also three other output ports allowing users to access the DC bus voltage and also connect the external braking resistor to the drive. The latter is used to

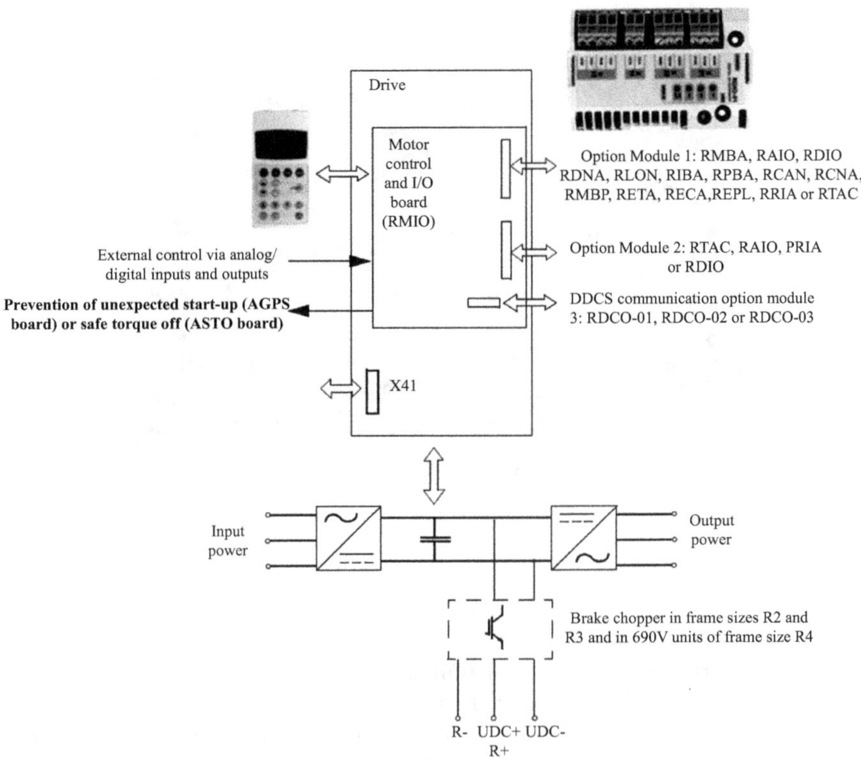


*Figure 3.22 Main circuit of ACS800 drive*[1]

consume large regenerated braking power with the goal avoiding higher thermal tensions applied to the drive.

It is very important to know that there are some safety rules to follow. For example, the drive should always be installed inside a grounded metal frame (see Figure 3.23) to prevent the external noise which might affect the control-level measurements. In addition, the life-threatening risk is reduced due to the fact that the frame voltage is deliberately brought to zero by connecting a ground wire to the body. A very important practical aspect is that if you use a computer to analyze the sampled motor signals, computer supply plug must be different from that of drive. Otherwise, you will observe a very undesirable signal to noise ratio in your sampled signals. Another alternative is to use three-phase chock.

Generally, the installation steps are as follows:

- Identify the frame size based on the user's manual.
- Select the required cables including the power and control cables depending on the motor–drive ratings the environmental conditions.

[1]https://library.e.abb.com/public/0cb03a89ae71bba5c1257b97004fdb01/EN_%20ACS800_01_HW_K_A4.pdf (access date: October 2016).

*Figure 3.23 Drive-shielded frame*

- Check for the availability of the entire necessary module in the drive box.
- Prepare a chock to prevent intercircuital interference between the drive and the data acquisition power supplies.
- Install the drive inside the metal frame.
- Connect the grounding cable of the frame.
- Connect and shield power cables. It is essential to use shielded power cables specifically where a long cable should be placed between the motor and the drive. This is the major requirement to surpass the radio frequency emissions radiated from the drive signals.
- Follow the cabling instruction to avoid improper installing angle of the cables at joints.
- Connect an external braking resistor if you might have a regenerative braking operation. This mostly happens when another drive which acts as the motor load is connected in parallel to the main drive. In this case, two drives share the DC bus voltage.

Finally, the motor is connected to the drive as shown in Figure 3.24. A set of fuses should be necessarily utilized as an electric safe guard in case if any undesirable over-currents take place (see Figure 3.24). The positive terminal of the braking resistor is regularly connected to the positive terminal of the DC bus, and a built-in terminal is embedded into the drive for the negative terminal. Both the motor and the drive bodies must be grounded together.

Usually two types of reference values are implemented in every drive, speed and torque reference values. Depending on the application in which the motor–drive system is used, different reference values might be chosen. Then, the rest of the operation is handled by the drive for controlling the motor speed or torque. It should be noted that precise determination of the motor quantities such as the stator inductance and the resistance are the essential task for presetting the drive.

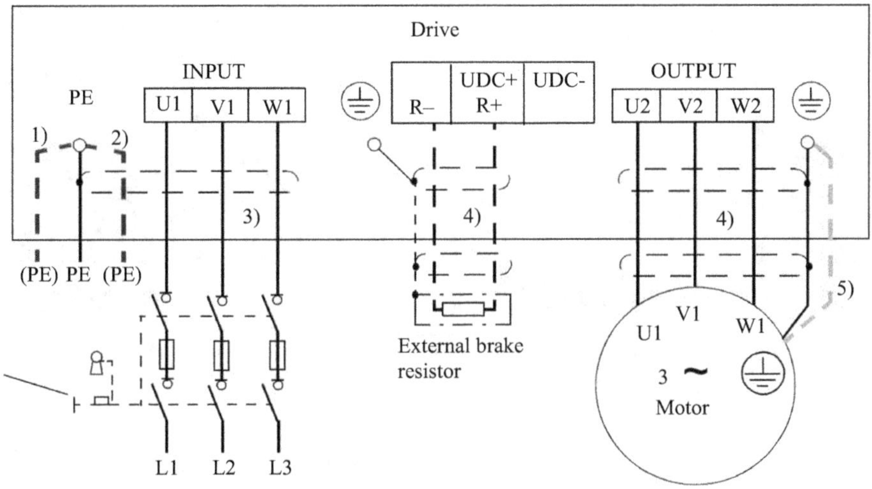


*Figure 3.24 Motor–drive connections*[2]

Otherwise, the drive might fail to estimate the motor flux, torque and speed accurately. As a result, all the control process should be questioned.

Another must-remember point is the acceleration profile adjustment which takes place at the very beginning step of parameter determination for the drive. Usually two general choices are available, linear and user-defined curves. The state of the art is to use a linear acceleration or even deacceleration profile. However, various control strategies certainly lead to a different profile of the motor quantities. This makes the diagnosis procedure a very complex task. Investigating the motor–drive behavior in the faulty modes is still an open area of research and readers are referred to the author's publications to find the potential research areas.

### *3.5.4 Motor load*

This is one of the most demanding aspects required to be as precise as possible if an accurate diagnosis investigation is really needed. First, it is the motor load level which affects all the motor quantities, so the necessity of presence of an accurate equipment for applying loads is one of the concerns. In addition, there are several types of equipment functioning as a load for any type of motors including:

- DC machines
- AC machines controlled by a drive
- advanced programable dynamometers.

These are the general types of loads functioning in different ways but fulfilling the same goal which is fixing the motor load level at a specific value or changing the

[2]https://library.e.abb.com/public/0cb03a89ae71bba5c1257b97004fdb01/EN_%20ACS800_01_HW_K_A4.pdf (access date: October 2016).

*Figure 3.25 Motor-generator mechanical connection using a coupler*

load based on a specific requirement such as oscillating load. Now, we are going to discuss each of the available options one by one from the simplest to the most advanced one sorted above.

#### 3.5.4.1 DC machines

This type of loads includes DC generators which are mechanically coupled to the shaft of induction motors, using a coupling device as shown in Figure 3.25. The more reliable the coupler is, the less the amount of unwanted oscillations caused by improper placement of the motor or the generator on the bench will be. It is highly recommended that the motor and the generator be aligned along their shaft. Otherwise, not only do undesirable oscillations harm the diagnosis procedure, but they also cause future mechanical defects if the system is used for a long time. DC machines are probably the simplest apparatus by means of which the motor load can be controlled. The underlying idea is to make the machine operate as a generator while it supplies a resistive load like what is shown in Figure 3.26. Significantly, the field and armature windings of the unitized DC machine should have a shunt nature as a series-connected topology suffers from the lack of field strength during the full-load operation. As a result, increasing the load, which is in turn achievable by adding a parallel resistive branch illustrated in Figure 3.26, leads to a dramatic reduction of the field of a series-connected topology; hence, the generator starts to be demagnetized. This means that the DC machine would not be able to compensate for the increased load level, and the total system load drops down the desired value. This only happens if the rated power of the DC machine is almost close to that of the motor. To come up with a solution, a considerably larger DC machine is recommended. Otherwise, a shunt topology should be used.

One of the drawbacks associated with the mentioned approach of load control is that the load level indicator is the induction motor current or speed while what really matters in fault diagnosis applications is the motor slip. Nevertheless, the proposed system seems to be very simple and straightforward while some precautions in terms of the appropriate load-level control should be always kept in mind. The shaft speed can also be measured by a tachometer.

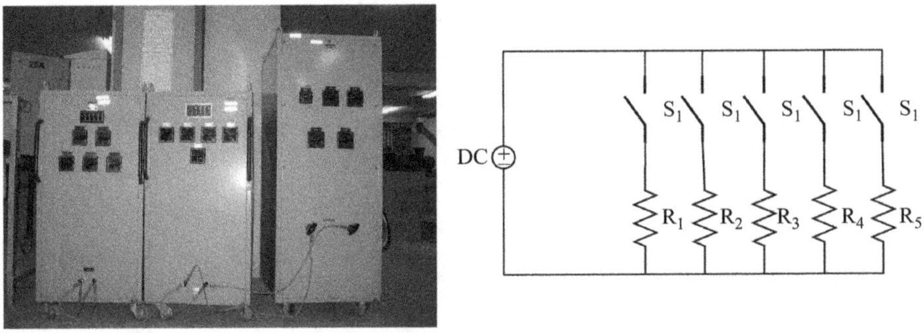


*Figure 3.26 Resistive load*

The proposed scheme is considered as of the unsafe approaches if an amateur practitioner uses the equipment. The reasons are the naked terminals of both the induction and DC motor along with the fact that the load should be changed manually by switching the resistors. Therefore, the user should be alert enough to avoid undesirable risks.

#### 3.5.4.2 AC machines controlled by a drive

This scheme of loading the induction motor is very straightforward, useful, safe enough, but a little costly (see Figure 3.27). The required ingredients of the scheme are as follows:

- tested induction motor as the main machine
- another AC machine, preferably a synchronous one, as the generator acting as a load
- two drives
- fuses.

The whole system is supplied by only one three-phase network connected to the machines and the drives through a set of fuses preventing the electrical stresses in the network side if any over currents happen in the system. Note that the fuses are supposed to be there any time. The three-phase fuse set should be used; moreover, two other fuses should connect the positive and the negative terminals of the DC buses of two drives. A choke is also required to prevent high-frequency PWM signals from circulating through the network. Drive#1 controls the induction motor while Drive#2 is connected to the AC machine, acting as the load, whether synchronous or asynchronous. The second motor–drive system should be set at torque control mode while the first system can be used in either torque or speed control mode depending on the test requirement. Preferably, the power, speed and torque ratings of the second motor–drive system which is used to act as the load should be similar to that of the induction motor. Otherwise, the AC machine power should be necessarily larger than that of the induction motor. Moreover, if there is so much difference of speed between the machines, a gearbox whose main operation is to change the speed while ideally transferring the same power is used to couple the shafts mechanically.

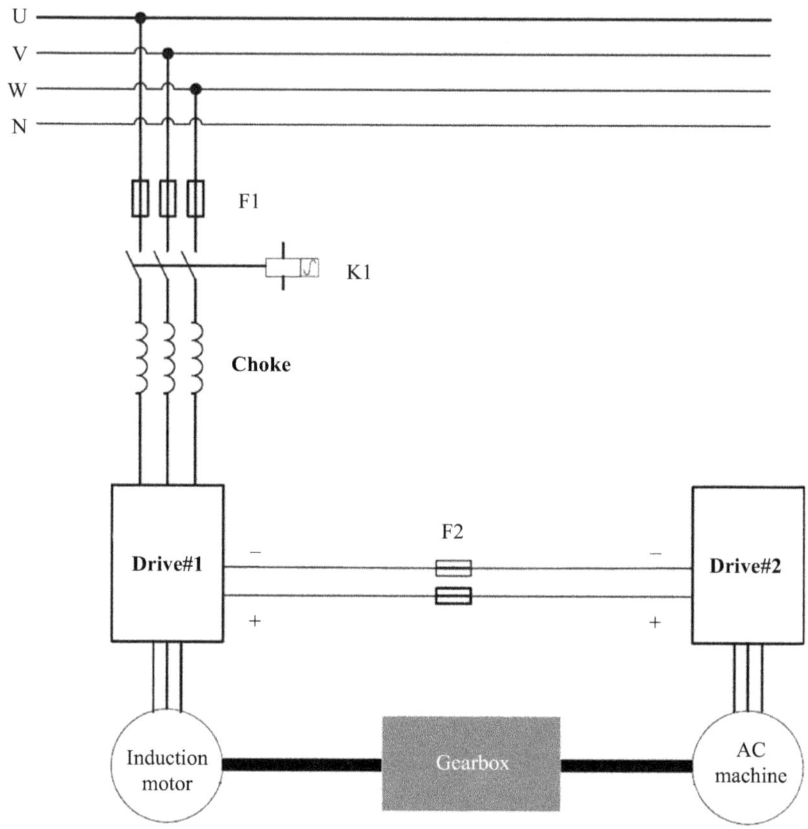


*Figure 3.27 Machine-drive systems with a common DC bus*

Due to its unique capability in accurate control of the torque or speed quantities, the closed-loop strategy should be hired in the load side. Otherwise, it is not guaranteed that a constant load is applied to the shaft if an open-loop strategy is used. So make sure to prepare a drive which is capable of handling closed-loop strategies, no matter if it is of a DTC or FOC nature, both work well. To reduce the budget, look for a drive with a sensor-less control included. Otherwise, a resolver or encoder should be prepared as well.

Whenever a single drive is connected to the network and the load is provided by the first scheme, the DC machine, all the required currents including the transient one come from the network and if a large power motor is utilized, other network-connected utilities might be affected harmfully due to the loading effect of the motor test setup. However, the second scheme, two motor–drive systems, is as demanding as the first scheme as the induction motor current is partly supplied by the AC machine which operates as a generator. So the current is circulated among two machines, and the network is only responsible for the power supply of drives. The situation is somewhat different in transient start-ups while once the system comes to a stable point, the network normally operates with a minimum current defined by the drives.

A programable drive#2 which facilitates applying a nonconstant load to the system is recommended. This kind of drive is useful when phenomena such as oscillating loads are studied. Otherwise, studies should be limited to a constant but altering load level which means a part of actual operations are neglected. So the generality of the investigations directly depends on the capability of drives. This is an important feature which we are aiming at in this book. With the use of the DC machine scheme, it is almost impossible to apply an oscillating load. This is another shortcoming of the first scheme highlighting it as an inefficient way of tackling fault diagnosis matters. To put it differently, it is not mainly proposed to go through the first scheme of motor loading.

On the other hand, the second scheme does not essentially ask for an auto-transformer connecting the network to drive systems. The reason is that the drive itself takes care of stabilization of its own terminal voltages and eliminates the need for an extra stabilizing tool. This is another advantage of the mentioned motor–drive systems in use. Furthermore, the braking resistor is not required anymore if the motor and AC machines powers are properly matched, or in the best case, the AC machine power exceeds that of the induction motor.

There is one important question "Is it possible to test a line-start induction motor, using the second scheme?" The answer is "YES." By removing the drive#1 and separately supplying the induction motor and the drive#2 to the network, a very fantastic line-start setup by means of which a pure constant load level can be easily applied is achieved. Do not forget to set the drive#2 at the torque control mode.

#### 3.5.4.3 Advanced programable dynamometers

This is certainly the most accurate and appealing way of applying various load levels to an induction motor under the test. Dynamometers generally operate on the basis of absorbing the energy of the motor shaft and acting as a load. There are able to operate at various torque–speed profiles providing the commanded mechanical load for motors. Sometimes, there are measurement devices implemented to measure speed and torque to provide a proper command while some of the types of dynamometers do not necessarily require a speed measure (Figure 3.28).

Regardless of what is connected to the dynamometer shaft, it operates as a load-producing equipment guaranteeing a specific torque on the shaft. On the basis of what structure or material produces the braking force acting as a load connected to the shaft, various dynamometers are available. Three important ones are as follows:

- **Hysteresis brake dynamometers**
  This is a perfect choice for experimenting motors ranging from fractions of a kilowatt to medium-power applications. A full range of motor speed including the free-run to locked rotor can be easily controlled by means of a hysteresis dynamometer. This is possible just because hysteresis dynamometers do not generally need a speed measurement to precisely output a torque. They are

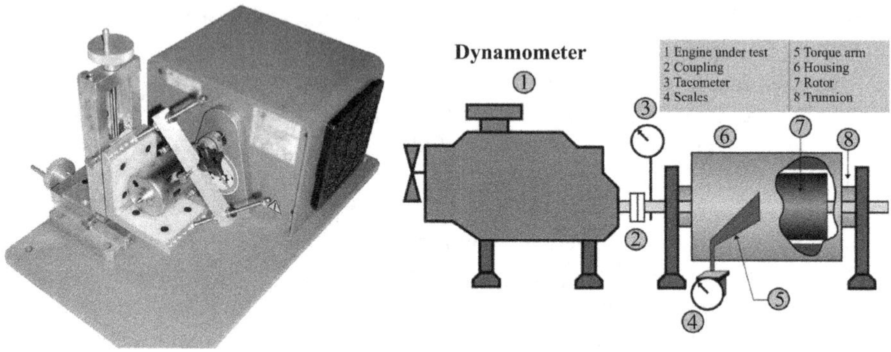


*Figure 3.28 Dynamometer*[3]

indeed accurate, and the maximum error, depending on the size and accuracy of the configuration, does not go beyond 1%. Considering the nature of hysteresis dynamometers, they can be programed to be even used in varying load applications.

- **Eddy-current brake dynamometer**
  The main feature of this type is the dependency of its torque to speed of rotation. It is something like fans in which the developed load is proportional to square of speed. So not all range of torque–speed profiles might be covered by this means. However, if a high-speed application should be tested, eddy-current dynamometers are the best.
- **Powder brake dynamometers**
  This type is usually recommended for a considerably high torque application such as traction motors while the corresponding speed level should not be so much high. Again, a very well-designed dynamometer of this type might reach a maximum error of 1% in the worst-case scenario.

Dynamometers usually come with a comprehensive measurement and analysis package called power and spectrum analyzers by means of which all the ordinary motor signals including the voltage, the current, the active and reactive power, the inductance and also the corresponding frequency and time-domain variations are explored. This gives users the opportunity to access the essential mediums of fault diagnosis procedure straightforwardly. However, such a measurement and loading equipment is supposed to be very expensive. An example will be provided later in this chapter.

### *3.5.5 Implementation of broken bar fault*

Initially, let us provide a complete 2D view of an induction motor crosssection shown in Figure 3.29(a). The stator and the rotor consist of 36 slots and 28 bars, respectively. Every 9 slots and 7 bars form pole out of 4 poles of the motor. Due to

[3]https://en.wikipedia.org/wiki/Dynamometer (access date: October 2016).

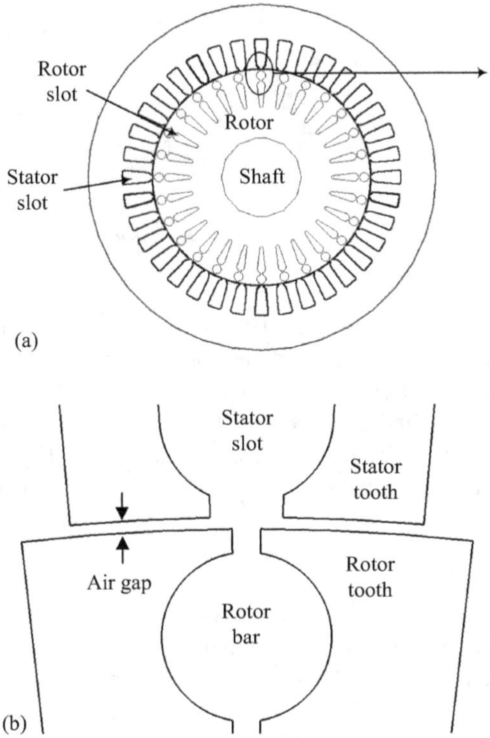


*Figure 3.29 (a) Squirrel-cage induction motor and (b) magnified stator-rotor teeth*

the symmetry of the rotor, it is not important which bar is assigned as the first bar of the rotor. Then, the other bars are numbered clockwise or counter clockwise from 2 to 28. The air-gap length of the motor is equal to 0.45 mm (Figure 3. 29(b)), and the bars possess one upper and one lower parts dealing with the motor start-up and steady-state operation, respectively.

The air gap is normally so small that even a very negligible amount of bearing defect or eccentricity fault leads to a considerable percentage of air-gap distortion causing an unbalanced motor operation. A very promising way of doing this is to fix the motor housing on the working bench and clamp the shaft from both ends and then try to loosen and open the plates. To perform it more smoothly, preheat the plates close to the bearing area to make sure the bearing nests are loose enough.

Prior to any fault implementation, the first step is to disassemble the motor by opening the two end plates. Preferably, start with the load-end plate. Make sure to clamp the rotor shaft and not to release it while opening two end plates. Even if you intend to release the shaft and put it on inner surface of the stator, do it safely to prevent the stator material from being damaged or rubbed. If not so, the stator would be apparently useless for future fault diagnosis attempts. Finally, take the entire rotor body out and put it somewhere safe. Make sure to do it gently with the

*Figure 3.30 Cage rotor*

goal of protecting the rotor material. By "material," we mean the silicon steel material, not the bar material because the bars are often intact and inaccessible while the outer layers of the silicon steel materials are of course in contact. The bearings are mainly attached and fixed to the two shaft ends. So leave them the same unless an eccentric rotor should be investigated (see Figure 3.30).

In this specific rotor, the bars are in the shape of a deep bar which consists of two parts, the upper and the lower cages, as shown in Figure 3.31. This makes the fault implementation somewhat tricky as there is not a uniform distribution of the bar crosssection across the rotor bar depth. So while applying the partial broken bar, it is not a 100% accurate practice. Nevertheless, applying a full broken bar is easy and precise. The main aspect of implementing the breakage is to know the exact dimension and shape of bars, as it is visible from outside of the rotor. Therefore, having the structural map of the motor in hand is essential.

Considering the double cage rotor bars shown in Figure 3.31, the maximum diameters of the upper and lower cages are equal to 6.2 and 5.4 mm, respectively. So, it is obvious that a full broken bar requires an elimination of a crosssection covered by a width 6.2 mm, while the partial broken bar only asks for a smaller portion of the mentioned value. The question is that how the removal of the crosssection is done. Actually, the goal is to partially or fully eliminate the bar current and prevent the corresponding bar from contributing to the magnetic field generation. How is it handled?

To apply broken bars fault to the rotor, the rotor bar should be drilled to prevent the current from passing through the bar from one end to the other end. In general, there are two main locations to make the desired hole (Figure 3.32):

- somewhere between two end-rings along the bar length
- right at the connection of bar and end-rings joints.

The latter is the most probable location of the fault in real motors (Figure 3.32(a)). The joint between bar and end-ring is not always as strong as it should be due to some internal cracks or bad welding or casting process. When the motor is subject to undesirable thermal or mechanical stress, the joint is broken and it is literally said that the bar is broken. It is not proposed to drill the hole at the joint of bar and end-ring as it probably damages the end-ring which leads to a worse fault level. First of all, it is better to make a very small hole on the rotor bar, somewhere between two end-rings, using a very thin drill. Do not put so much pressure on the

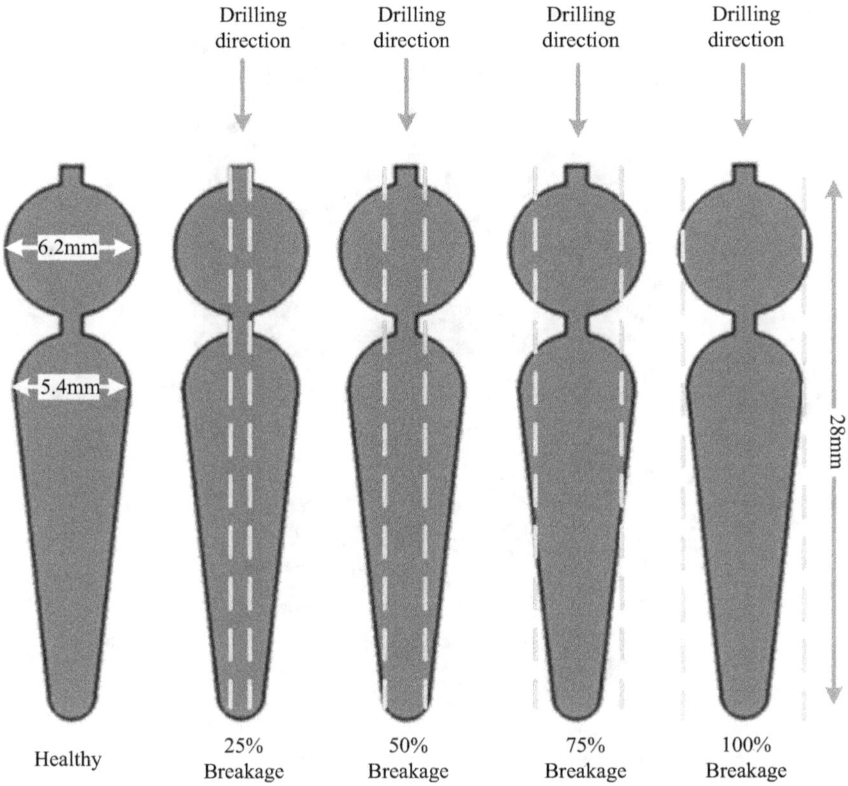


*Figure 3.31 Rotor bar slot and different degree of broken bar fault in one slot including partial and full broken bar*

drill because the bar is made of aluminum which is a really soft and easily drilled material. Keep in mind to use fixture to prevent the rotor from moving when it is drilled. Otherwise, the silicon steel material might be damaged. If so, not only the core losses but also the flux distribution will dramatically be increased. To make sure that you are just drilling the bar not the core, you have to keep eye on color of swarf coming out of the drill. If the color is bright, it is the aluminum which is drilled. Otherwise, if it is dark gray, you are drilling the core material. So you have to stop immediately. Be careful to drill the bar with a very high speed and low force level to guarantee that the drill would not brake and stick into the bar.

Depending on the diameter of drilling area, different levels of the breakage is applied. Figure 3.31 clearly illustrates four different possibilities, namely the 25%, 50%, 75% and 100% breakage.

The formal course of drilling action is to utilize a stand-up drill like what is shown in Figure 3.33. Do not lose the sight of the fact that the rotor should be certainly fixed by means of a fixture. Act very gently and smoothly! If you are going to investigate different fault levels ranging from partial to several full

*Figure 3.32 Drilling the bar (a) joint of bar and end-ring and (b) between the two end-rings*

broken bars, go ahead with the smallest fault level, for example 25% partial breakage. Then, move toward the higher levels to keep the tested rotor less damaged. Note that the broken bars fault is an irreversible kind of fault leading to a permanent damage.

A very critical point is to get more than one rotor ready for experiments as you might also intend to study more than one breakage at different poles. So one single rotor is not enough to fulfill the requirements.

Partial and full breakages are two types of broken bars. If it is a partial breakage, it is called partial broken bar in which a fraction of healthy current passes through the bar. By increasing the level of partial breakage, the bar resistance is increased up to almost infinity. Sometimes, there is full broken bar. This is simply known as broken bar fault. It is mostly considered that the resistance of a broken bar is infinite and no current passes through it, although it is not infinite when the bar is partially broken. What really happens is a bit different due to the existence of interbar currents. Although the current might not be able to pass through one bar in full breakage case, it might enclose its pass through cracks existing in core between

*Figure 3.33 Stand-up drill*

two or more bars. This phenomenon has not yet been studied deeply, but the effects of interbar currents have been observed in different tests. On the basis of this claim, the resistance of a broken bar is not infinite at all. If you need to simulate the broken bar fault, you might choose a larger value compared to that of the healthy motor. This value will be discussed in the next chapters. Sometimes, more than one broken bar, for example 2, 3 or more, might exist or happen in a rotor. The broken bars could be in adjacent or nonadjacent locations. So the broken bars fault is categorized into three main groups as follows:

- partial
- adjacent
- nonadjacent.

Figure 3.34 shows partial, 1, 2, 3, 4 adjacent broken bars. As the entire rotor circumference cannot be shown in 3D figures, the nonadjacent broken bars located at different poles are shown in Figure 3.35 by means of a 2D representation. P1, P2, P3 and P4 stand for first, second, third and fourth poles, respectively. Each small gray circle is the symbol of one bar. Red color means that the bar is broken. The number of nonadjacent broken bars might vary from 2 to larger numbers.

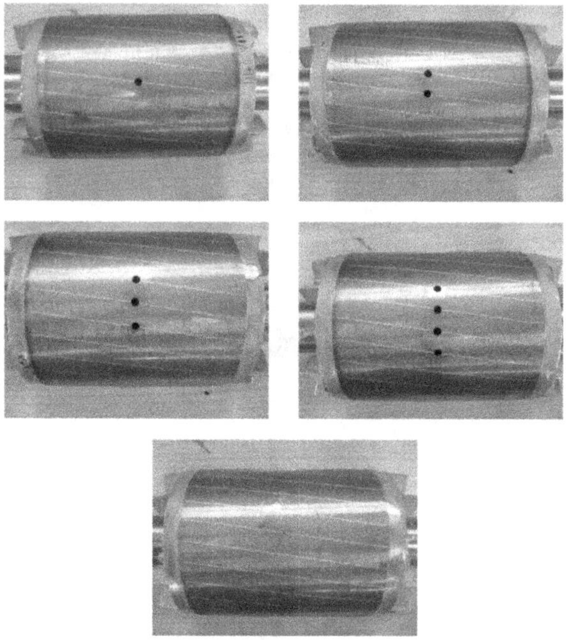

*Figure 3.34 Different types of broken bars*

(Instead of color, special signs must be used to have clear figures in the black and white hard copy of the book.)

As you notice, there could be various degrees, numbers and locations of the broken bars. So three totally different concepts, briefly addressed before, are introduced as follows:

- **Detection:** Monitoring incipient broken bars fault occurrence
- **Determination:** Monitoring number of broken bars
- **Fault-location:** Monitoring location of the broken bars.

Combination of "detection," "determination" and "fault-location" procedures builds a very comprehensive advanced industrial concept which is called "Diagnosis." The same is valid for the other types of motor faults. Diagnosis is not a specific process of monitoring for electric machines. It is a very complicated process which is required in all engineering and medical applications to provide a safe and reliable operating condition not only for devices but also for technicians.

Taking the possibility of having different levels and locations of the fault into account, one can easily understand the struggling of the diagnosis procedures based on which users, researchers and technicians are trying to detect, determine and even locate the fault. It is a really overwhelming task unless a clear interpretation of faults and their effects on the motor signals is prepared. That is why a very basic to advanced discussion of every influential factor of motors, drives and faults have been and will be discussed in this book. Ignoring one important aspect, something

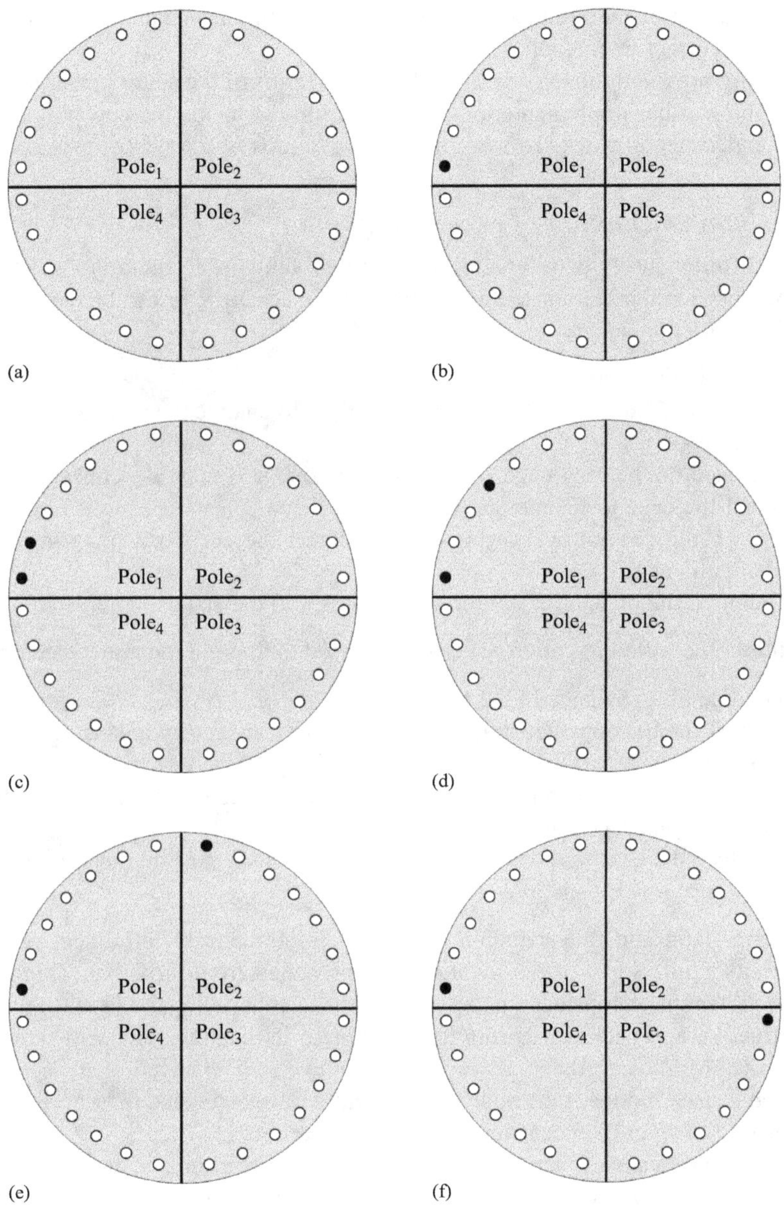


*Figure 3.35 Broken bar (bb) locations (a) healthy, (b) 1 bb, (c) 2 adjacent bbs (case 1), (d) 2 bbs at half pole-pitch distance (case 2), (e) 2 bbs at one pole-pitch distance (case 3) and (f) 2 bbs at two pole-pitch distance (case 4). (bb: Broken Bar)*

like the location of the fault, would probably result in unsatisfactory outputs leading to an incorrect maintenance process.

Having provided the experimental details in terms of the setup preparation and also the broken bar implementations, now it is time to tackle the eccentricity fault and its experimental sides. To learn more, have a look at the next subsection.

### *3.5.6 Implementation of eccentricity fault*

So far, different ways of producing eccentricity fault in a laboratory scale have been proposed in the literature. For example, static eccentricity can be the output of the following approaches:

- to use a faulty bearing
- to design and build specific plates enabling us to change the vertical position of the bearings
- to replace the bearing cage by an eccentric one and then use concentric rings around the cage to fill the empty spaces
- to place the concentric bearing cage between the eccentric rings inside and outside the cage
- to remove the plates, fix the rotor on stands and the stator on adjustable stand

In addition, the following approaches can be used to apply a dynamic eccentricity:

- to couple an unbalanced load to motor
- to chisel the bearings housing and fill the empty spaces by means of eccentric rings
- to chisel the bearings housing and make an eccentric space and fill the empty spaces by means of noneccentric rings
- to replace the bearing by bearings with a larger inner race diameter and fill the empty space by means of eccentric rings.

Using one static and one dynamic approach simultaneously introduces a mixed eccentricity fault. Some of the mentioned approaches are destructive. This means that the motor structure is damaged permanently after applying the fault. In the case of destructive approaches, returning the motor to its healthy state will be almost impossible. Moreover, the fault level cannot even be changed if a destructive approach is used. Some of the approaches require a sophisticated measurement tool to measure the level of eccentricity applied to bearings.

We decide to stick to an approach which reduces the number of disadvantages mentioned above. Mainly, the approach should not be destructive and not require a specific measurement tool. So various types and levels of eccentricity can be applied and eventually removed from the motor to get it back to a healthy operation. It is an extension of the forth dynamic approach. For this purpose, instead of the main bearing with the code 6309, another bearing called 6011 whose inner race diameter is 10 mm larger and outer race diameter is 10 mm smaller than that of 6309 is utilized. Placing the new bearing helps us with two empty spaces inside and

(a)

(b)

*Figure 3.36 Experimental implementation of eccentricity fault*

outside it and allows us to fill both spaces by means of rings leading to the possibility of both the static and dynamic eccentricities. So the steps are as follows:

- to install a concentric ring inside and an eccentric ring outside the bearing to produce a static eccentricity. The fault severity is controlled by means of the level the eccentricity of the outer ring.
- to install an eccentric ring inside and a concentric ring outside the bearing to produce a dynamic eccentricity. The fault level is controlled by means of the level of eccentricity of the inner ring.
- to combine the previous two approaches to apply a mixed eccentricity fault.

Finally, the initial healthy motor can be assembled again using the replacement of faulty bearings by the healthy one. A very clear illustration of the mentioned steps is provided in Figure 3.36. The challenging part of the implementation is to appropriately prepare the eccentric rings with an exact level of eccentricity. As the

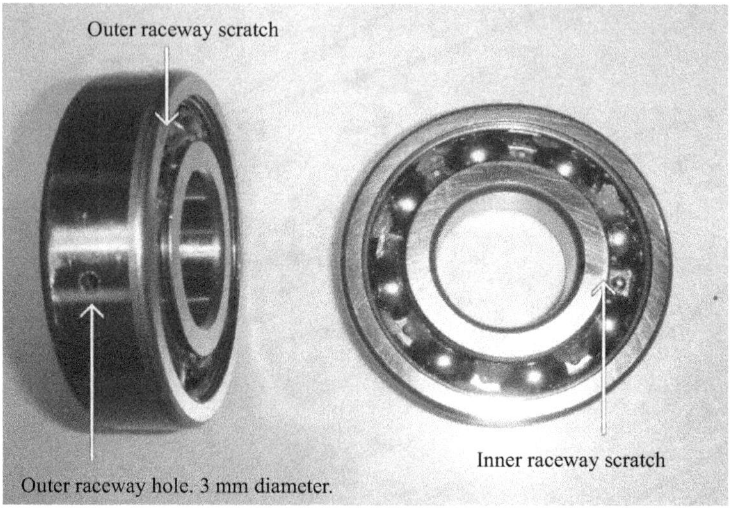


*Figure 3.37 Experimental implementation of bearing fault*

air-gap length is very small, a very accurate tool such as a computerized numeric control machine might be required. This is probably an expensive but appreciable practice.

Now, let us switch to bearing fault and a very simple but intuitive ways of applying defects to it (see Figure 3.37). Simply drill the inner or outer race to produce the inner race or outer race fault. To apply a ball defect, it is better to apply a crack or breakage to one of the balls. So it requires disassembling the bearing.

### *3.5.7 Implementation of interturn short-circuit fault*

In a healthy condition, the only accessible parts of the turns or coils are the end terminals allowing us to connect them to the supply while a short-circuit fault stands in need of new electrical connections as shown in Figure 3.38. These new connections are definitely different from the terminals and provide a current pass through somewhere in the middle of one of the windings, a, b or c to the same winding, another winding or even the ground (motor housing). Respectively, they are called interturn, phase-to-phase and phase-to-ground fault. Therefore, to apply a short-circuit fault, the insolation of at least two turns should be partly removed to access the turns. So it is certainly a destructive task and needs a careful attention. Even after removing the insolation, an additional resistance shown by $R_f$ should be used to limit the circulating fault current to a safe value. Actually, if the fault connections are attached to each other, using a low resistance wire, the fault current in the new phase might go beyond control, and the phase winding(s) burns completely. In this case, the motor will be useless.

Depending on the fault severity, location and type, different values of $R_f$ are demanded. For example, when dealing with phase-to-ground short-circuit fault, probably the larger $R_f$ should be used as it is often the electric path with the smallest

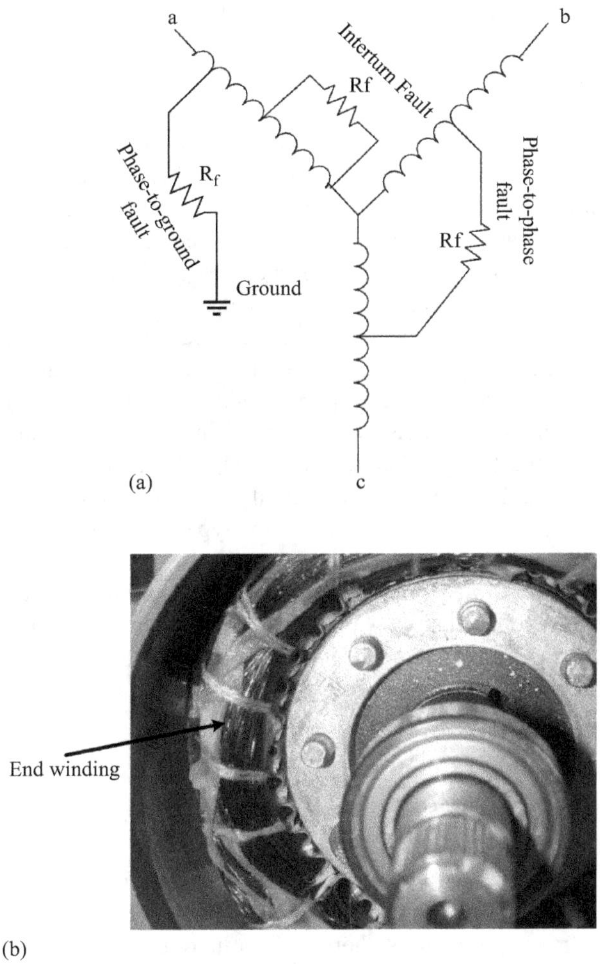


*Figure 3.38 (a) Schematic implementation of short-circuit fault and (b) motor end winding*

resistance. Hence, a larger circulating current will flow for a given path voltage. However, the interturn fault, which is usually the most possible case, is of the smallest current, so the corresponding $R_f$ might also be equal to zero, depending on the insulation quality. This is actually the insulation and partly the copper quality which define the maximum allowable circulating current and subsequently the $R_f$ value. In fact, the more the insulation and copper endurance is against the temperature rise caused by a large circulating current, the smaller the value of $R_f$ needs to be. Therefore, higher levels of fault might also be studied. High-quality insulation also makes it possible to sample a larger time period giving the opportunity to have a better spectral resolution. $R_f$ should be a high-power resistance which is capable of tolerating large power dissipation. Then, $R_f$ is to be electrically connected to the turns which are planned to be short-circuited. As the turns are not accessible through the stator stack length, the only part facilitating the process is the end-winding areas coming out from the stack.

In practice, as investigating real large current faults is impossible due to the fact that the motor will be permanently destructed, the trend is to only analyze lower levels of fault to first find proper indicators and then check how the proposed indicator depends on the fault increase or the other influential factors. The other faults are treated the same while some of them such as eccentricity fault are of a reversible nature pointing it as a motor friendly type of fault.

The aim of this section is to deliver the real sense of fault diagnosis and their laboratory-scale implementations. On the other hand, as the very beginning step of the diagnosis procedure, signals should be prepared for processing step. This is why measuring, or in other words, sampling motor signals including thermal, magnetic, electrical and mechanical signals are of a great interest as they are the mediums reflecting the possible faulty behavior. Motor signals such as voltage and current and their corresponding temporal and spectral variations are kind of signatures for any motor and differ depending on the motor type, the fault type, the processed signal and also the utilized processor.

Basically, an accurate measurement of proper motor signals is the most important concern of any diagnosis procedure. Therefore, we decide to establish a useful and comprehensive analysis of various types of sensing and sampling motor signals. To this end, move toward the next section which discusses signals and the corresponding sensor implementation.

## *3.5.8 Signals and sensors*

### 3.5.8.1 Voltage

One of the most well-known and significant motor signals is probably the voltage which is responsible for providing the amount of torque required for rotation. Although in line-start applications, there is no point of generally using the voltage as a signal for the diagnosis purposes, in inverter-fed applications it plays a vital role in defining some specific aspects of faults such as justifying the possible differences between various supply modes. However, the voltage might also be useful in a line-start application if calculation of quantities such as winding inductances is the main goal. Moreover, the time-domain samples of the line voltage are simply available in the I/O port of drives or by means of built-in packages or interfaces developed for a drive. Therefore, in drive-fed motors, using additional instruments is not recommended, but in case a line-start application is under the test or there is no access to drive interfaces, one might use a "Voltage Transducer (VT)" specifically developed for this kind of applications (see Figure 3.39). VTs are promising tools to measure and output a scale voltage exactly similar to their inputs which is the motor terminal voltage. A common structure of VTs is illustrated in Figure 3.39(a).

Any VT consists of a transformer used for isolating the primary and the secondary windings to increase the safety and also possibility of the measurement, using a small output voltage. The input voltage might be very high. A restricting

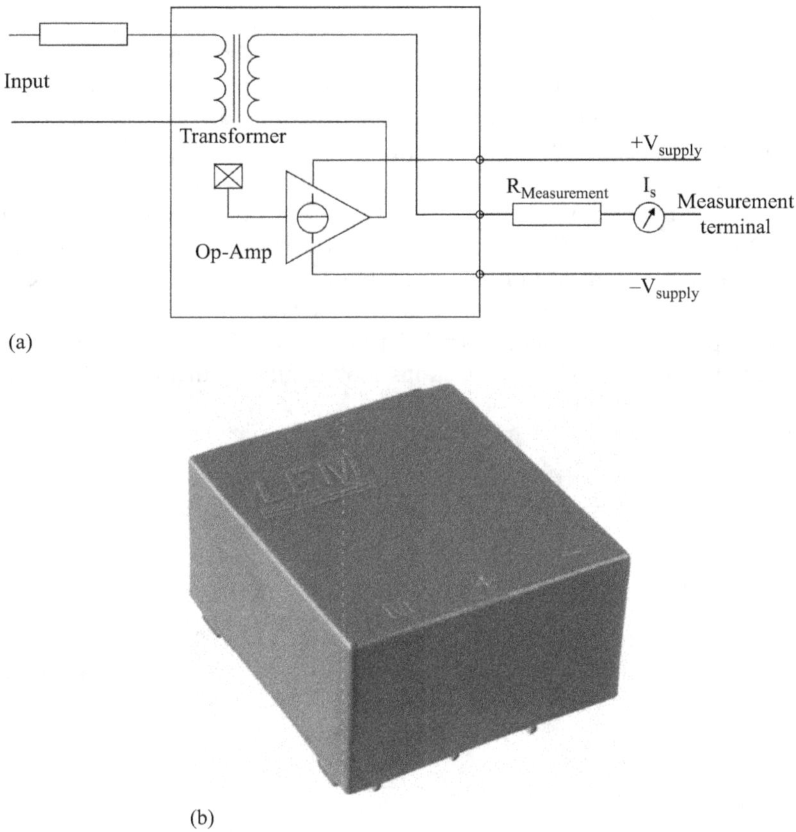


*Figure 3.39 (a) VT circuit and (b) LEM VT*

resistance connects the terminals to the VT and it should have a considerably large value to avoid the loading effect of the VT. Then, the output of transformer is connected to an op-amp for outputting a scaled voltage measured in the M (measurement) terminal. There is also a series-connected measurement resistance to prevent the output current from running above fractions of an ampere. If the output of voltages of transformer becomes larger than the supply voltage of the OP-AMP, the output is saturated and will represent a flat-topped signal. So, make sure the maximum terminal voltage does not go beyond the rated limits of the VT.

Using VTs is very simple and straightforward, and the only necessary additional thing is a DC voltage source to supply the OP-AMP. It is worth noting that the supply provides a positive–negative polarity voltage. Otherwise, one side of the signal, either the positive or negative side, is cut off in the output. The LEM company sells a wide variety of VTs as well as current transducers (CT) ranging from very small to medium ratings (see Figure 3.39(b)). Every single product of this company has its own specific datasheet in which the proper usage is provided.

#### 3.5.8.2 Current

As a principle ingredient of diagnosis procedure, the current should be considered as the most tackled and useful signal in any procedure. Both the time and the frequency analysis of this practically appreciated signal have been matter of various investigations, so an abbreviated term called MCSA which stands for "motor current signature analysis" is usually assigned to the process of investigating the fault, using the motor current. The widespread use of the motor current is underlined by the fact that it reflects almost all the essential fault behavior and also it is the easiest and safest signal in terms of sampling. Unlike the motor voltage which might be very dangerous if precautions are not taken into account, the current sampling is painless. The only required instrument is a CT, especially a Hall-effect sensor (see Figure 3.40). The connections look similar to that of a VT except the

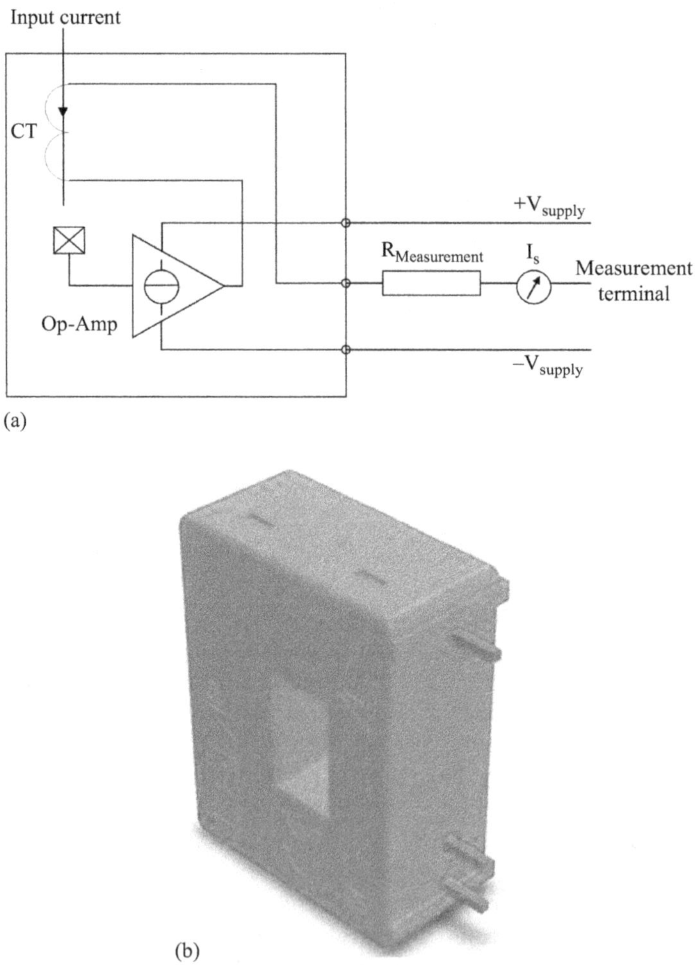


*Figure 3.40 (a) CT circuit and (b) LEM CT*

input connections. Instead of a simple isolated transformer, a Hall-effect sensor, which does not require an electric connection in the primary, is implemented. The primary is the motor terminal wires passing through the middle open space of the CT shown in Figure 3.40(b). Using the magnetic induction on the secondary winding of the sensor, the transformer output is passed to the OP-AMP and then scaled to the output which is the same as that of VT.

#### 3.5.8.3 Torque

In general, there are two types of torque-measurement sensors as the following:

- invasive
- noninvasive.

Invasive type of measurements is normally coupled to the motor shaft mechanically; hence, there is always a mechanical coupling which might affect the motor behavior in a bad way. In fact, there is the possibility of the signals used for the diagnosis being affected. Therefore, it is called "invasive." The terminology of "invasive measurement" used in this book is way broader than what is usually referred to in the literature. In fact, any device or even device placement revealing any kind of potentially harmful or disturbing effect might be included in this category. Otherwise, it is called a "noninvasive" technique. Definitely, a noninvasive method should be practically preferred to an invasive one. However, the corresponding noninvasive sensors (see Figure 3.41(b)) which are principally based on the Hall-effect do not cover a wide range of applications. So it is recommended to prepare a mechanically coupled sensor (see Figure 3.41(a)) if a larger range of torque and speed variations should be investigated.

One of the disadvantages of the mechanically coupled sensors is the presence of a mechanical connection between the motor and the sensor shafts. Thus, if eccentricity is investigated, there is a possibility of a bended sensor shaft as well. This not only applies asymmetry in a long-term use, but also might reduce the accuracy of the sensor. However, both types are usually robust enough to withstand sever applications up a certain point. Regardless of the type, the associated error of torque sensors today does not usually surpass 1% which is an acceptable range for the diagnosis purposes.

Both types consist of a scaled electrical output to hand in the motor torque instances mostly in an analogue regime. Actually, as they mostly use analogue devices, the output torque has a continuous nature while digital torque sensors providing described signals are also available. Close attention should be paid to the selection of a digital torque sensor in terms of the sampling rate. The higher the sampling rate is, the higher the resolution and the number of observable harmonic orders will be in the torque spectrum.

#### 3.5.8.4 Speed

Encoders and tachometers are two mediums of measuring the motor speed, each having its own advantages and disadvantages (see Figure 3.42). Encoders are mechanically coupled devices while tachometer should not necessarily be

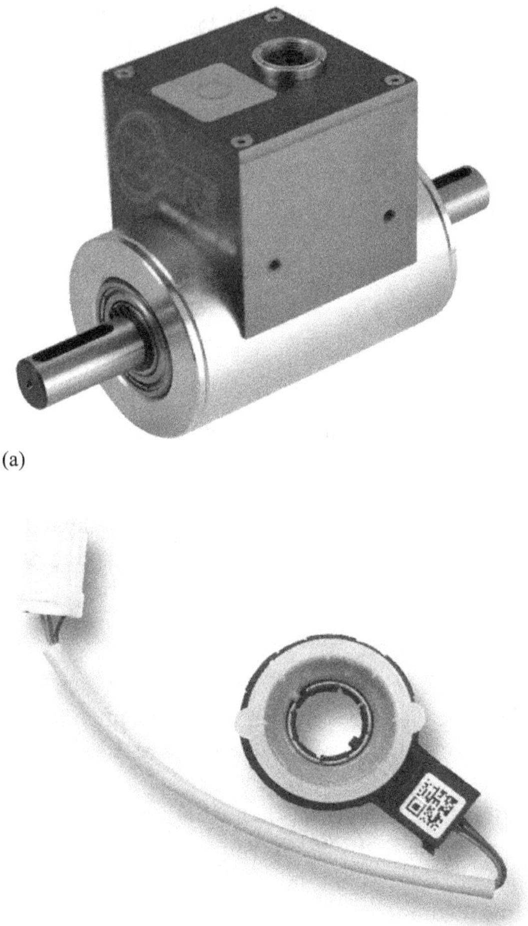

(a)

(b)

*Figure 3.41 Torque sensors (a) invasive method and (b) noninvasive method*

connected mechanically. What makes a difference between these two types is the way they are used to measure the speed. The following are the main differences:

- Encoders are mechanically connected to the motor shaft; hence, are considered as an invasive method of measurement. On the other hand, tachometers work on the basis of laser light which sends and receives high-frequency waves reflected from the shaft. Normally, unlike encoders, tachometers do not provide us with an output port of speed measurement, and they only consist of a digital indicator indicating an average value of speed.
- Encoders are of course capable of sampling the speed variation while tachometers often return one single number in average. So if the goal is to merely define the steady-state speed of the motor shaft, tachometers are the best choices. Otherwise, in case of a need for an accurate time-dependent step-by-step speed development of the shaft, encoders should be used undoubtedly.

(a) (b)

*Figure 3.42 Speed sensors (a) encoder and (b) tachometer*

Discussing the details of operating principles of sensors is not a part of this book, and the only focus is on their application in accurate signal measurement. For more information about the principles, readers are referred to datasheets distributed by companies.

### 3.5.8.5 Flux

Magnetic flux density is the key agent of transferring power from the stator to the rotor. Without a magnetic coupling through the motor air gap, electromagnetic energy conversion is stopped and the motor operation fails. This is why it is sometimes considered as the heart of any electrical machine. On the other hand, although the original signal of supplying the motor power is the terminal voltage, what handles the rotation is the magnetic force or torque developed by the magnetic flux density. Moreover, previous mathematical and simulation-based developments proved that any fault behavior, either in the stator or the rotor, is somehow transferred to the rotor or the stator only by means of the air-gap magnetic flux. The mentioned point necessitates the utilization of an approach to access the magnetic flux at least at air-gap level. To this end, two approaches, namely an FE-based simulation and an experimental sensor, are available. The first approach is the subject of the next chapters while here we are going to introduce the second approach.

Observing an induction motor structure, whether wound or cage rotor, one will simply understand that in an ordinary motor, there is no way of accessing the air gap as motors are totally encapsulated by the housing and there is no access inside. So the trick to access the air-gap quantities is to use an invasive technique called “search coil” shown in Figure 3.43. The approach is called invasive as the motor plates should be opened, and then the search coil is placed around one of the stator teeth.

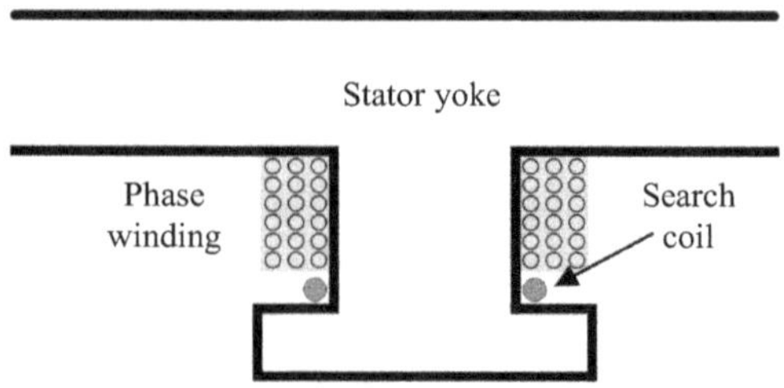


*Figure 3.43 Search coil*

The larger the number of coils is, the larger the output voltage will be. However, even one turn works very well. Be careful not to increase the number of turns so much that it impacts the flux distribution of the main coils. The search coil provides two open terminals whose differentiated voltages should be equal to its induced voltage on it. As the search coil is installed as close as possible to the air gap, the most dominant neighboring magnetic field is that of the air gap. According to the Faraday's law, while the magnetic field rotates and passes through the coil, a specific level of EMF is induced and measured at the terminals. This is a highly invasive but very useful technique. The magnetic flux density of the other parts of the motor including the stator tooth, the stator yoke, the rotor tooth and also the rotor yoke is not measurable at least by means of ordinary approaches. It is not also possible to implement such a device in a laboratory scale. This is why researchers rely on 2D or 3D FE simulations to study behavior of faulty motors. Furthermore, in the case of a cage rotor, rotor quantities are actually beyond reach, so the mentioned sensors will be useless unless a preimplemented sensor, manufactured by factories, is used.

A very important point is to get familiar with the issues associated with magnetic characteristics of faulty induction motors. This is not achievable unless a very comprehensive FE model of motor is investigated. In fact, any posterior analysis including thermal or loss characterization highly depends on an accurate magnetic flux distribution prediction and FE approach makes this possible. Therefore, one complete chapter will be devoted to basics, formulations, implementation and postprocessing the FE analysis of healthy and faulty induction motors. Along with the other chapters, the chapter related to the FE analysis definitely discriminates this book from similar books in which a shallow study of the motor quantities is addressed without providing justifications.

#### 3.5.8.6 Vibration

Practically, vibration analysis is the core diagnosis approach of induction motors. Vibration in vertical, horizontal and axial directions returns astounding information on the motor behavior. Most of the time, it is by mistake assumed that vibration analysis only deals with mechanical defects and has nothing to with electrical faults. However, any type of fault, as what was previously mentioned, produce a specific oscillating trend causing monotonic mechanical pulsation of the motor body. Even the short-circuit fault which is a kind of naturally electrical fault also

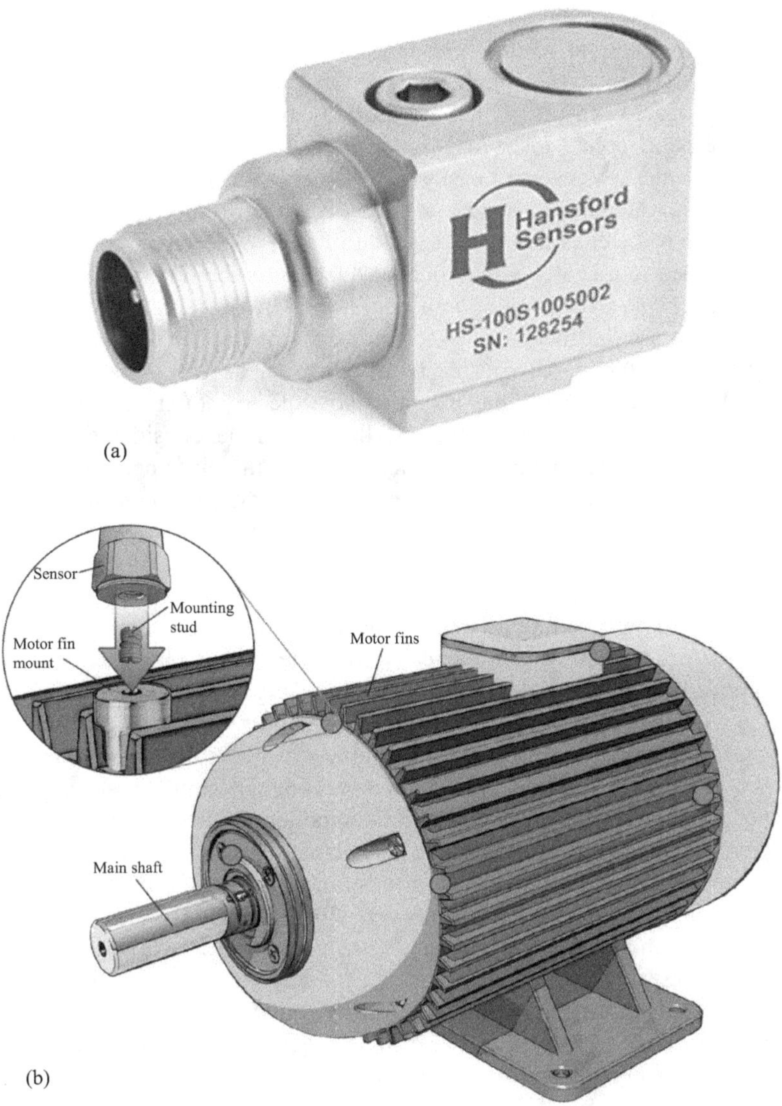


*Figure 3.44 (a) Vibration sensor and (b) possible installation locations*[4]

produces a pulsating air-gap flux density leading to a mechanical vibration. This is exactly what is sensed by a vibration sensor and returned as an electrical signal. A typical vibration sensor, along with the possible installation locations, is shown in Figure 3.44. Accelerometer is the conventional name of vibration sensors. They

[4]http://www.imi-sensors.com/ (access date: October 2016).

are magnetically mounted devices which can be rotated by user around the motor to capture vibration patterns of different parts of the motor.

There are types of two implementations of accelerometers, temporary and permanently located sensors, which are used in easily accessible and totally inaccessible operating areas, respectively. In other words, if the motor should be installed once and not accessed easily later on, it is usually preferred to install a permanently mounted wired sensor instead of a rotating accelerometer. If more than one accelerometer is mounted or used to monitor the vibration, data from all sensors is passed to a switch which is connected to a monitor or any kind of indicator. Then, sensors data is monitored one by one by switching between the sensors. Depending on the location of sensors on the motor housing, shaft, plates or even bearings, different patterns each revealing an aspect of faulty motor are achieved. It should be noted that it is also possible to mount a sensor internally inside the motor housing. Nevertheless, this might be considered an invasive technique and is not recommended. Having said that, the placement and location of sensors itself is a cause of decision change of a condition monitoring procedure. If so, vibration analysis should be included in the category of invasive approaches as the way it is implemented affects the diagnosis procedure. However, once the sensors are implemented and fixed, outputs should be the same for a given faulty condition. So in this sense, it is a noninvasive technique. Equally important, vibration sensors are always mechanically connected to the motor.

Vibration sensors are comparably more expensive than VTs or CTs. Therefore, it might not be affordable by academic research centers whose budgets are limited. In this case, it is proposed to use a search coil and measure the air-gap flux density. Then, the radial force can be calculated and assigned as a vibration producing component, using (3.38). As the tangential component of the air-gap flux density is almost zero, this leads to an acceptable approximation of the radial force which is the main reason of the vibration. The other option is to use an accurate FE approach to simulate the vibration signals. However, vibration analysis has been used for several decades and is still of a great interest in industries.

Sound (noise) is a direct consequence of any kind of vibration. So in an environment that there is no additional noise or sound-producing factor except the tested motor, it is a premium approach to the analysis of the fault as it certainly has a noninvasive nature. The corresponding sound or noise sensors are in the form of antenna absorbing the noise or any sound disturbance produced by the fault. The significant requirement of any sound sensing is the presence of a noise-free room (antenna room) in which the motor should be tested. Although the required room should not be very large, it is a very dedicated and expensive facility not found easily everywhere (see Figure 3.45). Sound analysis technique is a very promising way of diagnosing motor faults.

#### 3.5.8.7 Temperature

Temperature sensing is one of the common monitoring techniques of electrical machines including motors and generator. Depending on the machine and fault type, different parts of the machine are subjected to a thermal tension requesting for

*Figure 3.45 Antenna room*

a precaution in terms of possible future defects. In a healthy induction motor, end windings are the highly heated parts due to the higher leakage inductance while in a faulty motor, every part experiencing a dramatically increasing saturation level is the target of thermal tensions. For example, when a broken bar or short-circuit fault occurs, the area neighboring the fault reveals a considerably high temperature leading to an unbalanced temperature distribution of the machine. An eccentricity fault introducing a rotating saturated region called UMP point applies a total temperature rise of the motor. However, the other types of fault are also factors for generally increasing the motor temperature compared to a healthy operation. Temperature monitoring is not usually performed as a sole-task, and it is combined with some other techniques such as the vibration analysis to return helpful information on motor operation. Moreover, the operating mode and the motor load level should be first defined in any thermal analysis as these are the principle factors in determining the temperature.

There are two basic monitoring approaches dealing with the thermal analysis of an induction motor:

- measuring local temperatures
- measuring a bulk image or motor temperature.

The first approach, which is a kind of invasive technique, requires several thermocouples (see Figure 3.46) connected or installed at target part of the motor body including end windings, stator core, rotor core, housing and bearings, etc. This approach is based on a well-known effect called Seebeck which is the direct conversion of a difference in temperature between to materials to an electric signal. It literally means that an electrical equivalent value is assigned to heat existing at a joint of two wires with different materials (Figure 3.46(a)). Two metals face the same heat source while the rate of temperature increase is totally different as the materials are not the same. Considering that any temperature change leads to a flow

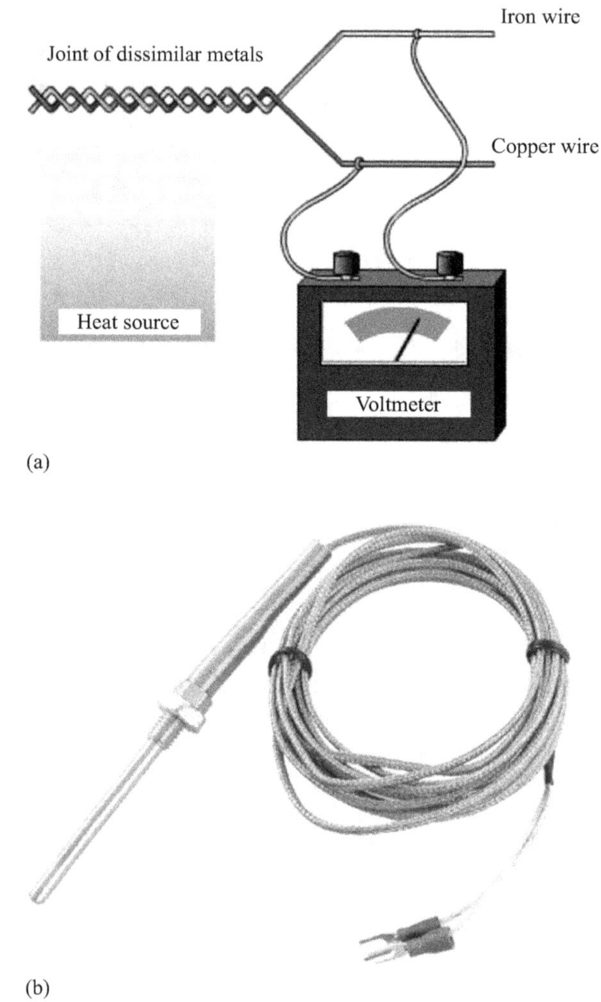


*Figure 3.46 Thermocouple (a) circuit and (b) real one*[5]

of electrons and assuming two materials with a great difference in their heat transfer capability, an electric voltage difference between two terminals in the measurement side takes place. This is what should be equivalently measured instead of real temperature difference. As thermocouples are very handy and small devices, they can be easily bonded on every part of motors body. This makes them one of the interesting ways of exact temperature monitoring.

Although thermocouples are very easy to use, their life time is somehow short compared to the other types of sensors discussed so far. The accuracy of the device

[5]http://www.kalkaheater.com/product/thermocouples (access date: October 2016).

might be affected by thermal tensions existing in the test environment. So calibrating the device and making sure if the device works very well should be one of the ongoing steps of using them.

Unlike the local thermal analysis, a global value indicating a total temperature change is sometimes assigned to the motor. The best candidate to reflect the change is the coolant temperature which is usually the air flowing in the air gap or close to it. So one can use a thermometer, regardless of type, to measure the time-dependent change of the motor temperature. This technique has nothing to do with an exact diagnosis procedure as the global temperature variation might be the product of any other unknown reason. Therefore, it is not proposed to go this way. Instead, utilizing a thermography camera is highly recommended (see Figure 3.47). It is a potential device to return a very discriminative temperature distribution of motors. In the case of stationary faults including the stator short circuit and also static eccentricity, it is expected to have a very clear representation of the fault location, and the detection is performed very well. However, if a rotating fault exists, it will not be easily detectable. As the thermal time constant is always larger than the magnetic time constant, incipient faults might not be detected by thermal analysis.

Considering the comments in terms of the motor signals and the corresponding sensing devices, all the sensors return electrical values which are relevant to the measured quantity. The products are electrical signals which should analyzed to conduct a diagnostic task. Therefore, another significantly important step is the so-called "data acquisition" which is introduced below. This step comes ahead of any signal processing and fault diagnosis step as the latter cases are impossible to be done unless a signal is in hand.

### *3.5.9 Data acquisition*

We prefer to introduce a very useful general application data acquisition (DAQ) device called "AdvanTech PCI-1710 HG" which enables us to sample several signals at the same time with difference qualities. A typical device is shown in Figure 3.48.

The specifications of the mentioned DAQ device are as follows:

- 16 single-ended or 8 differential or a combination of analog inputs
- 12-bit A/D converter, with up to 100-kHz sampling rate
- programable gain
- automatic channel/gain scanning
- onboard FIFO memory (4,096 samples)
- two 12-bit analog output channels (PCI-1710/1710HG only)
- 16 digital inputs and 16 digital outputs
- onboard programable counter
- board ID™ switch.

A very compelling aspect of the device is the number of input channels which is equal to 16 if a single-ended topology is used. On the other hand, if a noisy environment exists, the differential topology consisting of 8 channels is recommended.

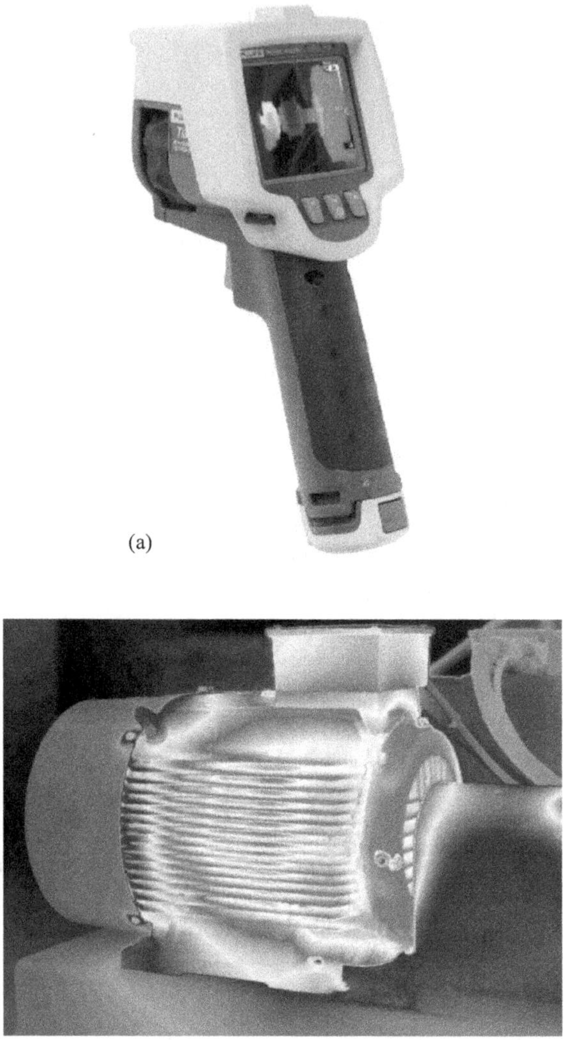

(a)

(b)

*Figure 3.47 (a) Thermography camera[6] and (b) motor temperature distribution[7]*

The analogue to digital convertor implemented in this device is capable of sampling at a maximum rate of 100 kHz indicating that a maximum sampling frequency of 6,250 Hz can be assigned to every channel in a single-ended topology. In the case of differential inputs, 12,500 Hz is the target. The corresponding sampling frequencies cannot be increased over the mentioned values unless

[6]http://www.conrad-electronic.co.uk/ce/en/product/100926/Fluke-Ti32-Infrared-Thermal-Imaging-Camera (access date: October 2016).

[7]http://iactthermography.org/ (access date: October 2016).

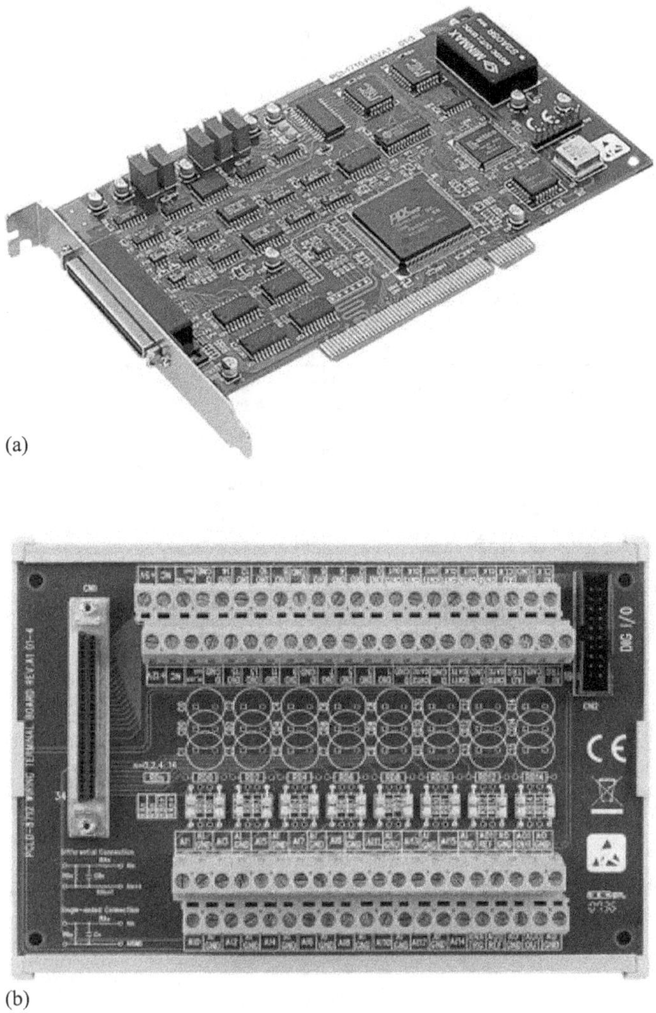


(a)

(b)

*Figure 3.48 AdvanTech PCI-1710 HG (a) main board and (b) terminal box*

some of the channels are free and not used. Then, the sampling frequency of used channels may increase if needed. The more the sampling frequency is, the more information preserved in a sampled signal will be. According to the Nyquist's law, the sampling frequency should be at least equal to twice the maximum frequency which should be included in the spectrum. Considering this fact and also taking the spectral resolution into account, it is suggested to take the safe side and go beyond the Nyquist's law and increase the sampling frequency as much as possible. The upper limit is usually forced by the data storage and also real-time analysis capability. Most of the time, it is favorable to have an online diagnosis technique which is able to detect incipient fault. In this case, there should be a

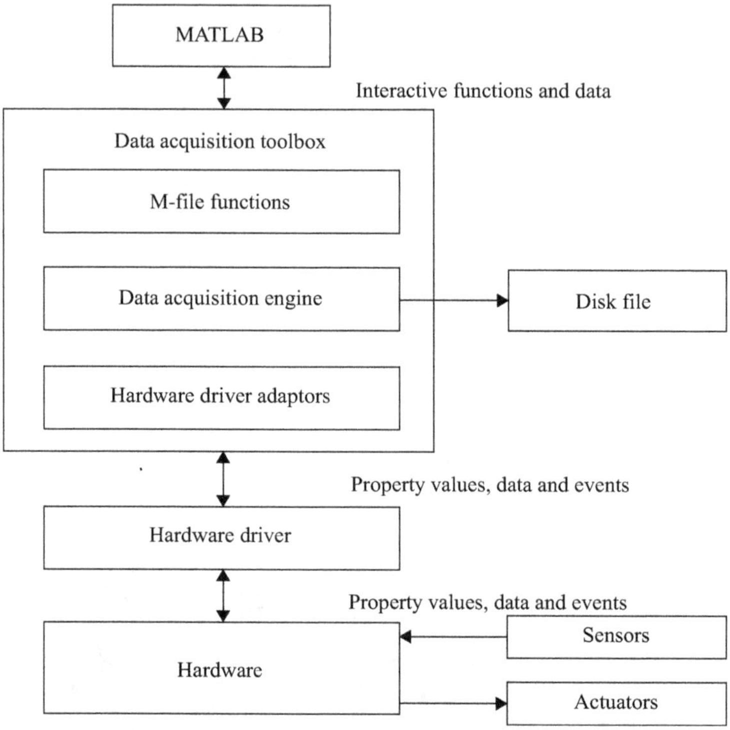


*Figure 3.49 MATLAB data acquisition routine*

trade-off between the accuracy and the required sampled data. In an offline case, users might even sample tons of data to increase the accuracy of the future investigations.

#### 3.5.9.1 MATLAB® code for an AdvanTech device

AdvanTech boards always come with a terminal box allowing us to connect the sensors outputs to the AdvanTech board installed inside a personal computer. The interface between the terminal box and the board is implemented by a 68-pin parallel connector shown in Figure 3.48(b). The whole system is controlled by the interfaces developed by the company or simply using MATLAB. Fortunately, MATLAB provides a fantastic DAQ interface not only for the AdvanTech products but also for a variety of other DAQ devices. Below is the routine which is used by MATLAB to input or output signals (see Figure 3.49). The interfaces taking care of MATLAB/AdvanTech interactions include m-files, data acquisition engines and hardware-driver adaptors. The last two aspects are themselves controlled by m-files or even SIMULINK® files.

This routine represents the possibility of inputting and outputting data while the focus of our work is on the inputting process. Assuming an inputting process, the steps are as follows:

- connect the sensors outputs to the terminal box
- connect the 68-pin parallel connector to the main AdvanTech board
- before going through defining a DAQ hardware, check if the hardware is available or not using the following command in the command line:

```
% show and save list of available and installed vendors
vendor=daq.getVendors
```

- open an m-file in a MATLAB environment and start typing the following code:

```
% Create a session for the required vendor
session = daq.createSession('advnatech')
% Add an input digital channel
ch1=addDigitalChannel(session,'DeviceID',
  'ChannelID', 'InputOnly');
% Specify channel specifications
ch1.TerminalConfig = 'Differential';
ch1.Range =[-10.0 10.0];
% Specify duration of acquisition
 session.DurationInSeconds = 2;
% Specify the sampling frequency or rate per second
session.Rate = 10000;
% Inquire about the limit of sampling rate determined by
  hardware
session.RateLimit
% start session
data = startForeground(s);
```

The steps have been explained as comments shown by percent symbols. Simply, one can call an AdvanTech Package and set the options according to the requirements and then start sampling input signals, using MATLAB. A real-time practice of data acquisition toolbox is also implementable by means of MATLAB/SIMULINK blocks. Moreover, the above-mentioned process shows a digital input scheme while the analogue one is also applicable if needed. For more details, readers are referred to the Mathwork website (*https://www.mathworks.com/help/daq/functionlist.html* (access date: October 2016)).

Some technical points while using any acquisition hardware or device should be taken into account:

- Do not touch the input terminals with a bare hand. Static electricity might decalibrate the device.

- Calibrate the device before using it.
- Scale the input signals applied to the terminal box to a measurable range of the DAQ device, i.e., [−10, 10] V for AdvanTech PCI-1710 HG. In the situation that inputs run over the mentioned range, the sampled signal's top will be cut off and look like a flat-topped curve. Furthermore, relatively large inputs, those which are higher than the tolerable range of the device, might lead to a complete device failure by introducing insulation failure.
- When connecting more than one input, if the number of inputs does not surpass the number of channels, assign distanced channels to inputs to reduce the possibility of loading effects of channels onto each other's signals. However, all the channels should be isolated in the manufacturing process.

What has been already discussed in terms of the DAQ devices is focused on a multitask hardware built by the AdvanTech Corporation. This device does not come with a set of sensors such as VTs or CTs, and one should first prepare the required sensors and then integrate them with a DAQ board. This means that although the provided measurement and acquisition setup components are very handy and useful, they might not perfectly match each other in terms of impedances, calibration, input/output ranges and so forth. So another comparable option is to use an integrated digital oscilloscope accompanied by a matched set of current and voltage probes. Tektronix is the best example of manufacturer of this kind. It provides high resolution multichannel digital oscilloscopes; moreover, very accurate current and voltage probes at different ratings are also available. The only drawback is related to the variety of type of probes (sensors) which does not cover the mechanical, magnetic and thermal sensors. So if one looks for a full control of the DAQ devices, the first approach is recommended while the second one is probably easier to use.

After being sampled, all the required motor signals are passed to a processor which can be a simple personal computer or a separate DSP such as those developed by the Texas Instruments company. The diagnosis procedure normally ends with the signal processing and fault indicator extraction step. This is a major step through the final goal of the fault diagnosis. Thus, it will be deeply discussed in Chapter 7. For now, it is assumed that the corresponding prerequisites are ready to go further through introducing an integrated fault diagnosis scheme below in the next subsection.

### *3.5.10 Overall scheme of the conventional cabled diagnosis system implementation*

Taking all the mentioned aspects of a practical implementation of fault diagnosis procedure into account, a general scheme shown in Figure 3.1 is introduced as the final solution for the experimental setup. In a few words, the following step-by-step procedure should be followed up to the end point which is extracting fault information, using a time, frequency or time-frequency analysis (Figure 3.50).

- If the motor is a wound rotor topology, short circuit the rotor windings.
- Connect the tested induction motor to the three-phase network/Drive, using a switch to take care of both the line-start and inverter-fed modes.

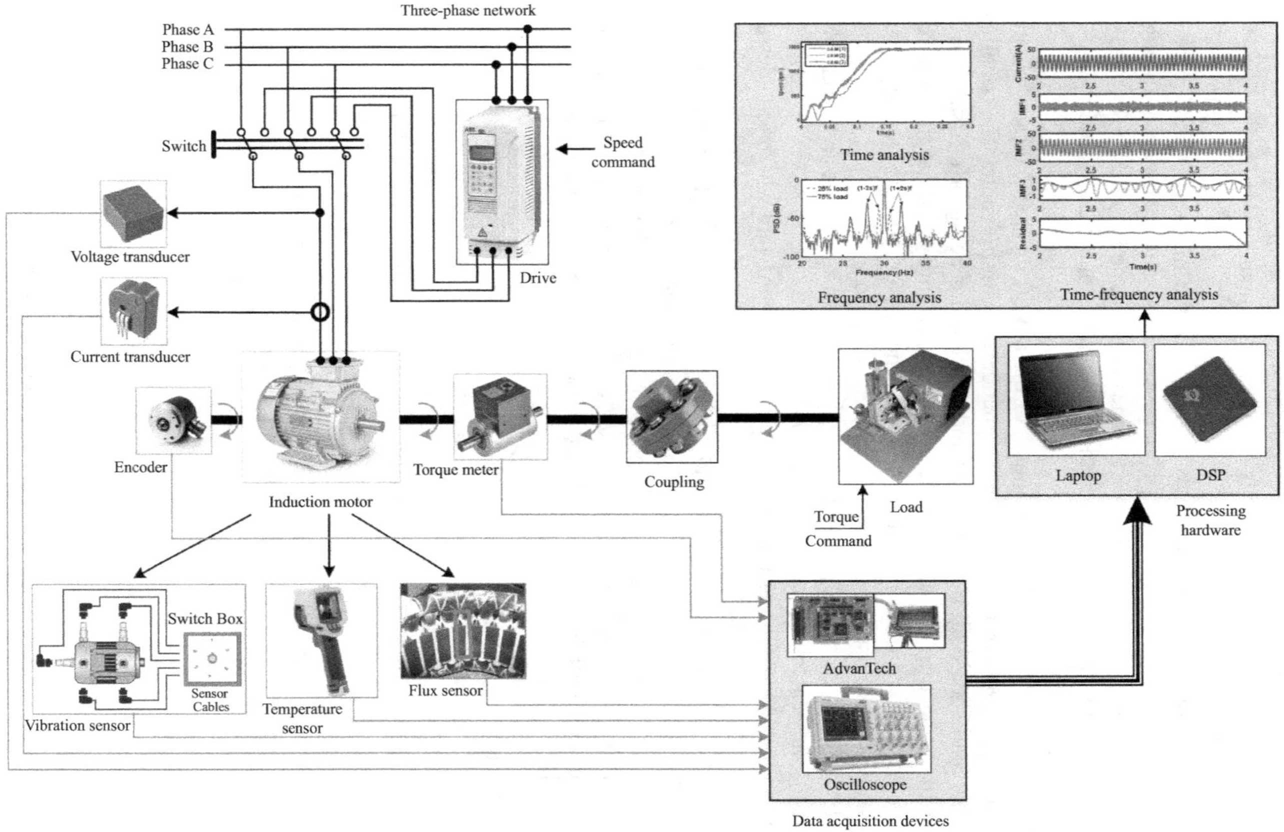


*Figure 3.50 Scheme diagram of a conventional cabled system implementation*

- The switch is used to switch between the line-start and inverter-fed modes.
- Connect the drive input to the three-phase network. One might also connect a three-phase choke to reduce the noise applied to the network.
- Connect the drives output terminals to the switch input terminals.
- Use shielded cables.
- Connect the switch output terminals to the motor terminals.
- Determine one of the phases as the diagnosis phase and then pass it through the CT. It is proposed to roll it several times around one leg of the CT to make sure the best primary coupling is made between the phase cable and the CT.
- Connect the test phase to the VT.
- Prior to connecting the electrical inputs of the motor, mechanical parts should be handled very well. It means that the encoder, the torque meter, the coupling and the load should all be connected and checked for being safe first.
- Prior to connecting the drive to the circuit, make sure to adjust the required initial parameters including the rise-time, the motor resistance, the motor inductance and most importantly the operating mode.
- Apply the temperature, flux and also vibration sensors as desired and then route all the sensing signals to the DAQ devices.
- Connect the DAQ devices outputs to a processing hardware which includes an ordinary CPU or a DSP.
- Obviously, the fault should be already applied to the motor. Furthermore, the speed and torque adjustments are supposed to be done.
- Finally, run the system and start sampling and analyzing the motor signals with the goal of extracting proper fault indicators. This step is the main focus of this book and will be further analyzed, discussed and explained in the next chapters.

One of the major aspects of the discussed scheme is the presence of a bunch of probably cumbersome wiring as all the signaling route is handled by means of cables. So if a long-distance analyzing center should be incorporated into the diagnosis procedure, this scheme, although provides a promising infrastructure, it will be somehow challenging in terms of signal routing. Instead, a wireless implementation of a condition monitoring system proposes a very tangible substitute for the conventional approach. The wireless condition monitoring is an almost new topic in the field, and not all researchers or even manufacturers have tried to address this approach. However, the existing literature clearly explains the idea.

### *3.5.11 Wireless condition monitoring setup*

Due to the recent developments achieved in the field of low-cost wireless sensors revealing a great ability in processing and communicating processed data, a couple of research attempts have been devoted to wireless implementation of a diagnosis setup. Consequently, online wired-monitoring systems, which indeed work well, might be totally replaced by wireless network alternatives. The main reason is probably the higher cost of shielded and isolated cable implementation while a wireless system does not require cabling at all. The only cost normally corresponds to the transmitters, receivers and likely the on-sensor processors if utilized. Another disadvantage associated with the conventional monitoring system is its inability to provide an opportunity for temporary or specialized condition monitoring practices.

Wireless sensor networks (WSN) commonly provide a cheaper system implementation along with a more flexible and relocatable monitoring system which does not restrict the location of system-utilizing technician. Nevertheless, no one can deny the existing shortcomings such as the potentially short range of operation and also possible electromagnetic interferences. On the other hand, WSNs have considerably constrained resources the significant of which is the associated battery life time. Regardless of its drawbacks, WSNs could definitely play a vital role in simplifying the diagnosis procedure.

In general, there are two kinds of topology for benefiting from a WSN:

- on-sensor processing
- off-sensor processing.

In the first topology, the signal processing, feature extraction and likely final fault diagnosis step are performed in the sensor itself. Therefore, sensors should provide an internally implemented microprocessor/microcontroller allowing processing the signals and then transmitting them to the monitoring unit. In the second topology, the sensor is just sensing and transmitting unprocessed signals to the receivers. Finally, the processing and feature extraction are handled by the monitoring unit which consists of a processing device such as DSPs. What is usually preferred is the first topology due to the fact that the processing unit is fixed and located close to the motor while the monitoring unit can be easily moving around the operating environment. If an off-sensor processing topology is used, the monitoring unit would probably be larger and more difficult to be flexibly moved. Besides, electromagnetic interference, if not dealt with properly, might harm the unprocessed signals and lead to a loss of valuable information during transmission of the signal. So the first topology is selected and shown in Figure 3.51.

For each sensor, one processing and transmitting unit is assigned while there can be only one receiver receiving all the data simultaneously by multiplexing among the inputs. The DAQ, signal processing and feature extraction is handled by a microcontroller or microprocessor, and the extracted features are sent to the transmitter. There are some types of microcontrollers such as Jennic JN5139 which is capable of multitasking. This means that they conduct all the tasks ranging from receiving sensor data to transmitting extracted features at the same time. So not only do they lead to a considerable save in the implementation space, but they also reduce the price of integrating devices. The proposed scheme in Figure 3.51 is applied to a line-start motor, but an inverter-fed application is also possible. To understand perfectly how a WSN works and is managed, a knowledge of microcontroller programing is also required. At this point, we do not aim at going through the details as it is not the main topic of this book, and it is assumed that readers have the required knowledge.

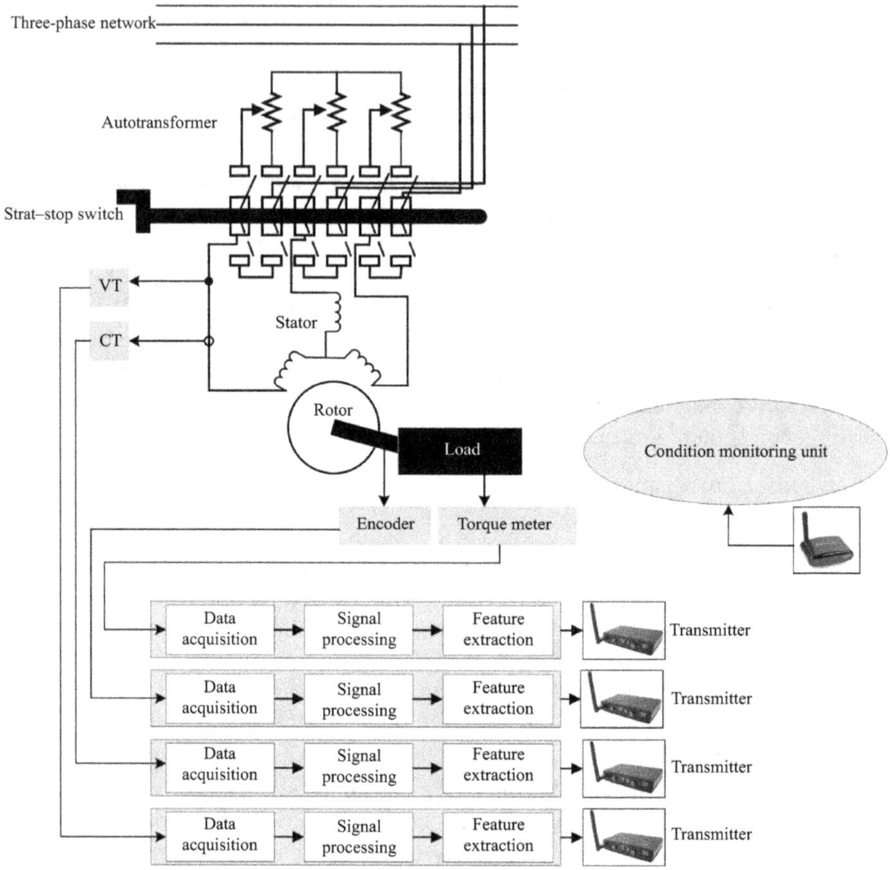


*Figure 3.51 Wireless sensor network-based condition monitoring setup*

A brief overview of the current chapter reveals the attempts to understand the fundamentals of fault occurrence and the way how fault affects the motor time-domain behavior. The attempts were totally devoted to explain the theoretical fault basis and consequently address the most significant time-domain variations by means of which a fault might probably be detected. Nevertheless, not all faults introduce a specifically detectable behavior, at least in a time domain. A step-by-step laboratory-scale implementation of various types of faults, along with the required details, was provided. By means of this knowledge, a simple but very useful setup of fault diagnosis based on which one can study different fault aspects is indeed achieved.

Bearing this in mind, in the next three chapters, we are going to target the best existing simulation approaches used for accurately modeling faults in induction machines. Not only the fundamental time harmonic components, but also all the possible spatial harmonic components are included and incorporated into the modeling process. The goal is actually to help researchers understand better how internal motor behaves, and it is not easily accessible by means of the existing experimental setups. For example, magnetic flux density at different parts of the motor or even the UMP is not the thing available at a cheap price in any laboratory or industry. Sometimes, it is even impossible unless an accurate modeling and simulation approach is hired. We believe no one's knowledge will get improved unless the upcoming information is studied and kept in mind. Therefore, it is highly proposed to study the following chapters first before attempting to read and understand the chapters coming after.

## References

[1] Momoh J. and Mili L. Dynamical models in fault tolerant operation and control of energy processing systems, *Operation and Control of Electric Energy Processing Systems*, 2010; Hoboken, NJ: Wiley-IEEE Press: 15–45.

[2] Anderson P.M. Sequence impedance of machines, *Analysis of Faulted Power Systems*, 1995; New York, NY: Wiley-IEEE Press: 183–230.

[3] Tavner P. Condition monitoring of rotating electrical machines, 2008, Stevenage: IET Press.

[4] Benbouzid M.E.H. A review of induction motors signature analysis as a medium for faults detection, *IEEE Transactions on Industrial Electronics*, 2000; **IE-47**: 984–993.

[5] Bellini A., Filippetti F., Tassoni C. and Capolino G.A. Advances in diagnostic techniques for induction machines, *IEEE Transactions on Industrial Electronics*, 2009; **IE-55**: 4109–4126.

[6] Erdman J.M., Kerkman R.J., Schlegel D.W. and Skibinski G.L. Effect of PWM inverters on AC motor bearing currents and shaft voltages, *IEEE Transactions on Industry Applications*, 1996; **IA-32**: 250–259.

[7] Thomson W.T. and Fenger M. Current signature analysis to detect induction motor faults, *IEEE Industry Applications Magazine*, 2001; **IAM-7**: 26–34.

[8] Kliman G.B., Koegl R.A., Stein J., Endicott R.D. and Madden M.W. Noninvasive detection of broken rotor bars in operating induction motors, *IEEE Transactions on Energy Conversion,* **EC-3**: 873–879.

[9] Bellini A., Filippetti F., Franceschini G., Tassoni C. and Kliman G.B. Quantitative evaluation of induction motor broken bars by means of electrical signature analysis, *IEEE Transactions on Industry Applications*, 2001; **IA-37**: 1248–1255.

[10] Filippetti F., Franceschini G., Tassoni C. and Vas P. Recent developments of induction motor drives fault diagnosis using AI techniques, *IEEE Transactions on Industrial Electronics*, 2000; **IE-47**: 994–1004.

[11] Jung J.H., Lee J.J. and Kwon B.H. Online diagnosis of induction motors using MCSA, *IEEE Transactions on Industrial Electronics*, 2006; **IE-53**: 1842–1852.

[12] Benbouzid M.E.H. and Kliman G.B. What stator current processing-based technique to use for induction motor rotor faults diagnosis? *IEEE Transactions on Energy Conversion*, 2003; **EC-2**: 238–244.

[13] Faiz J. and Ebrahimi B.M. A new pattern for detecting broken rotor bars in induction motors during start-up, *IEEE Transactions on Magnetics*, 2008; **MAG-44**: 4673–4683.

[14] Faiz J. and Jafari H. Accurate modeling of single-sided linear induction motor considers end effect and equivalent thickness, *IEEE Transactions on Magnetics*, 2000; **MAG-36**: 3785–3790.

[15] Ordaz-Moreno A., de Jesus Romero-Troncoso R., Vite-Frias J.A., Rivera-Gillen J.R. and Garcia-Perez, A. Automatic online diagnosis algorithm for broken-bar detection on induction motors based on discrete wavelet transform for FPGA implementation, *IEEE Transactions on Industrial Electronics*, 2009; **IE-55**: 2193–2202.

[16] Milimonfared J., Kelk H.M., Nandi S., Minassians A.D. and Toliyat H.A. A novel approach for broken-rotor-bar detection in cage induction motors, *IEEE Transactions on Industry Applications*, 1999; **IA-35**: 1000–1006.

[17] Hsu J.S. Monitoring of defects in induction motors through air-gap torque observation, *IEEE Transactions on Industry Applications*, 1995; **IA-31**: 1016–1021.

[18] Riera-Guasp M., Antonino-Daviu J.A., Pineda-Sanchez M., Puche-Panadero R. and Perez-Cruz J.A. General approach for the transient detection of slip-dependent fault components based on the Discrete wavelet transform, *IEEE Transactions on Industrial Electronics*, 2008; **IE-55**: 4167–4180.

[19] Faiz J., Ghorbanian V. and Ebrahimi B.M. EMD-based analysis of industrial induction motors with broken rotor bars for identification of operating point at different supply modes, *IEEE Transactions on Industrial Informatics*, 2014; **II-10**: 957–966.

[20] Bellini A., Filippetti F., Franceschini G. *et al.* On-field experience with online diagnosis of large induction motors cage failures using MCSA, *IEEE Transactions on Industry Applications*, 2022; **IA-38**: 1045–1053.

[21] Takbash A.M., Faiz J. and Ebrahimi B.M. Losses characterization in voltage-fed PWM inverter induction motor drives under rotor Broken bars fault, *IEEE Transactions on Magnetics*, 2013; **MAG-49**: 1516–1525.

[22] Faiz J., Ebrahimi B.M., Asaie B., Rajabioun R. and Toliyat H.A. A criterion function for broken bar fault diagnosis in induction motor under load variation using wavelet transform, 2007 International Conference on Electrical Machines and Systems (ICEMS), Seoul, 2007: 1249–1254.

[23] Ebrahimi B.M., Takbash A.M. and Faiz J. Losses calculation in line-start and inverter-fed induction motors under broken bar fault, *IEEE Transactions on Instrumentation and Measurement*, 2013; **IM-62**: 140–152.

[24] Faiz J., Ghorbanian V. and Ebrahimi B.M. A survey on condition monitoring and fault diagnosis in line-start and inverter-fed broken bar induction motors, 2012 IEEE International Conference on Power Electronics, Drives and Energy Systems (PEDES), Bengaluru, 2012,: 1–5.

[25] Abu-Elhaija W.S., Ghorbanian V., Faiz J. and Ebrahimi B.M. Impact of closed-loop control on behavior of inverter-fed induction motors with rotor

broken-bars fault, 2012 IEEE International Conference on Power Electronics, Drives and Energy Systems (PEDES), Bengaluru, 2012: 1–4.

[26] Faiz J., Ghorbanian V. and Ebrahimi B.M. A new criterion for rotor broken bar fault diagnosis in line-start and inverter-fed induction motors using Hilbert-Huang transform, 2012 IEEE International Conference on Power Electronics, Drives and Energy Systems (PEDES), Bengaluru, 2012: 1–6.

[27] Ghorbanian V. and Faiz J. A survey on time and frequency characteristics of induction motors with broken rotor bars in line-start and inverter-fed modes, *Mechanical Systems and Signal Processing*, 2015; **MSSP-54–55**: 427–456.

[28] Didier G., Ternisien E., Caspary O. and Razik H. Fault detection of broken rotor bars in induction motor using a global fault index, *IEEE Transactions on Industry Applications*, 2006; **IA-42**: 79–88.

[29] Schoen R.R., Habetler T.G., Kamran F. and Bartfield R.G. Motor bearing damage detection using stator current monitoring, *IEEE Transactions on Industry Applications*, 1995; **IA-31**: 1274–1279.

[30] Dorrell D.G., Thomson W.T. and Roach S. Analysis of airgap flux, current, and vibration signals as a function of the combination of static and dynamic airgap eccentricity in 3-phase induction motors, *IEEE Transactions on Industry Applications*, 1997; **IA–33**: 24–34.

[31] Cameron J.R., Thomson W.T. and Dow A.B. Vibration and current monitoring for detecting airgap eccentricity in large induction motors, *IEE Proceedings B – Electric Power Applications*, 1986; **EPA-133**: 155–163.

[32] Siddique A., Yadava G.S. and Singh B. A review of stator fault monitoring techniques of induction motors, *IEEE Transactions on Energy Onversion*, 2005; **EC-20**: 106–114.

[33] Ebrahimi B.M., Faiz J. and Roshtkhari M.J. Static-, dynamic-, and mixed-eccentricity fault diagnoses in permanent-magnet synchronous motors, *IEEE Transactions on Industrial Electronics*, 2009; **IE-56**: 4727–4739.

[34] Faiz J. and Tabatabaei I. Extension of winding function theory for nonuniform air gap in electric machinery, *IEEE Transactions on Magnetics*, 2002; **MAG-38**: 3654–3657.

[35] Faiz J., Ebrahimi B.M., Akin B. and Toliyat H.A. Finite-element transient analysis of induction motors under mixed eccentricity fault, *IEEE Transactions on Magnetics*, 2008; **MAG-44**: 66–74.

[36] Tabatabaei I., Faiz J., Lesani H. and Nabavi-Razavi M.T. Modeling and Simulation of a salient-pole synchronous Generator with dynamic eccentricity using modified winding function theory, *IEEE Transactions on Magnetics*, 2004; **MAG-40**: 1550–1555.

[37] Ebrahimi B.M., Javan Roshtkhari M., Faiz J. and Khatami S.V. Advanced eccentricity fault recognition in permanent magnet synchronous motors using stator current signature analysis, *IEEE Transactions on Industrial Electronics*, 2014; **IE-61**: 2041–2052.

[38] Faiz J., Ardekanei I.T. and Toliyat H.A. An evaluation of inductances of a squirrel-cage induction motor under mixed eccentric conditions, *IEEE Transactions on Energy Conversion*, 2003; **EC-18**: 252–258.

[39] Faiz F., Ebrahimi B.M. and Toliyat H.A. Effect of magnetic saturation on static and mixed eccentricity fault diagnosis in induction motor, *IEEE Transactions on Magnetics*, 2009; **MAG-45**: 3137–3144.

[40] Faiz J., Ebrahimi B.M., Akin B. and Toliyat H.A. Comprehensive eccentricity fault diagnosis in induction motors using finite element method, *IEEE Transactions on Magnetics*, 2009; **MAG-45**: 1764–1767.

[41] Blodt M., Granjon P., Raison B. and Rostaing G. Models for bearing damage detection in induction motors using stator current monitoring, *IEEE Transactions on Industrial Electronics*, 2008; **IE-55**: 1813–1822.

[42] Ojaghi M. and Faiz J. Extension to multiple coupled circuit modeling of Induction machines to include variable degrees of saturation effects, *IEEE Transactions on Magnetics*, 2008; **MAG-44**: 4053–4056.

[43] Thomson W.T., Rankin D. and Dorrell D.G. On-line current monitoring to diagnose airgap eccentricity in large three-phase induction motors-industrial case histories verify the predictions, *IEEE Transactions on Energy Conversion*, 1999; **EC-14**: 1372–1378.

[44] Faiz J., Ebrahimi B.M., Akin B. and Toliyat H.A. Dynamic analysis of mixed eccentricity signatures at various operating points and scrutiny of related indices for induction motors, *IET Electric Power Applications*, 2010; **EPA-4**: 1–16.

[45] Ojaghi M. and Faiz J. An experimental/simulation investigation to mixed eccentricity fault diagnosis of induction motors under DTC, 2014 IEEE International Conference on Industrial Technology (ICIT), Busan, 2014: 143–148.

[46] Abu-Elhaija W.S., Takbash A.M., Faiz J. and Ebrahimi B.M. Eccentricity fault diagnosis in induction motors using global processors — A review, *2012 IEEE International Conference on Power Electronics, Drives and Energy Systems (PEDES)*, Bengaluru, 2012: 1–6.

[47] Joksimovic G.M. and Penman J. The detection of inter-turn short circuits in the stator windings of operating motors, *IEEE Transactions on Industrial Electronics*, 2000; **IE-47**: 1078–1084.

[48] Gandhi A., Corrigan T. and Parsa L. Recent advances in modeling and online detection of stator interturn faults in electrical motors, *IEEE Transactions on Industrial Electronics*, 2011; **IE-58**: 1564–1575.

[49] Grubic S., Aller J.M., Lu B. and Habetler T.G. A Survey on testing and monitoring methods for stator insulation systems of low-voltage induction machines focusing on turn insulation problems, *IEEE Transactions on Industrial Electronics*, 2008; **IE-55**: 4127–4136.

[50] Schoen R.R., Lin B.K., Habetler T.G., Schlag J.H. and Farag S. An unsupervised, on-line system for induction motor fault detection using stator current monitoring, *IEEE Transactions on Industry Applications*, 1995; **IA-31**: 1280–1286.

[51] Thorsen O.V. and Dalva M. A survey of faults on induction motors in offshore oil industry, petrochemical industry, gas terminals, and oil refineries, *IEEE Transactions on Industry Applications*, 1995; **IA-31**: 1186–1196.

[52] Henao H., Demian C. and Capolino G.A. A frequency-domain detection of stator winding faults in induction machines using an external flux sensor, *IEEE Transactions on Industry Applications*, 2003; **EC-39**: 1272–1279.

[53] Su H. and Chong K.T. Induction machine condition monitoring using neural network modeling, *IEEE Transactions on Industrial Electronics*, 2007; **IE-54**: 241–249.

[54] Marques Cardoso A.J., Cruz S.M.A. and Fonseca D.S.B. Inter-turn stator winding fault diagnosis in three-phase induction motors, by Park's vector approach, *IEEE Transactions on Energy Conversion*, 1999; **EC-14**: 595–598.

[55] Bellini A., Filippetti F., Franceschini G. and Tassoni C. Closed-loop control impact on the diagnosis of induction motors faults, *IEEE Transactions on Industry Applications*, 2000; **IA-36**: 1318–1329.

[56] Arthur N. and Penman J. Induction machine condition monitoring with higher order spectra, *IEEE Transactions on Industrial Electronics*, 2000; **IE-47**: 1031–1041.

[57] Arkan M., Perovic D.K. and Unsworth P. Online stator fault diagnosis in induction motors, *IEE Proceedings – Electric Power Applications*, 2001; **EPA-148**: 537–547.

[58] Knight A.M. and Bertani S.P. Mechanical fault detection in a medium-sized induction motor using stator current monitoring, *IEEE Transactions on Energy Conversion*, 2005; **EC-20**: 753–760.

[59] Cruz S.M.A. and Cardoso A.J.M. Multiple reference frames theory: a new method for the diagnosis of stator faults in three-Phase induction motors, *IEEE Transactions on Energy Conversion*, 2005; **EC-20**: 611–619.

[60] Cruz S.M.A. and Cardoso A.J.M. Diagnosis of stator inter-turn short circuits in DTC induction motor drives, *IEEE Transactions on Industry Applications*, 2004; **IA-40**: 1349–1360.

*Chapter 4*

# Magneto-motive force waves in healthy three-phase induction motors

Alternating current (ac) windings in electrical machines basically have a twofold purpose. In electrical generators ac windings are place where electromotive force (EMF) should be induced. In electrical motors, primarily ac windings goal is to produce rotating magneto-motive force (MMF) wave. In any case, ac windings should be designed in such a manner that induced EMF or generated rotating MMF wave consists predominantly of the fundamental sinusoidal component. Therefore, the starting point of an induction machine study is the analysis of MMF waves in the air gap of such machines. This assumes knowledge of winding distribution along the air gap from stator as well as from rotor side.

The magnetic flux is analogous to the electric current. The MMF which sets up the magnetic flux is analogous to the EMF. The MMF is equivalent to a number of turns of wire carrying an electric current. If either the current through a coil (as in an electromagnet) or the number of turns of wire in the coil is increased, the MMF will be higher; and if the rest of the magnetic circuit remains the same, the magnetic flux increases proportionally [1].

At least two different approaches exist for describing winding distribution.

## 4.1 Current sheet concept

Suppose that exact winding distribution in stator and rotor slots can be taken into account and current sheet is defined. Let us analyze the simplest case where just one full-pitch coil with $N$ turns exists on the stator as shown in Figure 4.1(a). A full-pitch coil has pitch that is equal to the pole pitch, $y=\tau$. Let us assume that current $I$ flows through the coil. MMF in one slot is the product of number of conductors in that slot and current that is equal in all conductors. A current sheet, as the name implies, is a plane current flowing in one direction of the plane. By definition, current sheet $A$ is the ratio of the MMF of one slot, $F_S=N_SI$, and slot mouth opening $b$:

$$A=\frac{F_S}{b}=\frac{N_SI}{b}\,[\mathrm{A/m}] \tag{4.1}$$

where $N_S$ is the number of conductors in a slot ($N_S=N$ in this case). Current sheet distribution is depicted in Figure 4.1(b), as two rectangles of different signs as the

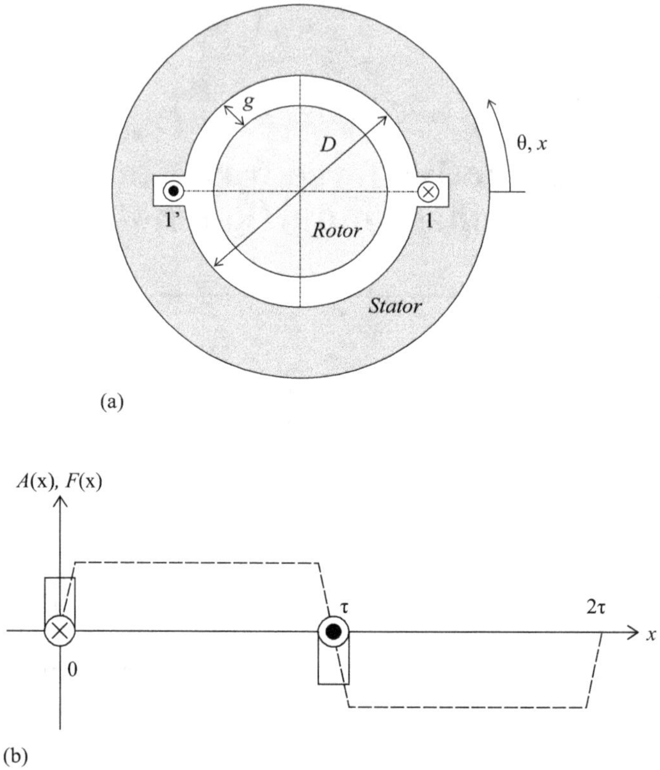


*Figure 4.1 (a) Cross-section view and (b) current sheet and MMF waveform around stator circumference in case of on full-pitch coil*

current in two slots is of different direction. The height of these two rectangles is given by (4.1). The MMF in the air gap due to the space distribution of windings (or current sheet) is also space-dependent. By definition, MMF on the place $x$ along the circumference of the machine is

$$F(x) = \int_0^x A(x)dx \tag{4.2}$$

where position $x = 0$ is defined as a place where current sheet distribution has maximal value. Previous expression was directly extracted from Ampere's circuital law. Trapezoidal MMF waveform along the machine circumference is also given in Figure 4.1(b). The peak value of MMF is obtained by substituting $x = b/2$ in (4.2):

$$F = \frac{NI}{2} \tag{4.3}$$

In the case where three series-connected identical coils (one of the phase windings) are placed in neighboring slots as illustrated in Figure 4.2(a), the current

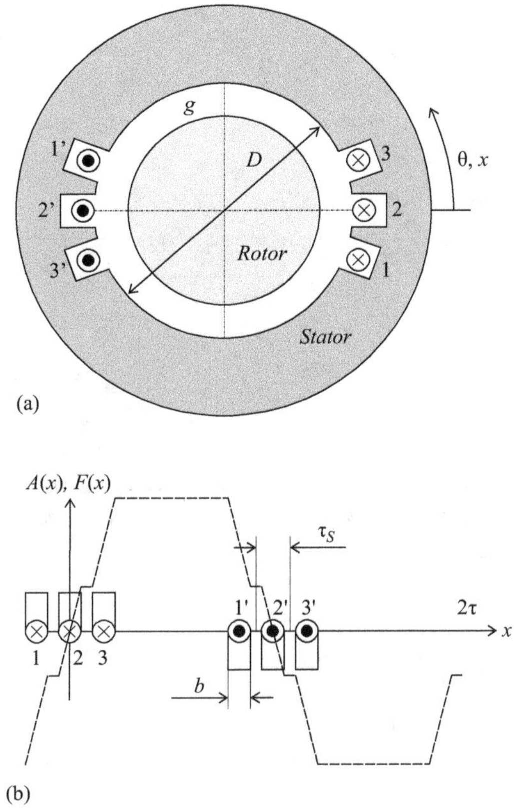


*Figure 4.2 (a) Cross-section view and (b) current sheet and MMF waveform along stator circumference for three full-pitch coils (one of the phase windings) connected in series*

sheet and MMF waves have somehow more complex shape (Figure 4.2(b)). Now, $N$ is the total number of turns per phase connected in series.

Current sheet could be alternatively defined as a ratio of slot MMF and slot pitch, and not slot mouth width. In this case, current sheet is

$$A = \frac{3N_S I}{3\tau_S} = \frac{N_S I}{\tau_S} = \frac{SN_S I}{\pi D}\,[\text{A/m}] \tag{4.4}$$

where $S$ is the number of stator slots and $\tau_S$ is the slot pitch or tooth pitch which are the same as displayed in Figure 4.3. Now the peak value of MMF wave for a machine with $p$ poles pair is as follows:

$$F = \int_0^{S\tau_S/2} A(x)dx = \frac{SN_S I}{\pi D}\int_0^{S\tau_S/2} dx = \frac{SN_S I}{\pi D}\frac{S\tau_S}{2} = \frac{SN_S I}{\pi D}\frac{1}{2}\frac{S}{2pm}\frac{\pi D}{S}$$
$$= \left(\frac{SN_S}{2m}\right)\frac{I}{2p} = \frac{NI}{2p} \tag{4.5}$$

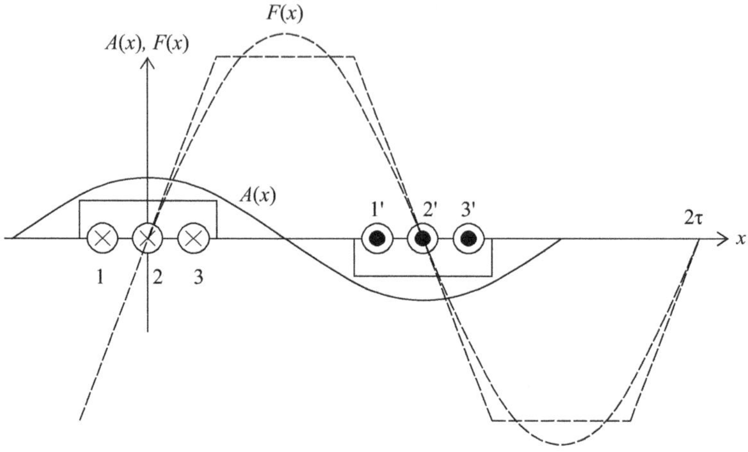


*Figure 4.3 Current sheet and MMF waveform along the stator circumference for three full-pitch coils connected in series (one of the phase windings)*

This is identical with (4.3), for $p = 1$, $N$ in (4.5) is the number of series turns per phase.

Obviously, current sheet and MMF waveform in Figure 4.3 are periodical. So, these functions can be expressed as the sum of a Fourier series elements. The fundamental harmonic approximation of both functions is given in Figure 4.3. These two harmonics are always shifted in space—MMF wave always leads one-half of pole pitch. Therefore, in general case, for fundamental harmonic, we have

$$A(x) = A_1 \cos\left(\frac{\pi}{\tau}x\right) \tag{4.6}$$

$$F(x) = \int_0^x A(x)dx = A_1 \frac{\tau}{\pi} \int_0^x \cos\left(\frac{\pi}{\tau}x\right) d\left(\frac{\pi}{\tau}x\right) = A_1 \frac{\tau}{\pi} \sin\left(\frac{\pi}{\tau}x\right)$$
$$= F_1 \sin\left(\frac{\pi}{\tau}x\right) \tag{4.7}$$

where

$$F_1 = \frac{\tau}{\pi} A_1 \tag{4.8}$$

Based on the following proportion,

$$x : 2\tau = \theta_{el} : 2\pi \tag{4.9}$$

$$\theta_{el} = \frac{\pi}{\tau} x \tag{4.10}$$

equation (4.6) can be given as a function of angular coordinate, electrical angle $\theta_{el} = p\theta_{meh}$:

$$A(\theta_{el}) = A_1 \cos(\theta_{el}) \tag{4.11}$$

$$F(\theta_{el}) = F_1 \sin(\theta_{el}) \tag{4.12}$$

**Remark 1:** *Fourier series consists of the sum of harmonic functions with different amplitudes, periods and phase positions. Amplitudes of those harmonic functions are coefficients of Fourier series. For functions that are periodic with fundamental period $2\pi$ and integrable over the range of $[-\pi,\pi]$ that coefficients are ($\nu$ is harmonic order; for $\nu = 0$ DC offset or mean value of function on that range is obtained):*

$$a_\nu = \frac{1}{\pi}\int_{-\pi}^{\pi} f(\theta)\cos(\nu\theta)d\theta, \quad \nu \geq 0$$

$$b_\nu = \frac{1}{\pi}\int_{-\pi}^{\pi} f(\theta)\sin(\nu\theta)d\theta, \quad \nu \geq 1$$

*The original $f(\theta)$ is as follows:*

$$f(\theta) = \frac{a_0}{2} + \sum_{\nu=1}^{\infty} [a_\nu \cos(\nu\theta) + b_\nu \sin(\nu\theta)]$$

*Often the exponential form of Fourier series is used which has a more concise form. Bearing in mind Euler's identity,*

$$e^{j\nu\theta} = \cos(\nu\theta) + j\sin(\nu\theta)$$

*Fourier series can be displayed in the following form,*

$$f(\theta) = \sum_{\nu=-\infty}^{\infty} c_\nu e^{j\nu\theta}$$

*where complex coefficient is:*

$$c_\nu = \frac{1}{2\pi}\int_{-\pi}^{\pi} f(\theta)e^{-j\nu\theta}d\theta$$

*The correlation between the previously introduced coefficients are given as follows:*

$$a_\nu = c_\nu + c_{-\nu} \quad \text{for} \quad \nu = 0, 1, 2, \ldots$$

$$b_\nu = j(c_\nu - c_{-\nu}) \quad \text{for} \quad \nu = 1, 2, \ldots$$

*In this chapter, the Fourier coefficients are obtained by utilizing the discrete Fourier transformation—FFT (Fast Fourier Transform) operator incorporated in software packet* MATLAB®.

**Example 4.1.** Find Fourier series of periodic function described graphically in Figure E4.1.1. This periodic function is often found in electrical machines.

Coefficient $a_\nu$ is

$$a_\nu = \frac{1}{\pi}\int_{-\pi}^{\pi} k(\theta)\cos(\nu\theta)d\theta = \frac{2}{\pi}\int_{-\pi/2}^{\pi/2} k(\theta)\cos(\nu\theta)d\theta$$

$$= \frac{2K}{\nu\pi}\int_{-\xi\pi/2}^{\xi\pi/2} \cos(\nu\theta)d(\nu\theta) = \frac{4K}{\nu\pi}\sin\left(\nu\xi\frac{\pi}{2}\right)$$

Coefficient $b_\nu$ is identically equal to zero.

Here, two different cases could be recognized regarding the value of coefficient $\xi$:

- For $\xi = 1$, coefficient $a_\nu$ is nonzero for odd-order harmonics, while for even-order, it is zero:

$$a_{\nu=1} = \frac{4}{\pi}K$$

$$a_{\nu=3} = -\frac{4K}{3\pi} = -\frac{1}{3}\frac{4}{\pi}K$$

$$a_{\nu=5} = \frac{4K}{5\pi} = \frac{1}{5}\frac{4}{\pi}K$$

In addition, the function has no DC offset:

$$a_{\nu=0} = 0$$

Hence, the original function could be described in the following manner,

$$k(\theta) = \frac{4}{\pi}K\cos(\theta) - \frac{1}{3}\frac{4}{\pi}K\cos(3\theta) + \frac{1}{5}\frac{4}{\pi}K\cos(5\theta) - \frac{1}{7}\frac{4}{\pi}K\cos(7\theta) + \cdots$$

and its development in Fourier series is depicted by Figure E4.1.2.

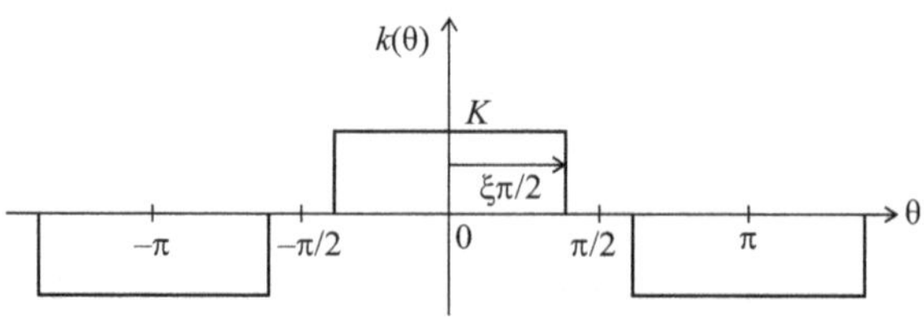


*Figure E4.1.1 Original function*

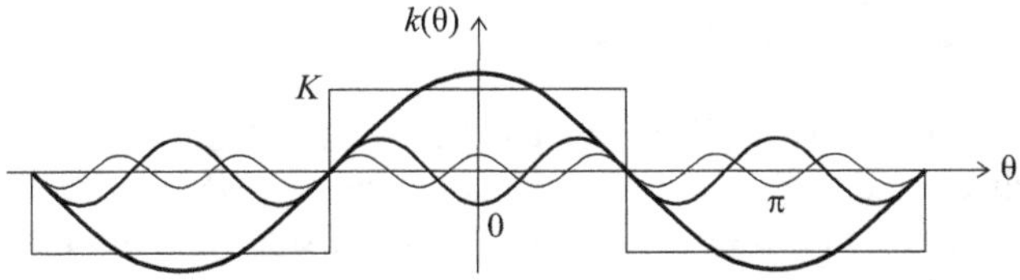


*Figure E4.1.2 Development of function in Fourier series for $\xi = 1$—fundamental, third and fifth harmonics*

- For $\xi < 1$, $a_\nu$ coefficient has the following form:

$$a_{\nu=1} = \frac{4}{\pi} \sin\left(\xi \frac{\pi}{2}\right) K$$

$$a_{\nu=3} = \frac{4}{\pi}\frac{1}{3} \sin\left(3\xi \frac{\pi}{2}\right) K$$

$$a_{\nu=5} = \frac{4}{\pi}\frac{1}{5} \sin\left(5\xi \frac{\pi}{2}\right) K$$

Moreover, the function has no DC offset:

$$a_{\nu=0} = 0$$

So the original function could be described as follows:

$$k(\theta) = \frac{4}{\pi} \sin\left(\xi \frac{\pi}{2}\right) K \cos(\theta) + \frac{4}{\pi}\frac{1}{3} \sin\left(3\xi \frac{\pi}{2}\right) K \cos(3\theta) + \frac{4}{\pi}\frac{1}{5} \sin\left(5\xi \frac{\pi}{2}\right) K \cos(5\theta) + \cdots$$

A factor occurred, along with the $\nu$th harmonic amplitude, is the so-called pitch factor ($\xi$ is equivalent to the shortening of the coil span):

$$k_{p\nu} = \sin\left(\nu\xi \frac{\pi}{2}\right) = \sin\left(\nu \frac{y}{\tau}\frac{\pi}{2}\right)$$

Therefore, in general, the harmonic content for $\xi < 1$ is identical with that of $\xi = 1$. However, for some specific value of $\xi$, some of the higher odd-order harmonics could be even totally eliminated. For example, for $\xi = 2/3$, the third, as well as all higher odd harmonics divisible by three (the so-called triplen harmonics), is identically equal to zero as the pitch factor for those harmonic components is equal to zero:

$$k_{p,\nu=triplen} = \sin\left(3\frac{2}{3}\frac{\pi}{2}\right) = \sin\left(9\frac{2}{3}\frac{\pi}{2}\right) = \sin\left(15\frac{2}{3}\frac{\pi}{2}\right) = \ldots = 0$$

## 4.2 Winding function concept

The winding function theory can be traced back to [2]. Contrary to the classical *d-q* model, this theory can take into account all winding MMF space harmonics in machines with small and uniform air gap. Now consider the simplest form of the stator phase winding which is a single concentrated coil with $N$ turns. Positions of coil's sides along the stator circumference are described by mechanical angles $\theta_1$ and $\theta_2$, i.e., $\alpha = \theta_2 - \theta_1$ which is coil pitch in mechanical radians as depicted in Figure 4.4. Assuming uniform small air gap length, neglecting the stator and rotor slots, assuming infinitely permeable iron as well as adopting point conductor approximation, Ampere's circuital law in common shape is as the following:

$$\oint_C \mathbf{H} \cdot d\mathbf{l} = \int_S \mathbf{J} \cdot d\mathbf{S} \tag{4.13}$$

and it's much simpler form is

$$H(\theta) \cdot g - H(0) \cdot g = n(\theta) \cdot i \tag{4.14}$$

where $H(\theta)$ is the radial component of magnetic field intensity in the air gap at position described by angle $\theta$, $g$ is the air gap length, $i$ is the coil current and $n(\theta)$ is the turn function [3]. The assumption of small air gap length means that the rotor radius is significantly larger than air gap length. On the other hand, this also means that tangential component of the magnetic field strength in the air gap is negligible compared with its radial component [4,5]. In expression (4.13), $C$ is contour 1-2-3-4 and $S$ is the area spanned by that contour.

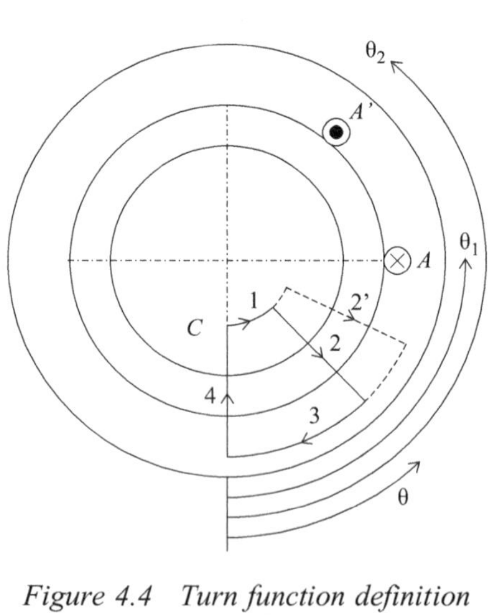


*Figure 4.4 Turn function definition*

From Figure 4.4, it is clear that by continuously changing the position $\theta$ of side 2 of the closed-loop $C$, turn function could be defined as follows:

$$n(\theta) = \begin{cases} N & \theta_1 < \theta \leq \theta_2 \\ 0 & \text{for other } \theta \end{cases} \tag{4.15}$$

By employing the Gauss law, we have

$$\oint_S \mathbf{B} \cdot d\mathbf{S} = 0 \tag{4.16}$$

and analyzing only electromagnetic processes in the air gap yields,

$$\mu_0 \oint_S \mathbf{H} \cdot d\mathbf{S} = 0 \tag{4.17}$$

or

$$\int_0^l \int_0^{2\pi} H \cdot r d\theta dz = 0 \tag{4.18}$$

Assuming that $H$ does not depend on the axial coordinate $z$, (4.18) leads to:

$$\int_0^{2\pi} H \cdot d\theta = 0 \tag{4.19}$$

Considering of (4.14) and (4.19) gives:

$$\int_0^{2\pi} \left[ H(0) + n(\theta) \cdot \frac{i}{g} \right] d\theta = 0 \tag{4.20}$$

$$H(0) \cdot 2\pi + 2\pi \frac{i}{2\pi g} \int_0^{2\pi} n(\theta) d\theta = 0 \tag{4.21}$$

$$H(0) = -\langle n(\theta) \rangle \cdot \frac{i}{g} \tag{4.22}$$

where $\langle n(\theta) \rangle$ is the mean value of the turn function. (4.14) and (4.22) yield,

$$H(\theta) \cdot g = [n(\theta) - \langle n(\theta) \rangle] \cdot i \tag{4.23}$$

where $H(\theta) \cdot g$ is MMF, and the last equation can be written as follows,

$$mmf = F(\theta) = [n(\theta) - \langle n(\theta) \rangle] \cdot i = N(\theta) \cdot i \tag{4.24}$$

where

$$N(\theta) = n(\theta) - \langle n(\theta) \rangle \tag{4.25}$$

is winding function [3]. Obviously, winding function is MMF per unit current. Henceforth, the terms "winding function" and "MMF per unit current" will be used interchangeably. Winding function could be alternatively written as follows:

$$N(\theta) = \begin{cases} N\left(1 - \dfrac{\alpha}{2\pi}\right) & \theta_1 < \theta \leq \theta_2 \\ -N\dfrac{\alpha}{2\pi} & \text{for rest of } \theta \end{cases} \tag{4.26}$$

Since infinite permeable iron is assumed, the principle of superposition could be adopted. So turn function or winding function of one-phase winding is simply the sum of adequate functions of coils that are constituents of the phase winding.

### *4.2.1 Concentrated full-pitch coil MMF*

For an $N$-turns full-pitch coil in a two-pole machine ($p = 1$) and for the adopted system of reference, based on Figure 4.5(a), turns and winding functions are as the following:

$$n_A(\theta) = \begin{cases} N & 0 < \theta \leq \pi \\ 0 & \text{for other } \theta \end{cases} \tag{4.27}$$

$$N_A(\theta) = \begin{cases} \dfrac{N}{2} & 0 < \theta \leq \pi \\ -\dfrac{N}{2} & \text{for other } \theta \end{cases} \tag{4.28}$$

Figure 4.5 illustrates the placement of the coil along the stator circumference, turn function profile, winding function profile as well as coefficients of the Fourier's series of winding function that are amplitudes of constituent space harmonics. Obviously, only odd higher space-harmonics are present in Fourier series of winding function. The fundamental space-harmonic amplitude is 6.37 turns (Figure 4.5(b)), i.e., $4/\pi$ higher than the winding function value, so the fundamental MMF wave amplitude is as follows:

$$F_{A\,\max\,1} = \frac{4}{\pi}\frac{N}{2}I_{\max} \tag{4.29}$$

where $I_{\max}$ is the amplitude of current passing the coil. Higher space-harmonic of order $\nu$ has $\nu$ times smaller amplitude than the fundamental one:

$$F_{A\,\max\,\nu} = \frac{4}{\pi}\frac{1}{\nu}\frac{N}{2}I_{\max} \tag{4.30}$$

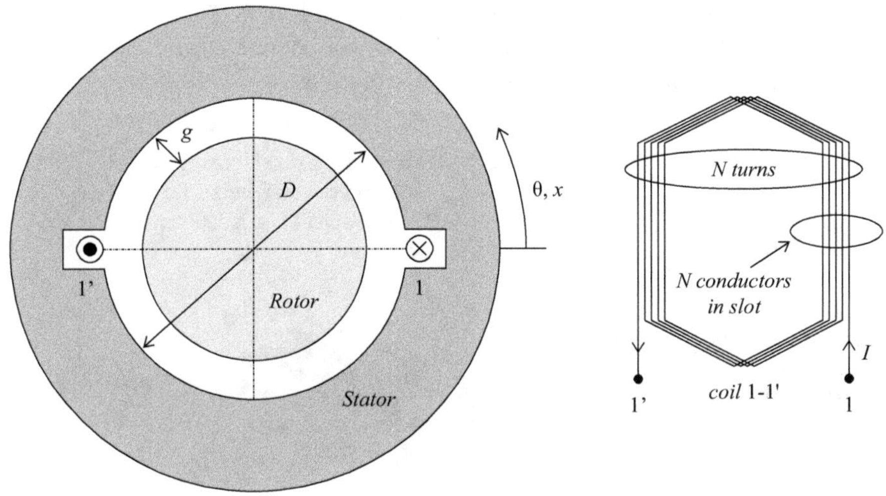


(a)

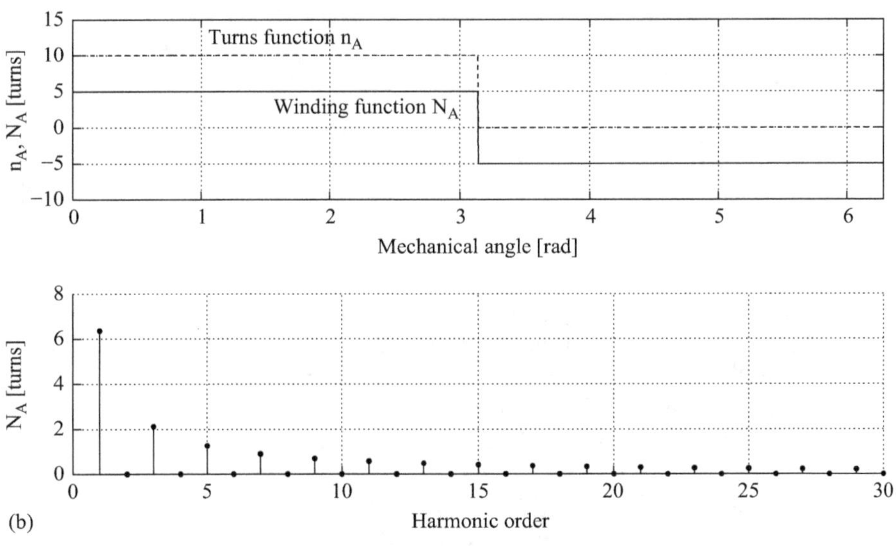


(b)

*Figure 4.5 (a) Crosssection view and (b) turn function, winding function and Fourier coefficients of winding function for a full-pitch coil in a two-pole machine (p = 1). Coil has N = 10 turns*

Turns and winding function for full-pitch coil as well as Fourier transform of winding function given in Figure 4.5 could be easily defined using the following short MATLAB script:

```
H=1024;                          % number of samples in 2*pi
dx=2*pi/H;                        % discretization step
nn=10;                            % number of turns in a coil
% na–turns function of phase (coil) A
i=1;
for teta=0:dx:2*pi-dx,
if((teta>0)&(teta<=pi)),
na(i)=nn;
else
na(i)=0;
end;
i=i+1;
end;
clear i

% Na–winding function of phase (coil) A

Na=na-mean(na);

% Fourier expansion

F_series=2*abs(fft(Na)/H);
stem(F_series(2:30),'filled')
```

In this section, the following two facts should be considered: (1) in order to obtain the exact discrete Fourier transform, the number of samples in $2\pi$ rad must be $H=2^n$ where $n$ is an integer and (2) the first element in Fourier transform is the coefficient equal to the mean value of the winding function. However, in this case, the number is already equal to zero, this element is not shown in Figure 4.1(b).

From Figures 4.1(b) and 4.5(b), it is apparent that the waveforms of MMF that are the result of the integration of current sheet and winding function are similar. The main difference is in the fact that the winding function has step rise in the center of the slot which is the result of earlier adopted point conductor approximation. On the other hand, the MMF has linear rise across the slot. However, definition of winding function could be easily adopted in order to take the linear rise of the MMF across the slot into account [6].

Figure 4.6(a) displays the crosssection view of the four-pole machine as well as turns and winding function for a full-pitch coil winding. Two full-pitched coils (in a four-pole machine, the full-pitch coil has a pitch $\alpha_{mech}=\pi/2$ rad) are

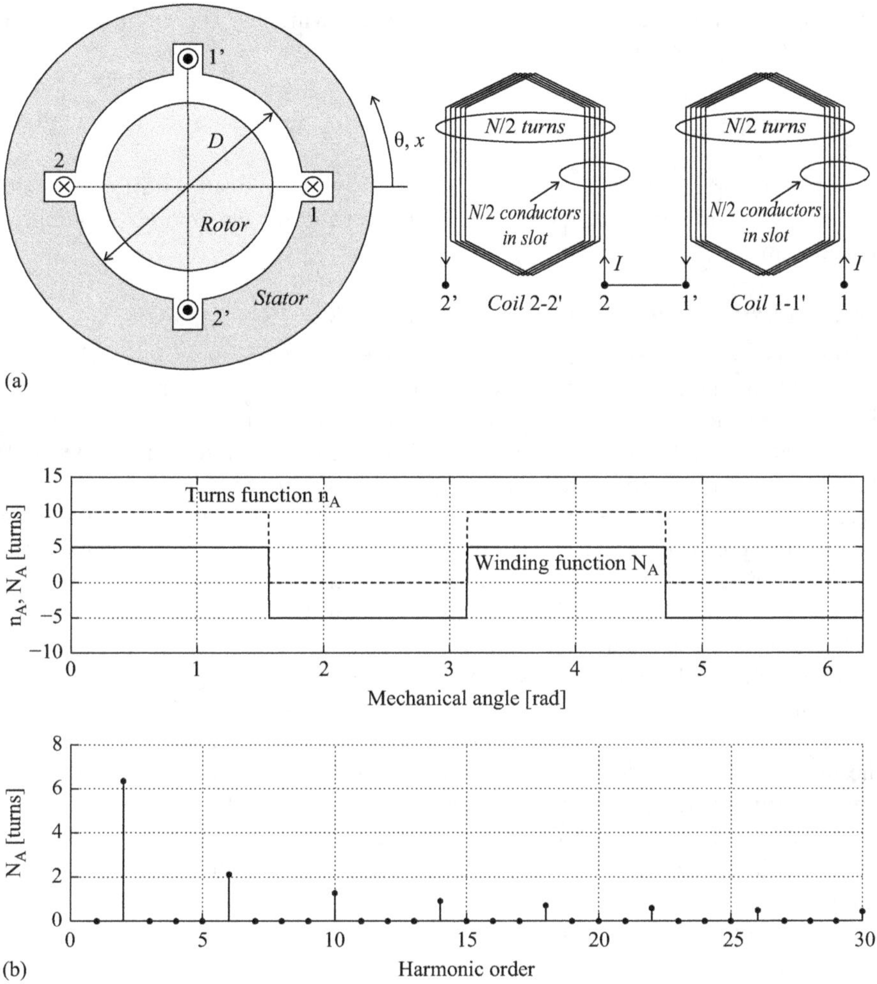


*Figure 4.6 (a) Cross-section view and (b) turns function, winding function and Fourier expansion of winding function of two series-connected full-pitch coils (1-1′ and 2-2′) in a four-pole machine ($p = 2$). Every single coil has 10 turns*

connected in series. The total turns and winding function of phase winding is simply the sum of turns and winding functions of every single coil in the winding. Now, the fundamental harmonic is second-order harmonic ($p$), the third harmonic is sixth-order harmonic ($3p$), etc. The amplitude of the fundamental MMF wave now is

$$F_{A\,\max 1} = \frac{4}{\pi}\frac{N}{2p}I_{\max} \tag{4.31}$$

where $N$ is the total number of turns in the phase winding. The amplitude of the higher order space harmonic components of $\nu$ is

$$F_{A\max\nu} = \frac{4}{\pi}\frac{1}{\nu}\frac{N}{2p}I_{\max} \tag{4.32}$$

In the above cases, turn and winding functions have been dealt with for the stator windings. All derived conclusions, however, are valid for the coils or windings in the rotor slots.

### 4.2.2 Distributed full-pitch phase winding

Fourier analysis showed that the air gap MMF produced by a full-pitch coil(s) consists of fundamental space-harmonic components as well as a series of all odd higher order space harmonic components. In the design of the ac windings, serious attempts are made to distribute the coils of the windings in such a manner as to minimize the higher order harmonic components and to produce an air-gap MMF wave consisting predominantly of the space-fundamental sinusoidal component. In this way, the machine is better utilized. Therefore, in practice, ac winding of induction machine is always designed as a distributed winding. Distributed phase winding of a three-phase machine occupies one-third of the stator slots along the circumference of the machine.

Figure 4.7(a) depicts such a one-phase winding in a two-pole machine when the stator has $S = 18$ slots. It means that one-phase winding (three series-connected full-pitch coils) occupies six slots, three under one and three under the other pole. The turn and winding function of phase $A$ winding could be easily obtained by summing turns and winding functions of individual coils. As Figure 4.7(b) illustrates, the resultant MMF profile of the distributed full-pitch winding has trapezoid shape which is much closer to the sinusoid than before (compare Figures 4.5(b) and 4.7(b)). This can also be concluded from the ratio of space harmonics amplitudes and fundamental harmonic amplitude. However, the amplitude of the resultant winding function is not three times greater than before, but somewhat smaller as it could be easily concluded observing the coefficients of the Fourier series. The reason for that is in the space displacement of individual coils. The winding distribution factor $k_d$, also known as a breadth factor, is ratio of the resultant MMF space-harmonic amplitude and sum of the MMF space-harmonics amplitudes of the constituent coils (ratio of vector and algebraic sum of constituent coil's MMFs):

$$k_{d\nu} = \frac{F_{A\max\nu}}{F_{A_1A_1'\max\nu} + F_{A_2A_2'\max\nu} + \cdots + F_{A_nA_n'\max\nu}} = \frac{F_{A\max\nu}}{n \cdot F_{coil\max\nu}} \tag{4.33}$$

Referring to Figures 4.5(b) and 4.7(b), this factor for the fundamental space harmonic ($\nu = 1$) is

$$k_{d1} = \frac{F_{A\max 1}}{3 \cdot F_{coil\max 1}} = \frac{18.3344}{3 \cdot 6.3662} = 0.96$$

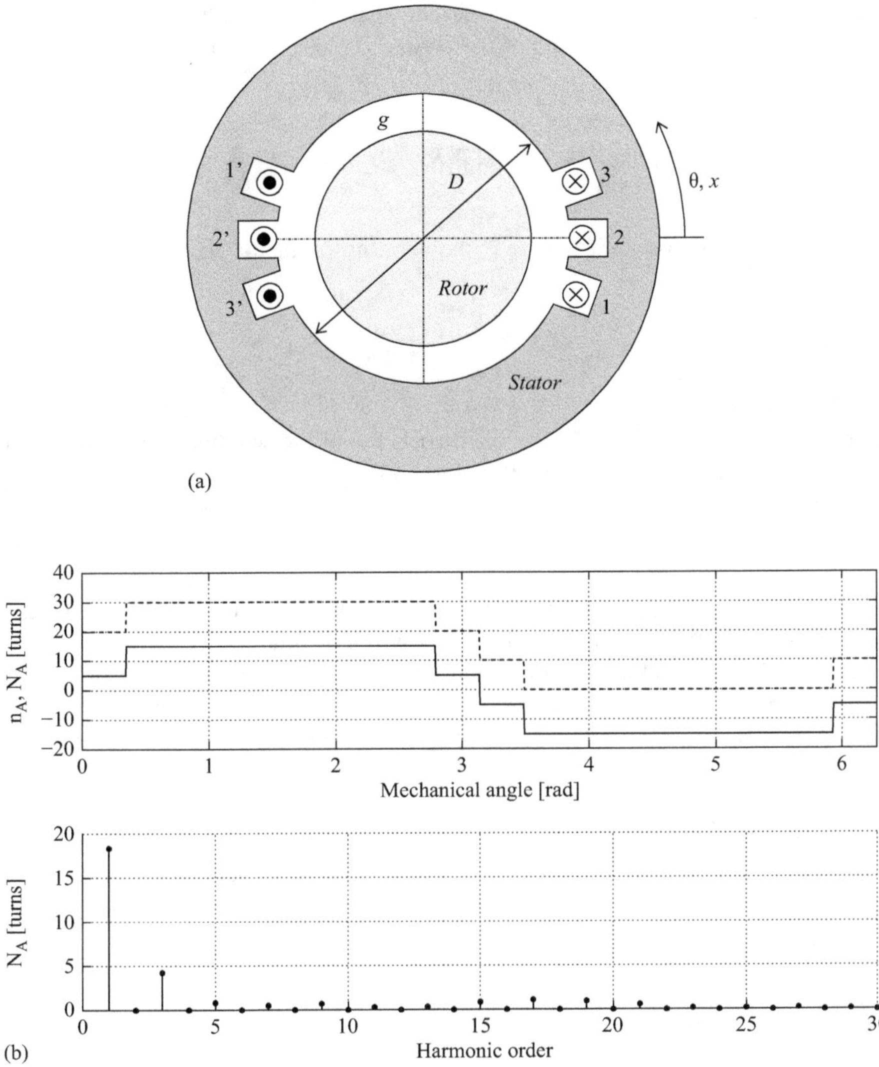


*Figure 4.7 (a) Cross-section view and (b) turns function, winding function and the Fourier expansion of distributed winding consist of three series-connected full-pitch coils (1-1′, 2-2′ and 3-3′) in a two-pole (p = 1) machine. Every single coil has $N_{coil}$ = 10 turns, so N = 30*

This is in good agreement with the factor obtained from the well-known equation for distribution factor [7]:

$$k_d = \frac{\sin\dfrac{z\theta_{el}}{2}}{z\sin\dfrac{\theta_{el}}{2}} \tag{4.34}$$

where $z$ is the number of slots per pole per phase and $\theta_{\text{el}}$ is the electrical angle between two slots, $\theta_{\text{el}} = p\theta_{\text{mech}}$. For the analyzed phase winding, $k_d$ is as follows:

$$k_d = \frac{\sin\frac{z\theta_{el}}{2}}{z\sin\frac{\theta_{el}}{2}} = \frac{\sin\frac{3\cdot 20^\circ}{2}}{3\sin\frac{20^\circ}{2}} = \frac{\sin 30^\circ}{3\sin 10^\circ} = 0.9598$$

For higher space harmonic of order $\nu$, the distribution factor is

$$k_{d\nu} = \frac{\sin\frac{z\nu\theta_{el}}{2}}{z\sin\frac{\nu\theta_{el}}{2}} \tag{4.35}$$

Considering the distribution factor, the distributed phase winding MMF amplitude for $\nu$th space harmonic could be defined as follows:

$$F_{A\max\nu} = \frac{4}{\pi}\frac{1}{\nu}\frac{k_{d\nu}N}{2p}I_{\max} \tag{4.36}$$

From Figures 4.7(b) and 4.2(b), it can be seen that the waveforms of MMF calculated by integration of current sheet and winding function are similar. Again, the main difference is that winding function has step rise in the center of the slot which is the result of earlier adopted point conductor approximation. On the other hand, MMF obtained from the integration of current sheet has linear rise across the slot.

### 4.2.3 Pulsating MMF

Winding function of $N$ turns full-pitch coil in two-pole machine $p = 1$ shown in Figure 4.5(b), could be resolved in Fourier series. As it is illustrated before, Fourier series is a sum of all odd harmonics

$$N(\theta) = \sum_{\nu=1,3,5,\ldots} N_{\nu\max}\sin(\nu\theta) \tag{4.37}$$

where

$$N_{\nu\max} = \frac{4}{\pi}\frac{1}{\nu}\frac{N}{2} \tag{4.38}$$

Neglecting all higher harmonics, this winding function could be approximated with the following expression:

$$N(\theta) = N_{1\max}\sin\theta \tag{4.39}$$

where $N_{1\max}$ is the amplitude of the fundamental harmonic:

$$N_{1\max} = \frac{4}{\pi}\frac{N}{2} \tag{4.40}$$

The same is true for Figure 4.7(b), namely (4.37) is valid also for this case, but the amplitudes of the constituent harmonics are something different,

$$N_{\nu\ \max} = \frac{4}{\pi}\frac{1}{\nu}\frac{k_{d\nu}N}{2} \tag{4.41}$$

where $N$ is the series turns number per phase (in all three coils) and $k_{d\nu}$ is the distribution factor. As it has been already mentioned, the winding function is MMF per unit current. So the waveforms of winding function in Figures 4.5(b), 4.6(b) and 4.7(b) is simultaneously the waveforms of MMFs if a dc of 1A passes the winding. If the dc rises to 2 A, MMF waveforms will be the same as winding function waveforms, but with double higher stairs. If ac with amplitude of 2A and frequency of 50 Hz passes the winding, the MMF will pulsate. When the current is maximal, MMF has the shape as sketched for winding functions. When the current is zero, MMF is also zero. If the current is maximal in other direction, MMF waveform is the mirror image of winding function. From the above explanation, it is clear that MMF in this case is pulsating. Figure 4.8 illustrates the waveforms of the pulsating MMF for the case of 3 turns full-pitch winding and ac of 2A amplitude. Therefore, MMF depends on two variables: angular coordinate and time.

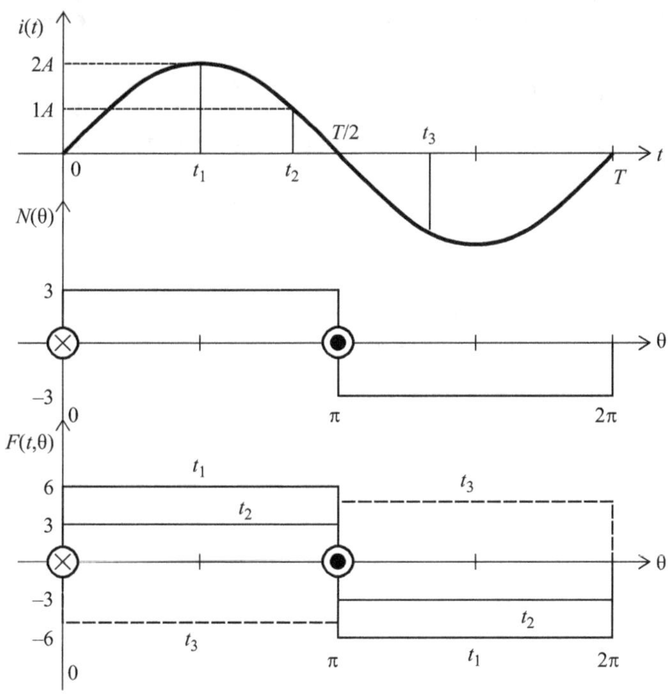


*Figure 4.8 Pulsating MMF: N = 3 turns full-pitch coil with ac of 2A amplitude*

Taking only the fundamental harmonic of winding function into account, pulsating MMF is as follows:

$$F_1(t,\theta) = N(\theta)\cdot i(t) = N_{1\,\max}\sin\theta\cdot I_{\max}\sin(\omega t) \tag{4.42}$$

Using trigonometric transformation (4.42) can be written as sum of two sinusoidal functions:

$$F_1(t,\theta) = \frac{N_{1\,\max}I_{\max}}{2}\sin(\omega t-\theta) + \frac{N_{1\,\max}I_{\max}}{2}\sin(\omega t+\theta) \tag{4.43}$$

$$F_1(t,\theta) = \frac{F_{1\,\max}}{2}\sin(\omega t-\theta) + \frac{F_{1\,\max}}{2}\sin(\omega t+\theta) = F_d(t,\theta)+F_i(t,\theta) \tag{4.44}$$

It means that the pulsating MMF can be resolved into two rotating MMF waves. One of them, called forward wave or direct wave (index *d*), rotates in anticlockwise direction and the other one, called backward or inverse wave (index *i*), rotates in opposite direction, as illustrated in Figures 4.9 and 4.10.

What is the speed of these waves in space? Let us look at forward wave. In different times, this wave will have fixed maximal value in different points along the stator circumference:

$$F_d(t,\theta) = \frac{F_{1\,\max}}{2}\sin(\omega t-\theta) = \frac{F_{1\,\max}}{2}\sin(\omega t-\omega_{\text{meh}}t) = \text{const}$$

$$\Rightarrow (\omega-\omega_{\text{meh}})t = \text{const} \tag{4.45}$$

Therefore, the mechanical angular speed of direct MMF wave, also known as *synchronous speed*, is:

$$\omega_{meh} = \omega_s = \omega = 2\pi f\left[\frac{\text{rad}}{\text{s}}\right] \tag{4.46}$$

That is also true of the speed of backward wave. In general case, for a machine with *p* pair of poles, the synchronous speed is:

$$\omega_s = \frac{2\pi f}{p}\left[\frac{\text{rad}}{\text{s}}\right] \tag{4.47}$$

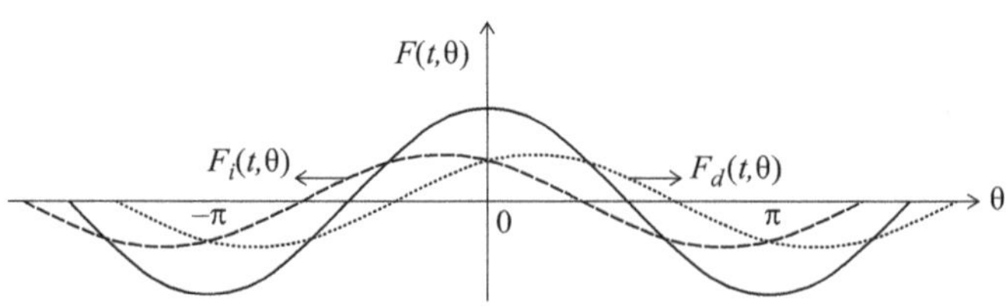


*Figure 4.9 Pulsating MMF at any point in space and any time is the sum of two rotating MMF waves with equal amplitudes: forward and backward rotating MMF waves*

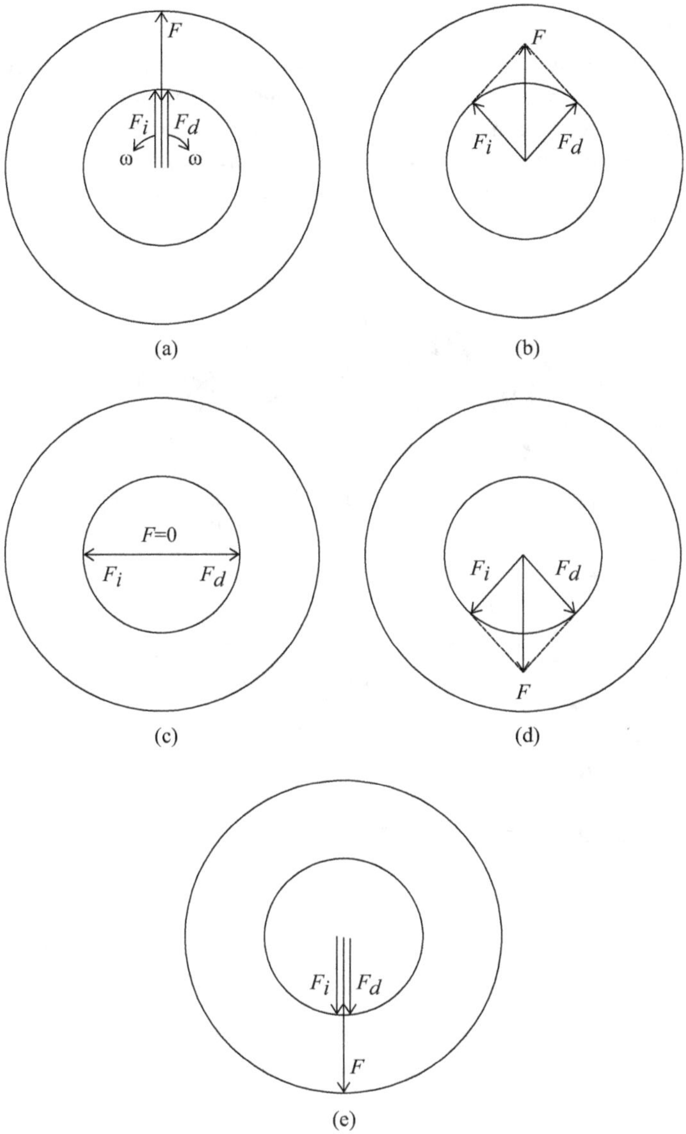


*Figure 4.10 Alternative view on pulsating MMF: pulsating MMF as a vector sum of two rotating waves also given in vector notation*

or

$$n_s = \frac{60f}{p} \left[\frac{\text{rev}}{\text{min}}\right] \tag{4.48}$$

where $f$ is the frequency of ac through the phase winding, i.e., frequency of the voltage supply.

### *4.2.4 Three-phase full-pitch (single-layer) winding*

A three-phase machine has three distributed windings on the stator. Phase windings are identical, but shifted in space by $2\pi/3p$ mechanical radians ($2\pi/3$ electrical radians). One-phase winding in a three-phase machine occupies one-third of the stator slots under one pole as shown in Figure 4.11(a). Winding functions for such a machine are provided in Figure 4.11(b). In order to obtain the resultant MMF

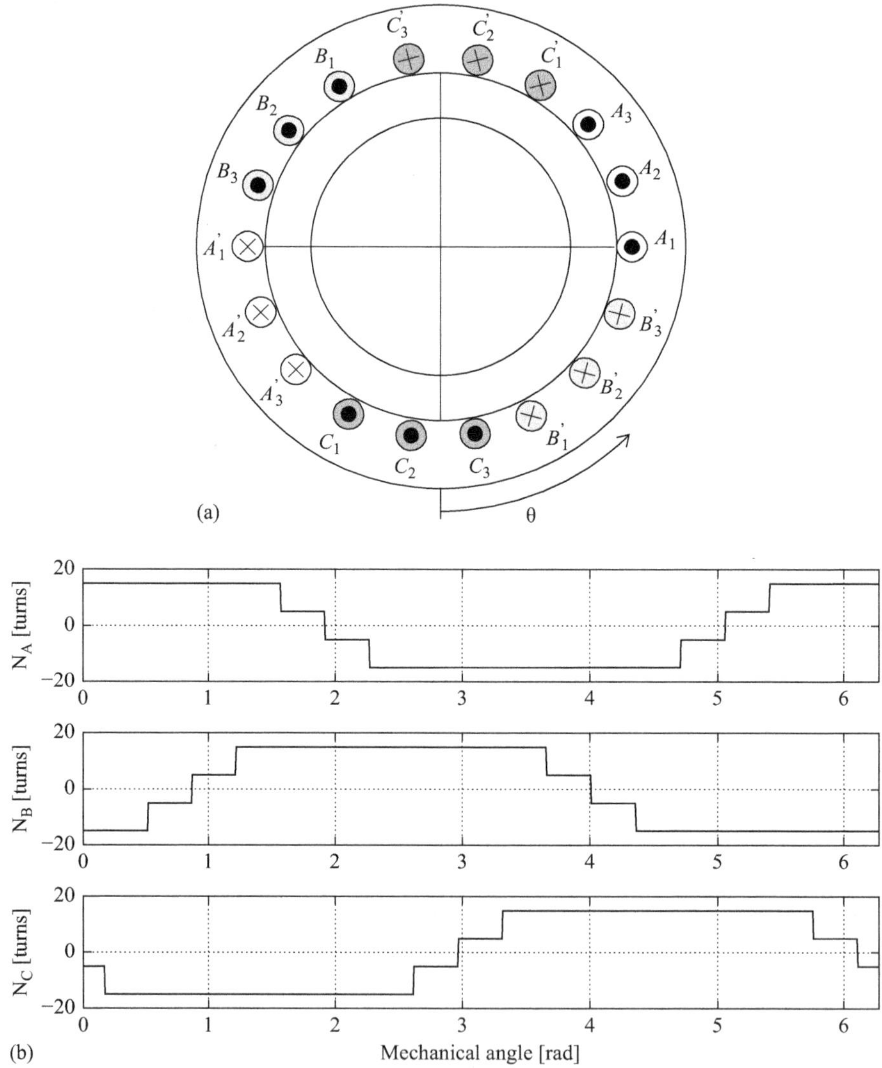


*Figure 4.11 (a) Cross-section view and (b) winding functions of phase windings. Every phase winding consists of three series-connected full-pitch coils. Two-pole machine, p = 1. Every single coil has 10 turns*

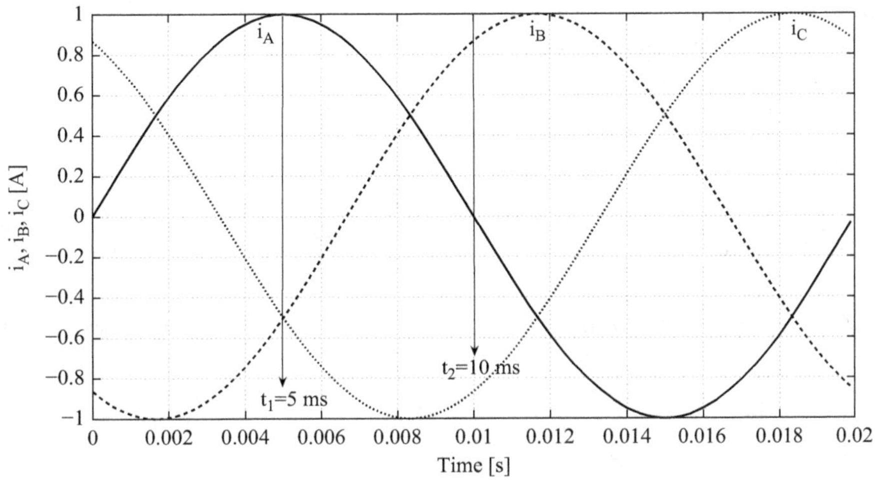


*Figure 4.12 Three-phase, 50Hz, unit currents: rms value of phase current are $1/\sqrt{2} = 0.707A$*

waveform, instantaneous value of phase currents must be taken into account. The phase-shifted current passes through the three-phase windings. At any time, for example $t_1 = 5$ ms (Figure 4.12), the phase current a is $i_a = 1$ A, while the other two-phase currents are $i_b = i_c = -0.5$ A. The resultant MMF waveform could be obtained by multiplying phase winding functions with adequate instant values of the phase currents:

$$F_{S\_t_1=5\ ms} = N_a \cdot i_a(t_1 = 5\text{ ms}) + N_b \cdot i_b(t_1 = 5\text{ ms}) + N_c \cdot i_c(t_1 = 5\text{ ms}) \tag{4.49}$$

i.e.,

$$F_{S\_t_1=5\ ms} = N_a - 0.5 \cdot N_b - 0.5 \cdot N_c \tag{4.50}$$

Similarly, for another time instance $t_2 = 10$ ms:

$$\begin{aligned} F_{S\_t_2=10\ ms} &= N_a \cdot i_a(t_2 = 10\text{ ms}) + N_b \cdot i_b(t_2 = 10\text{ ms}) \\ &\quad + N_c \cdot i_c(t_2 = 10\text{ ms}) \end{aligned} \tag{4.51}$$

i.e.,

$$F_{S\_t_2=10\ ms} = \frac{\sqrt{3}}{2}(N_b - N_c) \tag{4.52}$$

because $i_a(t_2 = 10\text{ ms}) = 0$, $i_b(t_2 = 10\text{ ms}) = \sqrt{3}/2$ and $i_c(t_2 = 10\text{ ms}) = -\sqrt{3}/2$, Figure 4.12.

Figure 4.13 displays the resultant MMF waveforms at these time instances. Obviously, the positions of the maximal values of resultant MMFs are shifted in space, i.e., the resultant MMF waveform is not fixed in space; actually it is rotating MMF.

The rotating speed could be easily found from the following considerations: at $t_1 = 5$ ms maximal value of the rotating MMF is in a position described by the axes of phase winding A, i.e., $\theta_{t1} = 20°$. At $t_2 = 10$ ms, the maximal value of the rotating MMF is at position $\theta_{t2} = 110°$. Rotating MMF speed, i.e., *synchronous speed*, is:

$$n_s = \frac{\text{angle}}{\text{time}} = \frac{110^\circ - 20^\circ}{t_2 - t_1} = \frac{90^\circ}{\frac{T}{4}[\text{s}]} = \frac{\frac{\pi}{2}[\text{rad}]}{\frac{T}{4}[\text{s}]} = 2\pi f \left[\frac{\text{rad}}{\text{s}}\right] = 60f \left[\frac{\text{rev}}{\text{min}}\right] \tag{4.53}$$

More generally, in case of machine with $p$ pole pairs, the synchronous speed is as the following:

$$n_s = \frac{2\pi f}{p}\left[\frac{\text{rad}}{\text{s}}\right] = \frac{60f}{p}\left[\frac{\text{rev}}{\text{min}}\right] \tag{4.54}$$

Hence, the resultant MMF wave rotates with the synchronous speed, having different waveforms in different times. However, regardless of the shape of MMF waveform in different times, Fourier transform of the MMF is always the same. Spectral content of the presented MMF waveforms is also shown in Figure 4.13.

The main difference in comparison with the phase windings MMF is that the resultant rotating MMF wave does not contain harmonic components with the order of three, six, nine, ... times the fundamental one. In other words, all MMF space-harmonics belong to the following series:

$$\nu = 6k + 1 \tag{4.55}$$

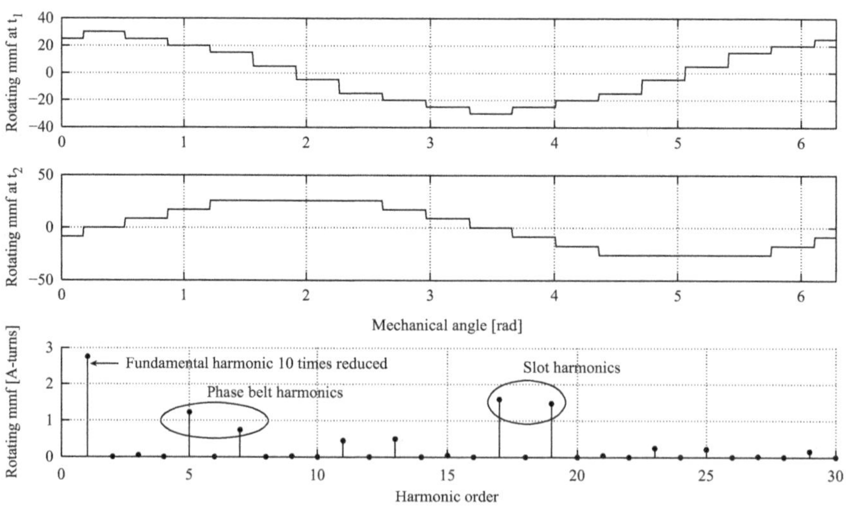


*Figure 4.13 Resultant rotating MMF wave in two different times, $t_1 = 5$ ms and $t_2 = 10$ ms and Fourier transform of MMF wave. Two-pole machine*

where $k = 0, \pm 1, \pm 2, \ldots$. It can be seen from (4.55) that the symmetrical three-phase winding supplied with the symmetrical three-phase voltage supply contains fundamental $\nu = 1$, fifth $\nu = -5$, seventh $\nu = 7$, eleventh $\nu = -11$, etc., space harmonics. The minus sign means inverse rotating MMF wave. The synchronous speed of $\nu$th space-harmonic is $\nu$ times smaller than the synchronous speed of the fundamental one:

$$n_{s\nu} = \frac{2\pi f}{\nu p}\left[\frac{\text{rad}}{\text{s}}\right] = \frac{60 f}{\nu p}\left[\frac{\text{rev}}{\text{min}}\right] \tag{4.56}$$

As seen in Figure 4.13, the higher space harmonics are fifth and seventh, the so-called *phase belt harmonics,* which are direct consequences of trapezoidal shape of the phase-winding MMF. However, the most significant higher space harmonics are *slot harmonics* of order $S/p \pm 1$. In the analyzed case, it is 17th and 19th space harmonics. They are direct consequences of the discrete nature of the winding, i.e., conductor placement in the slots. These harmonics in the best manner fill the "gaps" in the stepwise shape of the resultant MMF wave. Additionally, the rotating MMF space-harmonic amplitude is 1.5 times higher than the amplitude of the phase-winding MMF space harmonics:

$$F_{S\,\max\,\nu} = \frac{3}{2} F_{\text{phase}\,\max\,\nu} = \frac{3}{2}\frac{4}{\pi}\frac{1}{\nu}\frac{k_{d\nu}N}{2p} I_{\max} = \frac{3}{\pi}\frac{1}{\nu}\frac{k_{d\nu}N}{p} I\sqrt{2} \tag{4.57}$$

More generally, for a $m$-phase symmetrical winding distribution:

$$F_{S\,\max\,\nu} = \frac{m}{2} F_{\text{phase}\,\max\,\nu} = \frac{m}{\pi}\frac{1}{\nu}\frac{k_{d\nu}N}{p} I\sqrt{2} \tag{4.58}$$

### *4.2.5 Three-phase shorted-pitch coil (double-layer) winding*

An additional measure taken in order to further upgrade the rotating MMF waveform closer to the sinusoidal shape is the use of the short-pitch coils. However, in order to use short-pitch coils, the stator phase winding must be placed in two layers along the stator circumference, i.e., short-pitch coil must have one of its sides placed in the bottom layer in one slot but the other side should be placed in the top layer of other slot. In this way, double-layer winding is obtained. Using short-pitched coils, with adequate coil pitch, some of the higher space-harmonics in the resultant rotating MMF waveform could be canceled out or significantly attenuated.

Let us look at the following example. Figure 4.14 depicts double-layer three-phase stator winding. Every single coil in the phase windings is a short-pitch coil. Shortening the coil is one stator slot pitch. Now, instead of three coils in one-phase winding as shown in Figure 4.11(a), six shorted-pitch coils make one-phase winding. To compare with the previously analyzed case, every single coil should have one-half of the number of turns as before.

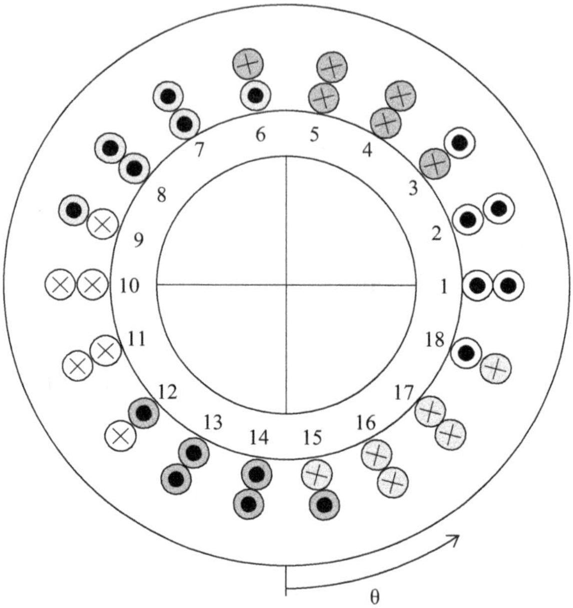


*Figure 4.14 Double-layer three-phase winding. Two-pole machine, p = 1. Coils are shortened for one stator slot pitch*

From phase winding in Figure 4.14 one could observe the following manner: the phase winding is organized in two layers, and every layer, observed individually, is a layer with full-pitch coils, but the two layers are shifted in space for one stator slot pitch—shortening the coil. Therefore, the resultant phase MMF, on the harmonic basis, could be seen as a vector sum of the layer's MMF. It is easy to introduce the other winding factor, *chord* or *pitch factor*, as ratio of vector and algebraic sum of layer's fundamental MMF wave [7]:

$$k_{p1} = \sin\frac{y}{\tau}\frac{\pi}{2} \tag{4.59}$$

or for any other higher space-harmonic $\nu$:

$$k_{p\nu} = \sin\nu\frac{y}{\tau}\frac{\pi}{2} \tag{4.60}$$

In previous expressions, $y$ is the coil pitch and $\tau$ is the pole pitch, both commonly given in the number of slots. By taking just the defined pitch factor into account, the amplitude of the rotating MMF space-harmonic in a distributed double-layer three-phase winding is as follows:

$$F_{S\max\nu} = \frac{3}{2}\frac{4}{\pi}\frac{1}{\nu}\frac{k_{p\nu}k_{d\nu}N_{\mathrm{phase}}}{2p}I_{\max} = \frac{3}{\pi}\frac{1}{\nu}\frac{k_{\nu}N_{\mathrm{phase}}}{p}I\sqrt{2} \tag{4.61}$$

where $k_\nu$ is the *winding factor* which includes distribution and pitch factor, $k_\nu = k_{p\nu} \cdot k_{d\nu}$. In many textbooks of electrical machines, the above expression is given in the following form:

$$F_{S\ \max\nu} = 1.35 \frac{N_{\text{eff}} I}{\nu p} \tag{4.62}$$

where $N_{\text{eff}} = k_\nu \cdot N_{\text{phase}}$ is the so-called number of effective turns per phase and where $I$ is the rms value of phase current. As $k_{p1} < 1$ for the shorted-pitch winding, it is implied that this measure has an impact on the fundamental MMF wave amplitude in such a way that this amplitude is somehow smaller than in full-pitch winding. However, the main result of the use of short-pitch coils is the attenuation of fifth and seventh space harmonics in the rotating MMF wave [8]. In fact, in order to cancel fifth harmonic the rotating MMF wave, the following condition must be satisfied:

$$k_{p5} = \sin 5\frac{y}{\tau}\frac{\pi}{2} = 0 \Rightarrow 5\frac{y}{\tau}\frac{\pi}{2} = \kappa\pi \Rightarrow y = \frac{2}{5}\kappa\tau \tag{4.63}$$

where $\kappa$ is an integer, $\kappa = 0,1,2,\ldots$ and $\kappa$ must be chosen in order that the coil pitch $y$ will be smaller than and close to the pole pitch expressed in the number of stator slots. For analyzed winding, $\tau = 9$ slots, so the reasonable solution for $y$ for $\kappa = 2$ is

$$y = \frac{2}{5} \cdot \kappa \cdot 9 \xrightarrow{\kappa=2} \frac{36}{5} = 7.2$$

As the coil-pitch must be an integer, possible solutions are $y = 7$ or $y = 8$. Obviously, for any choice, fifth space harmonic could not be eliminated, but for $y = 7$ it will be significantly reduced. Similarly, for canceling out seventh space harmonic, the following condition must be satisfied:

$$k_{p7} = \sin 7\frac{y}{\tau}\frac{\pi}{2} = 0 \Rightarrow 7\frac{y}{\tau}\frac{\pi}{2} = \kappa\pi \Rightarrow y = \frac{2}{7}\kappa\tau \tag{4.64}$$

$$y = \frac{2}{7} \cdot \kappa \cdot 9 \xrightarrow{\kappa=3} \frac{54}{7} = 7.71$$

Again, $y$ could be 7 or 8, and seventh space harmonic could be only attenuated, especially for $y = 8$, but not absolutely eliminated from the MMF waveform. If coils with 7 stator slot pitch are chosen, fifth harmonic will be attenuated more than seventh. Inversely, if coils with 8 stator slot pitch are chosen, seventh space harmonic will be attenuated more than fifth. In the analyzed case study, coils with 8 stator slot pitch are used (one stator slot pitch shortening), so seventh space harmonic will be attenuated much more than fifth, as could be easily concluded by comparison of the rotating MMF wave spectrum from Figures 4.13 and 4.15, which is given again in Figure 4.16. Usually, the coil pitch is chosen in order to attenuate

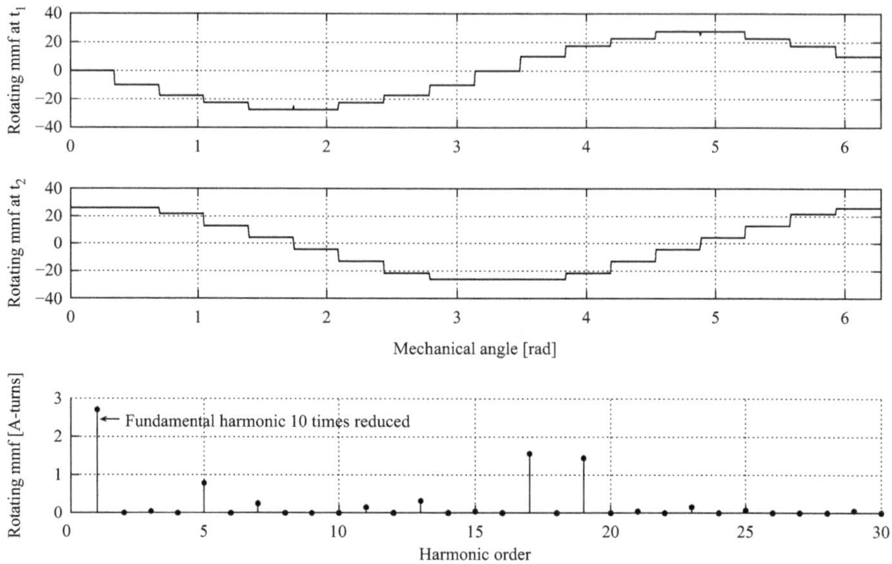


*Figure 4.15 Rotating MMF wave in two different instants of time and the corresponding Fourier's coefficients for machine from Figure 4.14. Two-pole machine. Every single coil has 5 turns*

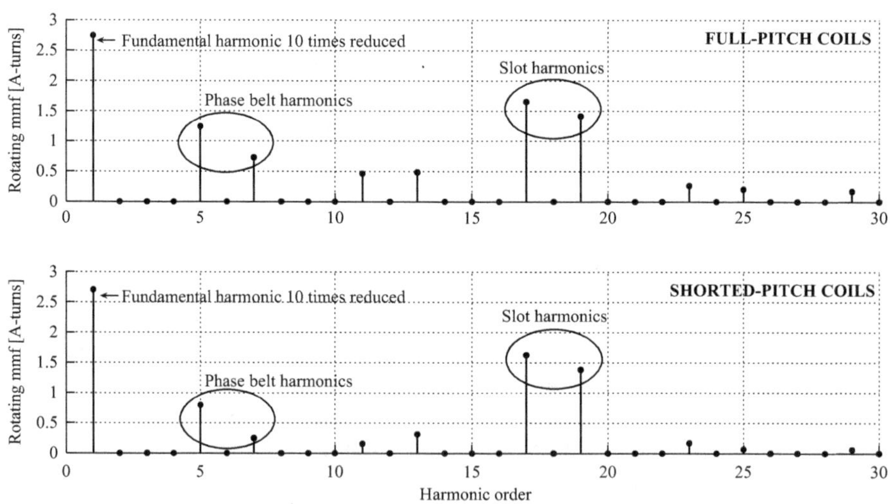


*Figure 4.16 Comparison of MMF wave spectrum for full and shorted-pitch winding from Figures 4.11 and 4.14*

simultaneously fifth and seventh space harmonic as much as possible, and the aim is fulfilled for the following coil pitch:

$$y = \frac{2}{6}\kappa\tau = \frac{1}{3}\kappa\tau \tag{4.65}$$

Along with the additional condition that $\kappa$ must be chosen in such a way that coil pitch $y$ is slightly shorter than the pole pitch, condition (4.65) could be alternatively defined as follows:

$$y = \tau - \frac{\tau}{6} = \frac{5}{6}\tau \tag{4.66}$$

As illustrated in Figure 4.16, shortening the coils has also positive impact on 11th and 13th harmonics. However, from Figure 4.16 it is observed that shortening the coils has no impact on the intensity of the slot harmonics. Why? Coil, regardless of its pitch, must begin in one slot and end in the other; therefore, the coil's pitch is an integral multiple of the slot pitch causing slot harmonics in the first place. It can be easily concluded from pitch factor for slot harmonics:

$$k_{p,\left(\frac{S}{p}\pm 1\right)} = \sin\left(\left(\frac{S}{p}\pm 1\right)\frac{y}{\tau}\frac{\pi}{2}\right) = \sin\left(\frac{S}{p}\frac{y}{\tau}\frac{\pi}{2} \pm \frac{y}{\tau}\frac{\pi}{2}\right) = \sin\left(\frac{S}{p}\frac{y}{\frac{S}{2p}}\frac{\pi}{2} \pm \frac{y}{\tau}\frac{\pi}{2}\right)$$

$$= \sin\left(y\pi \pm \frac{y}{\tau}\frac{\pi}{2}\right) \tag{4.67}$$

$$k_{p,\left(\frac{S}{p}\pm 1\right)} = \pm\sin\frac{y}{\tau}\frac{\pi}{2} \tag{4.68}$$

Obviously, this factor is the same as for the fundamental harmonic

---

**Example 4.2.** For an analyzed three-phase ($m = 3$) two-pole ($p = 1$) machine with $S = 18$ stator slots, calculate amplitudes of fundamental, phase belt and stator slot rotating MMF space harmonics for two different cases: (a) when it has full-pitch stator winding and (b) when it has stator winding with one slot shortening pitch. In both cases number of series turns per phase is the same, $N = 30$. Assume in both cases the amplitude of ac through the phase winding is 1A. Amplitude of $\nu$th MMF space harmonic is

$$F_{S\,\max\,\nu} = \frac{3}{2}\frac{4}{\pi}\frac{1}{\nu}\frac{k_{p\nu}k_{d\nu}N_{\text{phase}}}{2p}I_{\max}$$

where

$$k_{p\nu} = \sin\nu\frac{y}{\tau}\frac{\pi}{2}$$

$$k_{d\nu} = \frac{\sin\dfrac{z\nu\theta_{el}}{2}}{z\sin\dfrac{\nu\theta_{el}}{2}}$$

Number of slots per pole per phase is $z = S/(2pm) = 3$. Since machine has two poles, $\theta_{el} = \theta_{meh} = 360°/S = 20°$.

(a) Stator winding is made of full-pitch coils, $y = \tau$. A simple calculation leads to the following results:

| $\nu = 1$ | $\nu = 5$ | $\nu = 7$ | $\nu = 17$ | $\nu = 19$ |
|---|---|---|---|---|
| $k_{p1} = 1$ | $k_{p5} = 1$ | $k_{p7} = -1$ | $k_{p17} = 1$ | $k_{p19} = -1$ |
| $k_{d1} = 0.96$ | $k_{d5} = 0.22$ | $k_{d7} = -0.18$ | $k_{d17} = 0.96$ | $k_{d19} = 0.96$ |
| $F_{S\ max\ 1} = 27.50$ | $F_{S\ max\ 5} = 1.25$ | $F_{S\ max\ 7} = 0.73$ | $F_{S\ max\ 17} = 1.62$ | $F_{S\ max\ 19} = 1.45$ |

(b) Now coils are with shorted pitch. As shortening is one slot, $y = 8$ and $y/\tau = 8/9$. In this case, the following results are obtained:

| $\nu = 1$ | $\nu = 5$ | $\nu = 7$ | $\nu = 17$ | $\nu = 19$ |
|---|---|---|---|---|
| $k_{p1} = 0.98$ | $k_{p5} = 0.64$ | $k_{p7} = -0.34$ | $k_{p17} = -0.98$ | $k_{p19} = 0.98$ |
| $k_{d1} = 0.96$ | $k_{d5} = 0.22$ | $k_{d7} = -0.18$ | $k_{d17} = 0.96$ | $k_{d19} = 0.96$ |
| $F_{S\ max\ 1} = 27.08$ | $F_{S\ max\ 5} = 0.80$ | $F_{S\ max\ 7} = 0.25$ | $F_{S\ max\ 17} = 1.59$ | $F_{S\ max\ 19} = 1.43$ |

Comparing the obtained results, it is apparent that shortening pitch for one slot results in small attenuation of fundamental MMF; however, the main result is in the significant attenuation of fifth and especially seventh MMF space harmonic. On the other hand, slot harmonics are almost intact in this way. All obtained results are in a very good correlation with the results obtained from the Fourier transform, of the rotating MMF waveforms shown in Figure 4.16. If shortening pitch is two stator slots, the following results are achieved:

| $\nu = 1$ | $\nu = 5$ | $\nu = 7$ | $\nu = 17$ | $\nu = 19$ |
|---|---|---|---|---|
| $k_{p1} = 0.94$ | $k_{p5} = -0.17$ | $k_{p7} = 0.77$ | $k_{p17} = 0.94$ | $k_{p19} = -0.94$ |
| $k_{d1} = 0.96$ | $k_{d5} = 0.22$ | $k_{d7} = -0.18$ | $k_{d17} = 0.96$ | $k_{d19} = 0.96$ |
| $F_{S\ max\ 1} = 25.84$ | $F_{S\ max\ 5} = 0.22$ | $F_{S\ max\ 7} = 0.56$ | $F_{S\ max\ 17} = 1.52$ | $F_{S\ max\ 19} = 1.36$ |

where the fundamental harmonic is even smaller, but now fifth harmonic is attenuated much more than seventh. The smaller values of the slot harmonics are result of smaller value of fundamental one.

## 4.3 Rotating MMF wave—analytical approach

A general term for the rotational MMF of symmetrical three-phase machine fed by symmetrical three-phase voltage supply is easily reached also by following analytical procedures. Ac flowing through the phase A winding produces the following $\nu$th pulsating MMF harmonic:

$$F_{A\nu}(t,\theta) = N_{A\nu}(\theta)i_A(t) = \frac{4}{\pi}\frac{1}{\nu}\frac{N}{2p}\sqrt{2}I\cos(\nu p\theta)\cos(\omega t) = F_{\max\nu}\cos(\nu p\theta)\cos(\omega t) \tag{4.69}$$

that could be separated into the forward and backward MMF waveforms:

$$F_{A\nu}(t,\theta) = \frac{F_{\max\nu}}{2}\cos(\omega t - \nu p\theta) + \frac{F_{\max\nu}}{2}\cos(\omega t + \nu p\theta) \tag{4.70}$$

Phase winding B is space shifted for $2\pi/3p$ mechanical radians. Ac flows which is time shifted by $2\pi/3$ radians. Therefore,

$$F_{B\nu}(t,\theta) = N_{B\nu}(\theta)i_B(t) = F_{\max\nu}\cos\left(\nu p\left(\theta - \frac{2\pi}{3p}\right)\right)\cos\left(\omega t - \frac{2\pi}{3}\right) \tag{4.71}$$

$$F_{B\nu}(t,\theta) = \frac{F_{\max\nu}}{2}\cos\left(\omega t - \frac{2\pi}{3} - \nu p\theta + \nu\frac{2\pi}{3}\right) + \frac{F_{\max\nu}}{2}\cos\left(\omega t - \frac{2\pi}{3} + \nu p\theta - \nu\frac{2\pi}{3}\right) \tag{4.72}$$

For the third-phase winding (phase C):

$$F_{C\nu}(t,\theta) = N_{C\nu}(\theta)i_C(t) = F_{\max\nu}\cos\left(\nu p\left(\theta + \frac{2\pi}{3p}\right)\right)\cos\left(\omega t + \frac{2\pi}{3}\right) \tag{4.73}$$

$$F_{C\nu}(t,\theta) = \frac{F_{\max\nu}}{2}\cos\left(\omega t + \frac{2\pi}{3} - \nu p\theta - \nu\frac{2\pi}{3}\right) + \frac{F_{\max\nu}}{2}\cos\left(\omega t + \frac{2\pi}{3} + \nu p\theta + \nu\frac{2\pi}{3}\right) \tag{4.74}$$

The sum of three forward waveforms gives the forward waveform of the rotating MMF:

$$\begin{aligned} F_{d\nu}(t,\theta) = \frac{F_{\max\nu}}{2}\Bigg[&\cos(\omega t - \nu p\theta) + \cos\left(\omega t - \nu p\theta + (\nu - 1)\frac{2\pi}{3}\right) \\ &+ \cos\left(\omega t - \nu p\theta - (\nu - 1)\frac{2\pi}{3}\right)\Bigg] \end{aligned} \tag{4.75}$$

while the sum of backward waveforms gives the backward waveform of the rotating MMF:

$$F_{i\nu}(t,\theta) = \frac{F_{\max\,\nu}}{2}\left[\cos(\omega t + \nu p\theta) + \cos\left(\omega t + \nu p\theta - (\nu+1)\frac{2\pi}{3}\right) + \cos\left(\omega t + \nu p\theta + (\nu+1)\frac{2\pi}{3}\right)\right] \tag{4.76}$$

The fundamental MMF harmonic of phase windings, i.e., $\nu = 1$, results in forward rotating MMF waveform in which its amplitude is 3/2 times higher than the amplitude of constituent MMF waves while backward rotating MMF waveform does not exist:

$$F_{d,\nu=1}(t,\theta) = \frac{3}{2}F_{\max\,1}\cos(\omega t - p\theta) \tag{4.77}$$

$$F_{i,\nu=1}(t,\theta) = 0 \tag{4.78}$$

A similar situation could be analyzed for higher space harmonics. For example, for $\nu = 3$, there is no forward or backward MMF waveform.

$$F_{d,\nu=3}(t,\theta) = \frac{F_{\max\,\nu}}{2}\left[\cos(\omega t - 3p\theta) + \cos\left(\omega t - 3p\theta + \frac{4\pi}{3}\right) + \cos\left(\omega t - 3p\theta - \frac{4\pi}{3}\right)\right] = 0 \tag{4.79}$$

$$F_{i,\nu=3}(t,\theta) = \frac{F_{\max\,\nu}}{2}\left[\cos(\omega t + 3p\theta) + \cos\left(\omega t + 3p\theta - \frac{8\pi}{3}\right) + \cos\left(\omega t + 3p\theta + \frac{8\pi}{3}\right)\right] = 0 \tag{4.80}$$

For $\nu = 5$, there is no forward rotating MMF waveform:

$$F_{d,\nu=5}(t,\theta) = \frac{F_{\max\,5}}{2}\left[\cos(\omega t - 5p\theta) + \cos\left(\omega t - 5p\theta + 4\frac{2\pi}{3}\right) + \cos\left(\omega t - 5p\theta - 4\frac{2\pi}{3}\right)\right] = 0 \tag{4.81}$$

while backward rotating MMF exists:

$$F_{i,\nu=5}(t,\theta) = \frac{F_{\max\,5}}{2}[\cos(\omega t + 5p\theta) + \cos(\omega t + 5p\theta - 4\pi) + \cos(\omega t + 5p\theta + 4\pi)] \tag{4.82}$$

$$F_{i,\nu=5}(t,\theta) = \frac{3}{2}F_{\max\,5}\cos(\omega t + 5p\theta) \tag{4.83}$$

For $\nu = 7$, there is forward rotating MMF as follows:

$$F_{d,\nu=7}(t,\theta) = \frac{F_{\max 7}}{2}[\cos(\omega t - 7p\theta) + \cos(\omega t - 7p\theta + 4\pi) + \cos(\omega t - 7p\theta - 4\pi)] \tag{4.84}$$

$$F_{d,\nu=7}(t,\theta) = \frac{3}{2}F_{\max 7}\cos(\omega t - 7p\theta) \tag{4.85}$$

There is no backward rotating MMF waveform:

$$F_{i,\nu=7}(t,\theta) = \frac{F_{\max 7}}{2}\left[\cos(\omega t + 7p\theta) + \cos\left(\omega t + 7p\theta - \frac{16\pi}{3}\right) + \cos\left(\omega t + 7p\theta + \frac{16\pi}{3}\right)\right] = 0 \tag{4.86}$$

**Remark 2:** *The current in phase A is given by $i_A(t) = I_m sin(\omega t)$. If the current in phase B lags by $2\pi/3$ and the current in phase C lags by $4\pi/3$, there is a direct system of three-phase ac:*

$$i_A = I_m \sin(\omega t);\quad i_B = I_m \sin\left(\omega t - \frac{2\pi}{3}\right);\quad i_C = I_m \sin\left(\omega t - \frac{4\pi}{3}\right).$$

*It should be noted that lagging of currents is described by minus sign in analytical expressions of currents. On the other hand, phase A MMF in p pair poles machine is $N_A(\theta_{meh}) = N_m sin(p\theta_{meh})$. If phase B winding leads to space by $2\pi/3p$ mechanical radians and phase C winding leads to space by $4\pi/3p$ mechanical radians; then analytical expressions for MMFs are:*

$$N_A = N_m \sin(p\theta_{\text{meh}});\ N_B = N_m \sin\left(p\left(\theta_{\text{meh}} - \frac{2\pi}{3p}\right)\right);\ N_C = N_m \sin\left(p\left(\theta_{\text{meh}} - \frac{4\pi}{3}\right)\right).$$

*Now, the negative sign corresponds to leading of phases in space.*

*Described order of currents and windings gives forward rotating MMF wave.*

So the conclusion is as follows:

> The symmetrically constructed (identical phase windings with identical series number of turns per phase) and the symmetrically distributed three-phase winding (phase windings space shifted by $2\pi/3$ electrical radians) fed by symmetrical system of three-phase ac voltages (ac phase currents phase shifted by $2\pi/3$) produce symmetrical rotating MMF waves. The angular speed of this wave is known as synchronous speed.
>
> The described rotating MMF wave is not the result of mechanical rotation of any part of the electrical machine; rather, it is produced by the stationary phase windings powered by ac.

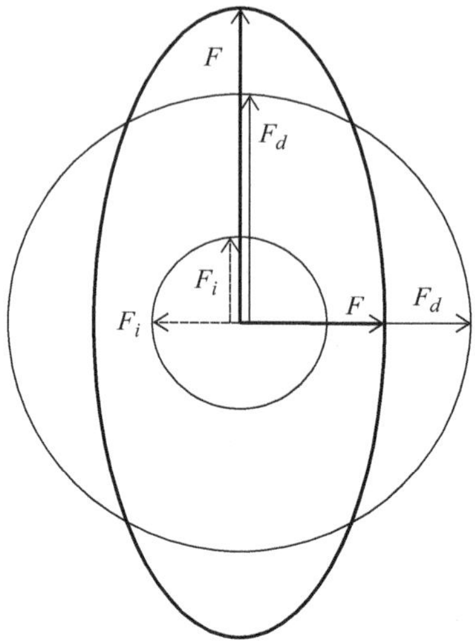


*Figure 4.17 The result of summing the forward and backward MMF waveforms of the same order is the rotational MMF wave of elliptical shape*

In addition to the fundamental MMF wave, there exist higher space harmonics of MMF that rotate forward or backward by angular speed which is $\nu$th times smaller than the synchronous speed. The order of harmonics are given by:

$$\nu = 6k + 1$$

where $k = 0, \pm 1, \pm 2, \ldots$ Noteworthy higher space harmonics are *phase belt harmonics*, i.e., fifth and seventh, as well as *slot harmonics* of order $S/p \pm 1$.

In case of any asymmetry in the machine as different number of turns in phase windings, different phase shift of phase windings, asymmetry in phase currents in magnitude or in phase shift, in addition to the direct rotating MMF wave, the backward rotating MMF wave always exists, i.e., $\nu$ could have value $\nu = -1$. As a result, a resultant elliptical MMF wave exists in the air gap as shown in Figure 4.17.

## 4.4 Fractional slot winding

The case in which only winding is embedded in the whole number of slots per pole per phase has been already analyzed. This winding is known as integral slot winding. However, there is also fractional slot winding. This is the case where the number of slots per pole per phase is not an integer. The manner of embodiment of this type of winding is numerous and so complex that it is beyond the scope of this book. Their common characteristics are unsymmetrical phase as well as rotating MMF wave, so the MMF spectral content could not be described by introduced rule $\nu = 6k + 1$, where $k = 0, \pm 1, \pm 2, \ldots$

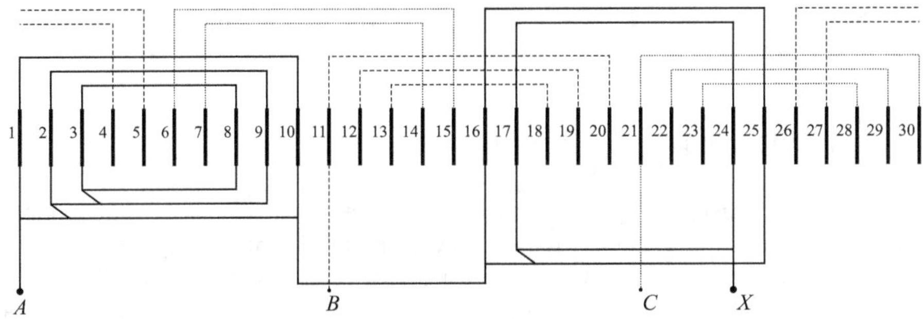


*Figure 4.18 Single-layer fractional slot winding (p = 2, S = 30, q = 2.5)*

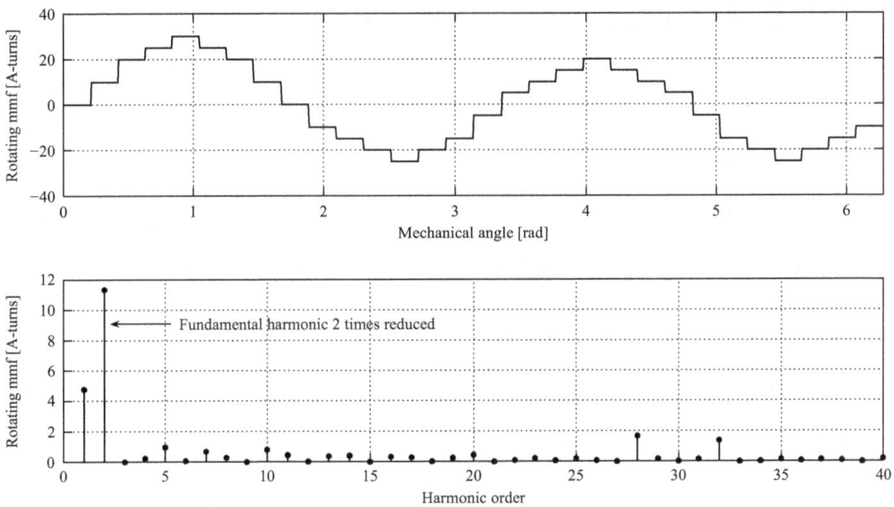


*Figure 4.19 Fractional slot winding rotational MMF waveform at any time and the corresponding spectral content*

As an example of such fractional slot winding in a three-phase four-pole machine inserted in $S = 30$ stator slots in a single layer is provided in Figure 4.18. Here, the number of slots per pole per phase is obviously a fractional number:

$$z = \frac{S}{2pm} = \frac{30}{2 \cdot 2 \cdot 3} = 2.5$$

The waveform of the rotating MMF at instant of time when $i_a = 1\text{A}$ and $i_b = -i_c = 0.5\text{A}$ is given in Figure 4.19, and it is asymmetrical. The same figure displays the spectral content of the MMF wave.

The noteworthy point is the existence of one sub-harmonic. Namely, the fundamental harmonic is second one of circa 24 A-turns amplitude. However, the first harmonic now exists with the amplitude of circa 5 A-turns. Additionally, there are now some even harmonics. However, the slot harmonics of order $S \pm p$, i.e., 28th and 32nd harmonics, which are rather prominent, exist in this case, too.

## 4.5 Wound rotor MMF space harmonics

In wound-rotor induction motor, the rotor carries three-phase winding with identical number of pole pairs as stator winding. Therefore, all the previously derived conclusions for the three-phase stator winding are also true for rotor winding. The main difference is in the fact that rotor currents are induced currents of frequency that is different from voltage supply frequency. Fundamental rotor current frequency is,

$$f_2 = s \cdot f_1 \tag{4.87}$$

where slip for fundamental MMF wave from stator side is,

$$s = \frac{\omega_s - \omega_r}{\omega_s} \tag{4.88}$$

or in general case

$$s_\nu = 1 - \nu(1 - s) \tag{4.89}$$

if of interest are rotor currents induced by $\nu$th MMF space harmonic wave from stator side. Hence, as it was the case for three-phase stator winding, rotor winding produces MMF waves of the same order,

$$\mu = 6q + 1 \tag{4.90}$$

where $q = 0, \pm 1, \pm 2, \ldots$ The most significant harmonics are phase belt harmonics and slot harmonics. The rotor slot harmonics of order $R/p \pm 1$ exist where $R$ is the number of rotor slots. If rotor winding is of fractional slot type then some additional harmonics could appear as it was explained previously.

## 4.6 Cage rotor MMF space harmonics

The MMF waves of symmetrical, integral slot three-phase stator winding of an induction machine has been already derived as follows:

$$F_s(t, \theta) = \sum_{\mu} F_{s\mu} \cos(\omega_1 t - \mu p \theta); \quad \mu = 6k + 1, \quad k = 0, \pm 1, \pm 2, \ldots \tag{4.91}$$

where $p$ is the number of pole pairs and $F_{s\mu}$ is the amplitude of the $\mu$th space harmonic. It is clear that besides the fundamental MMF wave ($k = 0$, $\mu = 1$), there exist waves with $5p$, $7p$, $11p, \ldots$ pair of poles even in the case of symmetrical machine. By assuming uniform air gap length, the flux density waves in the air gap produced by the stator windings have the same waveforms.

The winding function of multi-turn coil in slots described by angular coordinates $\theta_1$ and $\theta_2$ has been already defined by

$$N(\theta) = \begin{cases} N\left(1 - \dfrac{\alpha}{2\pi}\right), & \theta_1 \le \theta \le \theta_2 \\ -N\dfrac{\alpha}{2\pi}, & \text{for rest of } \theta \end{cases} \tag{4.92}$$

where $N$ is the number of turns and $\alpha$ is a coil pitch, $\alpha = \theta_2 - \theta_1$. In case of cage rotor, two nearby bars and ring segments between them form the rotor loop. Therefore, the rotor loop can be observed as a one-turn coil with pitch $\alpha = 2\pi/R$, where $R$ is the number of rotor bars. Hence, the winding function of the rotor loop 1, whose magnetic axis is in the center of the reference frame fixed to the rotor, is given by:

$$N_{\text{loop1}}(\theta_r) = \begin{cases} 1 - \dfrac{1}{R}, & -\dfrac{\pi}{R} \le \theta_r \le \dfrac{\pi}{R} \\ -\dfrac{1}{R}, & \text{for rest of } \theta_r \end{cases} \tag{4.93}$$

This function could be resolved by Fourier series as follows:

$$N_{\text{loop1}}(\theta_r) = \sum_{\nu=1}^{\infty} \frac{2}{\nu\pi} \sin\left(\nu\frac{\pi}{R}\right)\cos(\nu\theta_r) \tag{4.94}$$

The rotor current, which flows in this loop as a result of $\mu$th stator flux density space harmonic, produces the following MMF:

$$F_{\text{loop1}}(t, \theta_r) = \sum_{\nu=1}^{\infty} \frac{2}{\nu\pi} \sin\left(\nu\frac{\pi}{R}\right) I_{r\mu m} \cos(s_\mu\omega_1 t)\cos(\nu\theta_r) \tag{4.95}$$

where

$$s_\mu = 1 - \mu(1 - s) \tag{4.96}$$

and $s$ is the slip. The last equation could be resolved as

$$F_{\text{loop1}}(t, \theta_r) = \sum_{\nu=1}^{\infty} K_{\mu\nu}\left[\cos(s_\mu\omega_1 t + \nu\theta_r) + \cos(s_\mu\omega_1 t - \nu\theta_r)\right] \tag{4.97}$$

In the next rotor loop, with space displaced by $2\pi/R$, flows the same current and frequency, but shifted in phase by $2\pi\mu p/R$. This rotor loop produces the following MMF:

$$\begin{aligned} F_{\text{loop2}}(t, \theta_r) = \sum_{\nu=1}^{\infty} K_{\mu\nu} \Bigg[ &\cos\left(s_\mu\omega_1 t + \nu\theta_r - (\nu + \mu p)\frac{2\pi}{R}\right) \\ &+ \cos\left(s_\mu\omega_1 t - \nu\theta_r + (\nu - \mu p)\frac{2\pi}{R}\right)\Bigg] \end{aligned} \tag{4.98}$$

Summing the MMFs of all rotor loops, the resultant MMF of the cage rotor is [9]:

$$F_r(t,\theta_r) = \sum_{i=0}^{R-1}\sum_{\nu=1}^{\infty} K_{\mu\nu}\left[\cos\left(s_\mu\omega_1 t + \nu\theta_r - i\cdot(\nu+\mu p)\frac{2\pi}{R}\right)\right.$$
$$\left. + \cos\left(s_\mu\omega_1 t - \nu\theta_r + i\cdot(\nu-\mu p)\frac{2\pi}{R}\right)\right] \quad (4.99)$$

Let us analyze the last expression in the case where rotor currents are due to the $\mu$th stator flux density harmonic. By inspection of (4.99), the following rotor MMF waves exist:

for $\nu = -\mu p$, the first member of the sum is different from zero;
for $\nu = \mu p$, the second member of the sum is different from zero;

In the healthy induction machine, there are only stator MMF waves of order $\mu = 6k+1$, $k = 0, \pm 1, \pm 2, \ldots$. It means that the rotor cage produces MMF waves that are armature reaction to these MMF waves from the stator side. It should be noted that $\nu$ can be only positive integer, $\nu = 1, 2, 3, \ldots$.

However, there are cage rotor MMF waves in the following two cases:

for $\nu + \mu p = \pm\,\lambda R$ or $\nu = \pm\lambda R - \mu p$, where $\lambda = 1, 2, 3\ldots$ the first member of the sum is nonzero.
for $\nu - \mu p = \pm\,\lambda R$ or $\nu = \pm\lambda R + \mu p$, where $\lambda = 1, 2, 3\ldots$ the second member of the sum is nonzero.

Therefore, in addition to the fundamental cage rotor MMF wave which is the armature reaction to the stator MMF wave, there is also the so-called rotor slot harmonics of order $\lambda R \pm \mu p$, $\lambda = 1, 2, 3, \ldots$. These harmonics are the direct consequences of the space distribution of rotor bars, i.e., placement of rotor cage in slots. Consequently, the rotor currents caused by the $\mu$th stator flux density space harmonics generate the following MMF waves:

$$F_{r\mu}(t,\theta_r) = F_{r\mu 1}(t,\theta_r) + F_{r\mu sh1}(t,\theta_r) + F_{r\mu sh2}(t,\theta_r) \quad (4.100)$$

where

$$F_{r\mu 1}(t,\theta_r) = F_{r\mu 1}\cos(s_\mu\omega_1 t - \mu p\theta_r) \quad (4.101)$$

$$F_{r\mu sh1}(t,\theta_r) = F_{r\mu sh1}\cos(s_\mu\omega_1 t + (\lambda R - \mu p)\theta_r) \quad (4.102)$$

$$F_{r\mu sh2}(t,\theta_r) = F_{r\mu sh2}\cos(s_\mu\omega_1 t - (\lambda R + \mu p)\theta_r) \quad (4.103)$$

Using angular transformation,

$$\theta = \theta_r + \omega_r t = \theta_r + \left(\frac{1-s}{p}\right)\omega_1 t \quad (4.104)$$

these MMF waves, now observed from the stator side are

$$F_{r\mu 1}(t,\theta) = F_{r\mu 1}\cos(\omega_1 t - \mu p\theta) \tag{4.105}$$

$$F_{r\mu sh1}(t,\theta) = F_{r\mu sh1}\cos\left(\left(1 - \frac{\lambda R}{p}(1-s)\right)\omega_1 t + (\lambda R - \mu p)\theta\right) \tag{4.106}$$

$$F_{r\mu sh2}(t,\theta) = F_{r\mu sh2}\cos\left(\left(1 + \frac{\lambda R}{p}(1-s)\right)\omega_1 t - (\lambda R + \mu p)\theta\right) \tag{4.107}$$

Figure 4.20 shows the waveform of the fundamental cage rotor MMF wave ($\mu = 1$) at an instant of time assuming that the amplitude of the rotor loop current is 1 A. The rotor has $R = 40$ bars and the machine has one pair of poles, $p = 1$.

Figure 4.21 illustrates the spectral contents of MMF wave extracted from Figure 4.20. It is obvious that besides fundamental harmonic ($\mu p = 1$st), there exist first-order rotor slot harmonics, for $\lambda = 1$, $R - p$ and $R + p$, second-order rotor slot harmonics, for $\lambda = 2$, $2R - p$ and $2R + p$, etc. (i.e., 39th, 41st, 79th, 81st). This directly follows from (4.106) and (4.107).

Figure 4.22 displays the waveform of the cage rotor MMF, produced by rotor currents that are the result of $\mu = -5$ under the assumption that the amplitude of rotor loop current is 1 A. In this case, the number of rotor bars per one pole of harmonic field is less than before; as a result, more rugged waveform is obtained.

The spectral content of the waveform in Figure 4.22 is shown in Figure 4.23. It is obvious that in addition to the "fundamental" harmonic (now, $\mu p =$ fifth), the first-order rotor slot harmonics of the order $(R - 5p)$, $(R + 5p)$, $(2R - 5p)$, $(2R + 5p)$, i.e., 35th, 45th, 75th, 85th will appear. This result is consistent with

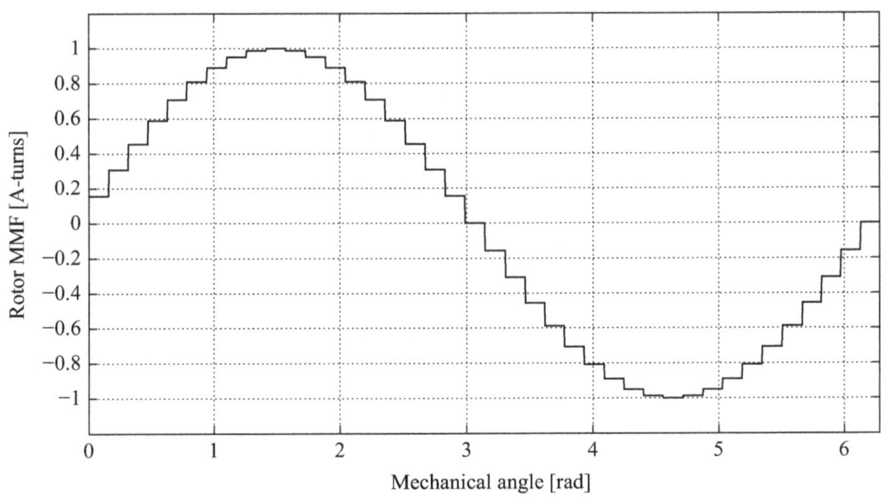


*Figure 4.20 Cage rotor MMF at an instant of time (p = 1, R = 40, μ = 1)*

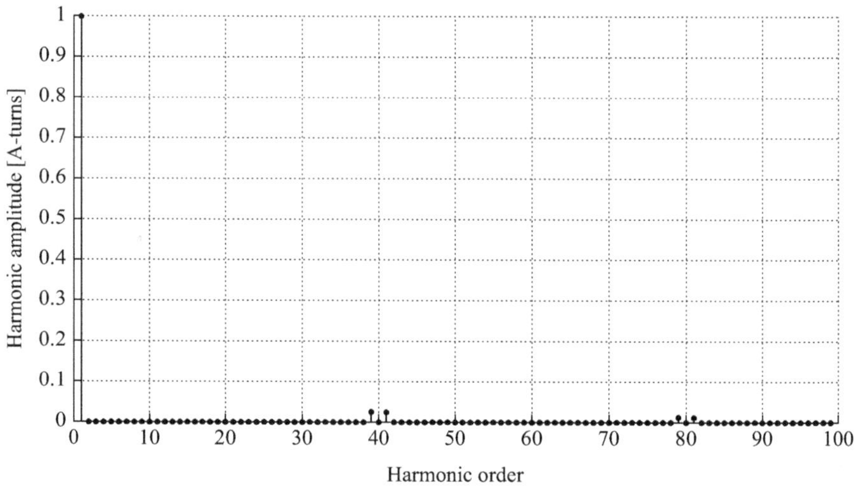


*Figure 4.21 Spectral content of cage rotor MMF given in Figure 4.20*

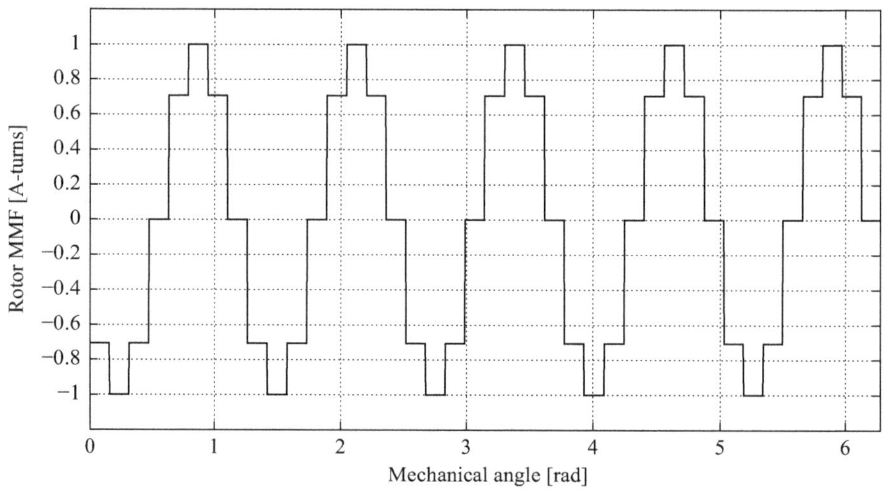


*Figure 4.22 Cage rotor MMF in an instant of time ($p = 1$, $R = 40$, $\mu = -5$)*

(4.106) and (4.107) as well. It is now clear that the amplitude of rotor slot harmonics is more pronounced in relation to the "fundamental" harmonic when compared to the situation in Figure 4.21.

However, the main conclusion can be derived from (4.105)–(4.107):

**The cage rotor of induction machine reflects all MMF space harmonics from stator side at the fundamental frequency $f_1$ and at frequencies $(1 \pm \lambda R(1 - s)/p)f_1$.**

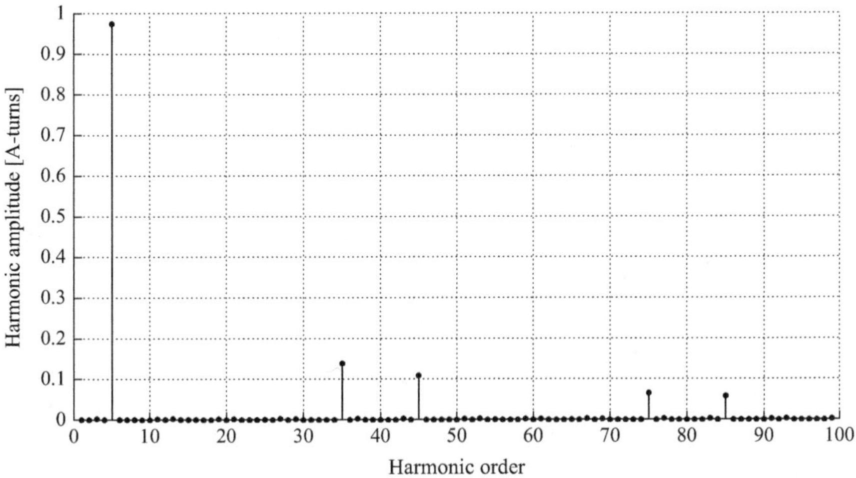


*Figure 4.23 Spectral content of cage rotor MMF given in Figure 4.22*

## References


[1] Encyclopaedia Britannica, https://www.britannica.com/science/magnetic-circuit#ref235761 (Access date: October 2016).

[2] Schmitz N.L. and Nowotny D.W. *Introductory Electromechanics*, 1965; The Ronald Press Company, New York.

[3] Lipo T.A. *Theory and control of synchronous machines*, ECE 511 lecture notes, University of Wisconsin-Madison.

[4] Serrano-Iribarnegaray L., Cruz-Romero P. and Gomez-Exposito A. *Critical review of the modified winding function theory*, Progress in Electromagnetics Research, 2013; **PER-133**: 515–534.

[5] Vukosavic S.N. *Electrical machines*, 2013; Springer-Verlag, New York.

[6] Joksimovic G., Djurovic M. and Obradovic A. *Skew and linear rise of mmf across slot modeling – winding function approach, 199; IEEE Transactions on Energy Conversion*, 1999, **EC-14**: 315–320.

[7] Fitzgerald A.E., Kingsley C. and Umans S.D. *Electric machinery*, 2003; McGraw Hill Professional, New York.

[8] Joksimović G. *AC winding analysis using winding function approach, International Journal of Electrical Engineering Education*, Manchester University Press, Manchester; Vol-48, 2011; 34–52.

[9] Joksimović G., Djurović M. and Penman J. Cage rotor MMF – Winding function approach, *IEEE Power Engineering Review*, 2001; **PER-21**: 64–66.

# Chapter 5
# Multiple-coupled circuit model of induction motors

There are a lot of different mathematical models which enable the analysis of an induction machine. Generally, they could be divided into static and dynamic models. Static models allow the analysis of static characteristics of induction motors. This model, for example, is the well-known single-phase equivalent circuit of induction motor. Dynamic models can provide time variations of motor currents, developed electromagnetic torque and rotor speed as outputs. Such models enable the analysis of induction machines in transient modes.

By appearance of static frequency converters, induction motor became very interesting for analysis from different point of views what is the result of its use as the most robust, reliable and simple ac motor type in different applications. On the other hand, condition monitoring as a discipline that has been widely used in the last two decades was an important incentive for the development of new motor models which enable the analysis of the motor in the time domain as well as its static characteristics of the asymmetrical operating modes caused by both external factors (complex periodic voltages) and internal factors (as a result of different faults or a design specific motor).

As regards the static models, in addition to the equivalent circuit of the motor, often a model based on rotating field theory is employed [1,2]. This model allows the analysis of static characteristics of the induction motor for any system of power supply and any stator winding configuration. The second model which is based on the theory of spatial vector allows the analysis of the magnetic field in the air gap at any asymmetrical configuration of stator windings [3,4]. The method introduced in [5] is a combination of the theory of rotating field and the theory of symmetrical components; however, it can only analyze star-connected induction motors.

In dynamic models, the most widely used model enabling the analysis of symmetrical induction motors is the well-known *d–q* model [6]. This model is based on the assumption of sinusoidal distributed windings in the rotor and stator so that it is impossible to directly analyze the operation of the motor with arbitrary distribution of windings. However, it is possible to apply *d–q* model for asymmetric modes analysis, through appropriate changes in parameters, in order to analyze the operation of the machine with a broken rotor bar, static and dynamic rotor eccentricity faults [7]. The results obtained in this way, however, show significant deviations from the experimental ones.

The second model which allows the dynamic analysis of the induction motor is slightly older and is based on the principle of mutually coupled electrical circuits. It enables the analysis of the motor by taking all the spatial harmonics of MMF in the machine into account, on the harmonic basis. This model is also primarily intended for the analysis of motor with symmetrical windings on the stator and rotor [8].

The model that enables the analysis of general machine with $m$ stator circuits (not windings) and $n$ rotor bars and allows dynamic analysis of the transient regimes of induction machine with any configuration of windings on the stator and the rotor, as with any profile of air-gap, is presented in [9]. This model is based on multiple-coupled circuit model, but with one distinct difference. The difference is the use of winding function theory that enables us to account for all spatial harmonics of MMFs in machine simultaneously. It means exact winding distribution and exact stepwise MMF distribution could be taken into account. In this chapter, the above-mentioned model is presented with some modifications in order to have more real picture of the machine. On the first place, skewed rotor bars are included into the model.

## 5.1 Model description

This model is based on the following assumptions which are common in analysis of induction motors:

- permeability of stator and rotor core is infinite;
- iron losses are ignored;
- concentrated parameters are used;
- rotor bars are electrically isolated from rotor core;
- windage and friction losses are neglected.

### *5.1.1 Electrical subsystem equations*

Cage rotor can be viewed as $n$ identical and evenly distributed rotor loops. The equations describing the general electric machine can be written as follows:

$$[U] = [R][I] + \frac{d[\Psi]}{dt} \tag{5.1}$$

$$[\Psi] = [L][I] \tag{5.2}$$

For induction machines with squirrel-cage rotor (5.1) and (5.2) can be broken down as follows:

$$\begin{bmatrix} [U_s] \\ [0] \end{bmatrix} = \begin{bmatrix} [R_s] & [0] \\ [0] & [R_r] \end{bmatrix} \begin{bmatrix} [I_s] \\ [I_r] \end{bmatrix} + \frac{d}{dt} \begin{bmatrix} [\Psi_s] \\ [\Psi_r] \end{bmatrix} \tag{5.3}$$

$$\begin{bmatrix} [\Psi_s] \\ [\Psi_r] \end{bmatrix} = \begin{bmatrix} [L_{ss}] & [L_{sr}] \\ [L_{rs}] & [L_{rr}] \end{bmatrix} \begin{bmatrix} [I_s] \\ [I_r] \end{bmatrix} \tag{5.4}$$

Then, they could be written as the following:

$$[U_s] = [R_s][I_s] + \frac{d[\Psi_s]}{dt} \tag{5.5}$$

$$[0] = [R_r][I_r] + \frac{d[\Psi_r]}{dt} \tag{5.6}$$

$$[\Psi_s] = [L_{ss}][I_s] + [L_{sr}][I_r] \tag{5.7}$$

$$[\Psi_r] = [L_{rs}][I_s] + [L_{rr}][I_r] \tag{5.8}$$

### 5.1.2 Mechanical subsystem equations

Rotor rotation depends on the load characteristics; therefore, the governing equations differ in various applications. It is assumed that the load torque which opposes the developed electromagnetic torque consists of an internal moment inertia of rotor and load, whose characteristics are known in advance explicitly. The well-known equation describing the rotor motion is:

$$T_{em} - T_L = J\frac{d\omega}{dt} = J\frac{d^2\theta}{dt^2} \tag{5.9}$$

where $T_{\text{em}}$ is the developed electromagnetic torque of the motor, $T_L$ is the load torque, $J$ is the rotor and load inertia referred to the rotor side, $\omega$ is the angular speed of the rotor and $\theta$ is mechanical angle in radians. Electromagnetic torque of the motor developed in the electromechanical conversion process is derivative of the magnetic coenergy regarding the spatial coordinate at constant stator and rotor currents:

$$T_{em} = \left(\frac{\partial W_{co}}{\partial \theta}\right)\bigg|_{I_s, I_r = const} \tag{5.10}$$

For the proposed linear magnetic circuit, i.e., infinite magnetic permeability of stator and rotor core, the magnetic coenergy is equal to the stored magnetic energy in the air gap as follows:

$$W_{co} = \frac{1}{2}\left[[I_s]^T [I_r]^T\right]\begin{bmatrix}[L_{ss}] & [L_{sr}] \\ [L_{rs}] & [L_{rr}]\end{bmatrix}\begin{bmatrix}[I_s] \\ [I_r]\end{bmatrix} \tag{5.11}$$

i.e.,

$$W_{co} = \frac{1}{2}[I_s]^T[L_{ss}][I_s] + \frac{1}{2}[I_s]^T[L_{sr}][I_r] + \frac{1}{2}[I_r]^T[L_{rs}][I_s] + \frac{1}{2}[I_r]^T[L_{rr}][I_r] \tag{5.12}$$

In case of uniform air gap for symmetrical machines, it is obvious that matrices $[L_{ss}]$ and $[L_{rr}]$ are constant and $[L_{rs}] = [L_{sr}]^T$. By substituting (5.12) into the equation for $T_{em}$, (5.10), we have

$$T_{em} = \frac{1}{2}[I_s]^T\frac{\partial[L_{sr}]}{\partial\theta}[I_r] + \frac{1}{2}[I_r]^T\frac{\partial[L_{sr}]^T}{\partial\theta}[I_s] \tag{5.13}$$

As $T_{\text{em}}$ is a scalar, the two terms of the right-hand side of (5.13) should be also scalars. On the other hand, transposition of the scalar is the scalar itself, and the following equation can be written:

$$[I_r]^T \frac{\partial [L_{sr}]^T}{\partial \theta} [I_s] = \left[ [I_r]^T \frac{\partial [L_{sr}]^T}{\partial \theta} [I_s] \right]^T \tag{5.14}$$

From matrix algebra, it is well known that

$$\begin{bmatrix} A^T & B^T & C \end{bmatrix}^T = \begin{bmatrix} C^T & B & A \end{bmatrix} \tag{5.15}$$

so

$$[I_r]^T \frac{\partial [L_{sr}]^T}{\partial \theta} [I_s] = [I_s]^T \frac{\partial [L_{sr}]}{\partial \theta} [I_r] \tag{5.16}$$

i.e., both members on the right-hand side of (5.13) are equal. Finally

$$T_{em} = [I_s]^T \frac{\partial [L_{sr}]}{\partial \theta} [I_r] \tag{5.17}$$

In an induction motor with mechanical asymmetries, which is the case in static or dynamic eccentricities, inductances of stator and/or rotor windings depend on rotor angular position, and the developed electromagnetic torque is calculated by taking the derivative of general equation for magnetic coenergy into account (5.12).

### *5.1.3 Model parameters*

#### 5.1.3.1 Inductances

The key parameters in any mathematical model of an electrical machine are inductances. In this model, the inductance of the windings in the machine, self or mutual, will be calculated using winding functions [10]. As shown in Chapter 4, winding $A$ having $N_A$ turns and carrying the current $i_A$ generates the MMF wave:

$$F_A(\theta) = N_A(\theta) \cdot i_A \tag{5.18}$$

where $N_A(\theta)$ is the winding function of winding $A$. This winding is shown as one coil in Figure 5.1. Assume that there is another winding, winding $B$ with $N_B$ turns, that is placed along the machine circumference on the totally arbitrary fashion. Considering stationary rotor, these windings are on the stator or on the rotor. In order to calculate mutual inductance between these two windings, magnetic flux-linkage of winding $B$ due to winding $A$ current should be calculated. The relationship between the magnetic flux, MMF and magnetic permeance is as follows:

$$\Psi = F \cdot \Lambda \tag{5.19}$$

The elementary magnetic flux passing through the air gap is

$$d\Psi = F_A(\theta) \mu_0 r l \frac{d\theta}{g_0} \tag{5.20}$$

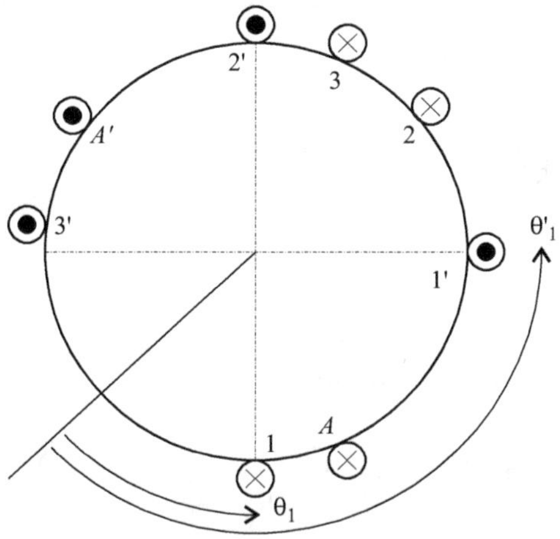


*Figure 5.1 Mutual inductance calculation*

All coils forming winding $B$ are shown and labeled in accordance with direction that is chosen as positive, Figure 5.1. It should be noted that it is necessary to have the same number of conductors that are labeled with $\times$ or with $\cdot$. The magnetic flux linkage of coil 1–1′ is as the following:

$$\Psi_{1-1'} = \frac{\mu_0 rl}{g_0}\int_{\theta_1}^{\theta_1'} F_A(\theta)d\theta \tag{5.21}$$

As turns function of coil 1–1′ has unit value on the angle range of $\theta_1 \le \theta \le \theta'_1$, while for all other angles it has zero value, previous expression could be given alternatively as:

$$\Psi_{1-1'} = \frac{\mu_0 rl}{g_0}\int_0^{2\pi} n_{B1}(\theta)F_A(\theta)d\theta \tag{5.22}$$

Similarly, flux-linkage of other coils in winding $B$ could be defined. If winding $B$ consists of $N_B$ coils, the total magnetic flux linkage of winding $B$, due to winding $A$ current is:

$$\Psi_{BA} = \sum_{i=1}^{N_B}\Psi_{i-i'} = \frac{\mu_0 rl}{g_0}\left(\sum_{i=1}^{N_B}\int_0^{2\pi} n_{Bi}(\theta)F_A(\theta)d\theta\right) = \frac{\mu_0 rl}{g_0}\int_0^{2\pi}\left[\sum_{i=1}^{N_B} n_{Bi}(\theta)\right]F_A(\theta)d\theta \tag{5.23}$$

The sum under the integral in (5.23) is turns function of winding $B$. Therefore, (5.23) will be as follows:

$$\Psi_{BA} = \frac{\mu_0 rl}{g_0}\int_0^{2\pi} n_B(\theta)F_A(\theta)d\theta \tag{5.24}$$

Mutual inductance between windings $B$ and $A$ is the ratio of winding $B$ flux-linkage and winding $A$ current. Hence,

$$L_{BA} = \frac{\Psi_{BA}}{i_A} = \frac{\mu_0 rl}{g_0} \int_0^{2\pi} n_B(\theta) N_A(\theta) d\theta \tag{5.25}$$

As from (4.25)

$$n_B(\theta) = N_B(\theta) + \widehat{n}_B \tag{5.26}$$

the mutual inductance will be as follows:

$$L_{BA} = \frac{\mu_0 rl}{g_0} \int_0^{2\pi} N_B(\theta) N_A(\theta) d\theta + \frac{\mu_0 rl}{g_0} \widehat{n}_B \int_0^{2\pi} N_A(\theta) d\theta \tag{5.27}$$

As winding function $N_A(\theta)$ is periodic function with zero mean value, the second integral in (5.27) is zero. Finally, mutual inductance is as the following:

$$L_{BA} = \frac{\mu_0 rl}{g_0} \int_0^{2\pi} N_B(\theta) N_A(\theta) d\theta \tag{5.28}$$

From (5.28), it is clear that the same expression is valid for mutual inductance between windings $A$ and $B$ as should be the case in linear magnetic circuit:

$$L_{AB} = \frac{\mu_0 rl}{g_0} \int_0^{2\pi} N_A(\theta) N_B(\theta) d\theta = L_{BA} \tag{5.29}$$

During the analysis, no assumption about the winding placement was made; both windings can be placed on the stator and on the rotor, or one winding could be on the stator and the other one on the rotor. Equation (5.28) is valid also for calculation of winding self-inductance:

$$L_{AA} = \frac{\mu_0 rl}{g_0} \int_0^{2\pi} N_A^2(\theta) d\theta \tag{5.30}$$

Thus, the general equation allowing the calculation of winding inductances in the machine is provided in (5.29). This equation calculates analytically in simpler cases or numerically using the standard numerical techniques when complicated forms of winding functions exist. In the following parts, both methods of calculation will be used.

- **CALCULATION OF STATOR WINDINGS INDUCTANCES**

Self- and mutual inductances of stator phase windings could be calculated by numerical integration of integrals that are parts of expressions (5.29) and (5.30) for the known phase winding scheme and geometrical details of the machine. The known winding scheme and the known number of turns in every coil of the phase winding mean the known turns and winding function of that phase winding.

It is worth noting that self-inductance of phase winding on the stator is due to the magnetic flux passing through the air gap as well as due to the leakage magnetic flux. Therefore, self-inductance of stator phase winding, i.e., $A$ is

$$L_A = L_{AA} + L_{\sigma A} = \frac{\mu_0 r l}{g_0}\int_0^{2\pi} N_A^2(\theta)d\theta + L_{\sigma A} \tag{5.31}$$

Leakage inductance $L_{\sigma A}$ could be calculated using the well-known methods described in many textbooks of standard electrical machines.

- **CALCULATION OF ROTOR WINDINGS INDUCTANCES**

In wound rotor induction machine type, self- and mutual inductances of rotor windings are calculated in the same manner as for the stator winding.

For a cage rotor with $R$ bars, the winding function of one rotor loop consisting of two neighboring bars and ring segments between them is:

$$N_{\text{loop}} = \begin{cases} 1 - \frac{1}{R}, & \theta_1 \leq \theta \leq \theta_2 \\ -\frac{1}{R}, & \text{for other } \theta \end{cases} \tag{5.32}$$

By substituting (2.32) into (2.30) rotor loop self inductance can be easily calculated. We get the following result:

$$L_{\text{self}} = \frac{\mu_0 r l}{g_0}\frac{2\pi}{R^2}(R-1) \tag{5.33}$$

Similarly, the mutual inductance between any two rotor loops as a result of magnetic flux that is closed through the air gap is

$$L_{\text{mutual}} = -\frac{\mu_0 r l}{g_0}\frac{2\pi}{R^2} \tag{5.34}$$

Taking the leakage inductances of the rotor bars $L_b$ and end-ring segments $L_{\text{ers}}$ into account, the rotor inductances matrix could be given as [8]:

$$[L_{rr}] = \begin{bmatrix} L_1 & L_2 & L_3 & L_3 & . & . & L_2 \\ L_2 & L_1 & L_2 & L_3 & . & . & L_3 \\ L_3 & L_2 & L_1 & L_2 & . & . & L_3 \\ . & . & . & . & . & . & . \\ . & . & . & . & . & . & . \\ L_3 & L_3 & L_3 & L_3 & . & . & L_2 \\ L_2 & L_3 & L_3 & L_3 & . & . & L_1 \end{bmatrix} \tag{5.35}$$

where

$$L_1 = L_{\text{self}} + 2(L_b + L_{ers}) \tag{5.36}$$

$$L_2 = L_{\text{mutual}} - L_b \tag{5.37}$$

$$L_3 = L_{\text{mutual}} \tag{5.38}$$

For the sake of completeness of the presented material, the relation between the rotor parameters in multiple-coupled circuit model and in the standard single-phase equivalent circuit of induction motor is provided. A cage rotor winding with $R$ rotor bars and $2R$ end-ring segments could also be observed as $R$-phase winding where one phase consists of one rotor bar and two associated end-ring segments. Using arguments of magnetic energy invariance, the equivalent inductance of one cage rotor "phase" is

$$L_{r\_\text{phase}} = L_b + \frac{L_{ers}}{2\sin^2\left(\frac{p\pi}{R}\right)} \tag{5.39}$$

and this refers to the stator side of $m_1$-phase machine as follows:

$$L_r' = \frac{4m_1}{R}\left(\frac{k_{w1}w_1}{k_{\text{skew}}}\right)^2 L_{r\_\text{phase}} \tag{5.40}$$

This is the inductance that occurs in one-phase steady-state equivalent circuit of the induction machine. More precisely, there it appears as leakage reactance:

$$X_r' = \omega_1 L_r' = 2\pi f_1 L_r' \tag{5.41}$$

The leakage inductances depend on the shape and dimensions of the rotor bars and the end-ring segments. For example, for the rotor bar and end-ring of shapes that are given in Figure 5.2, the appropriate leakage inductances are [11]

$$L_b = \mu_0 l_b \lambda_b \cong \mu_0 l_b \left(0.66 + \frac{2h_r}{3(d_1+d_2)} + \frac{h_{or}}{b_{or}}\right) \tag{5.42}$$

$$L_{ers} = \mu_0 l_{ers} \lambda_{ers} = \mu_0 \cdot \frac{\pi(D_{er}-b)}{R} \cdot \left[0.46 \cdot \log\left(\frac{2.35 \cdot (D_{er}-b)}{2a+b}\right)\right] \tag{5.43}$$

- **MUTUAL INDUCTANCE BETWEEN STATOR AND ROTOR WINDINGS**

In this case, two windings are on the opposite side of the air gap. Due to the rotor rotation, the mutual inductance between stator and rotor windings are spatially

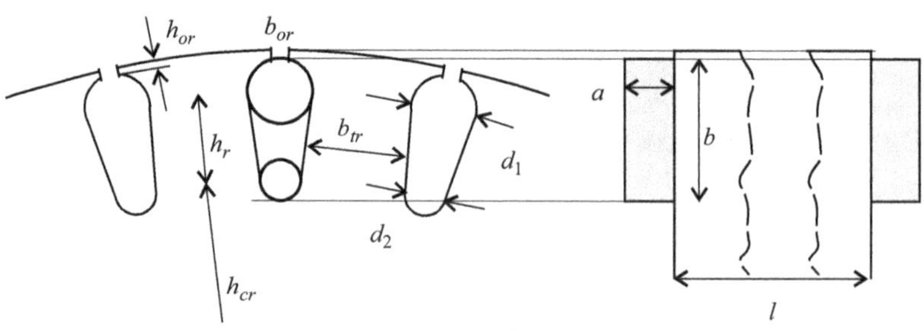


*Figure 5.2 Rotor slot details*

(time) dependent. In other words, during the mutual inductance calculation, the rotor position must be taken into account. Rotor rotation means rotation of all turns and winding functions that are defined for rotor windings. Therefore, calculation of mutual inductance means calculation of integral (5.29) for every different rotor position.

Let us look at the simple example of Figure 5.3 where one coil from the stator side and one rotor loop are involved [9].

Assuming that the pitch of the rotor loop is smaller than the pitch of the stator coil, during the rotor rotation, rotor loop can be found in four different positions with respect to the stationary coil on the stator side (Figure 5.3). Through knowing the winding functions of rotor loop and stator coil, the following equations could be easily derived:

$$L_{sirj}(\theta) = \begin{cases} \dfrac{\mu_0 r l N_{si}}{g_0}\alpha_{rj}\left(1 - \dfrac{\alpha_{si}}{2\pi}\right) & \theta_{i1} \le \theta_{j1} < \theta_{j2} \le \theta_{i2} \\ \dfrac{\mu_0 r l N_{si}}{g_0}\left(\theta_{i2} - \theta_{j1} - \dfrac{\alpha_{si}\alpha_{jr}}{2\pi}\right) & \theta_{i1} < \theta_{j1} < \theta_{i2} < \theta_{j2} \\ -\dfrac{\mu_0 r l N_{si}}{g_0}\dfrac{\alpha_{si}\alpha_{rj}}{2\pi} & \theta_{i1} < \theta_{i2} \le \theta_{j1} < \theta_{j2} \\ \dfrac{\mu_0 r l N_{si}}{g_0}\left(\theta_{j2} - \theta_{i1} - \dfrac{\alpha_{si}\alpha_{rj}}{2\pi}\right) & \theta_{i1} \le \theta_{j2} < \theta_{i2} < \theta_{j1} \end{cases} \tag{5.44}$$

However, in case of real stator phase winding that is of a more complicated form, the only reasonable manner for calculation of mutual inductance between the stator phase winding and rotor loop is numerical integration of expression (5.29). During this calculation one should bear in mind that rotor rotation also means rotation of the rotor winding functions, i.e., for one position of the rotor, one value of mutual inductance is obtained. For the next rotor position, the next value is obtained and so forth. The main idea is that one rotor revolution should be discretized in as many steps as it is meaningful. For example, one rotor revolution could be divided into 4,000 steps in such a way that the lookup table with 4,000 elements is obtained. In numerical procedure of integration of differential equations that describes the machine for every new rotor position, new value of mutual inductance could be pulled out from that lookup table.

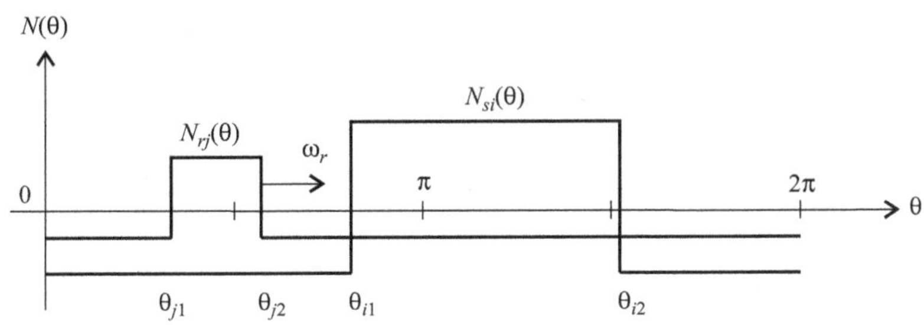


*Figure 5.3 Calculation of mutual inductance between one stator coil and one rotor loop*

In the case of healthy symmetrical machine, it will suffice to calculate one such lookup table that consists of mutual inductance between phase winding $A$ and rotor loop 1. All other mutual inductances could be obtained from this lookup table. For example, for the lookup table with 4,000 entries, the mutual inductance between phase $A$ and rotor loop 2, for one rotor position, is shifted for $4{,}000/R$ places regarding the mutual inductance between the same phase and rotor loop 1. The mutual inductance between phase winding $B$ and rotor loop 1 is shifted for $2\pi/3p$, i.e., $4{,}000/3p$ places and so forth, where $p$ is the number of pole pairs.

#### 5.1.3.2 Resistances

Matrix of stator winding resistances is simply diagonal matrix whose entries are stator phase winding resistances.

Matrix of rotor windings resistances $[R_r]$ is symmetrical and right cyclical as it is the matrix of rotor windings inductances $[L_{rr}]$.

Right cyclic matrix means that each row of that matrix could be obtained from previous row pushing that row for one place on right side and placing dropped out element in the first column of the new row. Dimension of this matrix is $R\times R$ and it is formed in the following manner. Let us look at the developed scheme of cage rotor winding illustrated in Figure 5.4.

Every loop (mesh) current is closed through the two bars and two ring segments. In addition, two neighboring mesh currents flow through one common bar in opposite directions. The matrix of rotor resistances will have the following entries,

$$[R_r] = \begin{bmatrix} R_1 & R_2 & 0 & 0 & . & . & R_2 \\ R_2 & R_1 & R_2 & 0 & . & . & 0 \\ 0 & R_2 & R_1 & R_2 & . & . & 0 \\ . & . & . & . & . & . & . \\ . & . & . & . & . & . & . \\ 0 & 0 & 0. & 0 & . & . & R_2 \\ R_2 & 0 & 0 & 0 & . & . & R_1 \end{bmatrix} \tag{5.45}$$

where $R_1 = 2(R_b + R_{ers})$ and $R_2 = -R_b$. Rotor bar resistance $R_b$ and ring segment resistances, i.e., $R_{ers}$, could be easily calculated from the known details about cage

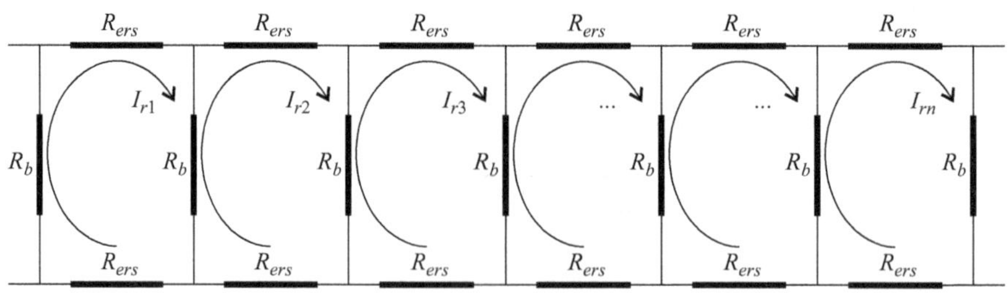


*Figure 5.4 Developed scheme of cage rotor winding*

rotor windings: cross-section area, length, specific resistance of cage rotor winding material, temperature, etc. Again, for the sake of completeness of the presented material, here will be given the relation between the rotor parameters in multiple-coupled circuit model and in the standard single-phase equivalent circuit of induction motor.

A cage rotor winding with $R$ rotor bars and $2R$ end-ring segments could also be considered as $R$-phase winding where one phase consists of one rotor bar and two associated end-ring segments. The relationship between the rms currents in the rotor bar and the end-ring segment in a machine with $p$ pair of poles is well known and given by [11]

$$I_b = 2I_{ers} \sin\left(\frac{p\pi}{R}\right) \tag{5.46}$$

The equivalent resistance of one cage rotor "phase" could be obtained from the condition of Joule power invariance:

$$R_{r\_\text{phase}} = R_b + \frac{R_{ers}}{2\sin^2\left(\frac{p\pi}{R}\right)} \tag{5.47}$$

This resistance referred to the stator side of $m_1$-phase machine is

$$R_r' = \frac{4m_1}{R}\left(\frac{k_{w1}w_1}{k_\text{skew}}\right)^2 R_{r\_\text{phase}} \tag{5.48}$$

where $k_{w1}w_1$ is the effective turns number of the stator phase winding and $k_\text{skew}$ is the skew factor for fundamental harmonic. For skewing of rotor bars for one stator slot pitch, what is the most common case in practice, this factor has the following value,

$$k_\text{skew} = \frac{Q_s}{p\pi}\sin\left(\frac{p\pi}{S}\right) \tag{5.49}$$

where $S$ is the number of stator slots.

Resistance calculated in (5.48) is resistance that occurs in a one-phase steady-state equivalent circuit of an induction machine.

## 5.2 Skewing of rotor bars

In the majority of induction machines and those with wound rotor and cage rotor, the rotor bars (slots) are skewed usually for one stator slot pitch. In this way, one will get calmer, quieter machines, more uniform torque, and the possibility of rotor locking will be largely eliminated, significantly reducing the torque dip which can cause "taxiing" of the rotor [12]. On the other hand, skewing of rotor bars has also negative effects such as developing less electromagnetic torque in the entire speed range due to the additional reactance. Moreover, as one of the negative consequences of skewing is the emergence of the so-called interbar currents, or currents to be closed between two bars through the lamination of the rotor magnetic core [13]. However, the positive effects of skewing are more significant than the negative ones, and skewing is applied to the vast majority of induction machines. It should be noted that skewing has effects only on the mutual inductance between stator and rotor windings.

Now skewing of rotor bars is taken into account in the calculation of stator-rotor mutual inductance. For reasons of clarity, (5.29) is rewritten as follows:

$$L_{ij}(\theta) = \frac{\mu_0 rl}{g_0} \int_0^{2\pi} N_i(\theta) N_j(\theta) d\theta \tag{5.50}$$

where indices $i$ and $j$ are related to $i$th stator winding and $j$th rotor winding. Validity of (5.50) is restricted when rotor bars (stator slots) are unskewed. It might be rewritten as follows:

$$L_{ij}(\theta) = l \cdot L_{ij}{}'(\theta) \tag{5.51}$$

where $L_{ij}{}'(\theta)$ is the mutual inductance between the stator winding and rotor loop, per unit of length, in the case of unskewed rotor bars. When the rotor bars are skewed uniformly, the mutual inductance between the stator winding $i$ and rotor loop $j$ can be defined per unit length as follows [14]:

$$L'_{ij\_skewed}(\theta, \xi) = L'_{ij}\left(\theta \pm \xi \frac{\gamma}{l}\right) \tag{5.52}$$

where $\xi$ represents the length along the axial direction of the rotor, $\xi \in (-l/2, l/2)$ and $\gamma$ is the mechanical angle of skewing in radians. When $\xi$ increases and the skewing of rotor bars is in a positive direction of $\theta$, then the sign in brackets of (5.52) is positive. Equation (5.52) can be interpreted as follows: in any radial crosssection of the machine, the mutual inductance per unit length, between stator winding $i$ and rotor loop $j$, has the same shape, but it is displaced in space, as shown in Figure 5.5.

Total mutual inductance in the case of skewed rotor bars is determined when (5.52) is integrated along the axial axis of the rotor:

$$L_{ij\_\text{skewed}}(\theta) = \int_{-\frac{l}{2}}^{\frac{l}{2}} L'_{ij\_\text{skewed}}(\theta, \xi) d\xi \tag{5.53}$$

It is apparent that the total mutual inductance between the stator and rotor winding, obtained by described method becomes function of one variable.

In other words, an obvious two-dimensional problem is reduced to one dimension, i.e., we again get the inductance as a function of spatial coordinate only. The described method can be applied to any case of a linear skewing of slots within the electrical machine, whether it is on the stator or rotor side. This approach takes all the spatial harmonics into account in contrast to a number of other well-known

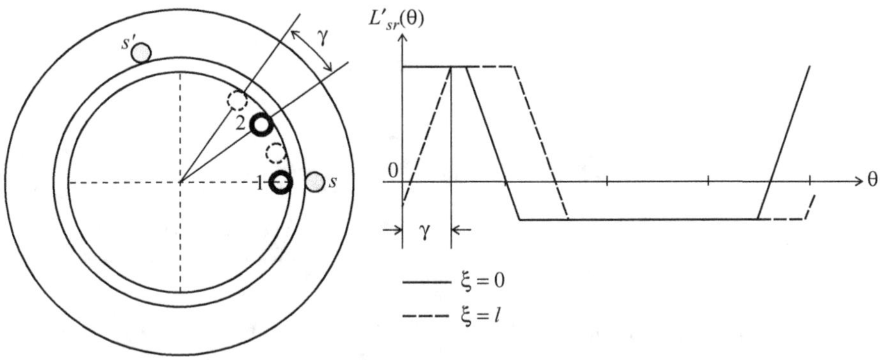


*Figure 5.5 Cross-section of machine for $\xi = 0$ and $\xi = 1$ and appropriate mutual inductance per unit length dependences, between stator coil s–s′ and rotor loop 1 (between the rotor bars 1 and 2)*

methods [8], which use the so-called integral factor on the harmonic basis and each harmonic of mutual inductance multiplied by them.

Figure 5.5 illustrates the effect of skewing of rotor bars on mutual inductance per unit length waveforms for $\xi = 0$ and $\xi = l$.

## 5.3 Linear rise of MMF across slot

Winding function approach is originally based on point conductor approximation. In order to consider the linear rise of MMF across the slot, winding function definition could be slightly changed as it is explained below [14]. Winding function of coil in which the effect of linear rise of MMF across the slot opening is considered is shown in Figure 5.6.

It is simple to propose a definition of winding function in this case. Referring to Figure 5.6, the winding function is as follows:

$$N_i(\theta) = \begin{cases} \dfrac{N_i}{2\pi\delta}[2\pi(\theta - \theta_{i1}) + \delta(\pi - \alpha_i)], & \theta_{i1} - \dfrac{\delta}{2} \le \theta \le \theta_{i1} + \dfrac{\delta}{2} \\ N_i\left(1 - \dfrac{\alpha_i}{2\pi}\right), & \theta_{i1} + \dfrac{\delta}{2} \le \theta \le \theta_{i2} - \dfrac{\delta}{2} \\ N_i\left[\left(1 - \dfrac{\alpha_i}{2\pi}\right) + \dfrac{\theta_{i2} - \theta}{\delta} - \dfrac{1}{2}\right], & \theta_{i2} - \dfrac{\delta}{2} \le \theta \le \theta_{i2} + \dfrac{\delta}{2} \\ -N_i\dfrac{\alpha_i}{2\pi}, & \text{for other values of } \theta \end{cases} \tag{5.54}$$

where $\delta$ is the slot opening width in radian.

Now, the mutual inductance between the coil and the rotor loop changes. There are two time more different positions between the coil and rotor loop.

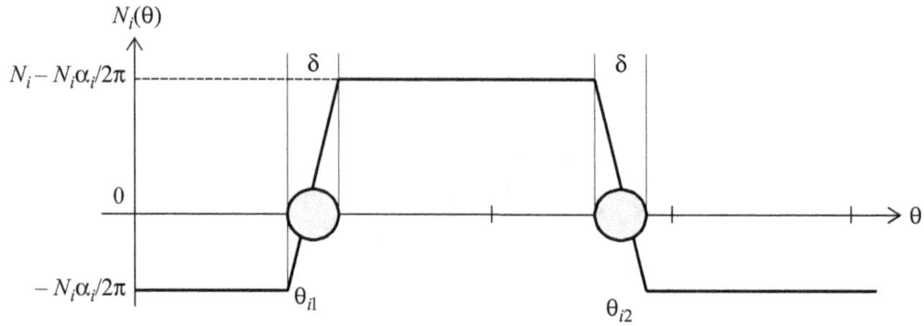


*Figure 5.6 Winding function of single coil. Linear rise of MMF across the slot opening is taken into account*

Mutual inductance equation as a function of rotor position in angular coordinate $\theta$ is as follows:

1. $\theta_{j1} < \theta_{j2} \leq \theta_{i1} - \dfrac{\delta}{2}$

$$L'_{sr}(\theta) = \frac{\mu_0 r}{g_0}\left(-\frac{N_i\alpha_i}{2\pi}\alpha_j\right)$$

2. $\theta_{j1} < \theta_{i1} - \dfrac{\delta}{2} < \theta_{j2} \leq \theta_{i1} + \dfrac{\delta}{2}$

$$L'_{sr}(\theta) = \frac{\mu_0 r}{g_0}\left[\frac{N_i}{2\delta}(\theta_{i1} - \theta_{j2})^2 + \delta(\pi - \alpha_i)\left(\frac{\delta}{2} + \theta_{j2} - \theta_{i1}\right) + \frac{N_i\alpha_i\delta}{2\pi}\left(\frac{1}{2} - \frac{\pi}{4\alpha_i} - \frac{\theta_{i1} - \theta_{j1}}{\delta}\right)\right]$$

3. $\theta_{j1} \leq \theta_{i1} - \dfrac{\delta}{2} < \theta_{i1} + \dfrac{\delta}{2} < \theta_{j2}$

$$L'_{sr}(\theta) = \frac{\mu_0 r N_i}{g_0}\left[\theta_{i2} - \theta_{i1} - \frac{\alpha_i\alpha_j}{2\pi}\right]$$

4. $\theta_{i1} - \dfrac{\delta}{2} < \theta_{j1} \leq \theta_{i1} + \dfrac{\delta}{2} < \theta_{j2}$

$$L'_{sr}(\theta) = \frac{\mu_0 r N_i}{2g_0}\left[\alpha_j + \theta_{j2} - \theta_{i1} - \frac{\delta}{4} - \frac{\alpha_i\alpha_j}{\pi} - \frac{(\theta_{j1} - \theta_{i1})^2}{\delta}\right]$$

5. $\theta_{i1} + \dfrac{\delta}{2} < \theta_{j1} < \theta_{j2} \leq \theta_{i2} - \dfrac{\delta}{2}$

$$L'_{sr}(\theta) = \frac{\mu_0 r N_i}{g_0}\left(1 - \frac{\alpha_i}{2\pi}\right)\alpha_j$$

6. $\theta_{j1} < \theta_{i2} - \frac{\delta}{2} < \theta_{j2} \le \theta_{i2} + \frac{\delta}{2}$

$$L'_{sr}(\theta) = \frac{\mu_0 r N_i}{g_0}\left[\left(1 - \frac{\alpha_i}{2\pi}\right)\alpha_j + \frac{\theta_{i2} - \theta_{j2}}{2} - \frac{(\theta_{i2} - \theta_{j2})^2}{2\delta} - \frac{\delta}{8}\right]$$

7. $\theta_{j1} \le \theta_{i2} - \frac{\delta}{2} < \theta_{i2} + \frac{\delta}{2} < \theta_{j2}$

$$L'_{sr}(\theta) = \frac{\mu_0 r N_i}{g_0}\left[\theta_{i2} - \theta_{j1} - \frac{\alpha_i \alpha_j}{2\pi}\right]$$

8. $\theta_{i2} - \frac{\delta}{2} < \theta_{j1} \le \theta_{i2} + \frac{\delta}{2} < \theta_{j2}$

$$L'_{sr}(\theta) = \frac{\mu_0 r N_i}{g_0}\left[\frac{\delta}{8} - \frac{\alpha_i \alpha_j}{2\pi} + \frac{\theta_{i2} - \theta_{j1}}{2} + \frac{(\theta_{i2} - \theta_{j1})^2}{2\delta}\right]$$

However, it can be easily proved that the linear rise of MMF across the slot and skewing of rotor bars has a very qualitatively similar effect on the mutual inductance waveform between stator winding and rotor loop [14]. Therefore, it will suffice to take one of the two mentioned effects into consideration. The experience shows that it is always much easier to consider the effect of skewing of rotor bars than to modify winding function definition in order to take linear rise of MMF across the slot opening into account.

---

**Example 5.1** Let us analyze a four-pole induction motor with the following nameplate data:

11 kW, 400 V, Y, 50 Hz, $p = 2$,

The number of stator slots are $S = 36$ and the motor has $R = 30$ rotor bars. The stator phase winding consists of 12 coils, 3 coils per pole with 9 turns in one coil, i.e., $N = 108$ series turns per pole per phase. The coil pitch is shortened, i.e., $y = (7/9)\tau$.

Phase winding scheme under one pole is: 1-8′-2-9′-3-10′.

Additional machine parameters are

$R_{\text{phase}} = 0.296\ \Omega$, $D_{\text{inner\_stator}} = 146.36$ mm, $g_0 = 0.4$ mm, $l_{\text{axial}} = 172.42$ mm
$L_{\sigma s} = 2.92$ mH, $R_b = 64.5\ \mu\Omega$, $R_{ers} = 1.55\ \mu\Omega$, $L_b = 400$ nH, $L_{ers} = 5.73$ nH,
$J = 0.3$ kg m$^2$

Self- and mutual inductances of stator phase windings could be easily calculated for known phase winding functions through using winding function approach. It could be easily proved that these values are

$$L_{\text{self\_phase}} = 123.4 \text{ mH}, \quad L_{\text{mutual\_between\_phases}} = -58.4 \text{ mH}$$

Cage rotor inductances are (5.33) and (5.34):

$$L_{\text{self}} = 8.06\ \mu\text{H}, \quad L_{\text{mutual}} = -0.279\ \mu\text{H}$$

Mutual inductance between stator phase and rotor loop as well as their derivative with respect to the angular position (needed for electromagnetic torque calculation (5.17)) for three different cases regarding skewing of rotor bars are given in the following figures.

Figure E5.1.1 shows the case where skewing of rotor bars are not taken into account. The maximum value of the mutual inductance in this case is 0.225 mH. Derivative of this function with respect to the angular position is also given in the figure. Due to the point conductor approximation, this function is rugged stepwise function which results in high spikes in the time-domain representation of electromagnetic torque as it will be illustrated later.

Figure E5.1.2 displays the common case where rotor bars are skewed for one stator slot pitch, i.e., for $\gamma = 2\pi/36$ radians. It is clear that the mutual inductance and its derivative wave shapes are much closer to the real situation.

Figure E5.1.3 shows the case that is not real as rotor bars are skewed more than one stator slot pitch, i.e., skew in this case was $\gamma = 2\pi/24$ radians. This case is chosen just for an illustration of mutual inductance wave shape for higher value of rotor bars skew angle.

As an illustration of the significant improvement of the model by taking into consideration the skewing of rotor bars, Figure E5.1.4 shows the results for developed electromagnetic torque obtained from the numerical model. The abnormal high spikes in electromagnetic torque that are the results of discontinuities in the derivative of mutual inductance, Fig. E5.1.1, are obviously significantly

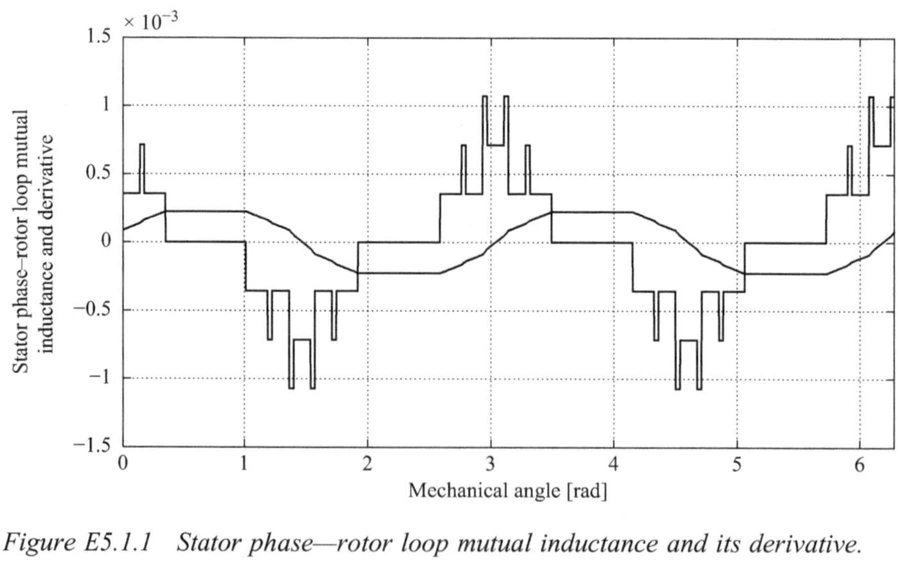


*Figure E5.1.1 Stator phase—rotor loop mutual inductance and its derivative. Skewing of rotor bars is not taken into account, $\gamma = 0$*

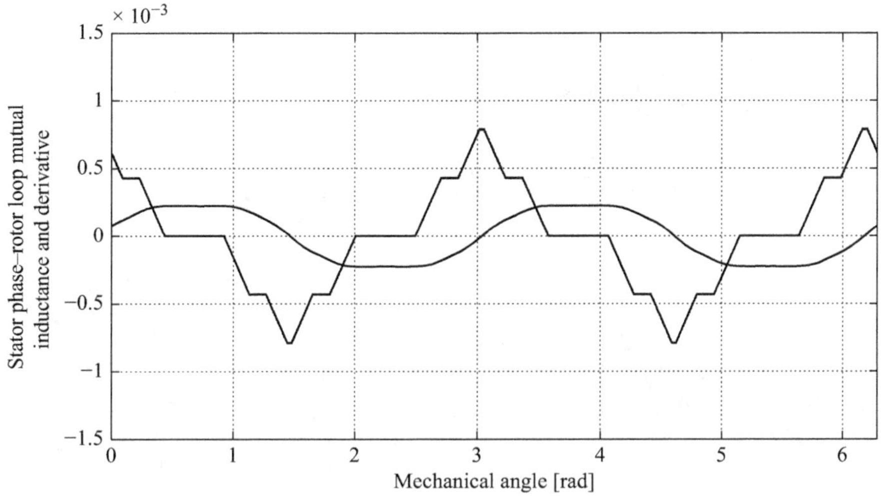


*Figure E5.1.2 Stator phase—rotor loop mutual inductance and its derivative. Rotor bars are skewed for $\gamma = 2\pi/36$ rad which is the common case—skewing for one stator slot width*

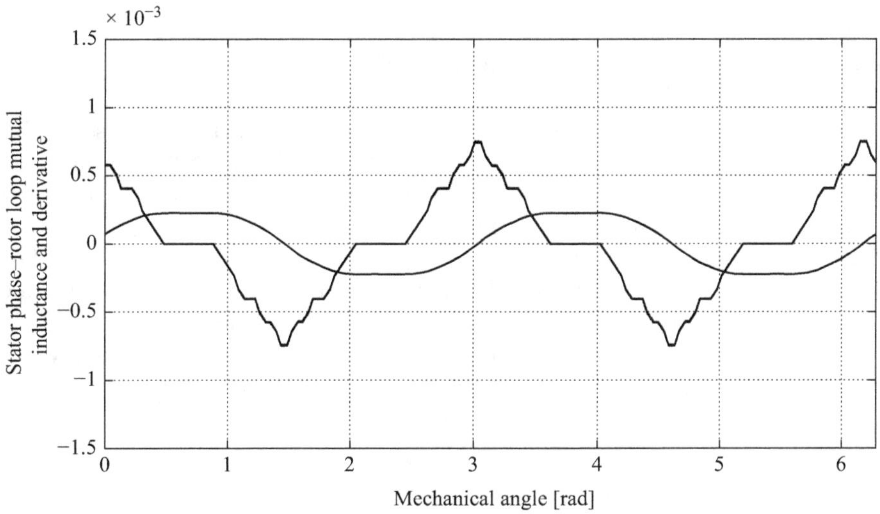


*Figure E5.1.3 Stator phase—rotor loop mutual inductance and its derivative. Rotor bars are skewed for $\gamma = 2\pi/24$ rad*

reduced. Due to the unreal high spikes in derivative of mutual inductance wave shape and, consequently, due to the high spikes in electromagnetic torque, speed-up time for machine with straight rotor bars is significantly shorter than in case with skewed rotor bars.

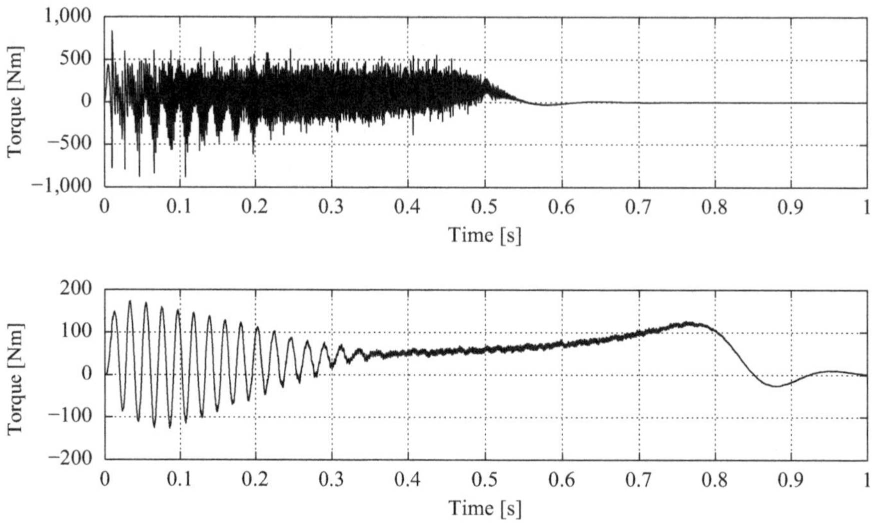


*Figure E5.1.4 Electromagnetic torque from the numerical model during the acceleration of an unloaded motor. Upside down: $\gamma = 0$ rad, $\gamma = 2\pi/36$ rad*

## 5.4 Solution of mathematical model

Equations (5.5)–(5.8) together with equations of mechanical part of the system, (5.9) and (5.17), provide a closed system of equations that fully describes an induction machine with known parameters. There are many numerical techniques for solution of such system of differential equations—system of nonlinear differential equations with time-varying coefficients: Runge–Kutta methods, Newton–Raphson method, etc. To solve such a system of equations, the most primitive method such as Euler method could be effectively used if step of integration is small enough.

The way to solve the above system of equations, using Euler method, will be shown in detail here. The system of (5.5)–(5.8), in addition to (5.9) and (5.17), is solved by fluxes and indirectly by currents:

$$\frac{d[\Psi_s]}{dt} = [U_s] - [R_s][I_s] \tag{5.55}$$

$$\frac{d[\Psi_r]}{dt} = -[R_r][I_r] \tag{5.56}$$

$$[I_s] = [L_{ss}]^{-1}\left[[\Psi_s] - [L_{sr}][K]^{-1}[L]\right] \tag{5.57}$$

$$[I_r] = [K]^{-1}[L] \tag{5.58}$$

where

$$[K] = [L_{sr}]^T[L_{ss}]^{-1}[L_{sr}] - [L_{rr}] \tag{5.59}$$

$$[L] = [L_{sr}]^T[L_{ss}]^{-1}[\Psi_s] - [\Psi_r] \tag{5.60}$$

$$T_{em} = [I_s]^T \frac{\partial [L_{sr}]}{\partial \theta}[I_r] \tag{5.61}$$

$$\frac{d\omega}{dt} = \frac{1}{J}(T_{em} - T_L) \tag{5.62}$$

$$\frac{d\theta}{dt} = \omega \tag{5.63}$$

The algorithm, by means of which the numerical integration of system of (5.55)–(5.63) could be performed, can be described in the following steps:

1. feeding all constant machine parameters as well as calculating all parameters that are not time- or space-dependent: $[R_s]$, $[R_r]$, and in case of uniform air gap $[L_{ss}]$ and $[L_{rr}]$;
2. calculating the mutual inductance between the stator phases and rotor loops $[L_{sr}(\theta)]$ as well as derivative of these functions with respect to angle $\theta$ $[DL_{sr}(\theta)]$;
3. defining initial conditions, i.e., $[\Psi_s(0)]$, $[\Psi_r(0)]$, $\theta(0)$, $\omega(0)$. Through these initial conditions, the initial conditions for currents $[I_s(0)]$ and $[I_r(0)]$ are known;
4. initializing $k$ and defining step of integration, $\Delta$ (sec);
5. defining phase or line voltages at $t=0$ if $k=1$, otherwise for $t=k\cdot\Delta$. Loading values for $[L_{sr}(\theta)]$ and $[DL_{sr}(\theta)]$ for $\theta=\theta(0)$ if $k=1$, otherwise for $\theta=\theta(k)$;
6. calculating the flux linkages, currents, torque, speed and rotor angular position according to the following iterative procedure:

$$[\Psi_s]_{k+1} = [\Psi_s]_k + \Delta \cdot \left[[U_s]_k - [R_s][I_s]_k\right]$$

$$[\Psi_r]_{k+1} = [\Psi_r]_k - \Delta \cdot [R_r][I_r]_k$$

$$[K]_k = [L_{sr}]_k^T[L_{ss}]^{-1}[L_{sr}]_k - [L_{rr}]$$

$$[L]_k = [L_{sr}]_k^T[L_{ss}]^{-1}[L_{sr}]_k - [L_{rr}]$$

$$[I_s]_{k+1} = [L_{ss}]^{-1}\left[[\Psi_s]_{k+1} - [L_{sr}]_k[K]_k^{-1}[L]_k\right]$$

$$[I_r]_{k+1} = [K]_k^{-1}[L]_k$$

$$T_{em_k} = [I_s]_{k+1}[DL_{sr}]_k[I_r]_{k+1}$$

$$\omega_{k+1} = \omega_k + \frac{\Delta}{J}(T_{em_k} - T_{L_k})$$

$$\theta_{k+1} = \theta_k + \Delta \cdot \omega_{k+1}$$

7. going back to step 5, loading new values for $[L_{sr}(\theta)]$ i $[DL_{sr}(\theta)]$ from the look-up tables for angle $\theta$ that is the result of previous iteration and so forth to the end of the predefined simulation time.

### 5.4.1 Stator phase windings connection

Multiple-coupled circuit model, along with the winding function approach for inductance calculation, is applicable to general induction machine with $m$ stator circuits and $n$ rotor bars [9]. All of them are assumed independent. Their mutual dependency could be established afterward, depending on the stator winding configuration and cage rotor configuration that could be irregular in case of some faults in rotor cage.

Mutual dependence between rotor currents in healthy rotor cage has been already established by defining rotor resistance matrix as well as defining matrix of inductance of cage rotor. The rotor currents are defined as mesh currents as every rotor loop current is connected with neighboring rotor loop current through the member $R_2 = -R_b$ in matrix of resistances or through $L_2 = L_{\text{mutual}} - L_b$ and $L_3 = L_{\text{mutual}}$ in inductance matrix.

Regarding the stator windings, currents in $m$ stator circuits have been already considered to be independent. However, in an actual machine, they are mutually dependent due to the different electrical connections. So the above model should be prepared to be used in two most common situations: wye and delta connections of stator windings.

#### 5.4.1.1 WYE connection

The described model is directly applicable to the case where the stator winding of induction machine has wye connection only in a symmetrical case and supplying the machine through a symmetrical three-phase (or multiphase) voltage. In this case, it is implicitly certain that the sum of stator currents is zero at any time. The input of the model is phase voltages.

As this model is developed to analyze general induction machines with $m$ stator circuits and $n$ rotor bars without any restrictions in terms of distributions and number of stator windings, symmetry or asymmetry of the machines, and air-gap profile, it is needed to make some modifications in the model as first Kirchhoff law could be explicitly defined. This is done in an original way in [9]. At the same time, the input of the model is line-to-line voltages that are only known. The details about the derivation and implementation into the model are provided.

Let us look at the system of equations that describes the three-phase induction machine:

$$\begin{bmatrix} U_a \\ U_b \\ U_c \end{bmatrix} = \begin{bmatrix} R_a & 0 & 0 \\ 0 & R_b & 0 \\ 0 & 0 & R_c \end{bmatrix} \begin{bmatrix} I_a \\ I_b \\ I_c \end{bmatrix} + \frac{d}{dt} \begin{bmatrix} \Psi_a \\ \Psi_b \\ \Psi_c \end{bmatrix} \tag{5.64}$$

$$\begin{bmatrix} \Psi_a \\ \Psi_b \\ \Psi_c \end{bmatrix} = \begin{bmatrix} L_{aa} & L_{ab} & L_{ac} \\ L_{ba} & L_{bb} & L_{bc} \\ L_{ca} & L_{cb} & L_{cc} \end{bmatrix} \begin{bmatrix} I_a \\ I_b \\ I_c \end{bmatrix} + \begin{bmatrix} L_{ar} \\ L_{br} \\ L_{cr} \end{bmatrix} [I_r] \tag{5.65}$$

where $L_{ar}$, $L_{br}$, $L_{cr}$ are $(1\times n)$ vectors. By subtracting each of the rows in (5.64) and (5.65), the following equations are obtained:

$$\begin{bmatrix} U_a - U_b \\ U_b - U_c \\ U_c - U_a \end{bmatrix} = \begin{bmatrix} R_a & -R_b & 0 \\ 0 & R_b & -R_c \\ -R_a & 0 & R_c \end{bmatrix} \begin{bmatrix} I_a - I_b \\ I_b - I_c \\ I_c - I_a \end{bmatrix} + \frac{d}{dt} \begin{bmatrix} \Psi_a - \Psi_b \\ \Psi_b - \Psi_c \\ \Psi_c - \Psi_a \end{bmatrix} \tag{5.66}$$

$$\begin{bmatrix} \Psi_a - \Psi_b \\ \Psi_b - \Psi_c \\ \Psi_c - \Psi_a \end{bmatrix} = \begin{bmatrix} L_{aa} - L_{ba} & L_{ab} - L_{bb} & L_{ac} - L_{bc} \\ L_{ba} - L_{ca} & L_{bb} - L_{cb} & L_{bc} - L_{cc} \\ L_{ca} - L_{aa} & L_{cb} - L_{ab} & L_{cc} - L_{ac} \end{bmatrix} \begin{bmatrix} I_a - I_b \\ I_b - I_c \\ I_c - I_a \end{bmatrix} + \begin{bmatrix} L_{ar} - L_{br} \\ L_{br} - L_{cr} \\ L_{cr} - L_{ar} \end{bmatrix} [I_r] \tag{5.67}$$

It is clear that new variables are obtained and, at the same time, line-to-line voltages are determined at the input of the model. By substituting the variables, the previous equations could be written as

$$\begin{bmatrix} U_{12} \\ U_{23} \\ U_{31} \end{bmatrix} = \begin{bmatrix} R_1 & -R_2 & 0 \\ 0 & R_2 & -R_3 \\ -R_1 & 0 & R_3 \end{bmatrix} \begin{bmatrix} I_1 \\ I_2 \\ I_3 \end{bmatrix} + \frac{d}{dt} \begin{bmatrix} \Psi_1 \\ \Psi_2 \\ \Psi_3 \end{bmatrix} \tag{5.68}$$

$$\begin{bmatrix} \Psi_1 \\ \Psi_2 \\ \Psi_3 \end{bmatrix} = \begin{bmatrix} L_{11} & L_{12} & L_{13} \\ L_{21} & L_{22} & L_{23} \\ L_{31} & L_{32} & L_{33} \end{bmatrix} \begin{bmatrix} I_1 \\ I_2 \\ I_3 \end{bmatrix} + \begin{bmatrix} L_{1r} \\ L_{2r} \\ L_{3r} \end{bmatrix} [I_r] \tag{5.69}$$

Regardless of whether the phase voltages are balanced or not, whether the windings are symmetrical or not, the sum of phase currents (it is at the same time the line currents in wye connection) must be zero any time. In other words, only two currents are independent. Therefore, (5.69) could be given as follows:

$$\begin{bmatrix} \Psi_1 \\ \Psi_2 \\ 0 \end{bmatrix} = \begin{bmatrix} L_{11} & L_{12} & L_{13} \\ L_{21} & L_{22} & L_{23} \\ 1 & 1 & 1 \end{bmatrix} \begin{bmatrix} I_1 \\ I_2 \\ I_3 \end{bmatrix} + \begin{bmatrix} L_{1r} \\ L_{2r} \\ 0 \end{bmatrix} [I_r] \tag{5.70}$$

It is obvious that the third row explicitly defines the condition that the sum of phase currents must be zero at any time. In addition, once again, inputs of the model are line-to-line voltages. As the rotor cage winding is short-circuited, these modifications from stator side do not implicate any modifications in the rotor side. If (5.68) and (5.70) are written as:

$$[U_{sl}] = [R_{s\ \mathrm{mod}}][I_s] + \frac{d[\Psi_s]}{dt} \tag{5.71}$$

$$[\Psi_{s\ \mathrm{mod}}] = [L_{ss\ \mathrm{mod}}][I_s] + [L_{sr\ \mathrm{mod}}][I_r] \tag{5.72}$$

where the subscript *mod* designates modified matrix, then iterative numerical procedure for integration of system of equations could be described as the following:

1. feeding all constant machine parameters as well as calculating all parameters that are not time- or space-dependent): $[R_{s\,\mathrm{mod}}]$, $[R_r]$, and in case of uniform air gap $[L_{ss\,\mathrm{mod}}]$ and $[L_{rr}]$;
2. calculating the mutual inductance between the stator phases and rotor loops $[L_{sr}(\theta)]$ as well as derivative of these functions with respect to angle $\theta$ $[DL_{sr}(\theta)]$;
3. defining initial conditions, i.e., $[\Psi_s(0)]$, $[\Psi_r(0)]$, $\theta(0)$, $\omega(0)$. From these initial conditions, the initial conditions for currents $[I_s(0)]$ and $[I_r(0)]$ are known;
4. initializing $k$ defining step of integration, $\Delta$ (sec);
5. defining phase or line voltages at $t=0$ if $k=1$, otherwise for $t=k\cdot\Delta$. Loading values for $[L_{sr}(\theta)]$,$[L_{sr\,\mathrm{mod}}(\theta)]$ and $[DL_{sr}(\theta)]$ for $\theta=\theta(0)$ if $k=1$, otherwise for $\theta=\theta(k)$;
6. calculating the flux-linkages, currents, torque, speed and rotor angular position according to the following iterative procedure:

$$[\Psi_s]_{k+1} = [\Psi_s]_k + \Delta\cdot\left[[U_{sl}]_k - [R_{s\,\mathrm{mod}}][I_s]_k\right]$$

$$[\Psi_r]_{k+1} = [\Psi_r]_k - \Delta\cdot[R_r][I_r]_k$$

$$\text{modification of } [\Psi_s]_{k+1} \quad \text{into} \quad [\Psi_{s\,\mathrm{mod}}]_{k+1}$$

$$[K]_k = [L_{sr}]_k^T[L_{ss\,\mathrm{mod}}]^{-1}[L_{sr\,\mathrm{mod}}]_k - [L_{rr}]$$

$$[L]_k = [L_{sr}]_k^T[L_{ss\,\mathrm{mod}}]^{-1}[\Psi_{s\,\mathrm{mod}}]_{k+1} - [\Psi_r]_{k+1}$$

$$[I_s]_{k+1} = [L_{ss\,\mathrm{mod}}]^{-1}\left[[\Psi_{s\,\mathrm{mod}}]_{k+1} - [L_{sr\,\mathrm{mod}}]_k[K]_k^{-1}[L]_k\right]$$

$$[I_r]_{k+1} = [K]_k^{-1}[L]_k$$

$$M_{em_k} = [I_s]_{k+1}[M]_k[I_r]_{k+1}$$

$$\omega_{k+1} = \omega_k + \frac{\Delta}{J}\left(M_{em_k} - M_{opt_k}\right)$$

$$\theta_{k+1} = \theta_k + \Delta\cdot\omega_{k+1}$$

7. going back to step 5, loading new values for $[L_{sr}(\theta)]$ i $[DL_{sr}(\theta)]$ from the look-up tables for angle $\theta$ that is the result of previous iteration and so forth to the end of the predefined simulation time.

It is important to note that in the above-described iterative procedure, there are now both matrices simultaneously: matrix $[L_{sr}(\theta)]$ and matrix $[L_{sr\,\mathrm{mod}}(\theta)]$.

#### 5.4.1.2 DELTA connection

In the case of delta-connection of the stator winding, any modifications to the original system of equations are not necessary except from the definition of the line currents through the phase currents as follows:

$$I_a = I_{ab} - I_{ca} \tag{5.73}$$

$$I_b = I_{bc} - I_{ab} \tag{5.74}$$

$$I_c = I_{ca} - I_{bc} \tag{5.75}$$

As it is known, the sum of phase currents of the stator windings coupled in a delta is zero in the case of a symmetric motor and a symmetric system supply voltage while in the case of any unbalance that is not the case. The phase voltages of the stator windings are also line-to-line voltages. Therefore, there is no need for any modification. On the other hand, regardless of whether the motor voltages are symmetrical or not, the condition in which the sum of the line currents is zero is already satisfied (5.73)–(5.75).

---

**Example 5.2** In order to illustrate the use and the power of the described multiple-coupled circuit model together with winding function approach, let us analyze an induction motor with the following nameplate data:

11 kW, 400 V, Y-connection, 50 Hz, $p=4$, $\cos\varphi=0.7$, $\eta=0.9$

The number of stator slots is $S=48$ and the motor has $R=30$ rotor bars. Stator phase winding consists of 16 coils, 2 coils per pole with 6 turns in one coil, i.e., $N=96$ series turns per pole per phase. Coil pitch is shortened, i.e., $y=(5/6)\tau$.

Phase winding scheme under one pair of poles is: 1-6′-2-7′-13-8′-12-7′.

Additional machine parameters are

$R_{phase}=0.222\ \Omega$, $D_{inner\_stator}=204.38$ mm, $g_0=0.4$ mm, $l_{axial}=201$ mm
$L_{\sigma s}=2.162$ mH, $R_b=77.4\ \mu\Omega$, $R_{ers}=4.57\ \mu\Omega$, $L_b=445$ nH, $L_{ers}=12$ nH,
$J=0.24$ kg m$^2$

Stator phase winding function is given in Figure E5.2.1.

Mutual inductance between the stator phase and rotor loop, as well as their derivative with respect to angular position which are required for electromagnetic torque calculation, is given in Figure E5.2.2 for the case of rotor bar skewing for one stator slot.

Figure E5.2.3 gives the content of rotor inductance matrix in 3D presentation. This 3D shape could be easily associated with the definition of this matrix, (5.35).

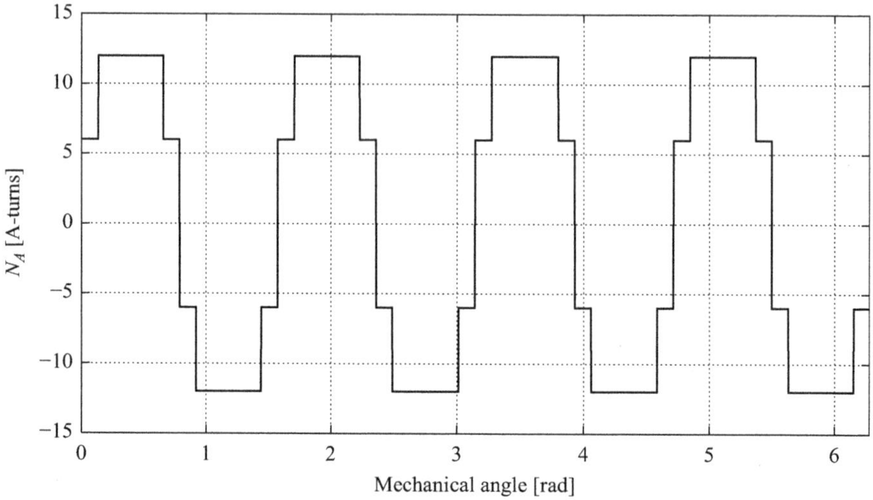


*Figure E5.2.1 Winding function of phase winding A*

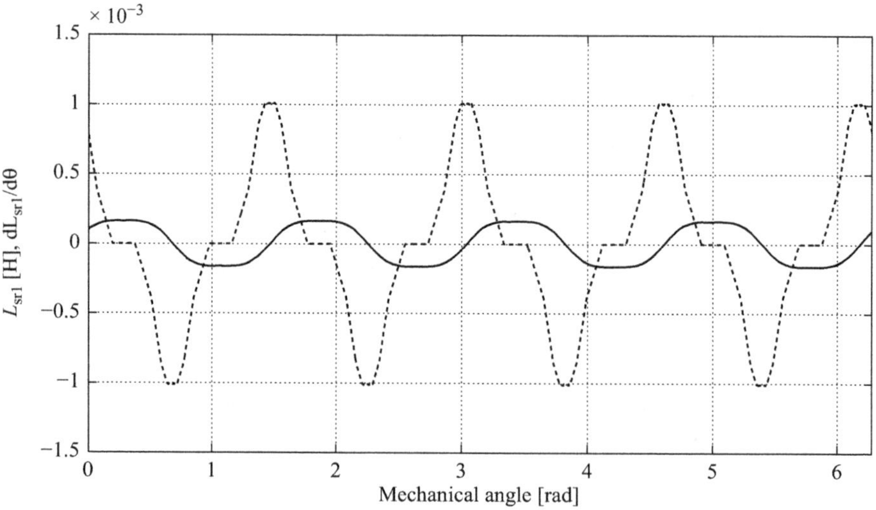


*Figure E5.2.2 Mutual inductance between stator phase and rotor loop and its derivative respective to the angle $\theta$. Rotor bar skew angle $\gamma = 2\pi/48$*

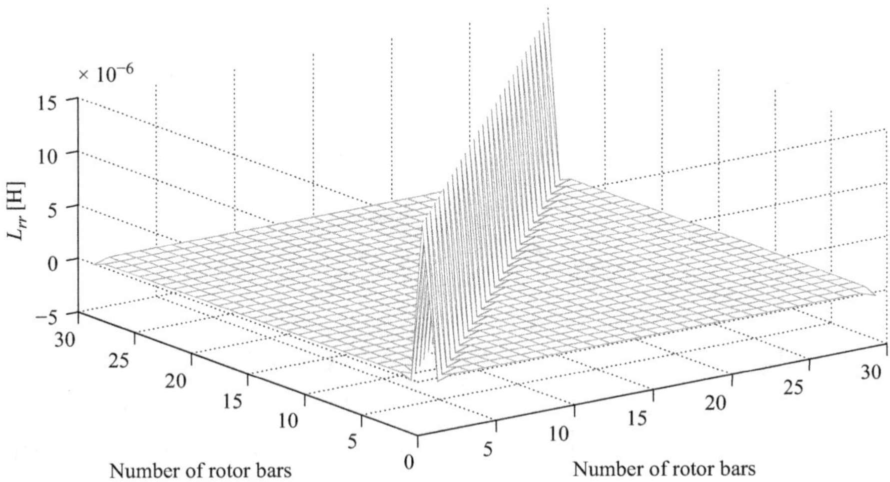


*Figure E5.2.3 A 3D presentation of the $L_{rr}$ matrix (5.35)*

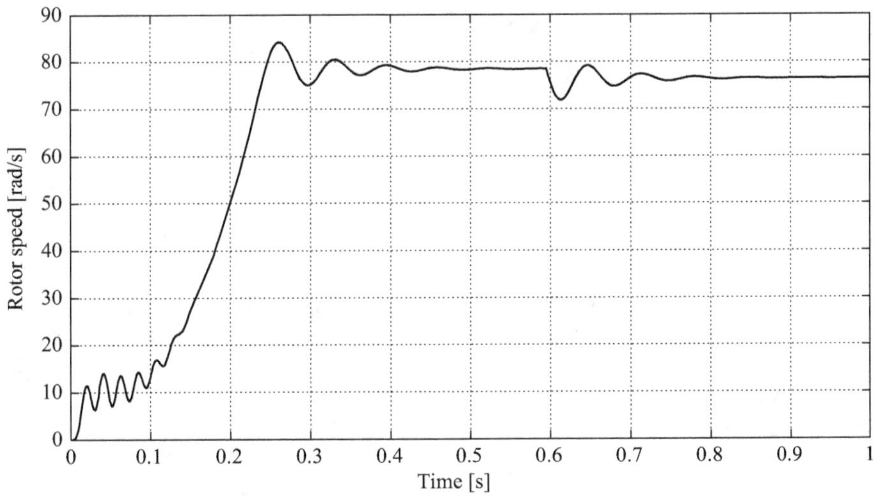


*Figure E5.2.4 Speed-up transients of analyzed motor. At t = 0.6s motor is instantly loaded with rated torque*

Figure E5.2.4 shows the rotor speed during the run-up transient of unloaded motor. Afterward, at $t = 0.6$ s, the motor is instantly loaded by rated torque. Figures E5.2.4–E5.2.9 show other values of interest including electromagnetic torque, stator phase currents, rotor loop current and rotor bar current during the same transient regime.

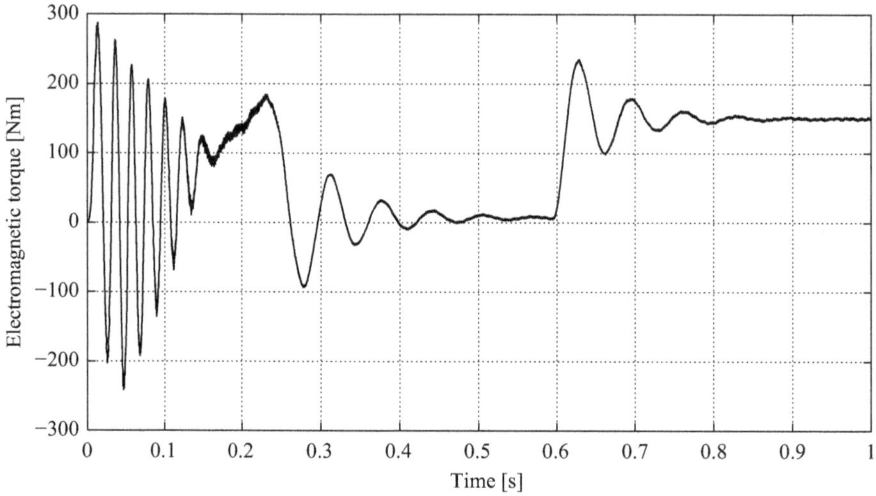


*Figure E5.2.5 Developed electromagnetic torque*

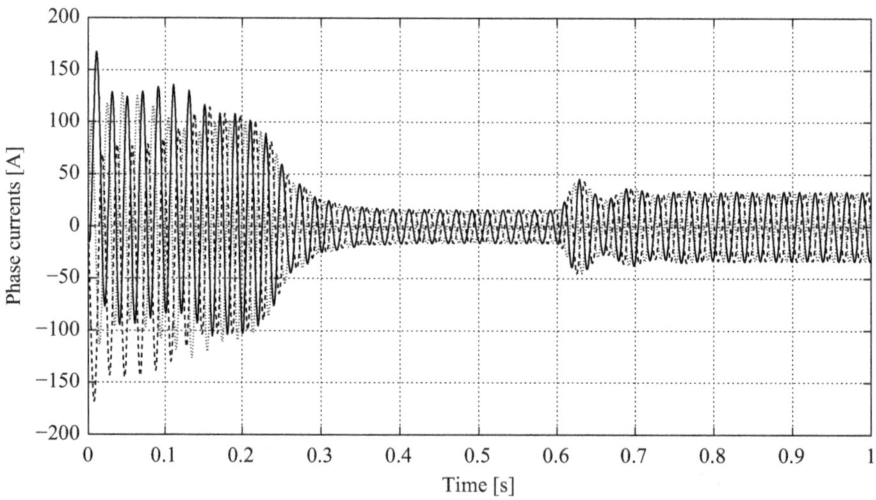


*Figure E5.2.6 Phase currents during the same transient conditions*

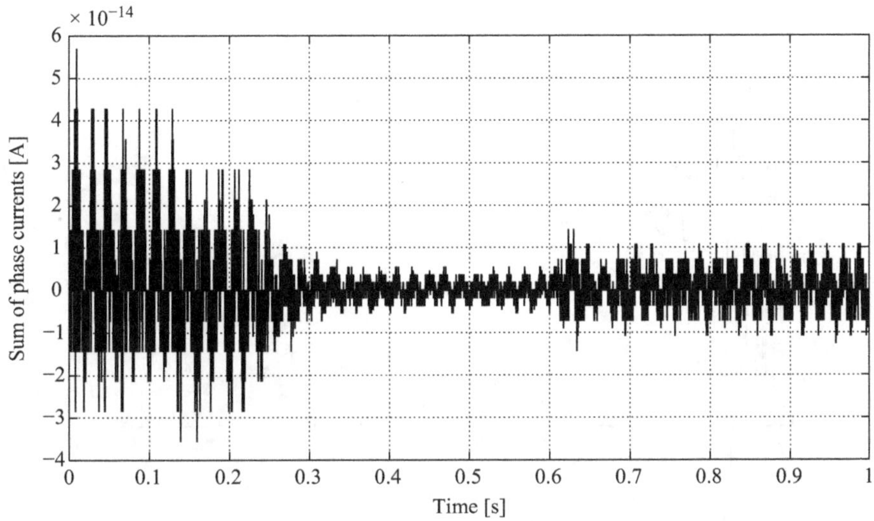


*Figure E5.2.7 Sum of phase currents are equal to zero in wye connection*

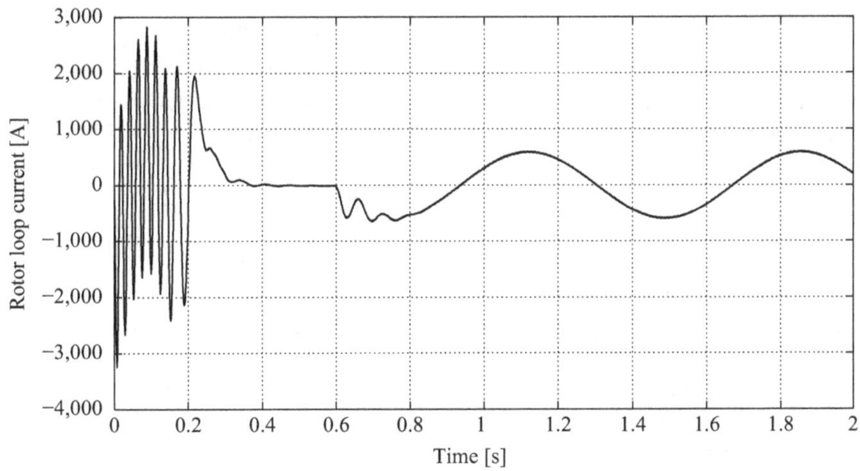


*Figure E5.2.8 Rotor loop current*

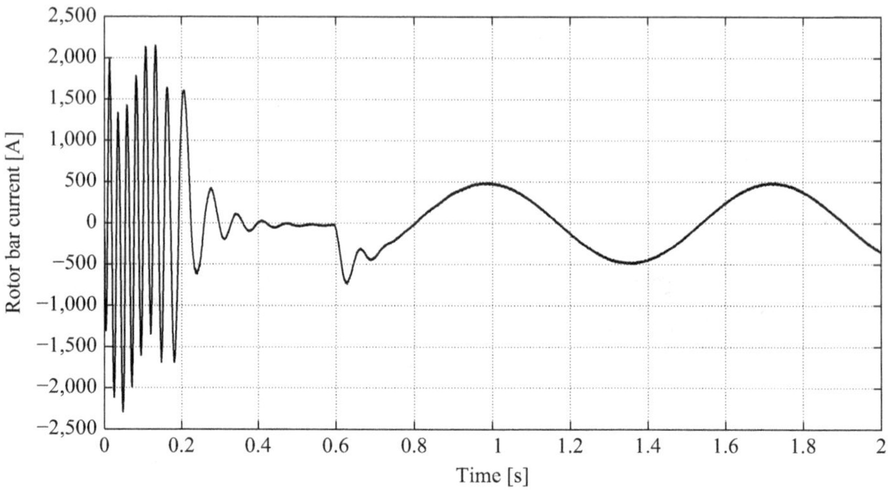


*Figure E5.2.9 Rotor bar current—difference of two side-up loop currents*

## 5.5 Modeling inductions motors with broken rotor bar(s)/end-ring(s)

Multiple-coupled circuit model which treats the currents in each rotor loop as an independent variable is the most appropriate model for analysis of the machine in this faulty regime. The most obvious way to model cage rotor induction motor with broken rotor bar(s), using multiple-coupled circuit model, along with winding function approach, is the approximation that resistance of faulty rotor bar(s) is higher than the resistance of other bars. In this way, there is no need for any additional changes in the model of healthy motor. This approach could be used for any number of broken rotor bars regardless of their relative position.

An alternative approach is to take real picture of the machine in case of broken rotor bar, i.e., to consider the fact that now in rotor cage there exists one rotor loop that is two times wider than others which have effect on the need for recalculation of self- and mutual inductances in comparison with healthy machine. This approach is viable; however, it is more time consuming, and finally, there is no significant difference between results obtained in one or another way.

If the rotor ring segment is broken, the simplest way to take this fault into account is to set appropriate rotor loop current to zero, Figure 5.4. In this way, there is no need for any additional changes also in the model of healthy motor. Moreover, this forced condition could be applied to any number of broken end-ring segments. Here will be presented some results obtained from multiple-coupled circuit model together with winding function approach for the motor whose details are provided in Example 5.2.

Figure 5.7 shows the rotor speed and developed electromagnetic torque in the case where one rotor bar is broken. This faulty regime is modeled by putting into the model the resistance of one rotor bar which is 100 times higher than other ones.

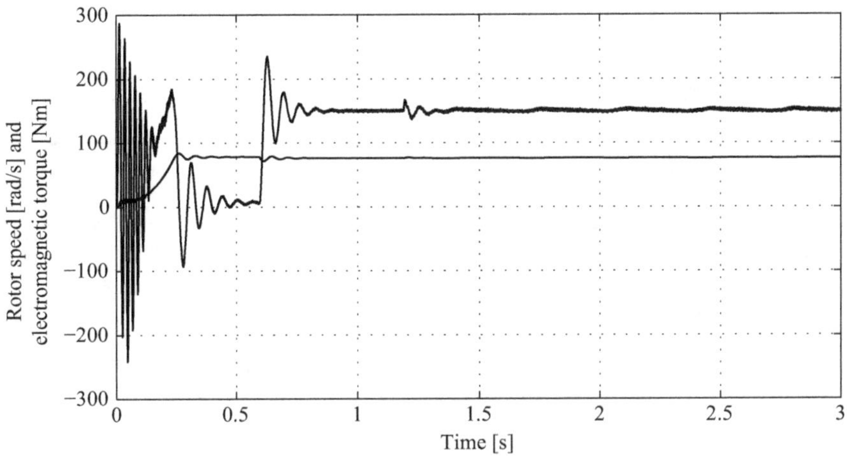


*Figure 5.7 Rotor speed and developed electromagnetic torque for fully loaded 50 Hz cage rotor induction motor with S = 48 stator slots, R = 30 rotor bars and p = 4 pole pairs @ s = 2.82%. One rotor bar broken at instant t = 1.2 s ($R_{bar\_broken} = 100 \times R_{b\_healthy}$)*

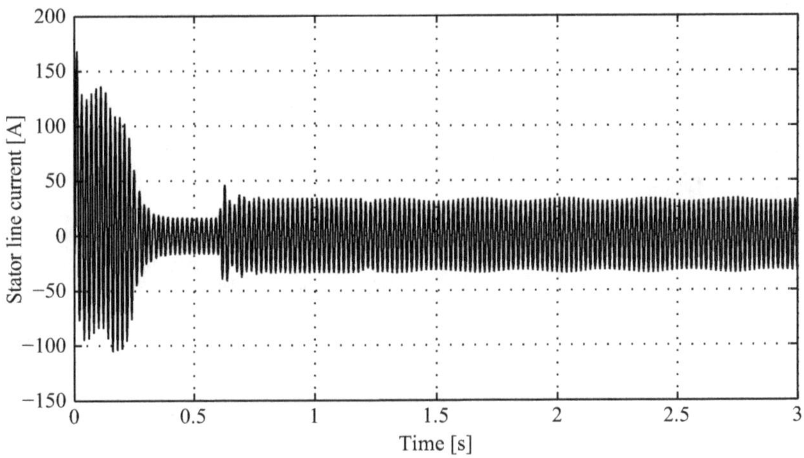


*Figure 5.8 Stator line current for fully loaded 50Hz cage rotor induction motor with S = 48 stator slots, R = 30 rotor bars and p = 4 pole pairs @ s = 2.82%. One rotor bar broken at instant t = 1.2 s ($R_{bar\_broken} = 100 \times R_{b\_healthy}$)*

The motor is speeded-up in no-load regime, rated torque is applied at 0.6 s and then one rotor bar is broken, at 1.2 s.

The main result that could be derived from Figure 5.7 is the fact that in steady state regime with broken rotor bar, there exists oscillating electromagnetic torque component whose frequency is $2sf_1$. Stator line current in this regime is given in Figure 5.8.

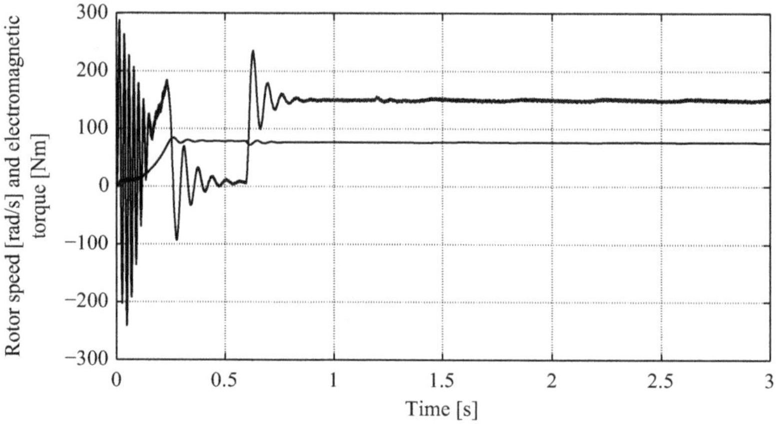


*Figure 5.9 Rotor speed and developed electromagnetic torque for fully loaded 50 Hz cage rotor induction motor with $S = 48$ stator slots, $R = 30$ rotor bars and $p = 4$ pole pairs @ $s = 2.57\%$. One rotor end-ring segment is broken at instant $t = 1.2$ s ($I_{r1} = 0$)*

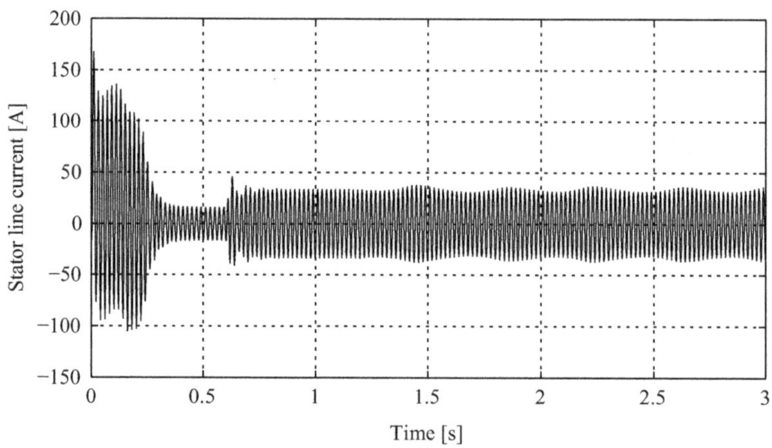


*Figure 5.10 Stator line current for fully loaded 50Hz cage rotor induction motor with $S = 48$ stator slots, $R = 30$ rotor bars and $p = 4$ pole pairs @ $s = 2.57\%$. One rotor end-ring segment is broken at instant $t = 1.2$ s ($I_{r1} = 0$)*

Figures 5.9 and 5.10 show the same variables in case where one rotor end-ring segment is broken. Modeling this regime is made by forcing that one rotor loop current is equal to zero.

## 5.6 Modeling induction motors with air-gap eccentricity

Air-gap eccentricity is one of the main faulted conditions of an induction machine. Air-gap eccentricity is recognized as being either static or dynamic in nature, and this distinction depends on the temporal properties of the air gap. Here will be presented the manner for modeling of these faulty regimes, using multiple-coupled circuit model along with the winding function approach. In order to calculate inductances of all windings in the machine when static or dynamic eccentricity is present, the modified winding function should be used as it is shown in [15].

### *5.6.1 Inductance calculation*

The nonuniform air-gap length in all rotating electric machinery with cylindrical stator and rotor can be described, taking the first two terms of Fourier series into account as [16],

$$g(\theta) \approx g_o - g_o \cdot \rho \cdot \cos\theta \tag{5.76}$$

where $\rho$ is the degree of eccentricity, $\rho \in (0,1)$ and $g_o$ is air-gap length in the machine with uniform air gap. Consequently, the inverse to the air-gap function is:

$$P(\theta) = \frac{1}{g_o}\sum_{i=0}^{\infty} P_i \cos(i\theta) \tag{5.77}$$

where

$$P_0 = \frac{1}{\sqrt{1-\rho^2}} \tag{5.78}$$

and

$$P_i = 2P_0\left(\frac{1-\sqrt{1-\rho^2}}{\rho}\right)^i \tag{5.79}$$

As shown in [15], the winding function of winding $x$, in the case of air-gap eccentricity, assuming infinite permeable iron, is given by:

$$N_x(\theta) = n_x(\theta) - \frac{\langle P \cdot n_x \rangle}{\langle P \rangle} \tag{5.80}$$

where $n_x(\theta)$ is turns function of winding $x$ and the operator $\langle f \rangle$ is defined as the mean value of function $f$ over $[0,2\pi]$. This winding function is similar to the equation in [17], which has been introduced as the winding function of a synchronous motor.

The mutual inductance between any $x$ and $y$ winding in the machine is

$$L_{xy} = \mu_0 rl \int_0^{2\pi} P(\theta) N_x(\theta) n_y(\theta) d\theta \tag{5.81}$$

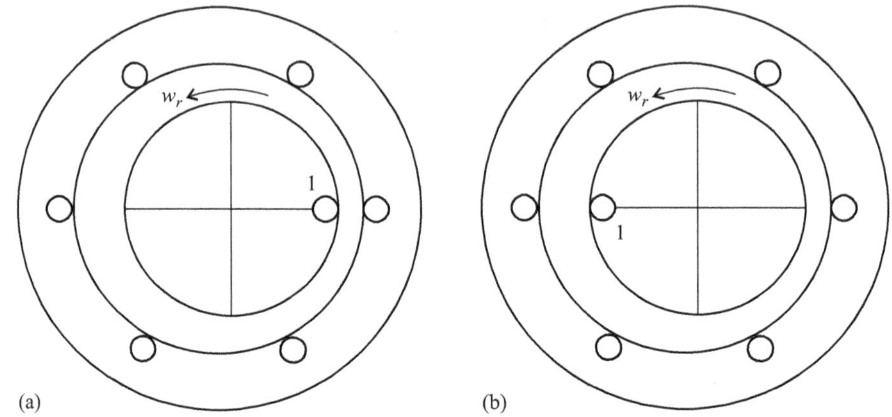


*Figure 5.11 Rotor bar position in machine with static eccentricity: (a) initial position and (b) position after half rotor revolution*

where $r$ is the mean air-gap radius and $l$ is axial length of the machine. As illustrated in [15], (5.81) gives $L_{xy} = L_{yx}$, in the case of a linear magnetic circuit. Skewing of rotor bars can be taken into account in the manner presented in [14]. Using expression (5.81) all inductances that describe induction machine, i.e., those inductances which comprise the system of (5.1)–(5.8) can be calculated.

## 5.6.2 Static eccentricity

Static eccentricity is the case where the air-gap length is not uniform around the rotor as shown in Figure 5.11. Bearing this in mind, all the machine inductances can be calculated by using some of the numerical techniques for integration of expression (5.81), for known machine geometry and winding details.

### 5.6.2.1 Calculation of matrix $[L_{ss}]$

As the air-gap picture does not change with rotor movement, the self- and mutual inductances of the stator phase windings will be independent of the rotor position. However, the self- and mutual inductances of the stator phases will be different for different phases. As an example, for the air gap minima position defined by $\theta = 0$ rad, for degree of eccentricity $\rho = 0.9$, for the first ten members in sum (5.77) and for the machine parameters given in paragraph 5.6.4, the following entries of the matrix $[L_{ss}]$ are obtained from (5.81) [16]:

$$[L_{ss}] = \begin{bmatrix} 0.4442 & -0.1576 & -0.1769 \\ -0.1576 & 0.4296 & -0.1832 \\ -0.1769 & -0.1832 & 0.3946 \end{bmatrix} [H]$$

For a uniform air gap, the matrix $[L_{ss}]$ is

$$[L_{ss}] = \begin{bmatrix} 0.2156 & -0.0910 & -0.0910 \\ -0.0910 & 0.2156 & -0.0910 \\ -0.0910 & -0.0910 & 0.2156 \end{bmatrix} [H]$$

Matrix $[L_{ss}]$ should be calculated only once and outside of the main iterative loop in the procedure of solving the system of differential equations describing the machine.

All inductances in this paragraph will be calculated for a degree of eccentricity $\rho = 0.9$. Although this seems to be an unreasonable high degree of eccentricity, it has been selected for the illustration of eccentricity effects in a numerical model.

#### 5.6.2.2 Calculation of matrix $[L_{sr}]$

The mutual inductance between the stator phase and the rotor loop, or inversely (in a linear magnetic circuit when $[L_{rs}] = [L_{sr}]^T$), is a function of rotor position as it is for a machine with uniform air gap. As the "picture" of the air gap does not change with the rotation of the rotor, only the function of mutual inductance between stator phase and one rotor loop, for example, the first must be known in the process of dynamic simulation. The mutual inductance between the stator phase and the second rotor loop could be obtained from the same curve for the angle which describes the position of the second rotor loop relative to the first rotor loop, i.e., for the angle $2\pi/R$ where $R$ is the number of rotor bars. This task can be solved outside the main iterative loop, making use of a one-dimensional lookup table, which contains the mutual inductance between the stator phase and the rotor loop for every rotor position. In this process, the integral in (5.81) must be solved for every rotor position with reasonable step size. During this process, the inverse air-gap function and winding functions of stator phases do not change, but the rotor loop turns function moves with the rotor.

Figure 5.12 shows the mutual inductance as a function of rotor position between stator phases and first rotor loop for the degree of eccentricity $\rho = 0.9$.

The position of the first rotor loop at $t = 0$ is defined with $\theta = 0$ rad (the air-gap length minima position), as it is the case for first stator slot where stator phase $A$ begins. One rotor revolution is discretized in 1,440 points. From the known matrix $[L_{sr}]$, the matrix of derivatives $d[L_{sr}]/d\theta$ in the expression for electromagnetic torque (5.10) can be determined.

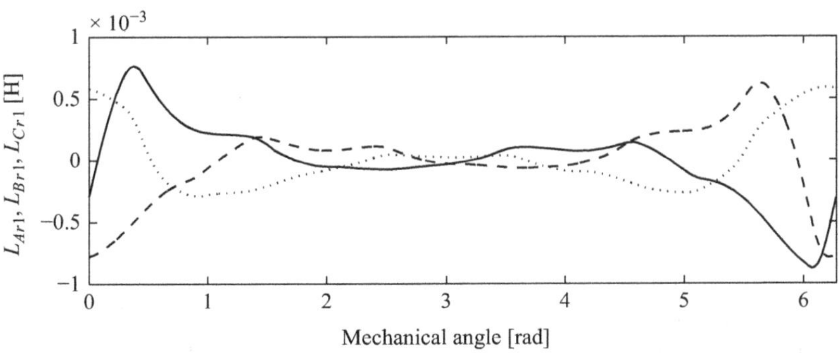


*Figure 5.12 Mutual inductance between stator phases and first rotor loop for $\rho = 0.9$*

———— $L_{Ar1}$ - - - - - - - $L_{Br1}$ . . . . . . . $L_{Cr1}$

#### 5.6.2.3 Calculation of matrix $[L_{rr}]$

The calculation of this matrix is the most difficult task in the modeling and simulation of static eccentricity in a squirrel-cage induction machine. As a consequence of static eccentricity, the rotor loop self-inductances and mutual inductances between rotor loops are functions of the rotor position. In the common squirrel-cage induction machines the rotor always has a relatively large number of bars and consequently the dimension of matrix $[L_{rr}]$ is large. For the machine analyzed, the dimension of the matrix $[L_{rr}]$ is $R \times R = 40 \times 40$. In the case of static eccentricity, each member of this matrix must be a vector of the length $N$, so a lookup table for this matrix must be three-dimensional with $R \times R \times N$ elements. Obviously, this approach highly consumes time and computer resources.

Alternatively, an analytical approach may be used. In this way, analytical expressions for the elements of the matrix $[L_{rr}]$ must be derived as well as analytical expressions for the derivatives of these elements. The calculation of elements of these two matrices must be incorporated into a main iterative loop for numerical solution of the system of differential equations of the machine. Equation (5.81) gives the self-inductance of the $i$th rotor loop as follows [16],

$$L_{ii} = 2\pi\mu_0 rl\langle P(\theta)\cdot n_i(\theta)\rangle\left[1-\frac{\langle P(\theta)\cdot n_i(\theta)\rangle}{\langle P(\theta)\rangle}\right] \tag{5.82}$$

where

$$\langle P(\theta)\rangle = \frac{P_0}{g_0} \tag{5.83}$$

and

$$\langle P(\theta)\cdot n_i(\theta)\rangle = \frac{P_0}{g_0 R}+\frac{1}{2\pi g_0}\sum_{k=1}^{\infty}\frac{P_k}{k}(\sin k\theta_{i+1} - \sin k\theta_i) \tag{5.84}$$

where angle $\theta_i$ defines the $i$th rotor loop position relative to the fixed stator point. Similarly, for mutual inductance between the $i$th and $j$th rotor loops, the following expression can be obtained:

$$L_{ij} = -2\pi\mu_0 rl\frac{\langle P(\theta)\cdot n_i(\theta)\rangle}{\langle P(\theta)\rangle}\langle P(\theta)\cdot n_j(\theta)\rangle \tag{5.85}$$

where

$$\langle P(\theta)\cdot n_j(\theta)\rangle = \frac{P_0}{g_0 R}+\frac{1}{2\pi g_0}\sum_{k=1}^{\infty}\frac{P_k}{k}(\sin k\theta_{j+1} - \sin k\theta_j) \tag{5.86}$$

For the known expressions of self- and mutual inductances, it is straightforward to define derivatives of these functions. As an example of calculation of matrix $[L_{rr}]$, Figure 5.13 shows the content of matrix $[L_{rr}]$ for two different rotor positions.

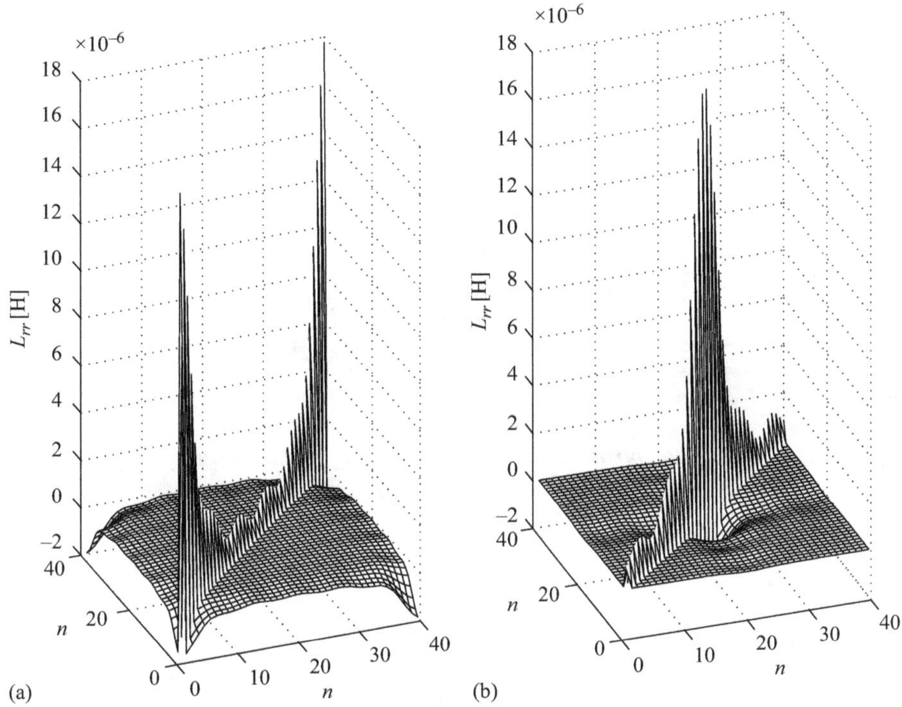


*Figure 5.13 Content of matrix [$L_{rr}$] for two different rotor position: (a) $\theta = 0$ and (b) $\theta = \pi$ rad. $\rho = 0.9$*

The degree of eccentricity is $\rho = 0.9$ and the first five members of the inverse air-gap function in (5.77) are taken into account.

### *5.6.3 Dynamic eccentricity*

Dynamic eccentricity is another faulty operating regime of the induction machine [16,17]. In this case, the air-gap length minima position moves along with the rotor as shown in Figure 5.14.

All inductances in the machine can also be calculated in this case by using expression (5.81), bearing in mind that the inverse air-gap function moves together with the rotor.

#### 5.6.3.1 Calculation of matrix [$L_{ss}$]

The self- and mutual inductances of the stator phases are functions of rotor position and this fact should be taken into account during the calculation of electromagnetic torque. For a known stator winding configuration, it is necessary to calculate the integral (5.81) for every rotor position during one revolution. The inverse air gap function value is different for different rotor positions. Therefore, it is expected to have a variable winding function of the stator phases upon changing the rotor position.

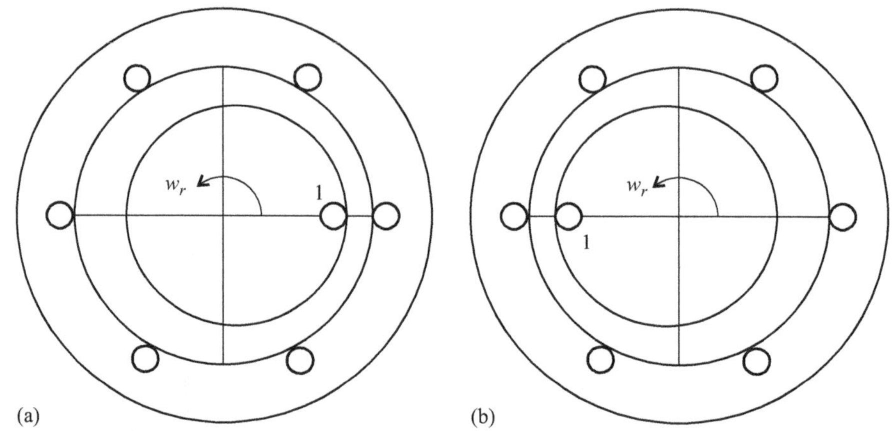


*Figure 5.14 Rotor bar position in a machine with dynamic eccentricity: (a) initial position and (b) position after half rotor revolution*

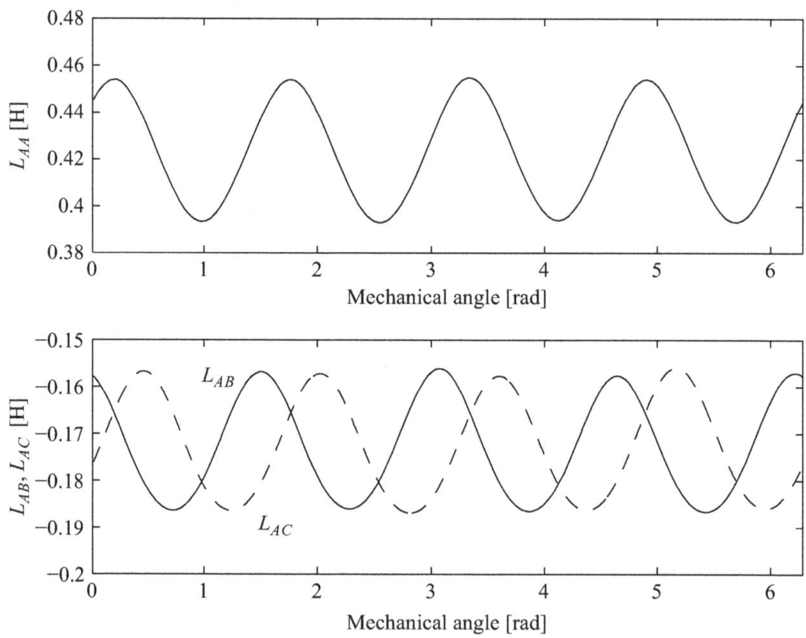


*Figure 5.15 Self- and mutual inductances of stator phases in case of dynamic eccentricity. Eccentricity degree $\rho = 0.9$*

Figure 5.15 shows the self-inductance of stator phase $A$ and the mutual inductances between stator phase $A$ and phases $B$ and $C$ for the degree of eccentricity $\rho = 0.9$. In the inverse air-gap function, the first ten members are taken into consideration.

It is evident that self- and mutual inductances are functions of rotor position and their mean values are significantly higher in comparison with the uniform air-gap case. In addition, it is obvious that mutual inductances between the different stator phases are different. For known self- and mutual inductances, their derivatives can be easily found as their presence is required for the calculation of electromagnetic torque.

#### 5.6.3.2 Calculation of matrix $[L_{\mathrm{sr}}]$

This matrix is of the same nature as in the case of static eccentricity. The principal difference is in the fact that the inverse air-gap function moves with rotor and should be considered during calculation of integral (5.81). Therefore, in the process of dynamic simulation, the function of mutual inductance between the stator phase $A$ and rotor loop 1, stator phase $A$ and rotor loop 2, …, stator phase $A$ and rotor loop $n$ must be known. This can be solved outside the main iterative loop, using a two-dimensional lookup table. The result of this procedure is matrix $[L_{sr}] = [[L_{Ar}]\ [L_{Br}]\ [L_{Cr}]]^T$. Submatrices $[L_{ir}]_{i=A,B,C}$ have dimension $R \times N$ where $N$ is the number of steps in $[0,2\pi]$ range (one rotor revolution). Figure 5.16 shows the mutual inductances between stator phase $A$ and rotor loops 1, 15 and 30 for the degree of eccentricity $\rho = 0.9$. The position of the first rotor loop at $t = 0$ is $\theta = 0$ rad (the air-gap length minima position at $t = 0$) as well as the position of the first stator slot where the stator phase $A$ begins. One rotor revolution is discretized into 1,440 points.

Figure 5.17 shows content of matrix $[L_{Ar}]$ as a 3D picture.

#### 5.6.3.3 Calculation of matrix $[L_{\mathrm{rr}}]$

During rotation, the rotor loops experience constant air-gap length; however, different rotor loops experience different air-gap lengths. Hence, matrix $[L_{rr}]$ should be calculated only once for the initial rotor position, outside of the main iterative loop. Figure 5.18 illustrates this and shows the content of matrix $[L_{rr}]$ for $\rho = 0.9$. The first ten terms are taken into account in (5.77).

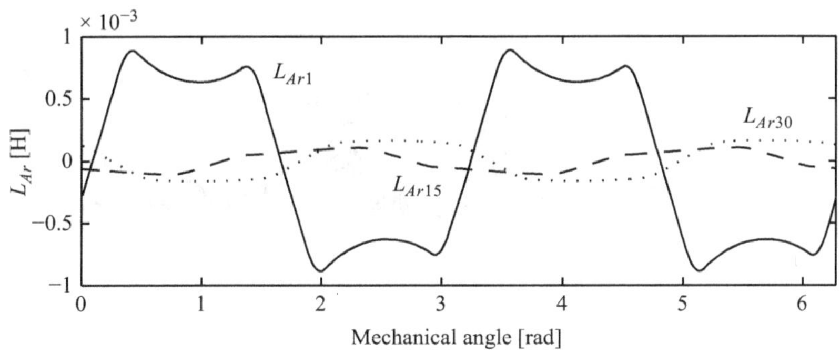


*Figure 5.16 Mutual inductance of stator phase* A *and some of the rotor loops for* $\rho = 0.9$

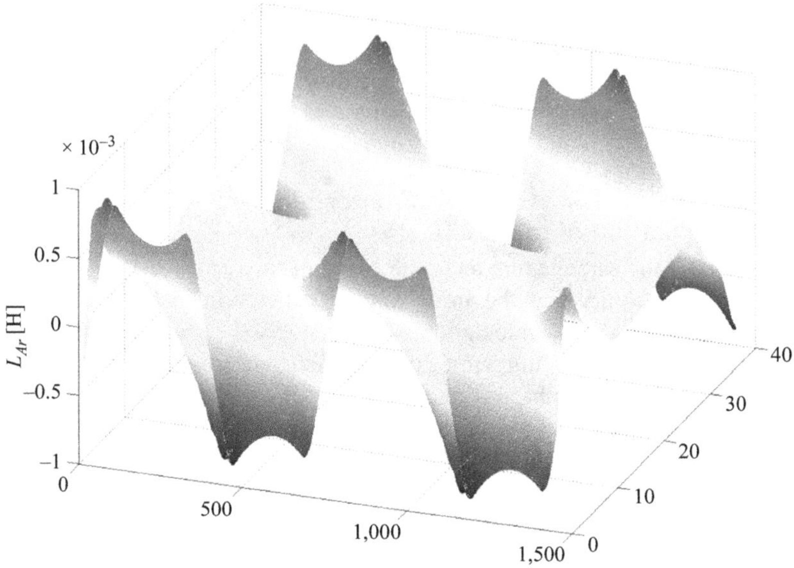


*Figure 5.17 Content of matrix $[L_{Ar}]$ for $\rho = 0.9$*

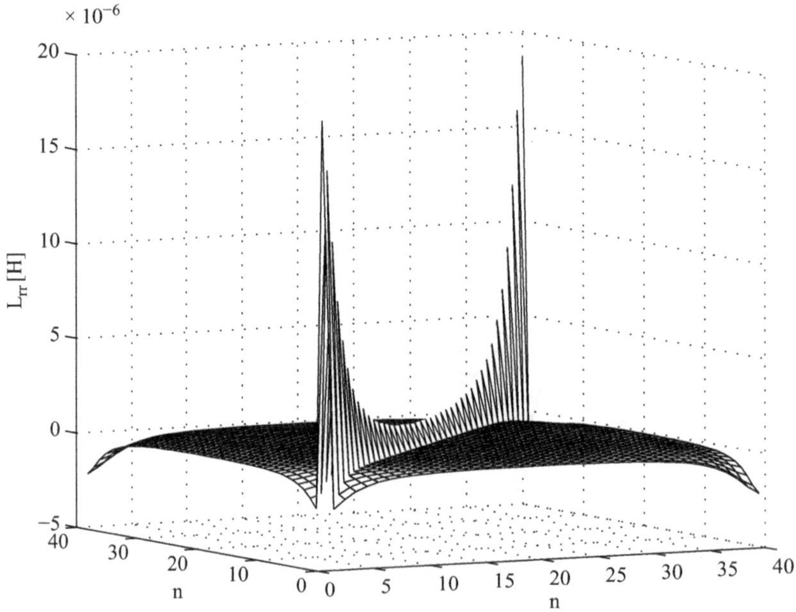


*Figure 5.18 Content of matrix $[L_{rr}]$ for $\rho = 0.9$*

### 5.6.4 Numerical model results

In order to illustrate the presented method of eccentricity modeling, an induction motor whose parameters are given below is modeled under static and dynamic eccentricity conditions. During the calculation of mutual inductances between the stator phases and rotor loops, rotor bar skewing is taken into account [14].

*Machine parameters*: $P = 15$ kW, $U = 415$ V, 50 Hz, Δ connection, $p = 2$ pair of poles, $l = 0.11$ m, $r = 0.082$ m, $S = 48$ stator slots, $n = 40$ rotor bars, $w = 28$ turns per coil, $g_0 = 0.8$ mm, $R_{s(\text{phase})} = 1.75\ \Omega$, $R_b = 30\ \mu\Omega$, $R_{ers} = 2\ \mu\Omega$, $L_b = 10$ nH, $L_{ers} = 2$ nH, $J = 0.0754$ kg m$^2$, $\gamma = 2\pi/48$ (angle of skewing of rotor bars).

*Winding scheme of phase A*:

A-1-16′-2-15′-3-14′-4-13′-25-40′-26-39′-27-38′-28-37′-X

Under eccentricity conditions, the run-up time of the machine depends on the position of air-gap flux density maximum at $t = 0$, i.e., on the instantaneous values of phase voltages at $t = 0$. Therefore, simulation was carried out for different instantaneous values of voltages at $t = 0$. However, no significant differences in results could be observed. Figure 5.19 shows the run-up of an unloaded healthy as well as machines with static and dynamic air-gap eccentricity for only one condition, where at $t = 0$, phase $A$ voltage had a maximum value. The run-up time of the machine with dynamic eccentricity differs slightly from the run-up time of a healthy machine and is a consequence of an additional reluctant torque component that compensates the inherently smaller induction torque component due to the higher air-gap length along the main part of machine circumference. The run-up time of the machine with static air-gap eccentricity is the longest, as expected.

Figure 5.20 displays stator phase currents for the fully loaded steady-state condition for a healthy machine and for static and dynamic air-gap eccentricity conditions for the same eccentricity degree, $\rho = 0.9$. In dynamic eccentricity conditions, the envelope of the phase current amplitudes coincides with the air-gap profile. In the static eccentricity conditions, the phase current is rich in high-frequency components.

Figure 5.21 shows the stator line currents for the same conditions as previously described.

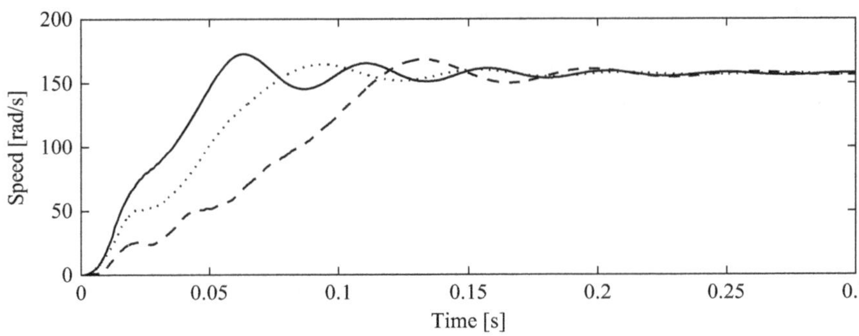


*Figure 5.19 Run-up of unloaded machine with uniform air gap (——), dynamic ($\rho = 0.9$)( . . . ) and static air-gap eccentricity ($\rho = 0.9$) ( - - -)*

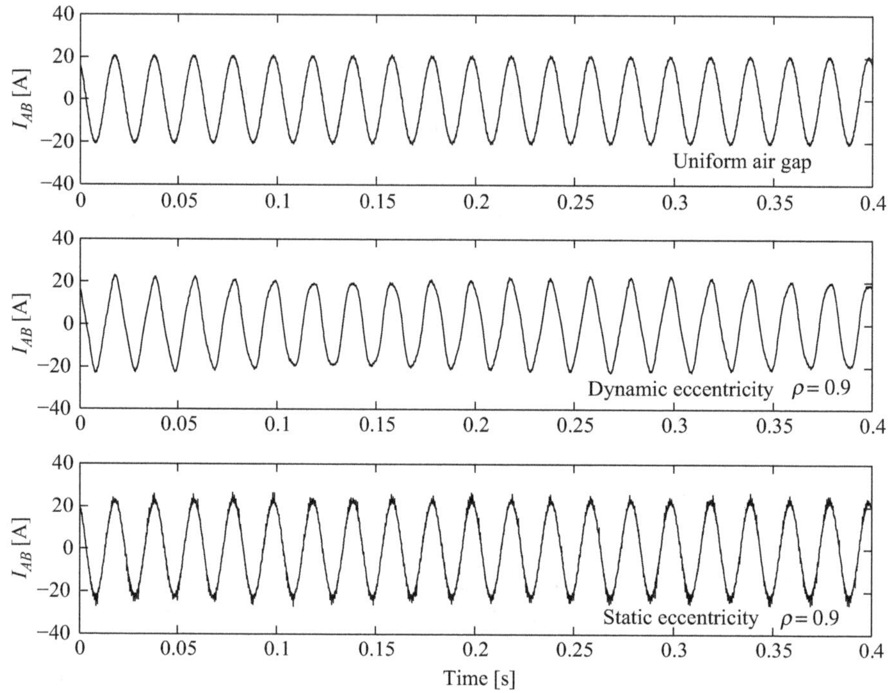


*Figure 5.20 Stator phase current for fully loaded machine in steady-state condition*

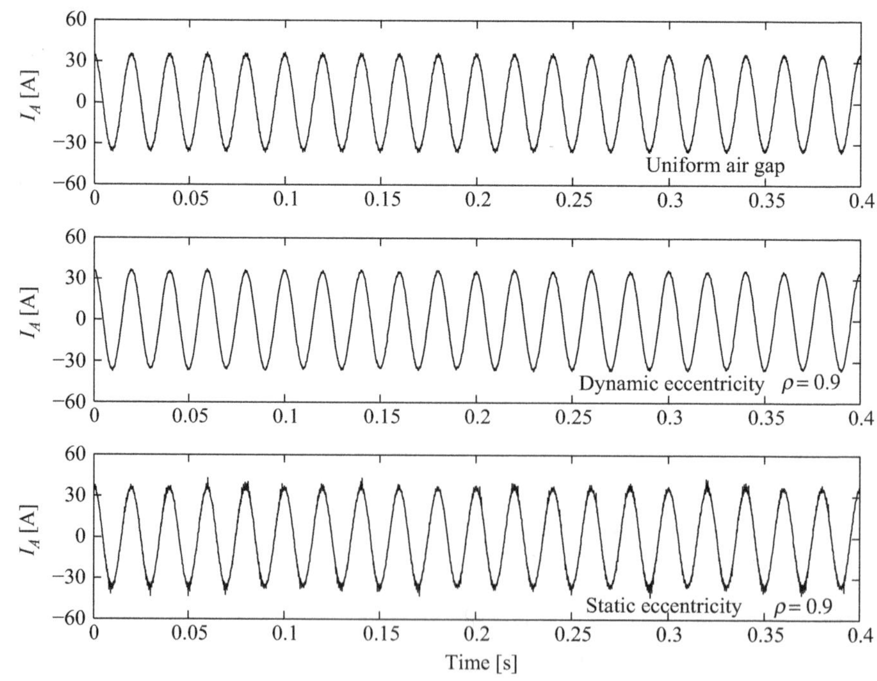


*Figure 5.21 Stator line current for fully loaded machine in steady-state condition*

## 5.7 Modeling induction motors with interturn short circuit in stator winding

A winding-function-based method could also be very effectively used for modeling of induction motors with interturn short circuit in the stator windings [18]. In the first instance, consideration is given to the simplest case where the coil has only one turn from which it is possible to draw some important conclusions. Figure 5.22, however, shows one-phase group of three coils. It is assumed that interturn short circuit arises between points *a* and *b*, as illustrated. It is evident that the path to the circulating current is closed. From simple theory, it is observed that the path *A*–*X* can be expanded into two independent circuits. It is obvious from Figure 5.22 that the two currents, the phase current and the current which flows in the short-circuit circulation current, produce opposite MMFs. Therefore, interturn short circuits have a cumulative effect in decreasing the MMF in the vicinity of the short-circuited turn(s). First, when a short circuit occurs, the phase winding has less turns, and therefore, less MMF. Second, short-circuit current MMF is opposite the MMF of the phase winding. Clearly, interturn shorts with more turns can be analyzed in a similar manner.

In most commercially available induction motors, coils are insulated from one another in slots as well as in the end winding region. Therefore, the highest probability for the occurrence of interturn is between turns in the same coil. Here, it is assumed that the interturn short circuit is between two turns in the same coil, and that one-half of the coil is short-circuited; this means that approximately 8% of turns of one-phase are short-circuited. Simulation was carried out for this condition.

As a consequence of the interturn short circuit, the MMF of the phase winding in which interturn short-circuit exists changes, as does the mutual inductance

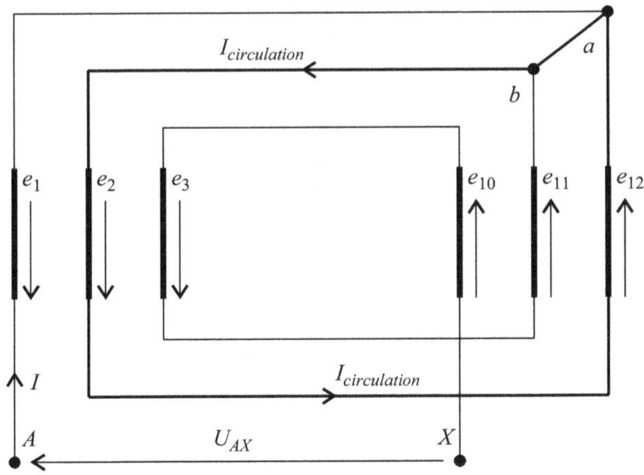


*Figure 5.22 Interturn short circuit between points a and b*

between that phase and all other circuits in the machine. In addition, a new "phase," short-circuited phase $D$ is introduced. It will be assumed that this phase has no conductive contact with other phases, but it is mutually coupled with all other circuits on both the stator and rotor sides. The relationship between the stator currents depends on the stator winding connection.

### *Wye Connection of Stator Windings*

When the stator windings and voltage supply are symmetrical, there is no requirement for any changes in the system of (5.1)–(5.8). In the case of unbalanced connections or an unbalanced supply, some changes must occur in order for the stator currents to satisfy Kirchhoff's current rule and to employ line-to-line voltages as inputs in the simulation model [9]. Applying these conditions to the case where the interturn short circuits occur, (5.3) becomes

$$\begin{bmatrix} u_{sa} \\ u_{sb} \\ u_{sc} \\ 0 \end{bmatrix} = \begin{bmatrix} R_a & -R_b & 0 & 0 \\ 0 & R_b & -R_c & 0 \\ -R_a & 0 & R_c & 0 \\ 0 & 0 & 0 & R_d \end{bmatrix} \begin{bmatrix} i_a \\ i_b \\ i_c \\ i_d \end{bmatrix} + \frac{d}{dt} \begin{bmatrix} \psi_{sa} \\ \psi_{sb} \\ \psi_{sc} \\ \psi_{sd} \end{bmatrix} \tag{5.87}$$

where $u_{si}$, $i = a, b, c$, are the stator line-to-line voltages. Of course, the zero term in the stator voltages vector describes the short-circuited "phase." Equation (5.4) is now

$$\begin{bmatrix} \psi_{sa} \\ \psi_{sb} \\ \psi_{sc} \\ \psi_{sd} \end{bmatrix} = \begin{bmatrix} L_{aa} & L_{ab} & L_{ac} & L_{ad} \\ L_{ba} & L_{bb} & L_{bc} & L_{bd} \\ L_{ca} & L_{cb} & L_{cc} & L_{cd} \\ L_{da} & L_{db} & L_{dc} & L_{dd} \end{bmatrix} \begin{bmatrix} i_a \\ i_b \\ i_c \\ i_d \end{bmatrix} + \begin{bmatrix} L_{ar} \\ L_{br} \\ L_{cr} \\ L_{dr} \end{bmatrix} [I_r] \tag{5.88}$$

where $L_{ir}$, $i = a, b, c, d$, are vectors of 1 by $R$ dimension. In order for the stator currents to satisfy Kirchhoff's current rule, (5.88) is written as follows:

$$\begin{bmatrix} \psi_{sa} \\ \psi_{sb} \\ \psi_{sc} \\ \psi_{sd} \end{bmatrix} = \begin{bmatrix} L_{aa} & L_{ab} & L_{ac} & L_{ad} \\ L_{ba} & L_{bb} & L_{bc} & L_{bd} \\ 1 & 1 & 1 & L_{cd} \\ L_{da} & L_{db} & L_{dc} & L_{dd} \end{bmatrix} \begin{bmatrix} i_a \\ i_b \\ i_c \\ i_d \end{bmatrix} + \begin{bmatrix} L_{ar} \\ L_{br} \\ 0 \\ L_{dr} \end{bmatrix} [I_r] \tag{5.89}$$

Hence, from (5.89), it is clear that the condition which must be forced in solving the above system of equations is:

$$\psi_{sc} = L_{cd} i_d \tag{5.90}$$

Figure 5.23 shows the stator star winding arrangement when an interturn short circuit occurs. As a consequence of the fault, we have now an independent circuit with fault current $i_d$.

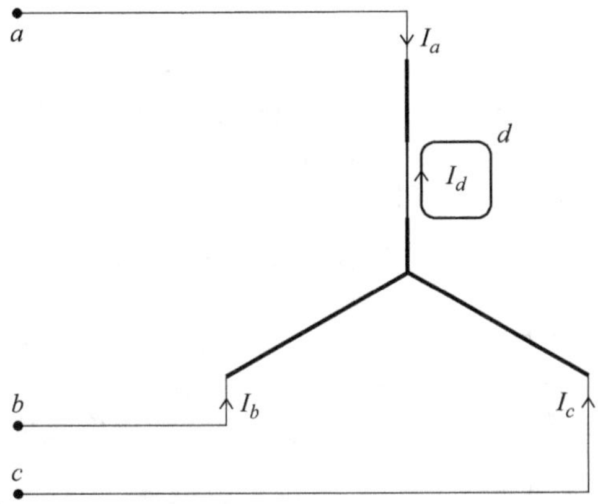


*Figure 5.23 Stator winding configuration with an interturn short circuit—wye connection*

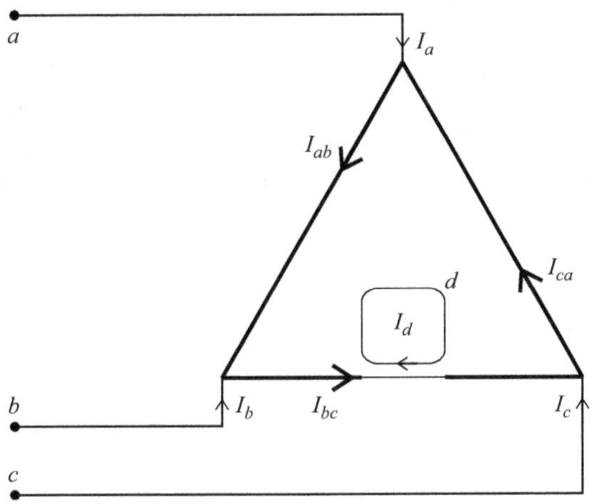


*Figure 5.24 Stator winding configuration with an interturn short circuit—delta connection*

***Delta Connection of Stator Windings***

When stator windings are delta connected, then no changes in the original system of equations are necessary. The inputs to the model are the line-to-line voltages which are the phase voltages. It is well known that under asymmetrical conditions in delta-connected windings, the sum of phase current is not zero; however, this is the case for the line currents.

Figure 5.24 shows an interturn short circuit occurring in a phase winding between terminal points $A$ and $B$. As before, a consequence of the fault is that now there exists an independent circuit with fault current $i_d$.

### 5.7.1 Numerical model results

In order to illustrate the presented method of interturn short-circuit modeling, an induction motor whose parameters are given below is modeled. During the calculation of mutual inductances between the stator phases and rotor loops, rotor bar skewing is taken into account [14].

$P_r = 3$ kW; $U_r = 415$ V; $f = 50$ Hz; $p = 3$; $n_r = 950$ rev/min; $I_r = 6.4$ A; delta connection; turns per coil $N = 77$; $N_{\text{slot}} = 77$; $S = 36$; $R = 32$; $J = 0.0258$ kg m$^2$; $r = 0.0681$ m; $l = 0.115$ m; $g_0 = 0.58$ mm; $R_{ph} = 5.7\ \Omega$; $R_b = 72.8\ \mu\Omega$; $R_e = 438$ n$\Omega$; $L_b = 30$ nH; $L_e = 6$ nH.
Resistance of short-circuit "phase": $R_d = (38/462)\cdot R_{ph} = 0.469\ \Omega$.

*Winding scheme of phase A*:

A-1-6′-12-7′-13-18′-24-19′-25-30′-36-31′-X

The motor is loaded with 30 N m under steady-state condition. Fault is made in such a manner that 38 out of 77 turns is short-circuited in one stator phase coil under one pole.

Figure 5.25 gives MMF of short-circuited phase $d$ and its harmonic content. As the analyzed motor is motor with $p = 3$ pole pairs, fundamental harmonic $\nu = 1$ means six poles. In case of short circuit in one coil, it means that phase $d$ also produces subharmonics. Figure 5.26 gives time series of rotor speed and developed electromagnetic torque in healthy motor. Figure 5.27 gives time series of rotor

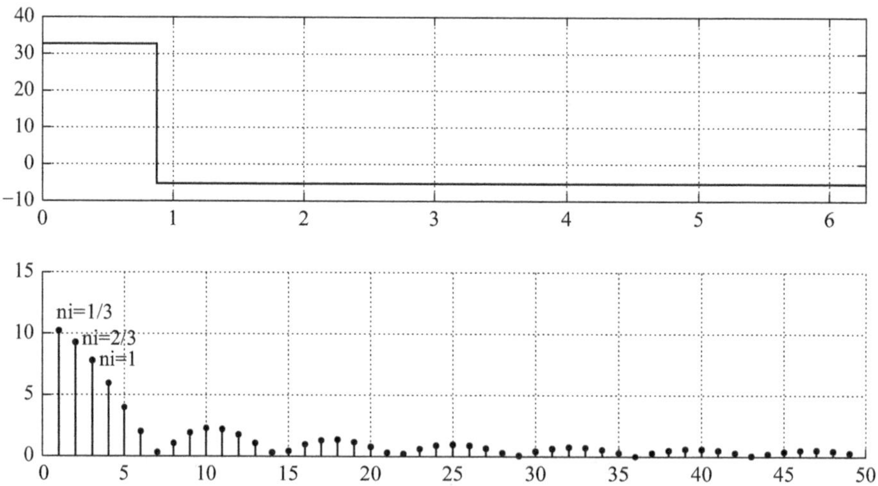


*Figure 5.25 Short-circuited "phase" d MMF*

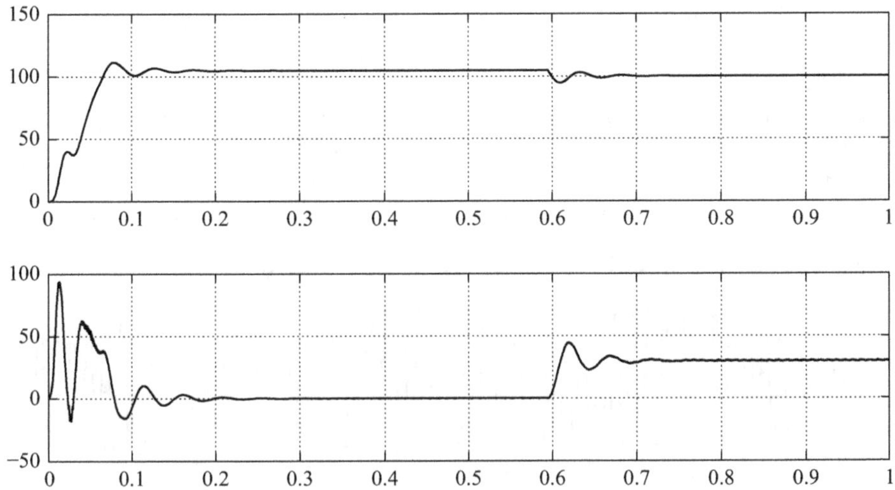


*Figure 5.26 Speed and developed electromagnetic torque—healthy motor*

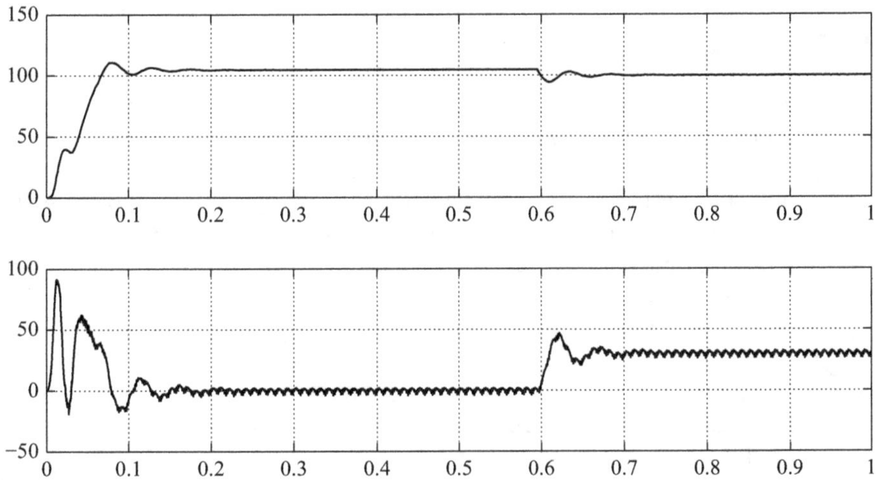


*Figure 5.27 Speed and developed electromagnetic torque—faulty motor*

speed and developed electromagnetic torque in faulty motor when interturn short-circuit exists in stator phase winding.

Figure 5.28 gives time series of stator line current $I_a$ and current in short-circuited phase $d$.

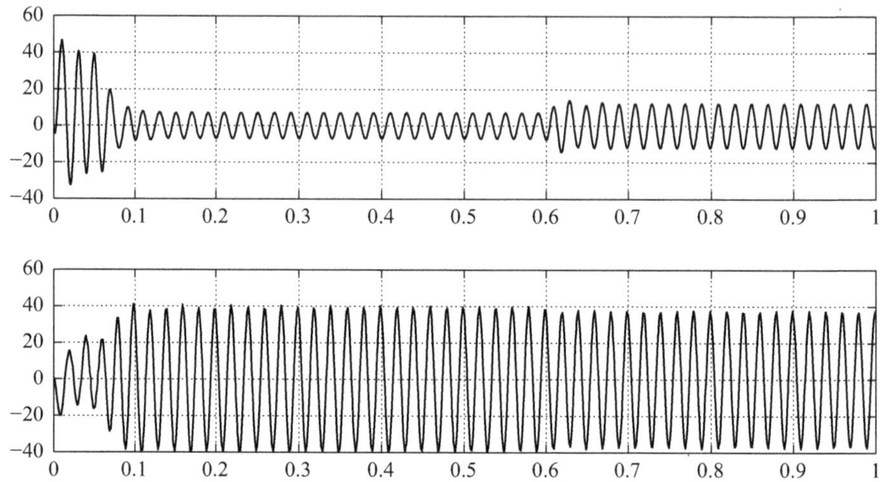


*Figure 5.28 Stator line current and current in short-circuit "phase" d*

## References


[1] Williamson S. Power-factor improvement in cage rotor induction motors, *IEE Proceedings B – Electric Power Applications*, 1983; EPA-130: 121–129.

[2] Williamson S. and Laithwaite E. Generalised harmonic analysis for the steady-state performance of sinusoidally-excited cage induction motors, *IEE Proceedings B – Electric Power Applications*, 1985; EPA-132: 157–163.

[3] Stepina J. Fundamental equations of the space vector analysis of electrical machines, *Acta Technica CSAV*, 1968; 13, 184–198.

[4] Stepina J. Matrix analysis of space harmonics of asymmetrical stator windings, *IEE Proceedings B – Electric Power Applications*, 1987; EPA-134: 207–210.

[5] Alwash J.H.H. Generalised approach to the analysis of asymmetrical three-phase induction motors, *IEE Proceedings – Electrical Power Applications*, 1995; EPA-142: 87–96.

[6] Krause P.C. *Analysis of Electric Machinery*. Piscataway, NJ: McGraw-Hill Book Company, 1996.

[7] Penman J. and Stavrou A. Broken rotor bars – their effect on the transient performance of induction machines, *IEE Proceedings—Electrical Power Applications*, 1996; EPA-143: 449–457.

[8] Fudeh H.R. and Ong C.M. Modeling and analysis of induction machines containing space harmonics. Part 1: Modeling and transformations. *IEEE Transactions on Power Apparatus and Systems*, 1983.

[9] Luo X., Liao Y., Toliyat H.A., El-Antalby A. and Lipo T.A. Multiple coupled circuit modeling of induction machines. *IEEE Transactions on Industry Applications*, 1995; IA-31: 311–317.

[10] Lipo T.A. Theory and control of synchronous machines, ECE 511 lecture notes, University of Wisconsin-Madison, 2000.

[11] Boldea I. and Nasar S.A. *The induction machine handbook*, Boca Raton, FL: CRC Press, 2002

[12] Alger P.L., *Induction machines*, Boca Raton, FL: Gordon and Breach Science Publishers, 1970.

[13] Meshgin-Kelk H., Milimonfared J. and Toliyat H.A. Interbar currents and axial fluxes in healthy and faulty induction motors, *IEEE Transactions on Industry Applications*, 2004; IA-40: 128–134.

[14] Joksimović G., Đurović M. and Obradović A. Skew and linear rise of MMF across slot modeling – winding function approach, *IEEE Transactions on Energy Conversion*, 1999; EC-14: 315–320.

[15] Faiz J. and Tabatabaei I. Extension of winding function theory for non-uniform air gap in electric machinery, *IEEE Transactions on Magnetics*, 2002; MAG-38: 3654–3657.

[16] Joksimović G. "Dynamic simulation of cage induction machine with air gap eccentricity", *IEE Proceedings – Electrical Power Applications*, 2005; EPA-152: 803–811.

[17] Toliyat H.A. and Nuaim N.A. Simulation and detection of dynamic air-gap eccentricity in salient-pole synchronous machines, *IEEE Transactions on Industry Applications*, 1999; IA-35: 86–91.

[18] Joksimović G. and Penman J. The detection of inter-turn short circuits in the stator windings of operating motors, *IEEE Transactions on Industrial Electronics*, 2000; IE-47: 1078–1084.

# *Chapter 6*

# Finite element implementation of induction motors in healthy and faulty conditions

## 6.1 Introduction

The finite element method (FEM) is indeed one of the most accurate techniques of simulating and approximating the electromagnetic and mechanical devices including motors and generators. FEM is based on numerical techniques applied to equations representing a real phenomenon which do not have an explicit analytical solution. Although FEM is a kind of approximation, it has proved to be an acceptably accurate practice. Hence, some applications including the fault diagnosis procedures have been so far addressed using FEM. Basically, a set of differential equations representing motor behaviors is solved using FEM. Some other alternative methods are the finite differential method, the finite volume method and the boundary element method. These techniques normally convert the original set of differential equations into an algebraic set of equations which can be easily solved in one or a couple of iterations.

Induction motors, whether healthy or faulty, are electromagnetic devices consisting of magnetically linear and nonlinear materials such copper, air and silicon steel, each must be treated separately regarding the magnetic characteristics of materials. On the other hand, esoteric geometry of electrical motors makes them a tricky device to be modeled and studied analytically. Sometimes, it is not even possible to extract analytical or even semianalytical equations. Therefore, the trend of modeling and simulating attempts has been switched to an FEM-based analysis which in fact returns experimentally proved accurate results [1–11]. This is the edge of knowledge in the field of research and development of fault diagnosis process; therefore, getting used to it is highly recommended [12,13].

Any diagnosis of an induction motor is grounded on the knowledge of magnetic field distribution in different parts of the motor. So becoming familiar with and understanding the healthy and faulty trends of field variations in time or space should be in fact the starting major step to be taken. Otherwise, a shallow inference of the topic is achieved. Therefore, we insist on understanding a numerical technique as accurate as the FEM by means of which all the spatial and time harmonics, along with their effects, are modeled. Putting forward some of the fundamental ideas behind the necessity of the FEM, the following influential factors which directly affect a faithful motor simulation analysis are given as the modeling capabilities of an FEM package:

- **Nonlinearity of silicon steel materials**: This is probably the most significant contributing factor influencing the accuracy of the electrical machine model [14–16]. The main theoretical premise behind this claim is that any machine operation fails if a linear material is used. It is the nonlinear material which takes care of a highly permeable environment for a smooth pass of the magnetic field. So, this study tries to offer some important insights into the concept of nonlinearity and the application of the FEM in accurate modeling of the mentioned phenomenon [17–20].
- **Nonsinusoidal distribution of windings and rotor bars:** Nonsinusoidal distribution of stator windings or the rotor bars is the common practice of making a path for the current to produce a rotating field. Although implementing a pure sinusoidal distribution is always respected, it is not practically feasible to the existing mechanical limitations in terms of windings, slot punching, etc. On the other hand, the more sinusoidal the distribution is, the smaller the tooth width should be. This increases the possibility of the highly saturated regions which are not desired. Regardless of the type of winding distribution, the FEM is capable of fully modeling a 2D or even 3D shape of slots, teeth, coils and phase windings. However, a major problem with the existing analytical or even semianalytical approaches is the lack of potential for taking care of the mentioned structural nonuniformity while the FEM does it very well [21–31].
- **Accurate material modeling:** Recent developments in the FEM-based material modeling reveals the fact that not only the silicon steel materials, specifically the corresponding single-value *B–H* curves, but also a complete hysteresis loop model of the magnetic materials besides the linear copper, aluminum and insulations can be accurately modeled using the FEM. The only requirement is factory data at different magnetic fluxes, temperatures and also frequencies. The rest is handled by the FEM carefully. These cannot be easily lifted by means of analytical solutions.
- **Structural deformation:** Induction motors are usually designed as a symmetrically, whether electromagnetically or mechanically, device. Faults, especially the mechanical faults, produce a type of deformation which is reflected in the motor electromagnetic quantities. For example, in the case of a misaligned rotor, there is no symmetry through the *z*-axis anymore; moreover, the magnetic field of the rotor parts closer to the stator body are way larger than that of the other parts. Previous studies have reported some attempts for modeling of the deformations analytically, but they fail to address an accurate modeling and so a diagnosis procedure. However, the FEM has proved to be the best approach to model the deformations [32–38].

There has been already little quantitative analysis of induction motors and their FEM-based simulations based on which one can understand how to implement the FEM analysis of a faulty motor from the scratch. So, this will be the focus of this chapter which aims at discussing the FEM implementation assuming that an

FEM package is available. However, better understanding of implementation, adjustment and result inference of a commercial package require a minimum knowledge of electromagnetism along with the principles of the FEM. Therefore, a useful discussion explaining the fundamentals of the governing equations and FEM implementation is provided first in this chapter. The following are included in this chapter:

- a comprehensive formulation of the Maxwell's equations [39]
- the finite element formulations [1,40]
- an efficient approach to the implementation of broken bar, eccentricity and short-circuit faults in an FEM package [41–45]
- a discussion of the adjustments in terms of selecting a proper time step and mesh size for simulations
- the inclusion of a magnetic field analysis of induction motors by providing two examples, a cage and a wound rotor (the dimensions and materials are also provided so that readers might implement the same models in their own packages)
    - the study of the flux density variations in different faulty conditions and various motor parts including the stator yoke, the stator teeth, the air gap, the rotor teeth and the rotor back-iron.

## 6.2 Electromagnetic field equations

Induction motors operate based on the induction or Eddy current phenomenon. Any electromagnetic device representing an induction-based operation can be formulated by means of eddy current formulation. By definition, electrical and magnetic fields are highly correlated in an eddy-current problem, so a change in one of the fields causes a change in the other one. The change is usually demonstrated in time; therefore, the time-derivatives and their numerical approximations should be always treated accurately as the accuracy of the final model directly depends on this. The time derivative, as there might also be a set of partial differential equations in this chapter, is shown by $\frac{\partial}{\partial t}$. It is better to keep in mind this convention because henceforth it is going to be used often. On the other hand, due to the time-dependent nature of any electrical machine, the time-derivative, at least in the case of motor current, flux and voltage, is certainly nonzero [1,2].

To initiate the discussion, Maxwell's equations omitted as the basis for any low- or high-frequency electromagnetic application are provided below. Before that, a very simple eddy-current problem mimicking the behavior of an electrical machine is illustrated to clarify what kind of materials and magnetic divisions are present (see Figure 6.1).

There are two types of regions, the so-called "conducting" and "nonconducting." By "conduction," magnetic conduction is meant, not the electrical conduction. The conducting region in Figure 6.1, shown by $\Omega_c$, is surrounded by a non-conducting region shown by $\Omega_n$. The conducting material is usually of the type

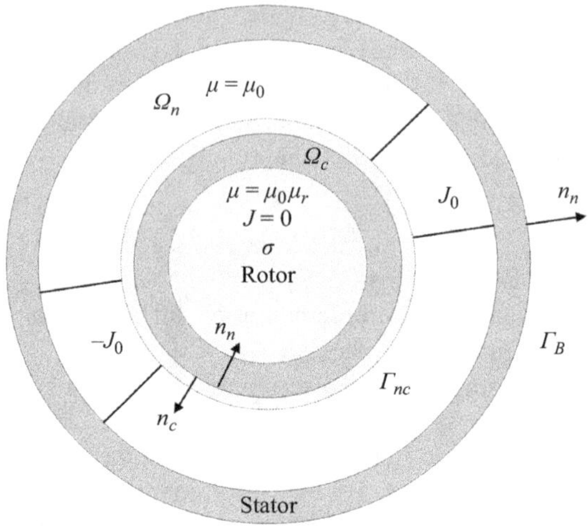


*Figure 6.1 Investigated eddy-current problem*

"silicon steel" and that of the nonconducting region is made of "air," "copper" or "aluminum." Depending on the type of material, of course from magnetic point of view, static and dynamic magnetic fields are presented in case there are nonconducting and conducting materials, respectively. So, the problem usually becomes more serious where a magnetically conducting material must be dealt with. Past decades have seen a rapid development of formulations the most reliable of which is the one developed by Maxwell in terms of expressing the relations inside and at the boundary of materials of the two types. The formulations naturally possess a differential type shown in (6.1) [2]:

$$\nabla \times \vec{H} = \vec{J}_0, \quad \text{in } \Omega_n \tag{6.1}$$

$$\nabla \cdot \vec{B} = 0, \quad \text{in } \Omega_n \tag{6.2}$$

$$\nabla \times \vec{H} = \vec{J}, \quad \text{in } \Omega_c \tag{6.3}$$

$$\nabla \times \vec{E} = -\frac{\partial \vec{B}}{\partial t}, \quad \text{in } \Omega_c \tag{6.4}$$

$$\nabla \cdot \vec{B} = 0, \quad \text{in } \Omega_c \tag{6.5}$$

$$\nabla \cdot \vec{J} = 0, \quad \text{in } \Omega_c \tag{6.6}$$

where $\vec{H}, \vec{B}, \vec{E}, \vec{J}$ and $\vec{J}_0$ are the magnetic field intensity, the magnetic flux density, the electric field intensity, the eddy-current density and the source current density, respectively. These are the main differential equations describing a low-frequency eddy-current problem. There are also some other relations called "constitutive relations" as follows:

$$\vec{B} = \begin{cases} \mu_0 \vec{H}, & \text{in } \Omega_n \\ \mu_0 \mu_r \vec{H}, & \text{in } \Omega_c \end{cases} \tag{6.7}$$

$$\vec{J} = \sigma \vec{E}, \quad \text{in } \Omega_c \tag{6.8}$$

The constitutive relationships can be expressed as follows:

$$\vec{H} = \begin{cases} \nu_0 \vec{B}, & \text{in } \Omega_n \\ \nu_0 \nu_r \vec{B}, & \text{in } \Omega_c \end{cases} \tag{6.9}$$

$$\vec{E} = \rho \vec{J}, \quad \text{in } \Omega_c \tag{6.10}$$

It should be noted that $\mu$ is the magnetic permeability, $\nu$ is the magnetic reluctivity, $\sigma$ is the electric conductivity and $\rho$ is the electrical resistivity of materials. Having discussed the basics of electromagnetism present in an electrical motor-like device, it is time to discuss how the regions with different types of materials react with respect to each other's magnetic and electric fields. This part critically traces the mutual effects and interactions. To this purpose, consider again the two regions $\Omega_c$ and $\Omega_n$ and their coupling at the interface $\Gamma_{nc}$. This is exactly where two materials are ideally separated from each other. The assumption is that the thickness of the interface, or technically the so-called "boundary," is almost zero; hence, there is nothing in between. On the other hand, the whole eddy-current problem is encapsulated inside a noneddy current region in which the magnetic flux density vanishes. So another boundary "$\Gamma_B$" is assigned to the problem to convey the idea of zero flux density outside the main problem. Although it might seem a very imaginary situation, it is often the case in any electrical machine analysis as the permeability of the stator core material which is the outer surface of any induction motor is considerably larger than that of the surrounding air. Practically, this leads to an almost zero flux leakage from the stator to the outer air environment. The result is the presence of a vanishing flux density toward the outer air space. During the modeling process, this phenomenon is applied by means of a zero-tangential flux density at the boundary. This makes the total flux density be zero at $\Gamma_B$. More useful discussions will be provided in the next sections of this chapter.

Previously, the tangential component of the flux density was targeted. It literally says that there should also be a normal component which builds the whole magnetic flux density vector together with the tangential component. The components of a vector are used to predict the electric and magnetic fields behavior at a boundary. They are identified as follows:

$$\vec{B} \cdot \vec{n} = -b \quad \text{or} \quad \vec{B} \cdot \vec{n} = 0, \quad \text{on } \Gamma_B \tag{6.11}$$

$$\vec{H}_c \times \vec{n}_c + \vec{H}_n \times \vec{n}_n = 0, \quad \text{on } \Gamma_{nc} \tag{6.12}$$

$$\vec{B}_c \cdot \vec{n}_c + \vec{B}_n \cdot \vec{n}_n = 0, \quad \text{on } \Gamma_{nc} \tag{6.13}$$

$$\vec{J} \cdot \vec{n}_c = 0, \quad \text{on } \Gamma_{nc} \tag{6.14}$$

where $\vec{n}$ is the outer normal unit vector, $\vec{H}_c$ is the magnetic field intensity in the conducting region, $\vec{H}_n$ is the magnetic field intensity in the nonconducting region, $\vec{n}_c$ is the outer normal vector of the conducting region, $\vec{n}_n$ is the outer normal vector of nonconducting region, $\vec{B}_c$ is the magnetic flux density in the conducting region and $\vec{B}_n$ is the magnetic flux density in nonconducting region. Note that $\vec{n}_n = -\vec{n}_c$.

The mentioned relations so far are kinds of general representations of a low-frequency and time-dependent electromagnetic field applicable to any type of applications among which induction motors is of a great interest. Henceforth, the focus will be switched to a more advanced discussion on the scalar and vectorized form of the formulations by incorporating a highly important term called "magnetic vector potential." Normally, all the 2D problems associated with an induction motor modeling are based on computing the magnetic vector potential, and then, the other magnetic and electric quantities are investigated.

## 6.3 Magnetic vector potential, magnetic scalar potential, current vector potential

The magnetic vector potential ($\vec{A}$) is formulated by

$$\vec{B} = \nabla \times \vec{A} \tag{6.15}$$

Equally important, (6.15) approves (6.2) according to the following relation which is held for any vector in space:

$$\nabla \cdot \nabla \times \vec{v} = 0 \tag{6.16}$$

Substituting (6.10) with (6.1) and (6.9), the partial differential equation relating the magnetic fields to an electric source is obtained as follows:

$$\nabla \times \left(\nu \nabla \times \vec{A}\right) = \vec{J}_0 \tag{6.17}$$

Moreover, using the Coulomb gauge:

$$\nabla \cdot \vec{A} = 0 \tag{6.18}$$

This guarantees the uniqueness of $\vec{A}$. If the electric source consisting of a current density $J_0$, only includes a $z$-direction component, the uniqueness of $\vec{A}$ is satisfied. Assuming a $z$-direct current density, the magnetic field intensity and flux density reveal a vector in $x$–$y$ plane.

$$\vec{J}_0 = J_0(x, y)\vec{e}_z \tag{6.19}$$

$$\vec{H} = H_x(x, y)\vec{e}_x + H_y(x, y)\vec{e}_y \tag{6.20}$$

$$\vec{B} = B_x(x, y)\vec{e}_x + B_y(x, y)\vec{e}_y \tag{6.21}$$

And the magnetic vector potential is:

$$\vec{A} = A_z(z)\vec{e}_z \tag{6.22}$$

Therefore,

$$\vec{B} = \nabla \times \vec{A} = \begin{vmatrix} \vec{e}_x & \vec{e}_y & \vec{e}_z \\ \dfrac{\partial}{\partial x} & \dfrac{\partial}{\partial y} & 0 \\ 0 & 0 & A_z \end{vmatrix} = \vec{e}_x \frac{\partial A_z}{\partial y} - \vec{e}_y \frac{\partial A_z}{\partial x} \tag{6.23}$$

The normal component of the magnetic flux density is set as:

$$\vec{B} \cdot \vec{n} = -b \xrightarrow{\text{yields}} \left(\nabla \times \vec{A}\right) \cdot \vec{n} = \nabla \cdot \left(\vec{A} \times \vec{n}\right) = -b \tag{6.24}$$

And

$$\nabla \cdot \left(\vec{n} \times \vec{A}\right) = b, \quad \vec{n} \times \vec{A} = \vec{\alpha} \quad \text{on } \Gamma_B \tag{6.25}$$

This condition is basically called a "Dirichlet boundary condition." The vector $\vec{\alpha}$ is not usually known, so the problem is dealt with by determining the value of $b$. $b$ is usually equal to zero on $\Gamma_B$. In a few words, the discussed electromagnetic system is represented by:

$$\nabla \times \left(\nu \nabla \times \vec{A}\right) = \vec{J}_0 \quad \text{in } \Omega_n \tag{6.26}$$

$$\vec{n} \times \vec{A} = 0 \quad \text{on } \Gamma_B \tag{6.27}$$

Despite the fact that $\vec{A}$ is the most addressed quantity in any FEM-based analysis, there is another formulation used for converting the vector potential to a scalar potential. To come up with the idea, the magnetic field intensity in the non-conducting region is reformulated as follows:

$$\vec{H} = \vec{T}_0 + \vec{H}_m \tag{6.28}$$

The curl of the imposed current vector potential $\vec{T}_0$ should be equal to the current density of the source, and the curl of the second right-hand side term is equal to zero.

$$\nabla \times \vec{T}_0 = \vec{J}_0 \tag{6.29}$$

$$\nabla \times \vec{H}_m = 0 \tag{6.30}$$

In the next step, $\vec{H}_m$ as the nonrotational (stationary) part of $\vec{H}$ is related to the scalar potential as the negative gradient of the magnetic scalar potential.

$$\vec{H}_m = -\nabla \Phi \tag{6.31}$$

On the other hand, $\nabla \times (\nabla \Phi) \triangleq \vec{0}$ holds for any scalar function. Therefore,

$$\vec{H} = \vec{T}_0 - \nabla \Phi \tag{6.32}$$

Equation (6.32) satisfies (6.1) in the nonconducting region. $\Phi$ is an abstract concept and usually named "reduced magnetic scalar potential." This arises from the fact that the source term is reflected in $\vec{T}_0$, so it is of a vector form which directly builds up the direction of $\vec{H}$. Using (6.7) we have:

$$\vec{B} = \mu(\vec{T}_0 - \nabla\Phi) \tag{6.33}$$

As the divergence of the magnetic flux density is zero, so

$$\nabla \cdot (\mu\nabla\Phi) = \nabla \cdot (\mu\vec{T}_0) \quad \text{in } \Omega_n \tag{6.34}$$

This is a very well-known equation called "generalized Laplace-Poisson" equation. Apart from the Dirichlet boundary condition, there is a Neumann boundary condition obtained by setting the normal component of the magnetic flux density on the boundary.

$$\vec{B} \cdot \vec{n} = 0 \xrightarrow{\text{yields}} (\mu\vec{T}_0 - \mu\nabla\Phi) \cdot \vec{n} = 0 \quad \text{on } \Gamma_B \tag{6.35}$$

Finally, the set of differential equations of a static magnetic (magneto-static) problem are summarized as follows:

$$\mu(\vec{T}_0 - \nabla\Phi) \cdot \vec{n} = 0 \quad \text{on } \Gamma_B \tag{6.36}$$

$$\nabla \cdot (\mu\nabla\Phi) = \nabla \cdot (\mu\vec{T}_0) \quad \text{in } \Omega_n \tag{6.37}$$

The term "magneto-static" is assigned to a problem in which there is neither mechanical motion nor a transient electromagnetic phenomenon. For instance, if an electrical machine should be analyzed in a steady-state regime in which there is no difference between different electric cycles, a magneto-static analysis is useful to return the flux distribution. However, if something like losses are to be computed, magneto-static solutions are useless even if the machine operates in a steady-state mode. It might also be noted that a magneto-static solution does not model an induction motor behavior efficiently and satisfactorily as it is a device with motion and an electromagnetic transient behavior. Even in one steady-state cycle, electrical and magnetic quantities contain nonconstant values. This is actually the nature of an eddy-current problem. In fact, the noticeable importance of an eddy current is the nature of induction caused by a dynamic field change. Evidently, if a constant magnetic field, whether spatial to temporal, exists, there will be no induction at all. Equations (6.3) and (6.4) clearly show the time-dependence of the eddy-current problem. The underlying argument in favor a magneto-static solution is that it can be easily solved in both time and space.

## 6.4 $\vec{T}$-$\Phi$ Formulation

Some main concepts including the $\vec{T}_0$, $\vec{A}$ and $\Phi$ have been already introduced. Along with the mentioned terms, an electric scalar potential shown by V denoting the difference of potential between two points in space is used to formulate a complete eddy-current problem. This task is set out with the aim of the pairs $(\vec{T}_0, \Phi)$ or $(\vec{A}, V)$ separately. Not both pairs are required at the same time. Bearing this in mind, the first formulation is a based on the pair $(\vec{T}_0, \Phi)$.

Using (6.3), (6.6) and (6.29), we have:

$$\nabla \times \vec{H} = \nabla \times \vec{T} \xrightarrow{\text{yields}} \nabla \times \left(\vec{H} - \vec{T}\right) = \vec{0} \tag{6.38}$$

This formulation directly satisfies (6.3) because for any scalar function $\varphi = \varphi(\vec{r}, t)$, the following equation holds:

$$\nabla \times \nabla\varphi = 0 \tag{6.39}$$

On the other hand,

$$\vec{E} = \rho\vec{J} = \rho\nabla \times \vec{T} \tag{6.40}$$

Substituting (6.40) and (6.7) with Faraday's law (6.4) leads to a partial differential equation:

$$\nabla \times \left(\rho\nabla \times \vec{T}\right) + \mu\frac{\partial\vec{T}}{\partial t} - \mu\nabla\frac{\partial\Phi}{\partial t} = -\mu\frac{\partial\vec{T}_0}{\partial t} \quad \text{in } \Omega_c \tag{6.41}$$

Then, using the Gauss's law,

$$\nabla \cdot \left(\mu\vec{T} - \mu\nabla\Phi\right) = -\nabla \cdot \left(\mu\vec{T}_0\right) \quad \text{in } \Omega_c \tag{6.42}$$

The solution of (6.41) and (6.42) gives the values for $\Phi, \vec{T}$.

## 6.5 $\vec{A}$–$V$ Formulation

Substitution of (6.15) with the Faraday's law results in:

$$\nabla \times \vec{E} = -\frac{\partial}{\partial t}\nabla \times \vec{A} = -\nabla \times \frac{\partial\vec{A}}{\partial t} \xrightarrow{\text{yields}} \nabla \times \left(\vec{E} + \frac{\partial\vec{A}}{\partial t}\right) = 0 \tag{6.43}$$

The curl-free vector field $\vec{E} + \frac{\partial\vec{A}}{\partial t}$ is then replaced by the gradient of the electric scalar potential:

$$\vec{E} + \frac{\partial\vec{A}}{\partial t} = -\nabla V \tag{6.44}$$

In a 2D problem, for example, a 2D induction motor model, $V$ can be set to zero, so

$$\vec{E} = -\frac{\partial\vec{A}}{\partial t} \tag{6.45}$$

On the basis of the evidence that the current density always has a $z$-direction in an induction motor model, the 2D problems are easily handled by the $\vec{A}$–$V$

formulation. The discussion is encouraged by substituting (6.45) and (6.15) with the (6.3):

$$\nabla \times \left(\nu \nabla \times \vec{A}\right) + \sigma \frac{\partial \vec{A}}{\partial t} = \vec{0} \tag{6.46}$$

where $\vec{A}$ is an unknown quantity. This is why only major equation is required to solve the $\vec{A}$–$V$ formulation. The foregoing discussion implies that (6.46) can also be implemented and solved in the frequency domain by simply replacing the term $\frac{\partial}{\partial t}$ by $j\omega$. This really makes the life easier if a static or even time-harmonic solution is enough to address the modeling task. The next step is to discuss how a magneto-static and eddy-current problem can be coupled in order to take the specifications of both types into account. In reality, the static magnetic field is the output of a current source, whether a real source or an eddy-current source. So coupling the magneto-static solutions with that of the eddy-current seems to be an essential task as both types are present in an induction motor. Having suggested the idea, let us move to the next section in which an inclusive debate on how these two systems are coupled is provided.

## 6.6 Coupled magneto-static and eddy current-field problem

Like the previous section in which two types of formulations were proposed to an eddy-current problem, two coupling methods are suggested as well. Due to the importance of $\vec{A}$–$V$ formulation, we first go through the corresponding topics and then change the strategy to a $\vec{T}$-$\Phi$ formulation. The following are some of the principles to keep in mind:

- $\vec{A}$ is a continues quantity. This means that the tangential and the normal components are continuous on $\Gamma_{nc}$.
- $\vec{A}$ is used throughout the conducting and nonconducting regions.
- $V$ is only used in the conducting regions.
- The normal component of the magnetic flux density is continuous.

Taking the mentioned facts into account and also bearing the previous discussions in mind, one ends up with the following formulation in terms of the magnetic vector potential held at the boundary of two regions.

$$\left(\nabla \times \vec{A}\right) \cdot \vec{n}_c + \left(\nabla \times \vec{A}\right) \cdot \vec{n}_n = \nabla \cdot \left(\vec{A} \times \vec{n}_c\right) + \nabla \cdot \left(\vec{A} \times \vec{n}_n\right) = \nabla \cdot \left(\vec{A} \times \vec{n}_c + \vec{A} \times \vec{n}_n\right) = 0 \tag{6.47}$$

Besides, the continuity of the tangential component of the magnetic flux density is added to the problem by means of a Neumann condition.

$$\left(\nu \nabla \times \vec{A}\right) \times \vec{n}_c + \left(\nu \nabla \times \vec{A}\right) \times \vec{n}_n = \vec{0} \tag{6.48}$$

The normal component of the current density has a vanishing trend on the boundary of two regions. We will now summarize the set of equations covered so far in terms of the $\vec{A}$–$V$ formulation of a magneto-static and eddy-current problem.

$$\nabla \times \left(\nu\nabla \times \vec{A}\right) + \sigma\frac{\partial \vec{A}}{\partial t} = \vec{0} \quad \text{in } \Omega_c \tag{6.49}$$

$$\nabla \times \left(\nu\nabla \times \vec{A}\right) = \vec{J}_0 \quad \text{in } \Omega_n \tag{6.50}$$

$$\left(\nu\nabla \times \vec{A}\right) \times \vec{n}_c + \left(\nu\nabla \times \vec{A}\right) \times \vec{n}_n = \vec{0} \quad \text{on } \Gamma_{nc} \tag{6.51}$$

$$\nabla \cdot \left(\vec{A} \times \vec{n}_c + \vec{A} \times \vec{n}_n\right) = 0 \quad \text{on } \Gamma_{nc} \tag{6.52}$$

$$\vec{n} \times \vec{A} = 0 \quad \text{on } \Gamma_B \tag{6.53}$$

In the frequency-domain, the following expressions are valid:

$$\nabla \times \left(\nu\nabla \times \vec{A}\right) + \sigma j\omega\vec{A} = \vec{0} \quad \text{in } \Omega_c \tag{6.54}$$

$$\nabla \times \left(\nu\nabla \times \vec{A}\right) = \vec{J}_0 \quad \text{in } \Omega_n \tag{6.55}$$

$$\left(\nu\nabla \times \vec{A}\right) \times \vec{n}_c + \left(\nu\nabla \times \vec{A}\right) \times \vec{n}_n = \vec{0} \quad \text{on } \Gamma_{nc} \tag{6.56}$$

$$\nabla . \left(\vec{A} \times \vec{n}_c + \vec{A} \times \vec{n}_n\right) = 0 \quad \text{on } \Gamma_{nc} \tag{6.57}$$

$$\vec{n} \times \vec{A} = 0 \quad \text{on } \Gamma_B \tag{6.58}$$

Having propounded the coupling between the magneto-static and eddy-current problems, using the $\vec{A}$–$V$ formulation, the $\vec{\boldsymbol{T}}$–$\boldsymbol{\Phi}$ formulation is going to be discussed. The underlying idea behind this formulation is the magnetic field intensity in the nonconducting and conducting regions. The corresponding values are $\vec{H} = \vec{T}_0 - \nabla\boldsymbol{\Phi}$ and $\vec{H} = \vec{T}_0 + \vec{T} - \nabla\boldsymbol{\Phi}$, respectively. On the boundary of the two regions, the tangential field intensity is set to be continuous. In addition, the tangential component of the current density is equal to zero on the boundary. Moreover, as the normal eddy-current density has a vanishing trend on the boundary, it satisfies the following relation:

$$\vec{J} = \nabla \times \vec{T} \xrightarrow{\text{yields}} \vec{J} \cdot \vec{n} = \left(\nabla \times \vec{T}\right) \cdot \vec{n} = \nabla \cdot \left(\vec{T} \times \vec{n}\right) = 0 \tag{6.59}$$

Furthermore, a Neumann boundary condition is applied for ensuring the continuity of the normal component of the magnetic flux density. In a few words, the $\vec{\boldsymbol{T}}$–$\boldsymbol{\Phi}$ formulation includes the following relations:

$$\nabla \times \left(\rho\nabla \times \vec{T}\right) + \mu\frac{\partial \vec{T}}{\partial t} - \mu\nabla\frac{\partial \Phi}{\partial t} = -\mu\frac{\partial \vec{T}_0}{\partial t} \quad \text{in } \Omega_c \tag{6.60}$$

$$\nabla \cdot \left(\mu\vec{T} - \mu\nabla\Phi\right) = -\nabla \cdot \left(\mu\vec{T}_0\right) \quad \text{in } \Omega_n \tag{6.61}$$

$$-\nabla \cdot \left(\mu\nabla\Phi\right) = -\nabla \cdot \left(\mu\vec{T}_0\right) \quad \text{in } \Omega_n \tag{6.62}$$

$$\left(\mu\vec{T}_0 - \nabla\Phi\right) \cdot \vec{n} = 0 \quad \text{on } \Gamma_B \tag{6.63}$$

$$\vec{T} \times \vec{n}_c = 0 \quad \text{on } \Gamma_{nc} \tag{6.64}$$

$$\left(\mu\vec{T}_0 + \mu\vec{T} - \mu\nabla\Phi\right) \cdot \vec{n}_c + \left(\mu\vec{T}_0 - \mu\nabla\Phi\right) \cdot \vec{n}_n = 0 \quad \text{on } \Gamma_{nc} \tag{6.65}$$

## 6.7 Transient-with-motion formulation

Undoubtedly, a real motor simulation requires this transient operation with motion, in which the voltages are the input. The above-mentioned current-based formulation is only useful under the following assumptions:

- The motor operates in a steady-state mode.
- The accurate current and advance angles are known not only for the stator but also for the rotor.

Otherwise, the current-based analysis is nonsense. In a real-world application, it is the voltage-based simulation which addresses the modeling challenges correctly. For the above-mentioned reason, any diagnosis procedure should be handled by a voltage-based simulation taking the electromagnetic and mechanical transients into account. Where the transient operation is under investigation, a time-harmonic model is definitely useless.

The foregoing discussion provides the electromagnetic equations corresponding to a "transient-with-motion-analysis" as follows [1]:

$$\nabla \times \left(\frac{1}{\mu}\nabla \times \vec{A}\right) + \delta\left(\frac{\partial \vec{A}}{\partial t} - \overrightarrow{\text{velocity}} \times \left(\nabla \times \vec{A}\right)\right) = \vec{J} \tag{6.66}$$

where the vector of velocity defines the relative speed of the rotor and the stator with respect to a reference frame which is usually the stator frame. If this formulation represents the electromagnetic equation of the stator, the velocity vector is equal to zero while that of the rotor is for sure nonzero. Then, the electrical connections relating the supply voltage to the electromagnetic equations through the current density are explained as follows:

$$\frac{V_b}{l} + \delta\left(\frac{\partial \vec{A}}{\partial t} - \overrightarrow{\text{velocity}} \times \left(\nabla \times \vec{A}\right)\right) = \vec{J} \tag{6.67}$$

where the amplitude of $V_b$ gradient in z-direction is $V_b/l$. The first term in the left-hand side of the above equation shows the current density provided by the supply voltage, the second term indicates the existence of an induced current and the third term corresponds to the EMF produced by the motion phenomenon. Considering the magnetic propagation equation,

$$\nabla \times \overrightarrow{\text{velocity}} \times \left(\nabla \times \vec{A}\right) = \delta\frac{V_b}{l} - \delta\frac{\partial \vec{A}}{\partial t} + \delta\ \overrightarrow{\text{velocity}} \times \left(\nabla \times \vec{A}\right) \tag{6.68}$$

As the above equation directly provides the current density, using the field equation, the electrical current is calculated by integrating it over the crosssection of conductors carrying the current. In a stationary reference frame, we have:

$$i = \iint \left( \delta \frac{V_b}{l} - \delta \frac{\partial \vec{A}}{\partial t} \right) dxdy \tag{6.69}$$

As seen, the conductors' current is related to the supply voltage through a resistance and an inductance. This is the equation expressing the current–voltage relation of one single conductor. So depending on the connection series or parallel connection, the voltages or the currents are added up to form a complete phase winding containing several coils and turns. Suppose that the coils $c_1, c_2, \ldots, c_n$ are connected in series to form a winding. All coils carry the total current, but the successive coils carry this current in the opposite direction. Two ends of the winding leave the magnetic field region, which is usually represented by an FE tool, and are connected to the voltage source which is the supply (Figure 6.2(a)). The resistance and inductance from these two ends are modeled by lumped resistance $R_{\text{ext}}$ and inductance $L_{\text{ext}}$. The applied voltage to the external terminals of

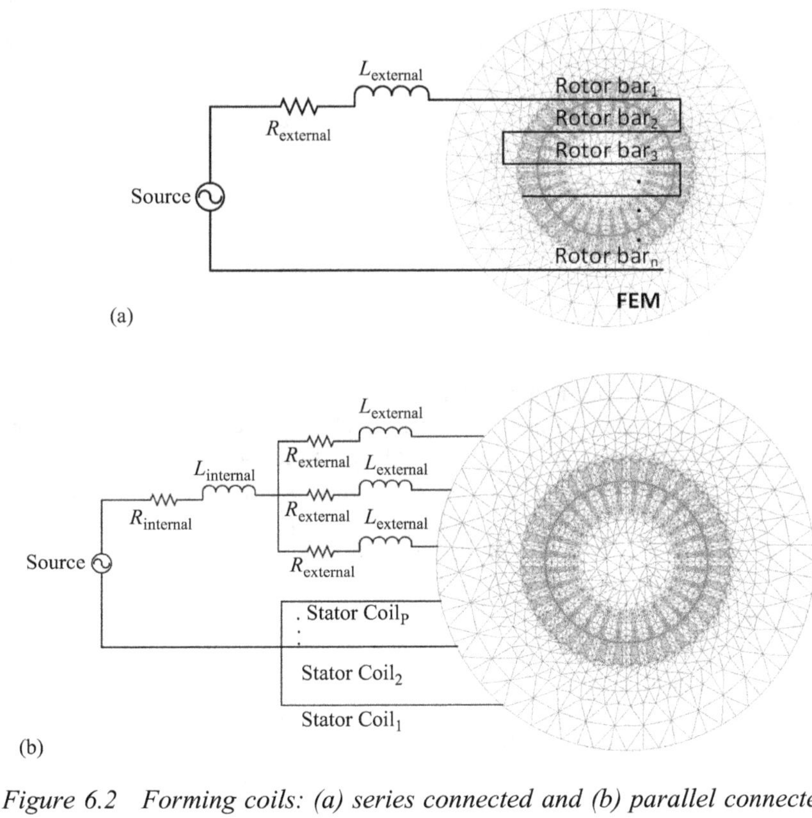


*Figure 6.2 Forming coils: (a) series connected and (b) parallel connected*

winding is $V_c$ and terminal voltage in the FE region is $V_t$. The voltages of the coils are as follows:

$$V_{t,c} = \sum_b d_b V_b \tag{6.70}$$

where $d_b$ is equal to $+1$ or $-1$ and it shows the polarity of coil b.

Therefore,

$$V_c = \{d_b\}_c^T \{V_b\}_c + L_{\text{ext}}\left(\frac{di_c}{dt}\right) + R_{\text{ext}} i_c \tag{6.71}$$

It means that this equation is used to relate the FE region with $V_b$ characteristic to external circuits and sources with $R_{\text{ext}}$, $L_{\text{ext}}$ and $V_c$. If the coils are connected in parallel, an individual circuit forms in the output terminals of the system. The general case consisting of connection of P coils in parallel with supply $V_s$, internal resistance $R_s$ and inductance $L_s$ has been illustrated in Figure 6.2. The governing RL equation is as follows:

$$V_s = R_S^T \sum i_c + L_s \sum \frac{di_c}{dt} + V_c \tag{6.72}$$

The developed electromagnetic torque is obtained as follows:

$$T_e = \left(\frac{L_{\text{ext}}}{\mu_0}\right) \int \left[ \left(\vec{B} \cdot \vec{n}\right)\left(\vec{r} \cdot \vec{B}\right) - 0.5B^2 (\vec{r} \times \vec{n}) \right] d\Gamma \tag{6.73}$$

where $\Gamma$ is the air-gap area of the motor. Finally, the electromechanical equation relating the developed electromagnetic torque to the angular speed of the rotor is as follows:

$$T_e - T_l = j\frac{d\omega_r}{dt} + B_m \omega_r \tag{6.74}$$

where $\omega_r$ is the angular frequency of the rotor.

The governing electromagnetic and mechanical equations of a rotating machine have been already explained regardless of the method by means of which the equations should be solved. Of course, we also noted the FEM as a promising tool to solve these kinds of equations and showed in Figure 6.2 that the electrical circuit can be connected to an FEM solution. In fact, we have a set of equations to be solved numerically. It is almost impossible to find an explicit analytical solution for the mentioned equations applied to a geometry such as the induction motors. In addition, the presence of the nonlinear materials which necessitates the use of an iterative approach to deal with them makes the problem more and more serious. Hence, a numerical and iterative routine is needed to tackle this problem. The FEM, of course, is one of the best methods which has been commercialized and used in industry not only for diagnosing but also for improving and optimizing designs.

## 6.8 Finite element method

FEM is used to find a numerical solution for electromagnetic field problems such as electrical machines regardless of the type and the structure. This concept was initially coined by Clough in 1960. Then, it found a widespread usage in different engineering problems ranging from architecture, civil, mechanical to electrical engineering. The FEM is very general in terms of applications, and applications which cannot be solved using the FEM are rarely found as the FEM does not care about the application. It deals with structures, materials as well as equations expressing the system or device behavior. The following are some of the main aspects of an FEM:

- handling very complex geometries,
- handling a wide range of problems such as solid mechanics, fluids, dynamics, heat problems and electrostatic problems
- handling complex restrains
- handling various loadings including nodal load, element load, time-dependent loading and frequency-dependent loading.

Although the FEM has a lot of astounding advantages, it does not return a closed-form solution which might be used for parameterized study of a device or structure. Once a solution is obtained, it is only valid for the parameters adjusted before the simulation process. The solution obtained from an FEM is only a numerical approximation of real solution. Moreover, as the FEM is implemented numerically, it has an inherent error corresponding to numerical solutions as well as the inaccuracy of structural modeling using meshes. Generally, it cuts the structure into several pieces called element or meshes. Then, it connects the elements at "nodes" and "edges" in a way that a group of several interconnected elements is constructed and mapped on the device structure in 2D or 3D. Most of the time, geometrical elements are in the shape of triangles and pyramids in a 2D and 3D problem, respectively. This process results in a set of simultaneous algebraic equations demonstrating the relations between the inputs and outputs of a problem. The set of equations can be linear which is usually the matter of a high-frequency application or nonlinear, which is mostly used in low-frequency applications such as electrical machines in which there are magnetically nonlinear materials like a silicon steel lamination used as the stator and rotor cores. The FEM uses a piecewise linear approximation of magnetic fields mapped on elements used to cover the device structure. By affixing elements and their corresponding field values obtained by solving the system of equations, an approximation of system output is obtained.

Figure 6.3 illustrates the general form of an FEM used in electromagnetism. Accordingly, drawing the geometry is the first step in defining the problem. To this end, a graphical user interface such as the "pdetool" developed by Mathwork can be used. Otherwise, a program should be developed to support this step. However, a graphical interface is not an essential part if one is expert enough to infer numerical

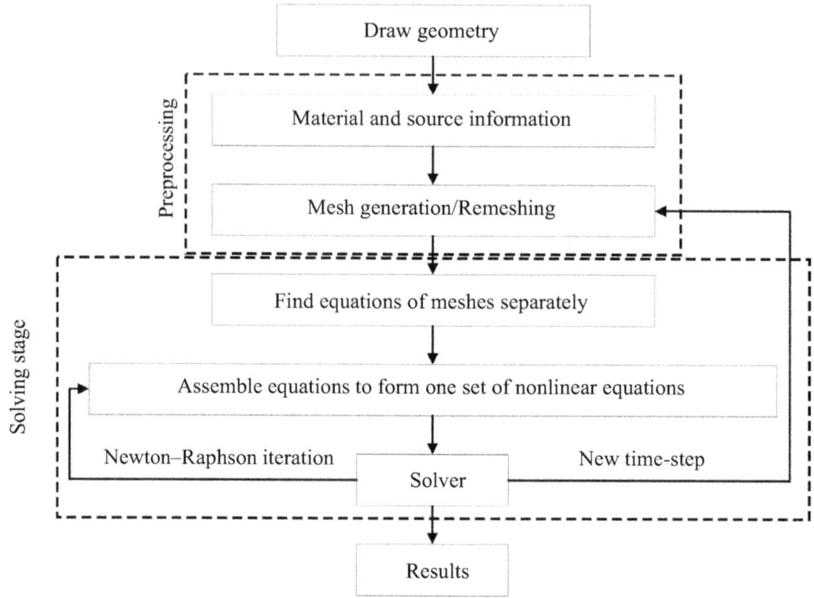


*Figure 6.3 FEM process*

data without it. Nevertheless, this is not an easy task and accessing an interface makes the life much easier.

If an interface is available, an example of the motor geometry can be drawn like what is illustrated in Figure 6.4. The stator and rotor cores consist of the segments with the same number of the stator slots and rotor bars, respectively. Although the segment-wise implementation of the components makes the solving process a little bit slower, it is usually recommended because different parts of the motor are fully separable in terms of materials as well as sources. This is a 2D representation while a comprehensive 3D analysis will result in a more accurate output. There has been an inclusive debate about whether 2D models work well in terms of predicting the motor quantities. The answer to these questions highly depends on the topology, the number of poles and the number of slots in both the stator and rotor sides. As a rule of thumb, if the skewing does not have a significant effect on the induction motor behavior or in the case that a thermal analysis is not targeted, a 2D model is satisfactory. Otherwise, a 3D model or a quasi 3D model should be utilized. Likewise, any motor fault leading to an asymmetry along the motor stack length, for example, the shaft misalignment, must be model in 3D.

### 6.8.1 Material modeling

The next step is to assign the materials to the problem. There are six types of materials used in an induction motor modeling as follows:

- copper for windings,
- silicon steel for stator and rotor cores

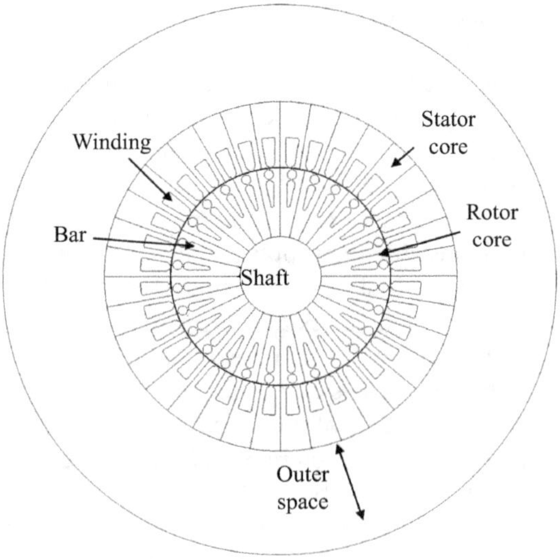


*Figure 6.4 Two-dimensional geometrical illustration of an induction motor with 36 stator slots and 28 rotor bars*

- aluminum for rotor bars
- air for the air gap and also the outer space of motors
- stainless steel for shaft
- paper for insulations.

Copper, stainless steel, aluminum, paper and air are the magnetically linear materials, while silicon steel reveals a highly nonlinear behavior with respect to the magnetic flux intensity. In fact, the linearity or nonlinearity of materials defines the variation of the magnetic permeability ($\mu = \mu_0 \mu_r$) as a function of the magnetic field intensity $H$ (Amps/m) (Figure 6.5). The fundamental curve of any nonlinear material is the one related to the Hysteresis $B$–$H$ curve which demonstrates a multivalue behavior. It means that there are two values of magnetic flux density for a given field intensity, depending on the increasing or decreasing slopes of the applied source to the magnetic problem. Actually, the field intensity increases upon increasing the input current. That is why the unit of $H$ is Amps/m revealing the dependency of $H$ on the input current signal provided by the voltage source.

For a nonlinear magnetic material, both $B$ and $H$ values are equal to zero if a nonmagnetized material is proposed. This point is called the initial magnetization point shown by "a" in Figure 6.5(a). Increasing the input current and consequently $H$ leads to an increase in the $B$ value up to the point "b" which is called the positive saturation point obtained by applying the maximum flux intensity. The initial parts of the dashed line (Figure 6.5(a)) are kind of linear representing the concept of linearity existing for a small value of the magnetic field. However, the trends changes to a highly nonlinear behavior while getting closer to the point "b." Then, the value of $B$ decreases as $H$ is reduced. The interesting point is that the curve does not necessarily follow the same path as it was built up, and a totally different $B$–$H$

curve with a higher magnetization level is generated by reducing $H$ up to the point "c" which is called the residual. This is the point in which $H$ is equal to zero, while the material is still magnetized to some extent. Depending on the material type, the residual point might be different. The process continues to build up the hysteresis $B$–$H$ curve enclosed by the points "gbcdeg," and the point "a" is normally bypassed after the first magnetic cycle.

The area of this produced curve is the main factor in determining the magnetic losses and increases by increasing the supply frequency. Although this is the reality of nonlinear materials, a closer look at the existing commercial packages shows that a multivalue implementation and simulation of this curve is still intact. There are two main reasons for this claim. First, it requires a high computational power. Second, there are convergence issues associated with solving multivalue nonlinear problems when it comes to the stage of numerical implementation of the Maxwell's equations. Therefore, the next discussion implies that a single-value implementation of the hysteresis curves is also possible and this is exactly what is used in commercial packages. A single-value curve (see Figure 6.5(b)) consists of a nonlinear $B$–$H$ curve usually modeled by several discrete points representing the curve. These points are collected by measurement setups and then imported into a package to mimic the behavior of the $B$–$H$ curve. The single-value $B$–$H$ curve is obtained by applying various induction levels, i.e., $H$, by means of an input current with different amplitude and then finding and connecting the positive saturation points of the obtained curves. The larger the number of samples is, the more accurate the modeled curve is. The slope of the curve is equal to the magnetic permeability $\mu$ which decreases up to the permeability of free space by increasing $H$. The permeability of free space $\mu_0$ is equal to $4\pi \times 10^{-7}$ and it is normally the basis of comparisons between materials. The $B$–$H$ curve of the free space is illustrated in Figure 6.5(c).

The larger the permeability is, the more the magnetic conductivity is. Thus, the initial linear region of the nonlinear $B$–$H$ curve is more conductive than the parts with a large $H$ value (over-saturated region). There is also a knee point which is the most challenging part of any nonlinear $B$–$H$ curve. As a technical issue, the number of measurements required for modeling the curve must be increased in the case of modeling the knee point. Otherwise, an unstable or inaccurate magnetic solution will be obtained. The reason is that the numerical implementation of a nonlinear problem is often handled by a Newton–Raphson method which is based on computing the derivative of the $B$–$H$ curve at different time steps. So increasing the number of samples in a highly nonlinear region is crucial for having an accurate derivative computation. There is also a minor nonlinear region called Rayleigh region. The available evidence seems to suggest that modeling the Rayleigh region does not impact the solution greatly as it only affects the problem regions with a relatively small field intensity levels. However, the electrical machines are designed to operate close to the knee point of the $B$–$H$ curve. The justifications are as the following:

- Below the knee point, power density decreases.
- Above the knee point, magnetic losses increase.

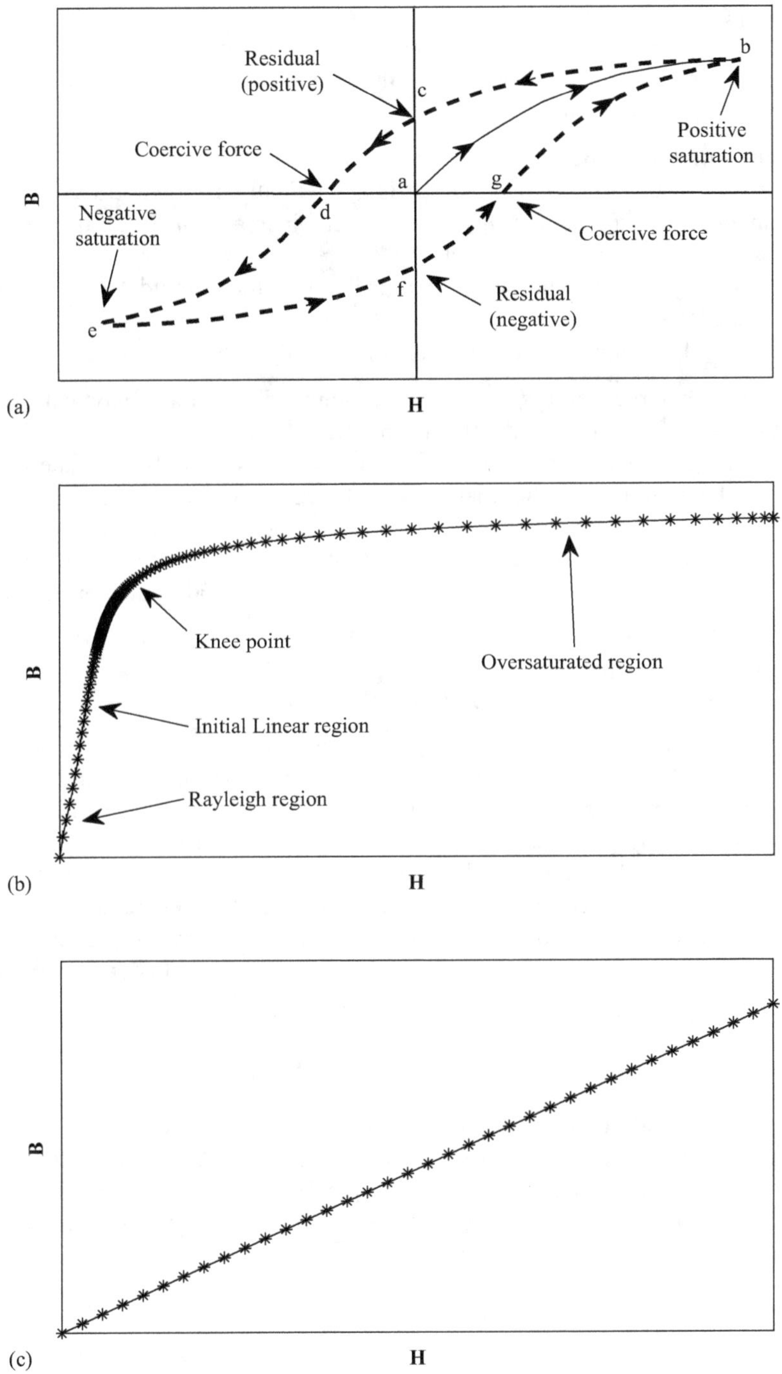


*Figure 6.5 (a) Multivalue nonlinear hysteresis curve, (b) single-value nonlinear B–H curve and (c) linear material*

So it is preferred to operate close to the knee point. During a simulation and Newton–Raphson process, interpolation is used to obtain the values of $B$ and $H$ between the measured data. Considering this, some rules should be followed when entering the measured data.

- Values of $B$ must be greater than or equal to zero.
- The differential permeability $dB/dH$ is theoretically such that $dB/dH$ is greater than or equal to $\mu_0$, forcing a monotonic increase in $B$ along the curve. Otherwise, an extremum is produced.
- $\mu_r > 1$ at all points (this is a theoretical result that should be automatically validated).
- The very first data point must be (0, 0).

For a better convergence, it is sometimes recommended that an analytical curve be fitted to the measured data and used for the next steps.

All the mentioned details in terms of the nonlinear materials associate with the silicon steel material. The other materials are linear and reveal a behavior similar to Figure 6.5(c). A constant slop of the *B–H* curve is the main characteristic and does not change as a function of the applied field intensity. They are easily modeled by means of a line passing through the origin with a slope equal to the permeability of the material.

### 6.8.2 Magnetic loss calculation

Assigning materials to a problem is the second major step of an FEM implementation. So if a highly variant thermal and electromagnetic problem is targeted, the materials properties should be updated for different operating conditions. For example, if the supply frequency is changed, the *B–H* curves and the corresponding loss values should also be updated. In fact, the *B–H* curves are not only the functions of the magnetic field intensity, they highly depend on temperature as well as the supply frequency. One of the significant frequency-dependent properties is the magnetic loss known as the combination of the hysteresis and eddy-current losses modeled as follows:

$$\text{Steinmetz Equation: Power loss density} = K_e(sfB)^2 + K_h f^\alpha B^\beta \tag{6.75}$$

where $K_e$ is the eddy-current loss coefficient, $s$ is the lamination thickness, $f$ is the frequency, $B$ is the magnitude of magnetic flux density, $K_h$ is the hysteresis loss coefficient, $\alpha$ and $\beta$ are the material-dependent coefficients expressing the dependency level of the hysteresis loss component to the frequency and the flux density. The values must be measured experimentally. However, there are companies providing the accurate data not only in terms of losses but also in terms of the *B–H* curves. On logical grounds, it is obvious that not all researchers have access to the corresponding measuring equipment, but they can buy the data. A typical variation of a silicon steel material versus the temperature and frequency

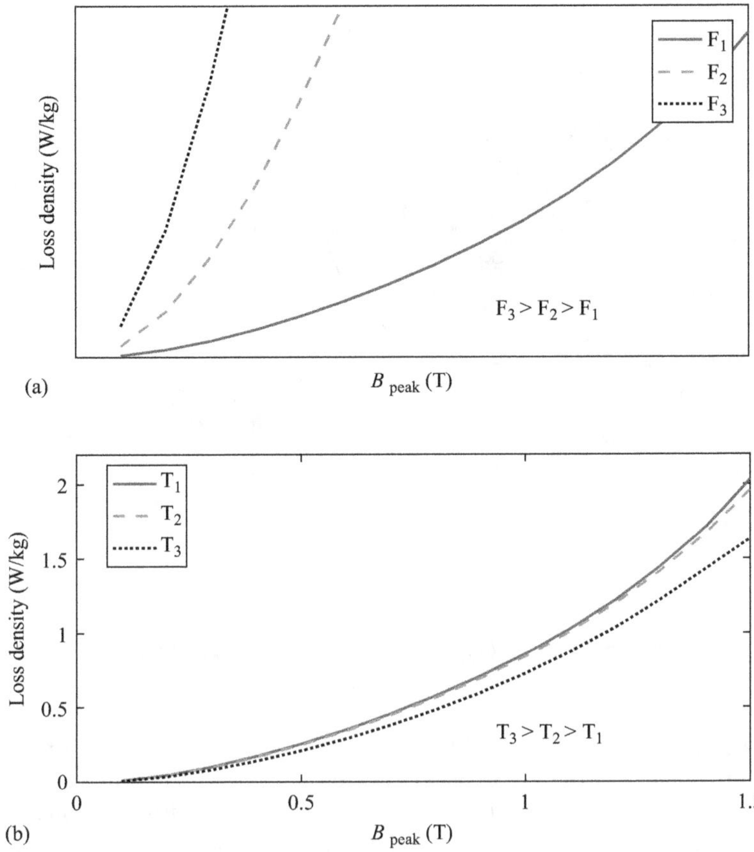


*Figure 6.6 (a) Frequency-dependent loss curves and (b) temperature-dependent loss curves*

change is shown in Figure 6.6. The following are the features of the corresponding curves:

- Increasing the frequency at a constant temperature increases the magnetic losses.
- Increasing the temperature at a constant frequency reduces the magnetic losses.
- Increasing the magnetic flux density increases the magnetic losses.
- The temperature-dependent variation of the magnetic losses is relatively smaller than that of the frequency-dependent variation.

The main theoretical premise behind the Steinmetz equation is the assumption of the linearity of the *B–H* curve. Although this is a source of uncertainty in loss characterization of electrical machines, there is an overwhelming evidence that this is the only available expression of the magnetic losses used so far. It consists

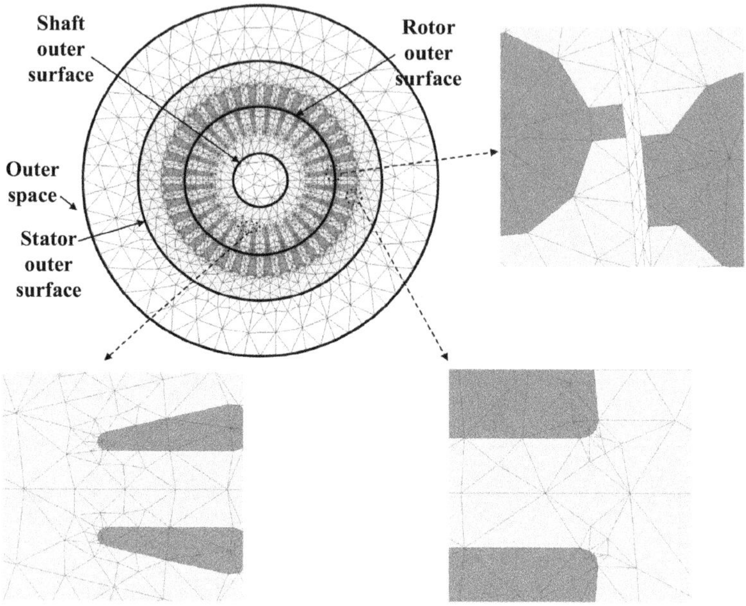


*Figure 6.7 Triangular finite element mesh applied to an induction motor problem*

of two components, namely the hysteresis and eddy-current components. The former is related to the area of the hysteresis curve shown in Figure 6.5(a), and the latter corresponds to the Ohmic losses produced by the currents induced by an alternative magnetic field inside the nonlinear material. To reduce the eddy-current losses, the materials are manufactured in a lamination form. That is why the stator and rotor stacks are made of several laminations sticking together. The larger the number of laminations is, the lower the eddy current is. In the existing commercial packages, the loss computation is handled internally as a post-processing step of simulations. The only task which should be done by the user is to enter the material, and loss data is the investigated material which does not exist in the database of the package. This is where the discussion provided so far plays an important role in understanding the corresponding details. The electrical loss calculation is simple and straightforward, so it is not addressed. For more information, refer to [1].

### *6.8.3 Mesh generation*

In any FEM package, the whole problem region is divided into many triangular elements called meshes (Figure 6.7). One mesh is a finite element in which the electromagnetic equations are set up and then assembled together in order to form a final linear or nonlinear set of equations relating the electromagnetic quantities of the meshes to each other as well as the boundary conditions. The problem of an electrical machine essentially requires a boundary condition unless a full model of the motor is modeled. This conveys the idea that there can be strategies to simulate a motor partially. The proper terminology used for this topic is the "use of symmetry" and will be explained further in this chapter.

The main rules to follow in any mesh generation task are as follows:

- Meshes should cover all the problem region.
- Materials should be assigned to all parts of the problem.
- Meshes for different parts should not necessarily have the same size.
- The jagged regions should consist of a higher mesh density containing a smaller mesh size.
- The larger the number of meshes is, the more accurate the solution is.
- The smoother the magnetic flux density of a region is, the smaller the number of meshes required for that region is.
- The outer space whose material is usually air should not necessarily have a refined mesh as the flux density remains inside the stator body and does not run away through the outer air due to the fact that the permeability of the silicon steel material is very higher than that of the air. Therefore, the magnetic quantities in the outer air gap are almost zero. Sometimes, the outer space is not even modeled because of the reason provided.
- The motor shaft, bars and windings which are made of stainless steel, aluminum and copper, respectively, act as an air-like linear region. Thus, a very smooth variation of the magnetic flux density is expected in those regions. As a result, the corresponding mesh density should not be high. It is worth noting that due to the shortcoming with the computational powers available in academic environment, distributing an optimized mesh refinement is a key point used for reducing the computational size of the problem as the dimensions of the final matrices to be solved directly depends on the number of meshes. Therefore, we insist on finding an optimum mesh refinement.
- A segment-wise demonstration of the motor topology helps assign specifically dedicated meshes to different parts. So it is proposed not to draw the geometry as a unit structure.
- Increase the air-gap meshes as much as possible. This is the crucial part of any rotating machine which contributes to the production of the electromagnetic torque.
- Geometry-wise, multiple layers of elements should be used in the air gap in order to properly model the large field gradients that might take place there. This can be done through the usages of mesh refinement and also the additional air-gap layers. In the case of latter option, at least two air-gap layers surrounding the rotor must be assigned. The inner layer contributes to the motion part which rotates with the same speed of rotor, and the outer layer is fixed. The air gap might also be split into more layers to increase the accuracy.
- When using a commercial FEM package, use the automatic mesh generator while assigning a maximum element/mesh size to the problem. Nevertheless, the meshes of the outer air box can still be very coarse (Figure 6.8).

*Figure 6.8 Refined meshes*

## 6.8.4 Set-up system of equations

The Galerkin method is the widely used one to solve the FEM equations [30]. It has higher convergence rate; therefore, it is popular compared to other techniques such as the variational methods. The method of weighted residuals, i.e., Galerkin, is applied as the following:

An operator:

$$\mathcal{L}(x) = 0 \tag{6.76}$$

is defined on the problem region, called $\Omega$, including a boundary condition C. Then, the solution is approximated by $\hat{x}$, so a residual is obtained as $x \neq \hat{x}$.

$$\mathcal{L}(\hat{x}) = \mathcal{R} \tag{6.77}$$

According to the method of weighted residuals, the integral of residual projection on a predefined function which is called weighting function is equal to zero over the problem region. For the FEM, it is preferred to assign a weighting function like FE shape function. The shape functions will be introduced later. This is specifically called the Galerkin method.

To better understand the discussion, we provide an example of the diffusion equation considering a known $z$-component of the magnetic vector potential, i.e., A. In a 2D Cartesian coordinates,

$$\frac{1}{\mu}\frac{\partial^2 A}{\partial x^2} + \frac{1}{\mu}\frac{\partial^2 A}{\partial y^2} = -J_0 + j\omega\sigma A \tag{6.78}$$

where $j\omega$ reveals the fact that a frequency-domain analysis is held. $J_0$ is the applied current density. As $A$ has only $z$-component, the vector form is reduced to a scalar form. Assuming an approximate solution of $A$ is equal to $\hat{A}$, the residual will be as follows:

$$\mathcal{R} = \frac{1}{\mu}\frac{\partial^2 \hat{A}}{\partial x^2} + \frac{1}{\mu}\frac{\partial^2 \hat{A}}{\partial y^2} - J_0 - j\omega\sigma\hat{A} \tag{6.79}$$

Therefore, the weighted function method requires that the integral of the weighted residual to be equal to zero.

$$\int \mathcal{R}W \, dxdy = 0 \tag{6.80}$$

Then,

$$-\iint W\left(\frac{1}{\mu}\frac{\partial^2 \hat{A}}{\partial x^2} + \frac{1}{\mu}\frac{\partial^2 \hat{A}}{\partial y^2}\right)dxdy + j\omega\sigma\iint W\hat{A}dxdy = \iint WJ_0dxdy \tag{6.81}$$

Integrating the first integral term by parts,

$$\iint W\left(\frac{1}{\mu}\frac{\partial^2 \hat{A}}{\partial x^2} + \frac{1}{\mu}\frac{\partial^2 \hat{A}}{\partial y^2}\right)dxdy = \iint \frac{1}{\mu}\left(\frac{\partial W}{\partial x}\frac{\partial \hat{A}}{\partial x} + \frac{\partial W}{\partial y}\frac{\partial \hat{A}}{\partial y}\right)dxdy - \oint \frac{1}{\mu}W\frac{\partial \hat{A}}{\partial \hat{n}}dc \tag{6.82}$$

The rightmost term is related to the boundary condition C. Now, it is time to break the integral operator to summations over very small areas which are the meshes covering the problem surface and the so-called finite elements.

$$\sum_M\left\{\frac{1}{\mu^e}\iint \frac{\partial W^e}{\partial x}\frac{\partial A^e}{\partial x} + \frac{\partial W^e}{\partial y}\frac{\partial A^e}{\partial y}\right\}dxdy + j\omega\sigma^e\iint W^eA^edxdy - \frac{1}{\mu^e}\frac{\partial \hat{A}}{\partial \hat{n}}\int W^edc$$
$$= J_0\iint W^edxdy \tag{6.83}$$

where $M$ is the number of meshes, "e" shows the element-wise implementation of the integral over the problem region, and $\frac{\partial \hat{A}}{\partial \hat{n}}$ is usually equal to zero in a 2D problem. Now, consider a mesh with a triangular shape whose vertices are highlighted by ($ijk$) indices (see Figure 6.9). A counter clockwise numbering system is adopted. For instance, the vertices $i$, $j$ and $k$ will be assigned to the numbers 1, 2 and 3, respectively. The vertices are the so-called "nodes" in an FEM problem and there are vector potential values as many as the number of the nodes in a problem. However, the vector potential is known for a couple of nodes using the boundary conditions, and the rest are unknown and should be calculated throughout the solving step.

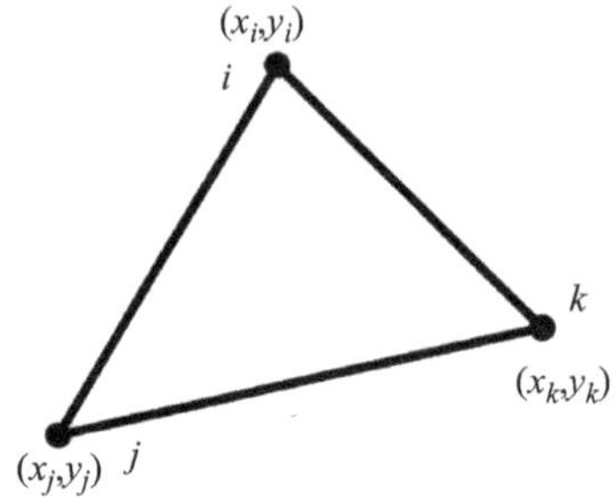


*Figure 6.9 First-order triangular mesh element*

The magnetic field distribution will be computed inside one element under the assumption of a predefined order of field variation. In other words, an approximation of the field distribution is computed considering a linear, quadratic or even higher order variation of the vector potential in one element or mesh. The mentioned order is called the "polynomial order" expressing the field distribution inside and on the boundary of one element in space. Previous research work seems to validate the view that a first-order polynomial order of 1, a linear approximation, works well if a reasonably large number of elements is assigned to the problem region. On the other hand, there is always a trade-off between the accuracy and speed of the simulation process. Increasing the number of meshes or the polynomial order leads to an increase in the simulation time while a better accuracy is achieved. The effect of increasing the polynomial order is greater than that of the number of meshes. Normally, a linear approximation is enough for a low-frequency application such as induction machines. Taking the mentioned points into account, the vector potential is approximated using the following linear expression:

$$\hat{A} = C_1 + C_2x + C_3y \tag{6.84}$$

where $C_1$, $C_2$ and $C_3$ are constants. In the first step, these constants should be determined utilizing the coordinates at vertices $i, j$ and $k$.

$$C_1 + C_2x_i + C_3y_i = \hat{A}_i \tag{6.85}$$

$$C_1 + C_2x_j + C_3y_j = \hat{A}_j \tag{6.86}$$

$$C_1 + C_2x_k + C_3y_k = \hat{A}_k \tag{6.87}$$

This provides a set of equations with three unknowns, $C_1$, $C_2$ and $C_3$ and results in the following solutions:

$$C_1 = \frac{\hat{A}_i\begin{vmatrix} x_j & y_j \\ x_k & y_k \end{vmatrix} + \hat{A}_j\begin{vmatrix} x_k & y_k \\ x_i & y_i \end{vmatrix} + \hat{A}_k\begin{vmatrix} x_i & y_i \\ x_i & y_j \end{vmatrix}}{2\Delta} \tag{6.88}$$

$$C_2 = \frac{\hat{A}_i \begin{vmatrix} 1 & y_k \\ 1 & y_j \end{vmatrix} + \hat{A}_j \begin{vmatrix} 1 & y_i \\ 1 & y_k \end{vmatrix} + \hat{A}_k \begin{vmatrix} 1 & y_j \\ 1 & y_i \end{vmatrix}}{2\Delta} \tag{6.89}$$

$$C_3 = \frac{\hat{A}_i \begin{vmatrix} 1 & x_j \\ 1 & x_k \end{vmatrix} + \hat{A}_j \begin{vmatrix} 1 & x_k \\ 1 & x_i \end{vmatrix} + \hat{A}_k \begin{vmatrix} 1 & x_j \\ 1 & x_j \end{vmatrix}}{2\Delta} \tag{6.90}$$

where $\Delta$ is the area of the triangle or mesh. So

$$\hat{A} = \frac{(a_i + b_i x + c_i y)\hat{A}_i + (a_j + b_j x + c_j y)\hat{A}_j + (a_k + b_k x + c_k y)\hat{A}_k}{2\Delta} \tag{6.91}$$

where

$$a_i = x_j y_k - x_k y_j \tag{6.92}$$

$$b_i = y_j - y_k \tag{6.93}$$

$$c_i = y_k - y_j \tag{6.94}$$

The linear functions of $xy$ which are coefficients of the vector potentials in (6.91) are called "shape functions." Finally, the magnetic vector potential is formulated as follows:

$$\hat{A} = \sum_{i=1}^{m} N_i(x, y)\hat{A}_i \tag{6.95}$$

where $m$ is the number of nodes in one element or mesh and $N$ is the shape function. As the name suggests, a linear distribution of the vector potential is obtained by means of (6.95). Therefore, any quantities such as the magnetic flux density, which is a derivative of the vector potential, will be constant over the surface of a mesh. As a result, a discrete distribution of the magnetic flux density is obtained. To deal with this issue, there are two general solutions:

- to increase the number of meshes in order to reduce the variation step of the magnetic flux density
- to use a higher order polynomial.

This topic matters in the case where a frequency analysis of the motor signals is targeted. Actually, the quality of a frequency response directly depends on the time-domain consistency and refinement. Although increasing the resolution of the quantities obtained by the FEM is time-consuming, it is highly recommended.

However, the shape functions satisfy the following rules:

- $N_i$ is equal to one at the node I, although it is equal to 0 at other nodes.
- The sum of the shape functions is equal to one.

Element-wise, and the nodal potentials are expressed as follows:

$$\hat{A}^e = \left(N_i^e, N_j^e, N_k^e\right) \begin{pmatrix} \hat{A}_i^e \\ \hat{A}_j^e \\ \hat{A}_k^e \end{pmatrix} \tag{6.96}$$

$$N_i^e = \frac{\left(a_i^e + b_i^e x + b_i^e y\right)}{2\Delta} \tag{6.97}$$

$$N_j^e = \frac{\left(a_j^e + b_j^e x + b_j^e y\right)}{2\Delta} \tag{6.98}$$

$$N_k^e = \frac{\left(a_k^e + b_k^e x + b_k^e y\right)}{2\Delta} \tag{6.99}$$

As mentioned before, the weighting functions are assumed to be the same as the shape functions. In addition, taking the derivatives of the vector potential leads to the following terms:

$$\frac{\partial \hat{A}}{\partial x} = \frac{1}{2\Delta}\left(b_i^e, b_j^e, b_k^e\right) \begin{pmatrix} \hat{A}_i \\ \hat{A}_j \\ \hat{A}_k \end{pmatrix} \tag{6.100}$$

$$\frac{\partial \hat{A}}{\partial y} = \frac{1}{2\Delta}\left(c_i^e, c_j^e, c_k^e\right) \begin{pmatrix} \hat{A}_i \\ \hat{A}_j \\ \hat{A}_k \end{pmatrix} \tag{6.101}$$

And

$$\frac{\partial W^e}{\partial x} = \frac{1}{2\Delta} \begin{pmatrix} b_i^e \\ b_j^e \\ b_k^e \end{pmatrix} \tag{6.102}$$

$$\frac{\partial W^e}{\partial y} = \frac{1}{2\Delta} \begin{pmatrix} c_i^e \\ c_j^e \\ c_k^e \end{pmatrix} \tag{6.103}$$

Therefore, the first term in the left-hand side of (6.83) becomes

$$\frac{1}{\mu^e} \frac{\partial W^e}{\partial x} \frac{\partial \hat{A}}{\partial x} + \frac{\partial W^e}{\partial y} \frac{\partial \hat{A}}{\partial y} \iint dxdy \tag{6.104}$$

$$\iint dxdy = \Delta \tag{6.105}$$

Then,

$$\frac{1}{\mu^e}\iint\left(\frac{\partial W^e}{\partial x}\frac{\partial \hat{A}^e}{\partial x}+\frac{\partial W^e}{\partial y}\frac{\partial \hat{A}^e}{\partial y}\right)dxdy$$

$$=\frac{1}{4\Delta\mu^e}\begin{pmatrix} b_i^2+c_i^2 & b_ib_j+c_ic_j & b_ib_k+c_ic_k \\ b_ib_j+c_ic_j & b_j^2+c_j^2 & b_jb_k+c_jc_k \\ b_ib_k+c_ic_k & b_jb_k+c_jc_k & b_k^2+c_k^2 \end{pmatrix}\begin{pmatrix}\hat{A}_i\\ \hat{A}_j\\ \hat{A}_k\end{pmatrix} \tag{6.106}$$

The coefficient matrix is called the stiffness matrix. The second term of (6.83) is calculated as follows:

$$j\omega\sigma^e\iint W^eA^e dxdy = j\omega\sigma^e\iint\begin{pmatrix}N_i^e\\ N_j^e\\ N_k^e\end{pmatrix}\begin{pmatrix}N_i^e & N_j^e & N_k^e\end{pmatrix}\begin{pmatrix}\hat{A}_i\\ \hat{A}_j\\ \hat{A}_k\end{pmatrix}dxdy$$

$$=\frac{\Delta j\omega\sigma^e}{12}\begin{pmatrix}2&1&1\\1&2&1\\1&1&2\end{pmatrix}\begin{pmatrix}\hat{A}_i\\ \hat{A}_j\\ \hat{A}_k\end{pmatrix} \tag{6.107}$$

Eventually, the right-hand side term of (6.83) leads to

$$J_0\iint\frac{(a_i+b_ix+c_iy)}{2\Delta}dxdy = J_0\frac{(a_i+b_i\bar{x}+c_i\bar{y})}{2}=\frac{J_0^e\Delta}{3}\begin{pmatrix}1\\1\\1\end{pmatrix} \tag{6.108}$$

where $\bar{x}$ and $\bar{y}$ are the coordinates of the center of mass of the mesh or the element.

Having discussed the FEM equations, the following details are taken into account:

- electric and magnetic material properties using $\sigma$ and $\mu$
- applied sources using $J_0$ which was previously related to a voltage source in order to build a transient-with-motion model
- the geometry using the meshes.

Therefore, an appropriate basis is prepared for understanding the future analysis. For more information about the FEM and its implementation, refer to [1]. The rest of details in terms of:

- the Newton–Raphson method
- an automatic mesh generation
- geometry implementation

and

- other refinements

are ignored as they are perfectly handled by the existing commercial packages. Nevertheless, some of the practical issues will be discussed shortly by giving an example of an industrial induction motor. The following are some of the possible fatal mistakes made by users:

- Element type is wrong. Often, a triangular mesh or element is suitable for low-frequency application
- Distorted elements
- The motion part is not modeled correctly
- Inconsistent units (this is a very important aspect when the user should define the inputs of the process).

## 6.9 Induction motor examples

In this section, an industrial induction motor is introduced, modeled and simulated by means of an FEM package. The corresponding motor parameters are listed in Table 6.1.

The motor dimensions are illustrated schematically in Figure 6.10. The stator winding layout is also shown in Figure 6.11. The rotor consists of a double-cage topology making it possible to have a better start-up by means of increasing the rotor resistance. The motor shaft, the stator coils and the rotor bars are made of stainless steel, copper and aluminum, respectively. Air gap as an essential part of the motor

*Table 6.1 Induction motor parameters*

| | |
|---|---|
| Power | 11 kW |
| Synchronous speed | 1,500 rpm |
| Synchronous frequency | 50 Hz |
| Voltage (rms) | 380 V |
| Current (rms) | 24 A |
| Number of poles | 4 |
| Rated slip | 2.9% |
| Power factor | 0.83 |
| Stack length | 160 mm |
| Number of stator slots | 36 |
| Number of rotor bars | 28 |
| Air gap length | 0.435 mm |
| Stator phase resistance | 0.87 Ω |
| Rotor bar resistance | 45 μΩ |
| End-ring resistance | 3.5 μΩ |
| Stator leakage inductance | 4.3 mH |
| Rotor bar leakage inductance | 5 nH |
| End-ring leakage inductance | 2 nH |
| Rotor inertia | 0.062502 |
| Skew angle | 22.87 mm (along the arc) |
| Number of turns | 27 |
| Silicon steel material | M19_29G |
| Winding layout | Single layer |
| Stator phase connection | Delta |

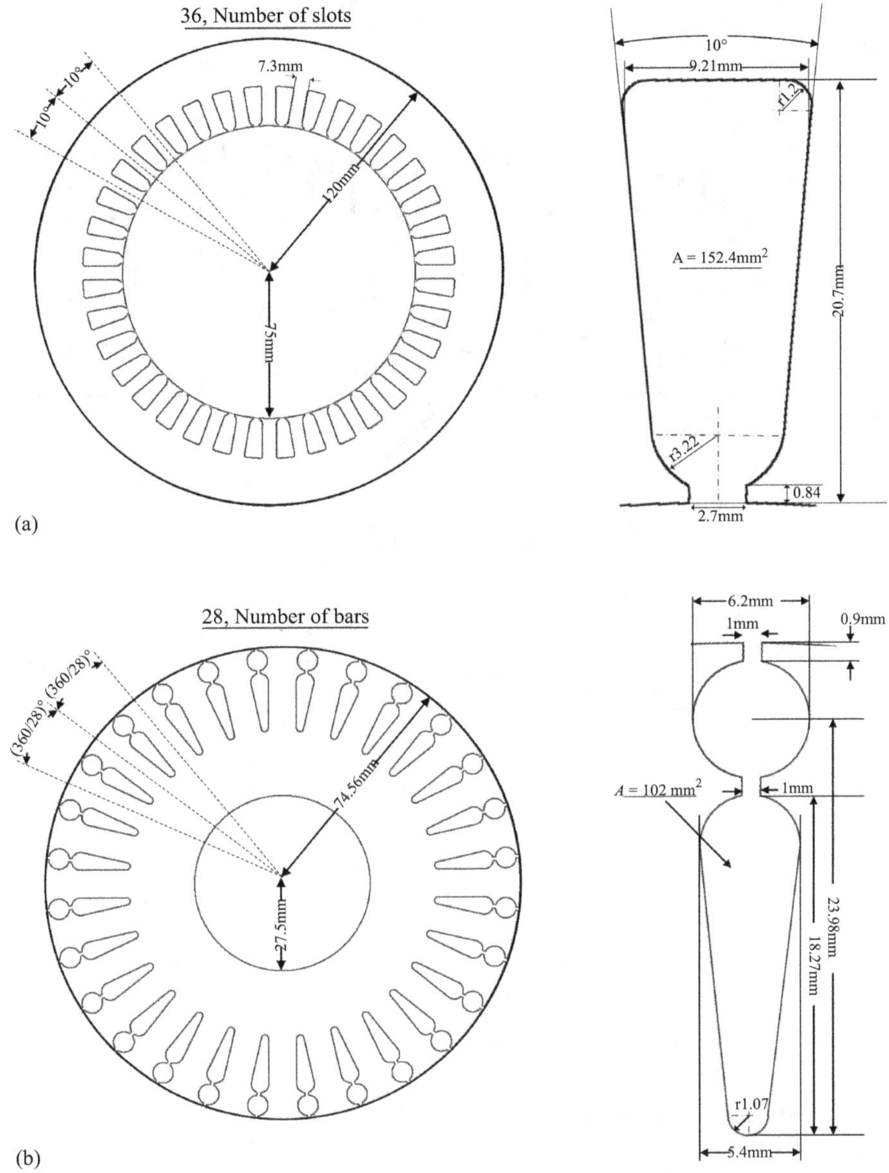


*Figure 6.10 Induction motor dimensions (a) stator and (b) rotor*

should be equal to the mentioned value, 0.435 mm. Otherwise, there can be a lot of difference between the experiment and simulation results. The underlying reason is that most of the input energy is stored in the motor air gap for developing the torque during the operation. On the other hand, a small change in the air-gap length highly affects the reluctance of the magnetic flux path to which the stored energy is related. So, the air gap should be necessarily the same as experiments. Note that a 2D

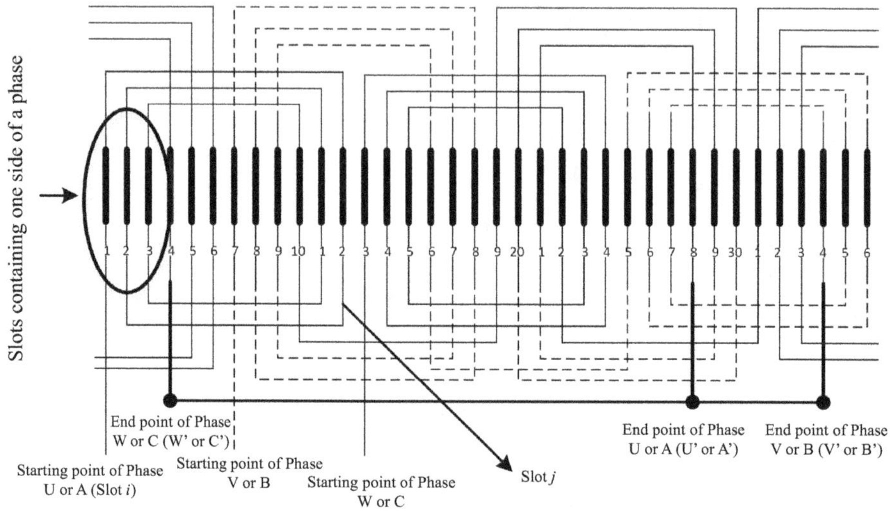


*Figure 6.11 Stator winding layout*

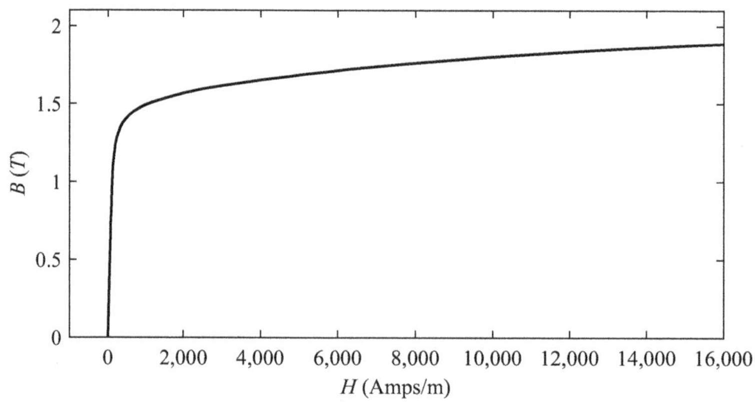


*Figure 6.12 M19_29G saturation profile*

implementation of the geometry is initially studied, and the skewing effect is ignored as it is a 3D phenomenon. Therefore, it is expected to have a higher order harmonic compared to experiments. A 3D problem is not only time-consuming but also sometimes impossible for most of the academic computational powers. The stator is composed of a single layer winding topology consisting of 27 turns in each slot. The stator slots are straight without any skew.

The silicon steel material used in this motor is M19_29G whose *B–H* curve is shown in Figure 6.12.

An outer air box with the radius of 175 mm is also assigned to the problem. This air box makes it possible to compute leakage flux running away from the stator outer surface. Normally, as the permeability of the silicon steel materials is way larger than that of the air, there will be no such flux. No boundary condition is

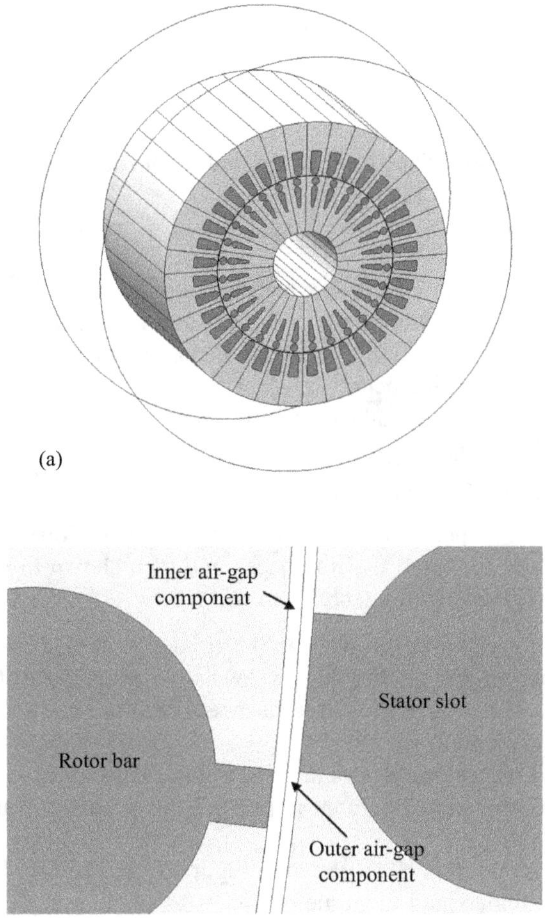


*Figure 6.13 (a) Entire motor geometry and (b) air-gap region*

required for this type of fully modeled motor problem if a relatively large air box is assigned. Otherwise, in case if there is not air box, the normal component of the magnetic flux should be set to zero on the outer surface of the stator. The situation is totally different in a partially modeled motor, and a proper discussion will be provided afterward. Moreover, the air gap has been split into two regions, a rotor-related component which rotates as the rotor rotates and a fixed stator-related component. Taking the above-mentioned points into account, the entire motor geometry is shown in Figure 6.13. The motion parts are as follows:

- the rotor core
- the shaft
- the bars
- the inner air-gap component.

The following adjustments are also applied to the problem:

- The motor is supplied by a system of three-phase voltages consisting of sinusoidal signals.
- The polynomial order is one.
- The maximum mesh size is set to 20 and 3 mm in order to investigate the difference between the results.
- The number of computer cores used for solving and remeshing steps is set to 2.
- The motor load is set to 70 N m which is the motor rated load level.
- The start time and the stop time are 0 and 300 ms, respectively.
- Two time-steps, i.e., 3 and 0.5 ms are studied as well in order to investigate the effect of time-step refinement.
- A transient-with-motion analysis is performed.

### *6.9.1 Healthy motor operation*

The simulations are done, and the magnetic flux distribution is investigated for the mentioned mesh sizes and the time steps and then shown in Figure 6.14. The following are the features extractable form this figure:

- Changing the mesh size, as well as the time step, leads to a change in the maximum and minimum flux density levels. As a result, all the motor quantities are affected. It is proposed to decrease both the mesh size and the time step as much as possible.
- Refining the meshes produces a smoother flux distribution.
- Reducing the time step has more impact on the simulation run time than the mesh size.
- A four-pole structure is clearly seen in Figure 6.14. The flux lines are the best candidates to understand the issue.
- No flux passes through the motor shaft.
- The stator yoke saturation takes place 45 mechanical degrees away from the stator tooth. The same thing is valid for the rotor.
- The air gap includes only the normal flux lines.
- There is a 1/2 and 1/4 symmetries associated with the flux distribution as well as the motor geometry.
- This means that only 1/4 of the motor is required to be modeled in a healthy condition. The presence of a fault changes the situations completely, and the user must probably model the entire motor.
- Considering a healthy motor, a 1/4 symmetry is applicable (see Figure 6.15). In this case, a boundary condition should be assigned to the surfaces which have been cut. The boundary condition is an odd periodic nature, meaning that this modeled motor pole will be repeated after a following opposite pole is connected to the modeled one. Using this context, if half of the motor is modeled, an even periodic boundary condition should be assigned.

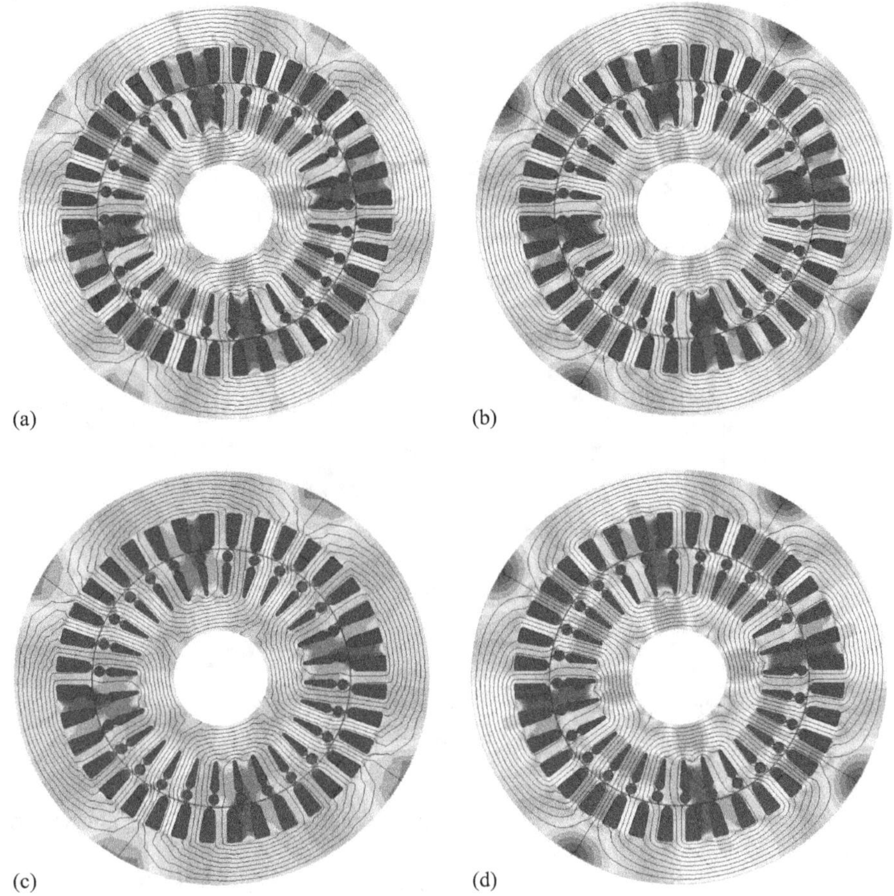

*Figure 6.14 Magnetic flux distribution of healthy induction motor in steady-state operation under the rated load. (a) Maximum mesh size = 20 mm and time step = 2 ms, simulation run time = 532 s, (b) maximum mesh size = 3 mm and time step = 2 ms, simulation run time = 882 s, (c) maximum mesh size = 20 mm, and time step = 0.5 ms, simulation run time = 1791 s and (d) maximum mesh size = 3 mm and time step = 0.5 ms, simulation run time = 3160 s*

Having applied the periodicity to account for the symmetry, the following adjustments applied to the input and outputs should be considered:

- The input voltage should be reduced to 1/4 of the rated value as there is 1/4 of the windings contribution to the model.
- The developed electromagnetic torque should be multiplied by 4.

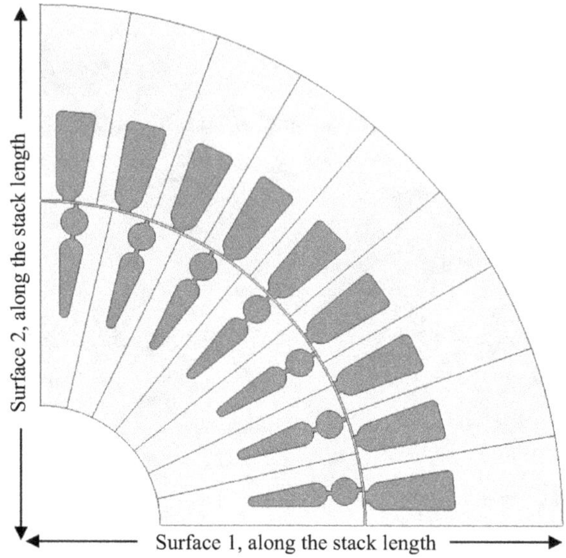


*Figure 6.15 1/4 motor symmetry*

- The motor current must not be changed.
- The motor flux linkage must be multiplied by 4.
- The motor speed remains the same.

Respecting the above-mentioned points, the FEM results are shown in Figure 6.16.

Simulating one cycle of the rotor quantities requires a relatively large stop time, 2,500 ms in this case. Therefore, this is only done for one of the refined cases, and the corresponding results are shown in Figure 6.17.

The illustrated graphs provide the following inferences:

- The transient current level shown in Figure 6.16(a) is more than that of the analytical and experimental solutions of the same motor provided in Chapter 3. This is due to the fact that the skewing effect has been ignored in the investigated 2D FEM model. It is noteworthy that the number of rotor bars is equal to 28. Hence, it is expected to have very much difference between the experimental and 2D simulation results compared to a motor with a larger number of rotor bars.
- On closer inspection, the smaller the mesh size as well as the time step is, the more accurate the results are. Therefore, the pair (mesh 3 mm, time step 0.5 ms) is chosen as the basis of comparison of various mesh size and time-step options. Taking this into consideration, the pairs (mesh 3 mm, time step 2 ms) and (mesh 20 mm, time step 2 ms) are not properly swipe the variation of the motor (Figure 6.16(b)).

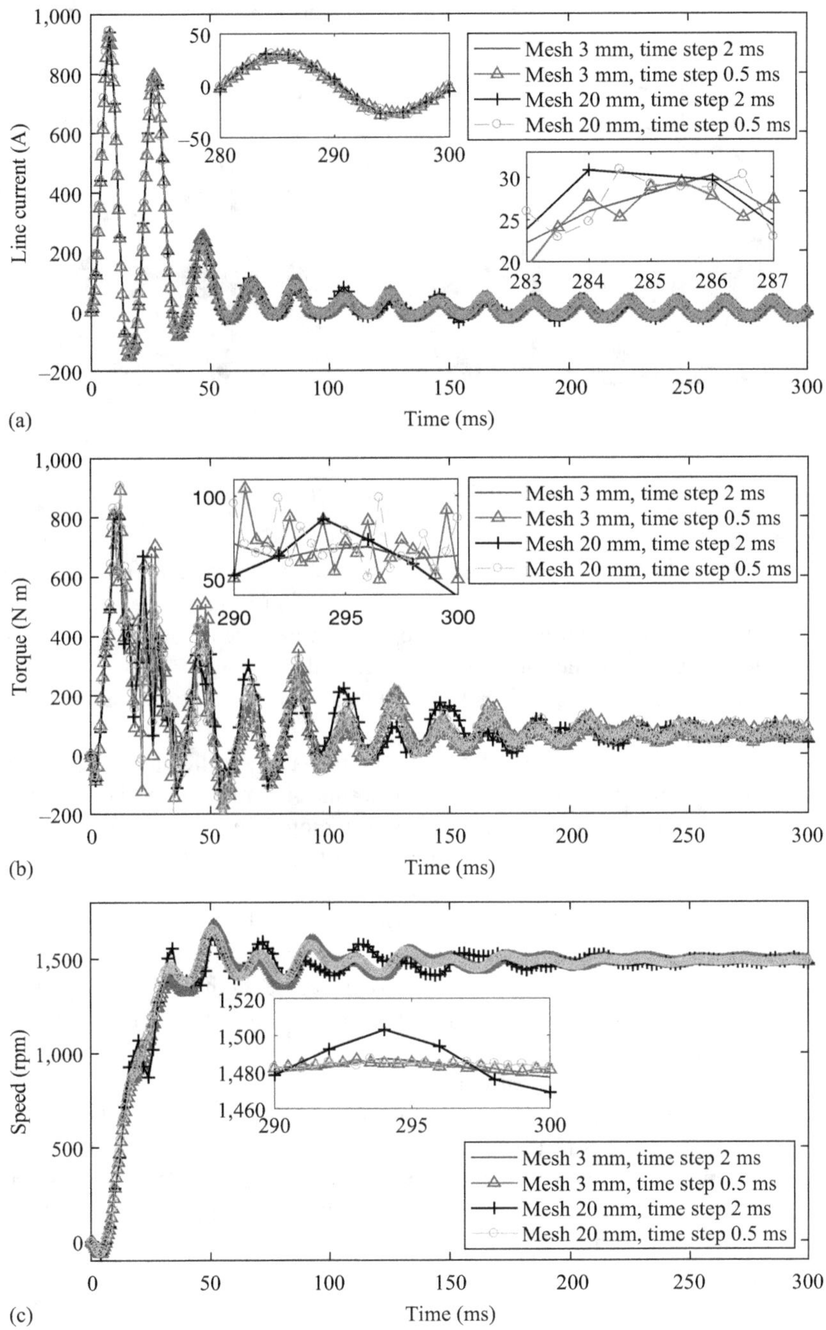


*Figure 6.16 Healthy motor quantities (a) line current, (b) electromagnetic torque, (c) speed and (d) flux linkage*

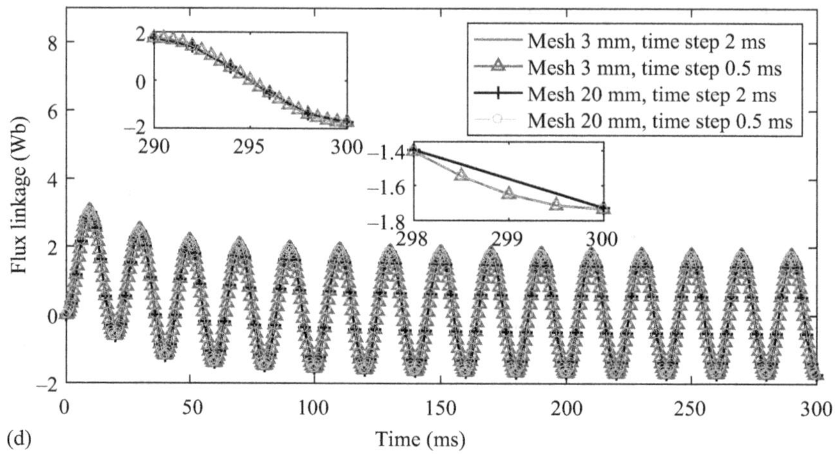


*Figure 6.16 (Continued)*

- Considering the motor speed (Figure 6.16(c)), the pair (mesh 20 mm, time step 2 ms) is out of the scope of an accurate simulation model because of a highly fluctuating speed signal leading to a possible interference with the fault components, specifically the motor broken bar fault.
- In terms of the speed signal analysis, the other speed-accuracy trade-offs work almost similar.
- Figure 6.16(d) is a clear proof of the claim that a smaller time step is more influential than a smaller mesh size in terms of a real motor behavior. This arises from a pure sinusoidal variation of the motor flux waveform for the pairs (mesh 3 mm, time step 0.5 ms) and (mesh 20 mm, time step 0.5 ms).
- The rotor electric cycle is way larger than that of the stator (Figure 6.17(a)). Therefore, processing the rotor signal during a diagnosis procedure requires a powerful computational power as the simulation should be run for at least a couple of seconds. This is also the case where a high-resolution frequency spectrum is needed. In this case, the motor signals must be sampled for more than a typical 5 s.
- The rotor current is not even comparable to that of the stator. An acceptable justification of this phenomenon is the presence of the rotor bars which are accounted for a one-turn coil while the stator consists of 27 turns in this case. On the other hand, the induction motor is a kind of rotating transformer. Therefore, simply using the governing equations of a transformer and multiplying the stator current by the inverse transformer ratio (1/27 in this case), a current level of almost 810 A is obtained. Considering the slip losses and the resistances of the rotor and the stator, the magnitude of almost 700 A which is observable in Figure 6.17(a) is justifiable.
- The bar voltage is almost equal to zero (Figure 6.17(b)).

For further information, the rotor circuit is shown in Figure 6.18.

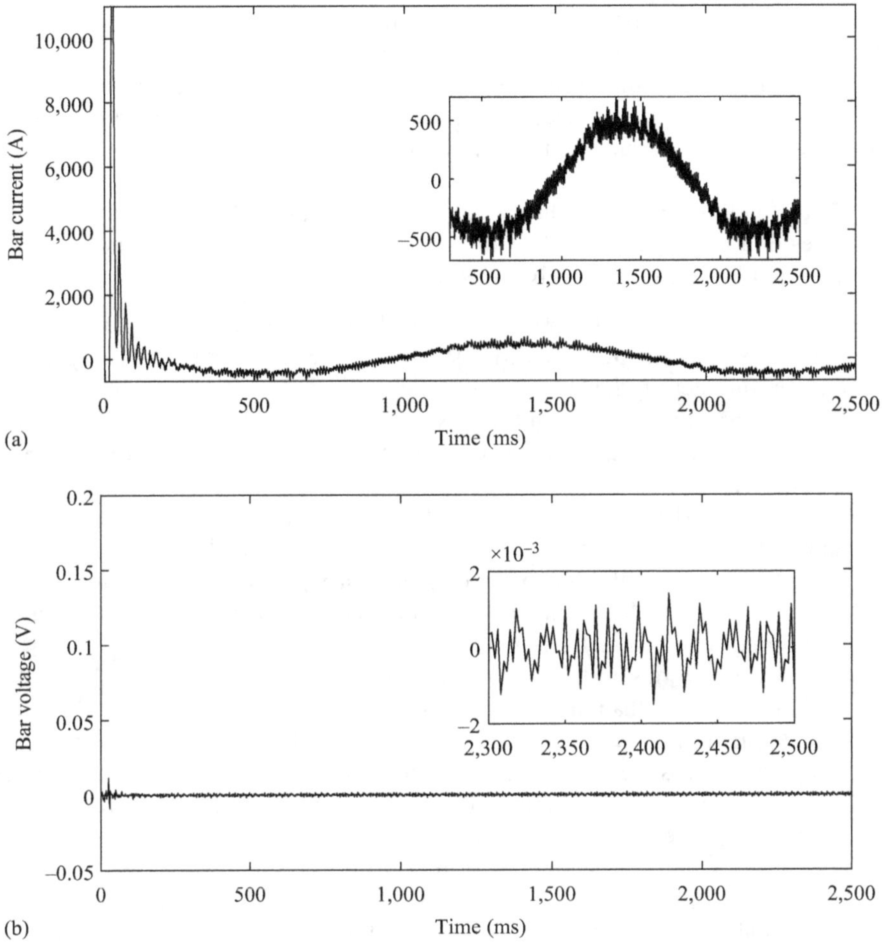


*Figure 6.17 Healthy rotor quantities (a) bar current and (b) bar voltage*

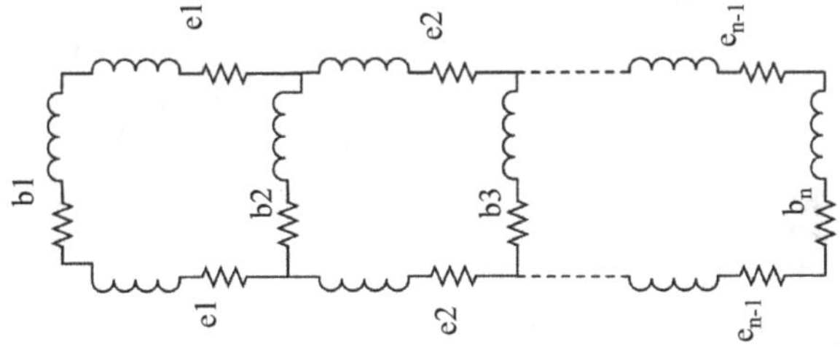


*Figure 6.18 Rotor connections*

All the signals investigated above are electric and mechanical signals while the distribution of the magnetic quantities is also important for the diagnosis purposes. However, magnetic quantities are not achievable easily or accurately by means of an analytical or semi-analytical model of a motor. The only accurate way of addressing magnetic issues, especially when it comes to computing the losses or local saturations, is the FEM.

Figure 6.19 demonstrates the spatial and temporal distribution of the magnetic quantities of a healthy motor. The values correspond to the mesh size and the time step of 3 and 0.5 ms, respectively. Accordingly,

- The stator tooth has the maximum available magnetic flux density in the motor. As the teeth are the narrowest parts, so the saturation level is higher for a given flux density level.
- The temporal distributions of the flux density of both the air gap and the rotor tooth level have pulsating shape revealing the fact that there is slotting effect in the motor.
- In Figure 6.19(b), the flux density of the stator tooth has a pulsating shape in addition to the rotor tooth and air gap.
- All the signals, whether temporal or spatial, are periodic. The magnitudes do not change from pole to pole or from cycle to cycle. This means that there is an electromagnetic symmetry in the motor.
- It is also interesting to know that the total magnetic flux density of the rotor back iron is larger than that of the stator yoke. On the other hand, the magnetic flux density of the rotor teeth is smaller than that of the stator teeth because the rotor experiences the same flux level as the stator while the rotor has a smaller number of teeth and therefore a larger tooth cross section. As a result, it is expected to have a smaller flux density for a given flux level.
- As a healthy motor is symmetrical in terms of the flux distribution, only half the mechanical cycle is shown in Figure 6.19(b). Illustrating a quarter of the cycle is also enough. However, when a fault exits, the whole motor should be illustrated.

A healthy induction motor is a highly symmetrical device, of course if the inherent asymmetry is ignored. Therefore, both the temporal and spatial distributions of all electrical and magnetic motor quantities should be a periodic signal in the steady-state mode. The transient mode is different as there are vanishing exponential components incorporated in the motor signals including torque, current, etc. Therefore, the periodicity is disturbed at the initial instants of the simulation process. Moreover, if the skewing effect is modeled correctly, smoother time-domain signals will be obtained. A useful implementation method of the skewing effect is going to be proposed at the end of this chapter where the short-circuit fault is discussed. It should be noted that ignoring the skewing effect does not change the symmetry of the healthy motor quantities. It might change the fluctuations and the corresponding magnitudes. However, if a fault takes place, the quantities such as the magnetic flux density and the current lose their periodic nature. This is the underlying reason for the necessity of modeling a complete motor, not a ½ or ¼

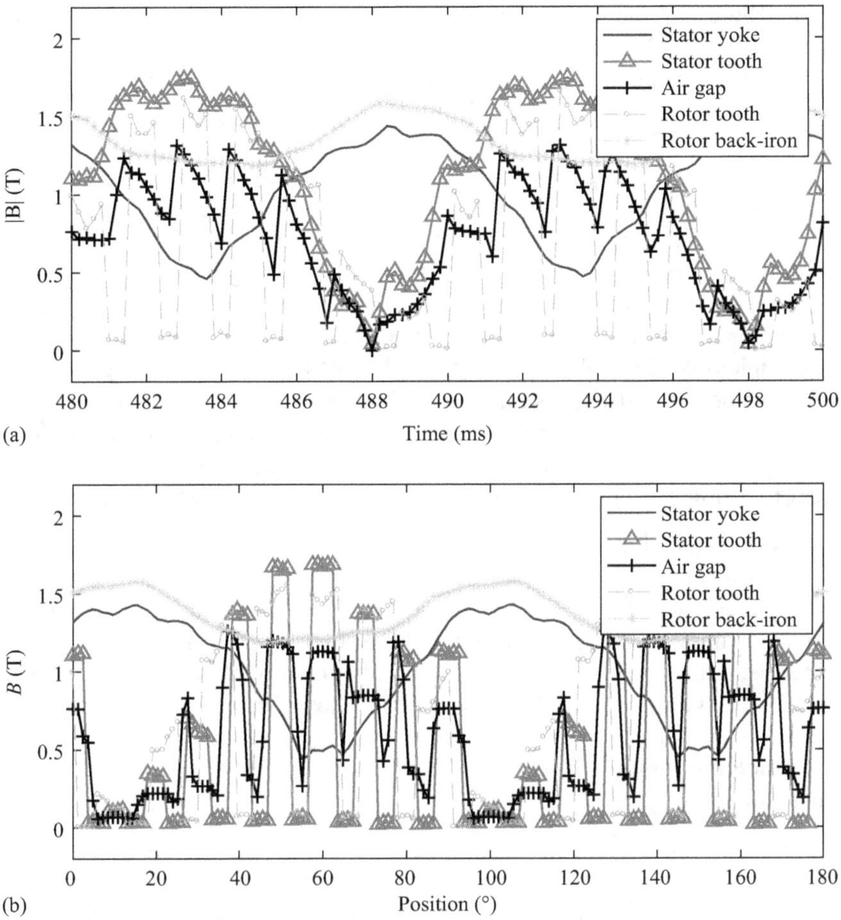


*Figure 6.19 (a) Temporal distribution of the magnetic flux density at a given spatial point and one electric cycle and (b) spatial distribution of the magnetic flux density at a given time step and half mechanical cycle*

symmetry model when a fault diagnosis process is targeted as there is no more symmetry, whether temporal or spatial, and different motor poles reveal various flux density levels. Thus, a complete motor model is used in this chapter in order to address the fault-related motor behaviors.

### 6.9.2 Broken bar motor operation

The broken bars fault affects the rotor bars in a way that the resistance of the broken bar increases [13,17,19,21]. So for a given EMF, the rotor current reduces depending on the increase in the bar resistance. Sometimes, the bar is fully broken, meaning that the bar current is almost zero and it is not able to produce an armature reaction. Therefore, the broken bar does not contribute to the motor torque. However, the motor load should be provided by means of the electromagnetic torque developed by the interaction of the rotor and stator fields. Therefore, the bars adjacent to the breakage area take the burden by tolerating a higher level of the current.

Modeling the broken bars fault, using the simulation processes, has always been a big question, especially in the case of a partial broken bar. However, the full broken bar fault can be somehow easily modeled by assigning a nonconducting material to the targeted bar or reducing the corresponding conductivity to almost zero. It is proposed to use the latter as the former option leads to unstable FEM solutions in some cases. To handle this step, the user-defined materials which are usually implemented in FEM packages should be used. This means the user is able to create his/her own material similar to the original bar materials, aluminum in the present case, but with a relatively smaller electric conductivity level. As an example of the approach, the following properties should be assigned to a material in order to mimic the characteristics of an aluminum which can be used as:

- relative magnetic permeability: 1
- magnetic coercivity: 0 A/m
- electric conductivity: 1 Siemens/m, 1.72 %IACS while the conductivity of a healthy aluminum bar is equal to 38,000,000 Siemens/m or 65.5 %IACS. Increasing the conductivity level allows a larger current to flow into the bar, so different levels of a partial broken bar can also be studied. A conductivity of 1 leads to an almost zero bar current.
- electric permeability: 1
- mass density: 2,707

The introduced approach is useful in a 2D or quasi-3D models in which the end-rings are not usually modeled. In a full 3D model motor, the broken bar can be modeled like what it is. In fact, a mechanical disjoint between the bars and end-rings is modeled. This is not our goal because handling a 3D model is sometimes impossible for most of the potential readers of this book.

The discussions presented above suggest that a high-resistance aluminum bar is initially defined by the user and assigned to the FE package. Then, the induction motor model is rebuilt using the defined high-resistance aluminum assigned to the broken bars.

Previously, it was mentioned that the number and the location of the broken bars are the influential factors in the diagnosis process. In this section, two following cases are simulated to check the effect of different possibilities compared to the healthy motor (see Figure 6.20).

- 2 adjacent broken bars in one pole
- 1 broken bar at each of the two adjacent poles

In the present cases, there are two broken bars in different locations around the circumference of the rotor. The corresponding broken bars are highlighted in white while the healthy bars are in dark gray. The black circle around the bars

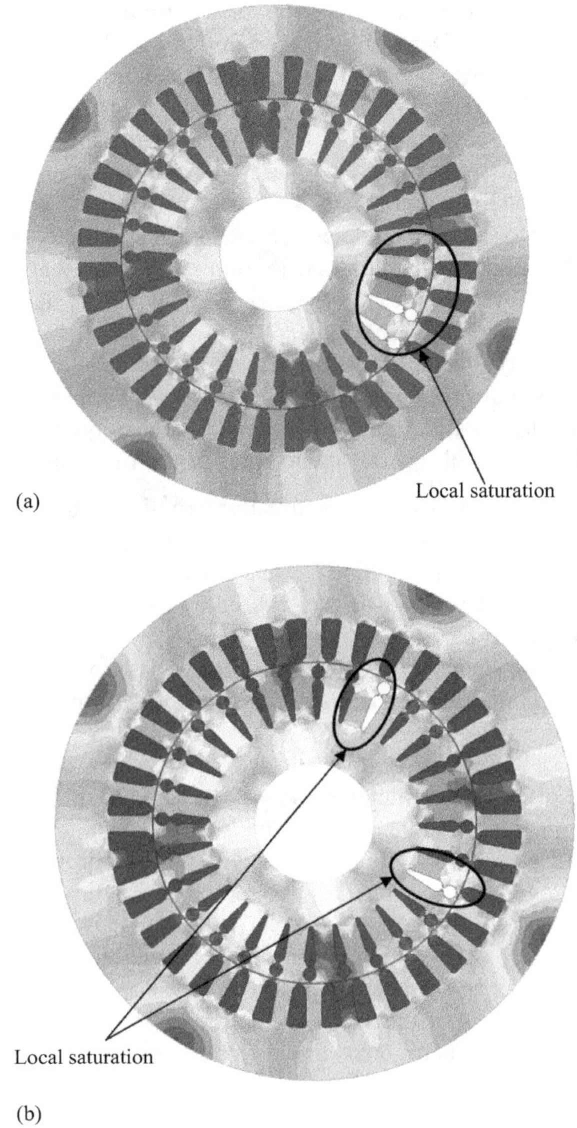


*Figure 6.20 (a) Two adjacent broken bars in one pole and (b) one broken bar in each of the two adjacent poles*

shows the local saturation regions caused by a higher current level of the bars adjacent to the breakage produced by removing the armature reaction of the broken bar from the vicinity of the adjacent bars. These local saturation regions, having a relatively larger magnetic flux density level, are the source of magnetic losses and also the thermal stress. Both the stator and the rotor are affected locally by the broken bars fault; hence, the local saturation is observable in both the stator and rotor teeth. Depending on the number or the locations of the broken bars, the local saturation level and so the produced unbalanced magnetic pull will be different. In fact, the presence of a broken bar introduces an unbalanced magnetic pull and increases the chance of having an eccentric rotor. However, the opposite is not generally true. This issue will be further analyzed in the next chapters. The broken bars fault causes an electromagnetic asymmetry which requires that the motor is fully modeled. So the use of symmetry is rejected in the case of broken bar fault as the flux distribution is not identical in all the motor poles while the underlying idea of the usage of symmetry is the presence of identical distributions in different poles.

Figure 6.20 clearly illustrates that the distribution of the motor magnetic and so electrical quantities will not be pole-wise or cycle-wise symmetrical. The numerical values of the flux distributions are compared in healthy and two faulty cases (see Figure 6.21). The significant characteristics of the figure are as follows:

- The healthy motor reveals a symmetrically periodic signal.
- The broken bars affect the flux density in a way that two consecutive poles have different flux amplitudes. The one including the broken bars has the largest magnitude of the flux density caused by the local saturation which is a result of the broken bars fault.
- None of the magnitudes of the flux density associated with four poles are the same as the flux distribution is distorted completely to the defect in the rotor. The same happens to the stator.
- According to Figure 6.21(b), two adjacent broken bars in one pole have the largest local effect on the motor behavior compared to the healthy case.
- Depending on the location of the fault, various fault intensities affect the motor. It will be shown in the next chapters that the broken bars existing in the opposite poles have the smallest magnitude of the flux densities among the other locations.
- Unlike the healthy motor in which the spatial distribution of the magnetic flux density always has upward or downward curves, alternating curves are observable in the envelope of the variations in the faulty motors (Figure 6.21(b)). This is the source of additional torque ripple introduced by the broken bars fault.

### 6.9.3 Eccentric motor operation

The eccentricity fault is a kind of defect which, in the first place, affects the motor geometry by disturbing the inherent symmetry existing in the healthy case [33,38,42]. Thus, what happens next should be logically an asymmetrical and unbalanced electromagnetic flux distribution. So the source of the unbalanced flux density is a mechanical defect. Like the broken bar fault, the motor must be modeled completely without assuming a symmetrical distribution of the magnetic flux. Considering that there are two types of an eccentricity fault including the static and dynamic types, both are discussed in this chapter.

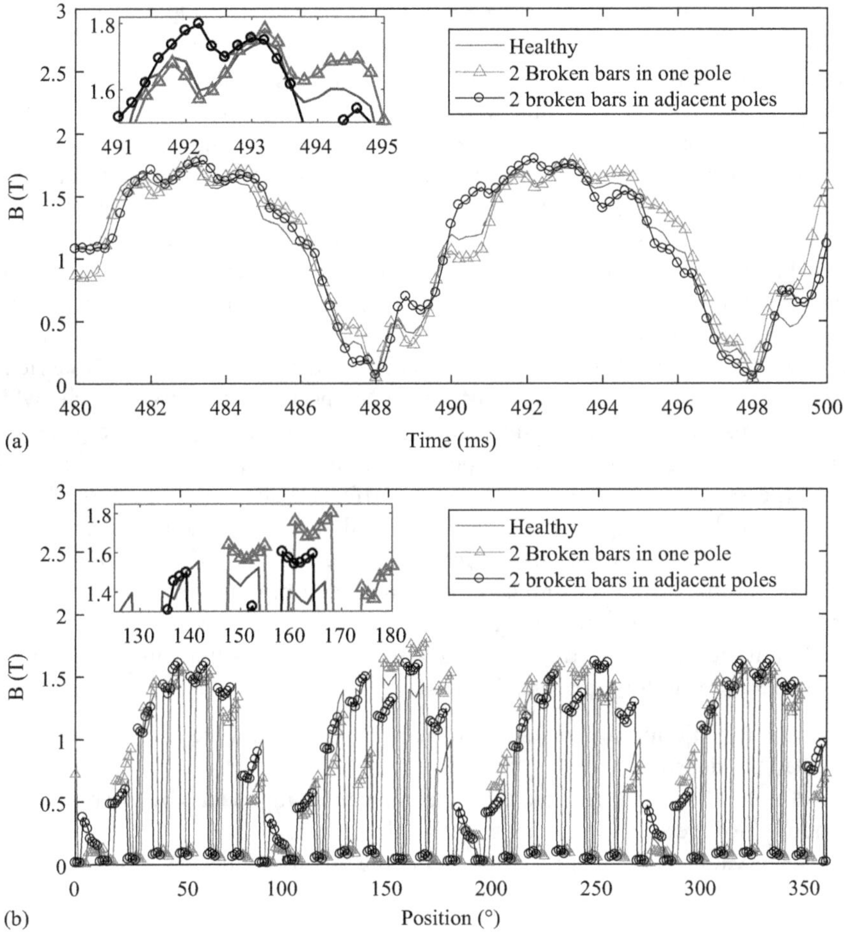


*Figure 6.21 (a) Time-domain variation of the magnetic flux density of one stator tooth and (b) spatial variation of the magnetic flux density at the rotor tooth level*

#### 6.9.3.1 Static eccentricity

First, static eccentricity is modeled using an FEM package. In this type of fault, the rotor is shifted toward the stator, so the stator and the rotor are not concentric any more. It should be noted that the position of the rotor center is fixed with respect to that of the stator; therefore, there is an unbalanced fixed air-gap length around the rotor circumference (see Figure 6.22). As can be seen, there is a middle air-gap circle around and concentric to the rotor center. This is the exact way of defining the two layers of the air-gap components mentioned earlier in this chapter. The inner layer is attached to the rotor and rotates with the rotor. The outer layer is a nonrotational layer. The rotor-concentric inner layer guarantees that there is no out-of-boundary meshes while the rotor rotates and requires remeshing. In fact, if the inner air-gap layer is concentric to the stator, the outer boundary of the inner layer will collide the stator inner boundary for a relatively large static eccentricity level, while the inner layer rotates with the rotor. That is why the inner layer should be fixed at the rotor center, not the stator center, to prevent this problem.

As another practically important point in terms of the simulation run time, an eccentric geometry totally changes the FEM stiffness matrix to an asymmetrical matrix which requires a long time to be solved compared to a geometrically symmetrical matrix. Therefore, it is expected to have a much larger simulation run time in the case of eccentricity fault. So if you notice such a problem, do not blame your model! In contrast, the simulation run time is very close to that of the healthy machine in the case of a broken bar motor because the motor is geometrically symmetrical. The stiffness matrix of an FEM model is a sparse matrix which usually represents a symmetrically distributed element leading to a very fast solving process using direct or decomposition approach. However, any factor disturbing the symmetry of the stiffness matrix affects the simulation run time.

It is proposed to apply a relatively large radius to the middle air-gap line and keep a balance between the air-gap lengths related to the inner and outer layers. Otherwise, if there is a very thin layer, whether inner or outer, the mesh generation process will encounter a problem or the generated mesh will not be optimum enough to increase the speed of the simulation process. For implementing a proper mesh generation, follow the rules below:

- Use automatic mesh generator for the regions where there is a smoothly variable flux density.
- Refine the meshes assigned to the air-gap components by increasing the number of elements or decreasing the element size.
- Use curvature refinement for improving the quality of meshes. This application can be found in any FEM package.

Figure 6.22 shows an example of the refined meshes around the air-gap region. The difference between the meshes is visible from this figure. The meshes are not

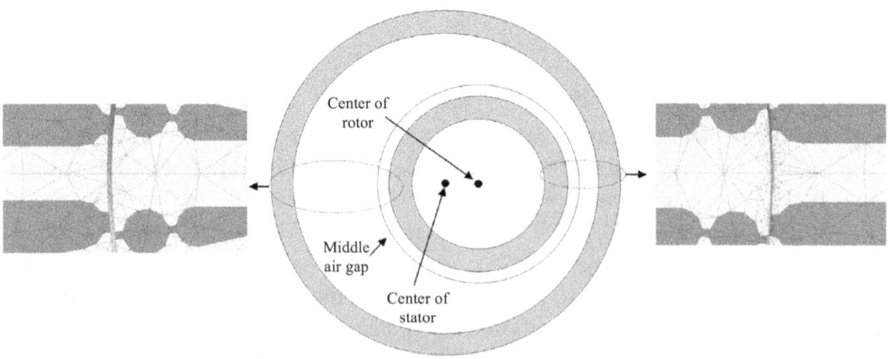


*Figure 6.22 Implementation of static eccentricity fault*

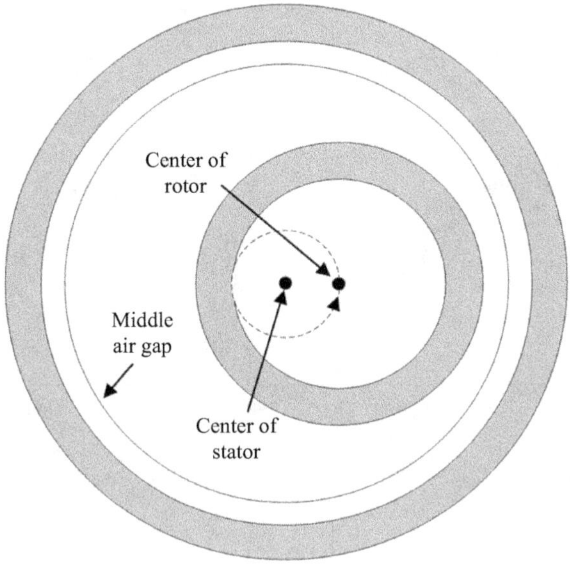


*Figure 6.23 Implementation of dynamic eccentricity fault*

identical in right and left side of the air gap as the geometry is not symmetrical. Moreover, the total air-gap length in the right side is relatively smaller than that of the left side because the rotor has been shifted toward the right end of the stator. The number of the utilized meshes in the right side is larger than that of the left side. In addition, the meshes related to the right side are more refined. Figure 6.22 is a snapshot of the initial mesh generated by the FEM package while the meshes are subjected to change where there is motion part in the problem.

### 6.9.3.2 Dynamic eccentricity

Dynamic eccentricity is a rotating type of eccentricity which introduces a variable air-gap length during the motion analysis of the faulty induction motor (Figure 6.23). Unlike the static eccentricity which requires a middle air-gap circle concentric to the rotor, the middle air-gap circle should be concentric to the stator in the case of a dynamic eccentricity. The reason is that the rotor center is not fixed at a point in space and rotates around the stator center. So the middle air-gap circles should be concentric to the stator, not the rotor. Otherwise, a mesh-generation problem caused by a collision of the middle air-gap circle, and the stator inner surface will occur during the simulation process.

Figure 6.24 shows time-domain variation of the magnetic flux density of one of the stator teeth upon changing the fault type from the static eccentricity to the dynamic eccentricity and then compares the results to that of the healthy motor. Moreover, the spatial distribution of the energy stored in the air gap is illustrated in Figure 6.24(b).

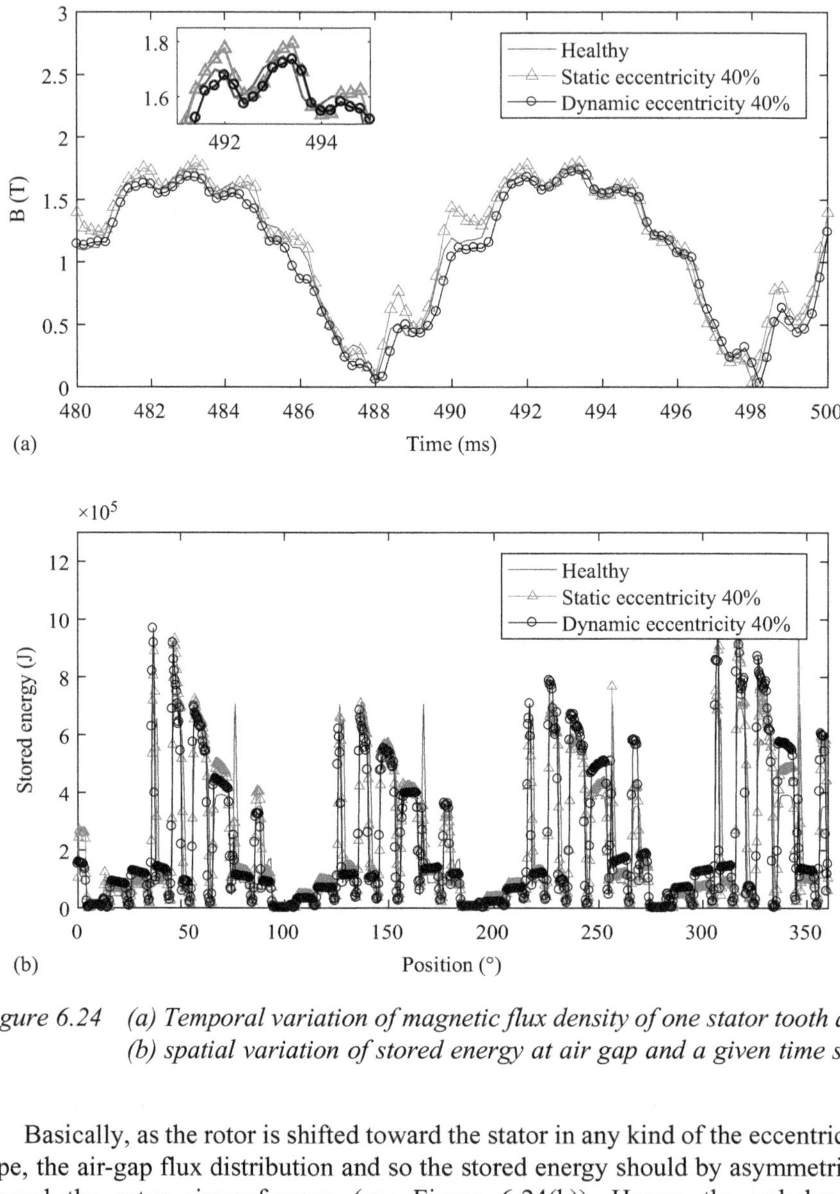


*Figure 6.24 (a) Temporal variation of magnetic flux density of one stator tooth and (b) spatial variation of stored energy at air gap and a given time step*

Basically, as the rotor is shifted toward the stator in any kind of the eccentricity type, the air-gap flux distribution and so the stored energy should by asymmetrical around the rotor circumference (see Figure 6.24(b)). Hence, the unbalanced magnetic pull is originated from this phenomenon and leads to a mechanically unbalanced motor operation. The stored energy is much more in the regions very close to the stator inner surface while the corresponding values are relatively small in the larger parts of the air-gap length. The healthy motor (see Figure 6.24(b), the solid line) possesses a completely balanced energy distribution with the magnitude of 7.1678e + 05. The corresponding values of the static and dynamic eccentricities are 1.0444e + 06 and 1.1504e + 06, respectively.

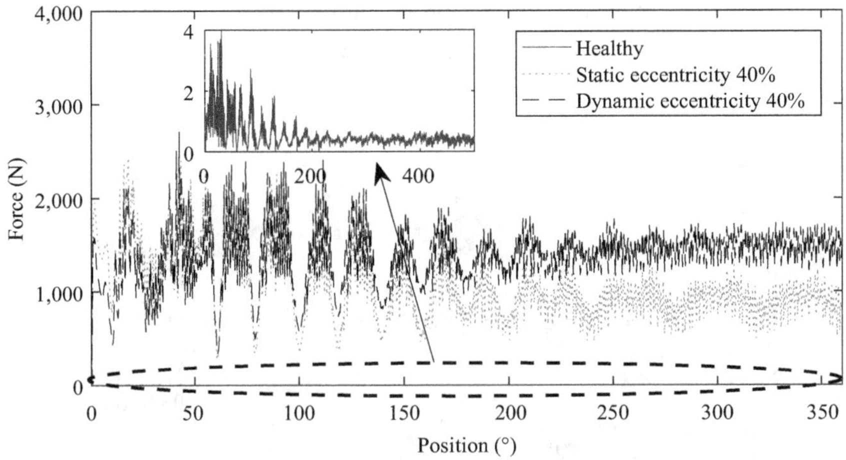


*Figure 6.25 Magnetic force applied to the rotor*

Figure 6.25 shows the net force applied to the rotor in the healthy, static eccentricity and dynamic eccentricity cases. On the basis of the values provided in this figure, the net force is almost zero for the healthy motor. There are some transients related to the motor start-up while the steady-state performance does not vary so much. In contrast to the healthy cases, the faulty motors reveal a relatively large net force applied to the rotor at different positions. The one corresponding to the static eccentricity has the largest value while that of the dynamic eccentricity consists of a smaller net force. This means that the static eccentricity generally imposes larger disturbance to the motor rather than the dynamic one considering the same fault level, 40% in this investigation. All the simulations which have been already done in this chapter were held at a full-load condition, 70 N m. If the motor load is changed, the absolute values of the illustrated temporal and spatial performance will be subject to a change. However, the trends and justifications are still valid even for smaller load levels. A general discussion on the effect of the load variations will be provided in the next chapters.

### 6.9.4 Short-circuited motor operation

Short-circuit fault can happen in a wound-rotor motor in the same fashion that the stator is affected [29]. So, in this subsection, we prefer to introduce a new motor which is of a wound-rotor type with the goal of investigating the short-circuit fault. To this purpose, the corresponding motor parameters are listed in Table 6.2.

The winding layouts corresponding to the stator and rotor parts are shown in Figure 6.26.

As this is a new type of an induction motor, different from what was previously studied, first the healthy operation is investigated in order to get better understanding of the variations of the motor signals. Then, the short-circuit fault is

*Table 6.2 Wound-rotor induction motor parameters*

| | |
|---|---|
| Power | 10 hp |
| Rated speed | 1,425 rpm |
| Synchronous frequency | 50 Hz |
| Voltage | 420 V |
| Current | 8.57 A |
| Number of poles | 4 |
| Rated slip | 5% |
| Stator phase resistance | 1.67 Ω |
| Rotor phase resistance | 0.3 Ω |
| Magnetizing inductance (stator reference frame) | 189 Ω |
| Rotor inertia | 0.21 kg/m$^2$ |
| Silicon steel material | M19_29G |
| Stack length | 103.9 mm |
| Number of stator slots | 36 |
| Number of stator parallel branches | 1 |
| Number of rotor parallel branches | 1 |
| Number of rotor bars | 28 |
| Stator inner diameter | 228.18 mm |
| Stator outer diameter | 353 mm |
| Rotor outer diameter | 227.15 mm |
| Shaft diameter | 120 mm |
| Winding factor (stator) | 0.95 |
| Winding factor (rotor) | 0.95 |
| Slot fill factor (stator) | 52.75% |
| Slot fill factor (rotor) | 31.10% |
| Air gap length | 0.5 mm |
| Coil half-turn length (stator) | 245.661 mm |
| Conductor type of stator | Copper |
| Number of stator slots | 48 |
| Stator slot opening | 2.5 mm |
| Number of rotor slots | 34 mm |
| Stator conductor diameter | 1.725 mm |
| Winding layout | Single layer |
| Connection | Grounded Y |
| Number of coils per pole per phase | 2 |
| Stator slot depth | 26.82 |
| Stator tooth width | 8.22 mm |
| Rotor slot opening | 2 mm |
| Rotor slot depth | 22.5 mm |
| Rotor conductor diameter | 1.9 mm |
| Number of rotor conductors in one slot | 6 |
| Number of coils per pole per phase | 3 |
| Rotor leakage inductance | 0.69 mH |
| Stator leakage reactance | 3.67 Ω |

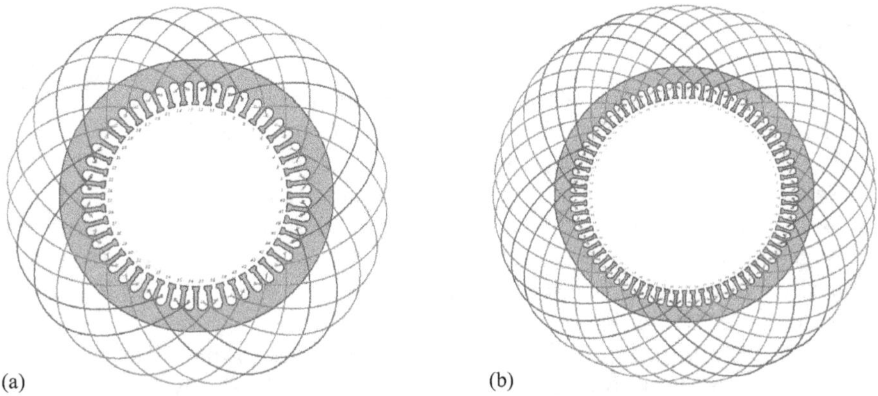

(a) (b)

*Figure 6.26 Winding layouts (a) stator and (b) rotor*

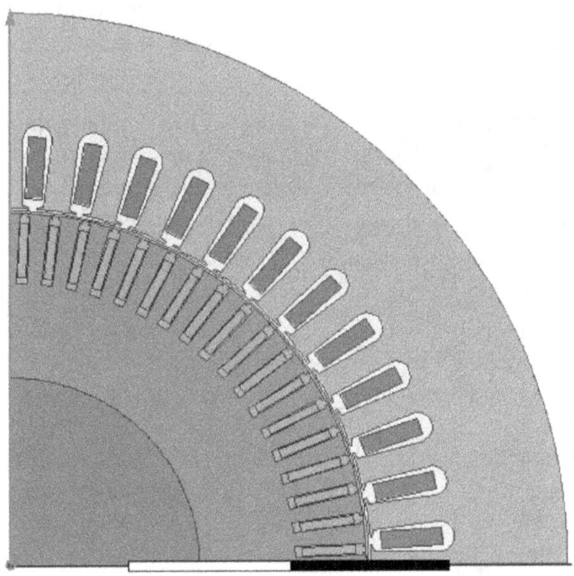


*Figure 6.27 Investigated motor topology*

analyzed. To this end, a 1/4 symmetry is used, and the motor is modeled as Figure 6.27.

After setting the rotor supply voltage to zero in order to create short-circuited rotor circuit, the stator supply voltage is applied. The current, the torque and the speed are illustrated in Figure 6.28. The following features are deducible from the figures:

- This motor has a smoother operation than the previously simulated one. The larger number of the rotor bars is the main reason for the claim.
- The motor speed signal does not reveal an over shoot.
- The motor torque represents a mostly positive signal increasing smoothly in the transient region and then decreasing to the steady-state point after 0.4 s. As the motor speed is an output of the motor torque interaction with the load, it is also expected to have a smooth speed variation unlike the previously studied motor with 28 rotor bars.
- Again, the rotor bar currents are very larger than that of the stator.
- The start-up current is almost five times more than the steady-state current. In induction motors, this is always a rule of thumb that the start-up current is five or six times more than the rated steady-state current.
- As a technical issue, to have an accurate motor performance prediction, a fully modeled skewing effect should be simulated. Instead, a motor with a relatively large number of rotor bars in which the skewing effect is negligible should be investigated in 2D.

FEM packages also provide a very promising tool to estimate the winding inductances which are very useful tools in analyzing an eccentric or short-circuited motor. Figure 6.29 shows an example of the motor inductances. $L_{aa}$ is the stator self-inductance, $L_{ab}$ is the stator–stator mutual inductance and $L_{au}$ is the stator–rotor mutual inductance.

Correspondingly, the following technical principles are always valid:

- The self-inductance of a winding or coil is always a positive term.
- The mutual inductance of two windings or coils which are located on one reference frame and have no variable relative position change is always a negative term.
- The mutual inductance of two windings or coils, for example, the stator and rotor windings, might have a sinusoidal wave including positive and negative terms, depending on the relative position of two windings.
- In a healthy motor, the average value of the inductances is a constant value in time domain while a shifted or oscillating average value is observed if an eccentricity or short-circuit fault occurs.

#### 6.9.4.1 Implementing skewing effect

In electrical machines with slots, the air-gap length is not basically equal to the difference between the stator inner radius and the rotor outer radius, but it is a part of the shortest available flux path inside a motor. This is a common mistake to consider the mentioned measure as the actual air-gap length of the motor. The air-gap length is in fact equal to the change in a distance that motor flux should travel in order to pass from the stator teeth to the rotor teeth and vice versa. This results in two outcomes: first, as the flux always looks for a path with the smallest magnetic

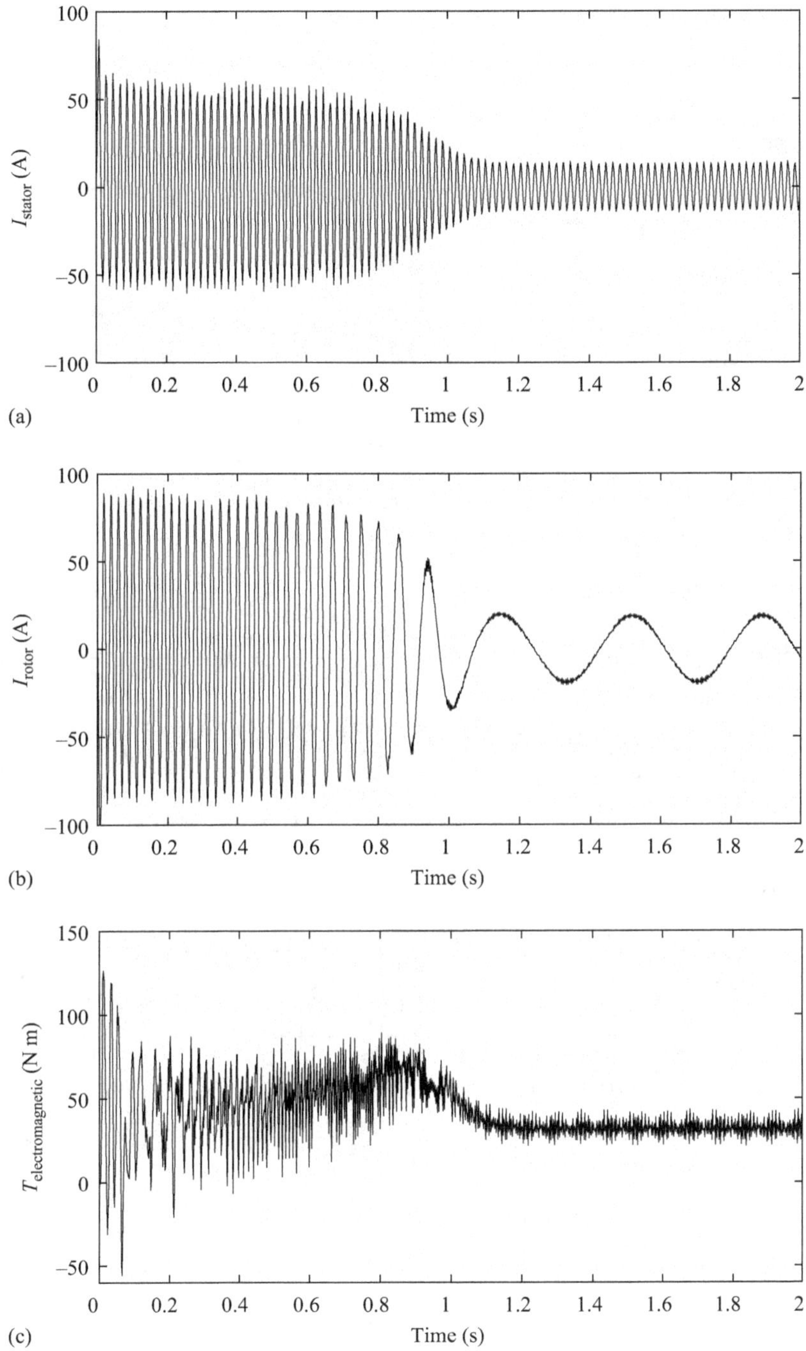


*Figure 6.28 Healthy motor operation (a) stator current, (b) rotor current, (c) torque and (d) speed*

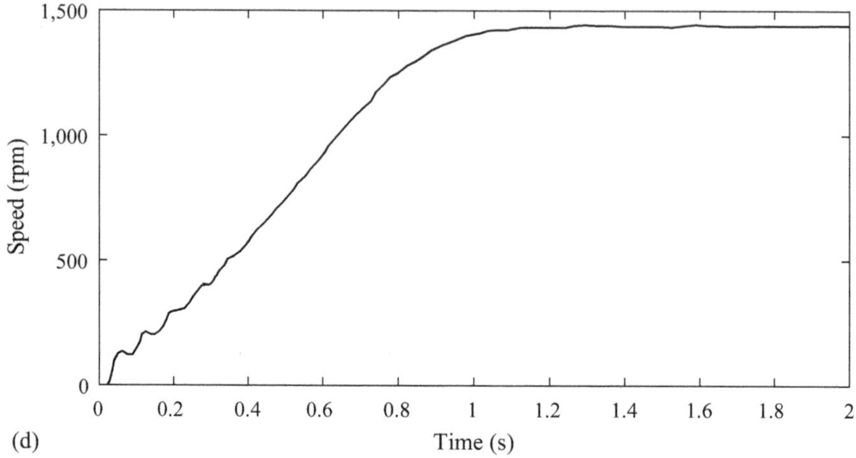


(d)

*Figure 6.28 (Continued)*

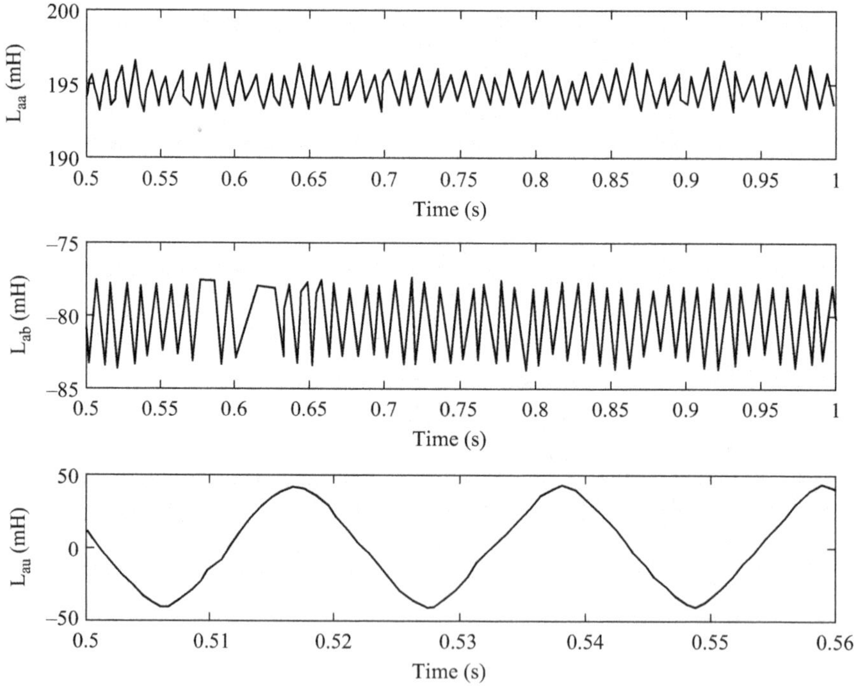


*Figure 6.29 Motor inductances obtained using FEM*

reluctance to pass through and because the stator teeth to the rotor teeth are the geometrically shortest path, flux concentrates around the teeth and results in a local saturation. Second, the torque ripple is the direct output of the first phenomenon which is caused by a highly nonuniform flux distribution in the regions close to the air gap. As a result, additional harmonics are introduced to the flux. On the other hand, the motor torque is proportional to the flux linkage of the stator and rotor and will be affected by high-order flux harmonics. Hence, a high-torque ripple is expected.

To have a more uniform flux path or in fact air-gap length, there are a couple of methods considered during the design process of induction motors:

- to use a proper stator–rotor slot combination
- to implement a double layer winding layout
- to change the slot shape
- to have totally closed stator and rotor slots
- to apply skew to the stator or rotor slots.

In the case of a skewed rotor, rotor slots experience a variable flux along the stack length. This leads to a considerable decrease in the motor torque ripple as the high-order harmonic components of the air-gap flux are weakened. However, the fundamental component is weakened as well. Therefore, the average torque becomes smaller by increasing the skew angle. The maximum appropriate skew angle is equal to the rotor slot pitch. Increasing the skew angle beyond the prescribed limit increases the torque ripple again.

To create the skew in induction motors, the rotor laminations are punched to each other with a slightly clock and counter clock-wise rotated center. Assuming that there are 201 laminations and 72 rotor bars, the maximum appropriate skew angle is 5 mechanical degrees. This means that each lamination must be rotated by 0.025 degrees with respect to the previous lamination attached to the rotor stack.

What was mentioned above suggests an experimental way of implementing the skew. The question is "How is the skew applied to a simulation model in an efficient way?" In order to account for the skewing effect, a quasi 3D model of an induction motor can be used. This includes multiple motor slices along the $z$-axis and the motor shaft (see Figure 6.30). In this approach, the motor is divided to $n$ slice along the motor stack length. Each slice has a length of (stack length/$n$), and the corresponding rotor is rotated by (360/$n$) degrees mechanically with respect to the previous slice. The stators are identical. Actually, the skew is modeled using nonskewed and small stack length motors operating in series. The larger the number of slices is, the more accurate the skewing effect is. This is a very useful method of incorporating the skewing effect into the simulation process in a reasonable period. Having applied this method, the simulated torque ripple might be reduced by factor of 5 or 6, depending on the skewing angle and the motor topology. The larger the number of the rotor bars is, the lower the skewing effect is.

To investigate the skewing effect, four different approximations of the skewed rotor are studied considering 3, 5, 8 and 10 slices, and their corresponding rotation angles are shown in Table 6.3.

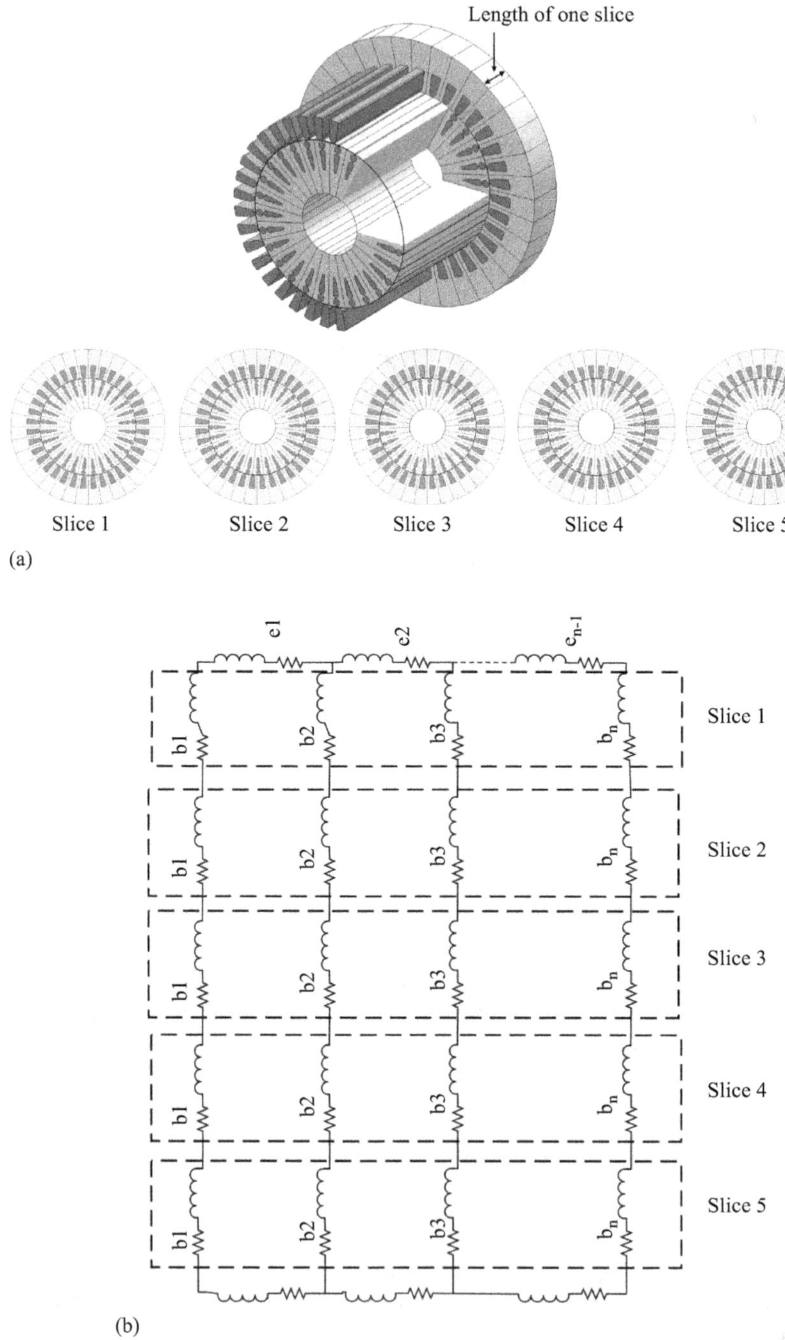


*Figure 6.30 (a) Multislice implementation of skewing effect and (b) rotor circuit*

*Table 6.3 Rotation angle of the rotor slices used for investigating the skewing effect*

| | | | | | | | | | | | |
|---|---|---|---|---|---|---|---|---|---|---|---|
| 3 slices | Slice number | 1 | | | 2 | | | 3 | | | |
| | Rotation angle | 0.83 | | | 2.5 | | | 4.17 | | | |
| 5 slices | Slice number | 1 | | 2 | | 3 | | 4 | | 5 | |
| | Rotation angle | 0.5 | | 1.5 | | 2.5 | | 3.5 | | 4.5 | |
| 8 slices | Slice number | 1 | 2 | 3 | 4 | 5 | 6 | 7 | | 8 | |
| | Rotation angle | 0.31 | 0.94 | 1.56 | 2.2 | 2.8 | 3.44 | 4.06 | | 4.7 | |
| 10 slices | Slice number | 1 | 2 | 3 | 4 | 5 | 6 | 7 | 8 | 9 | 10 |
| | Rotation angle | 0 | 0.5 | 1 | 1.5 | 2 | 2.5 | 3 | 3.5 | 4 | 4.5 |

*Table 6.4 Percentage of motor torque ripple*

| **Number of slices** | **1** | **3** | **5** | **8** | **10** |
|---|---|---|---|---|---|
| Torque ripple (%) | 25 | 17.9 | 13.3 | 10.1 | 9.6 |

The results of the simulations of the machines discussed above are listed in Table 6.4.

Accordingly, the torque ripple has a decreasing trend of variation as the number of slices increases. In Figure 6.31, the motor torque ripple is shown for two number of slices 1 and 10.

Now, the skewing effect is investigated using the spectral analysis of the motor current signal by means of the Fourier transform. To handle this step, two cases, S1 and S10, are intended. Figure 6.32 shows the current spectrum of the motors.

As the magnitude of the fundamental frequency, i.e., 50 Hz, is about 12 A, and the high-order frequency components have a relatively small magnitude, for the sake of clearness, the *Y*-axis is limited to smaller values so that the amplitude of the higher order harmonic components are observable. The third harmonic is not affected very much by imposing the skew to the rotor slots while the other dominant components including 235 and 335 Hz are greatly reduced, so it is expected to have a smaller torque ripples. Most of the time, applying the skew leads to a decrease in the torque development capability of the motor because not only the high-order components but also the fundamental one is affected in a bad way.

#### 6.9.4.2 Modeling rotor interturn fault

In practice, insulation fails to isolate the coils upon the occurrence of the interturn or any kind of short-circuit fault. The electrical resistance between two or multiple turns falls down from an infinite value to an almost zero value such that short-circuited turns are defected permanently. In this case, the short-circuited turns operate as a separate phase carrying an uncontrollably larger amount of current leading to a thermal stress of coils and windings. The corresponding resistance of

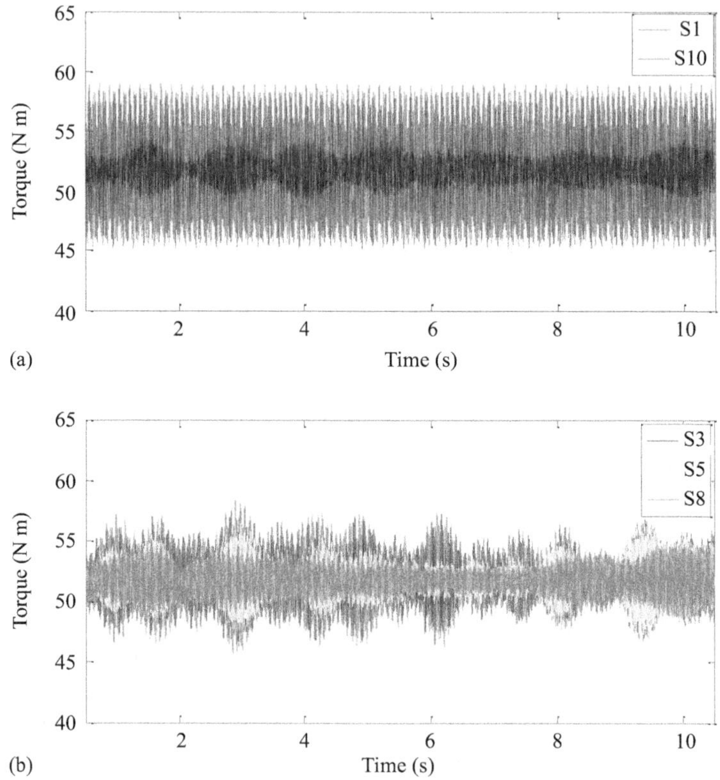


*Figure 6.31 Motor steady-state torque carrying the ripples (a) number of slices 1 and 10 and (b) number of slices 3, 5 and 8*

the mentioned phase is equal to the ratio of the shorted-circuited turns to the total number of turns multiplied by the resistance of a phase.

Figure 6.33 shows the rotor winding and 3 short-circuited turns applied to the rotor phase. It is noted that the investigated machine is a wound-rotor induction motor; hence, there is also a possibility of a short-circuit fault in the rotor as well as the stator. The points a and b are connected to each other directly without any intermediate resistance. Therefore, the faulty phase current is separated from that of the main phase. In order to account for this type of fault in an FEM package, the faulty phase must be modeled turn-wise. In fact, all the short-circuited turns should be modeled separately so that the connections a and b are available. Besides, the mesh size of the regions which correspond to the short-circuited turns, is better to be refined. The other slots can be kept intact compared to a healthy case. Actually, there is no need to model all the turns of all the slots.

The current of the short-circuited phase is shown for different number of faulty turns in Figure 6.34.

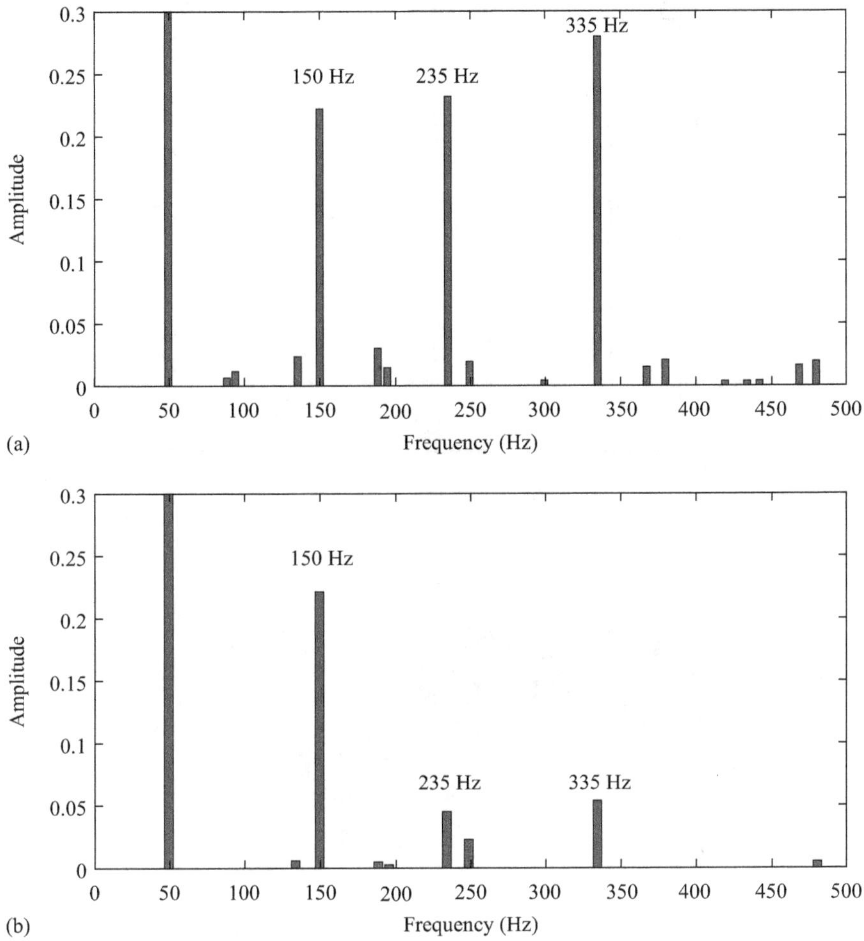


*Figure 6.32 Motor steady-state torque carrying the ripples (a) number of slices 1 and 10 and (b) number of slices 3, 5 and 8*

The important point is that the faulty phase current is not much different from the main phase current, and it is slightly larger. This means that a complete insulation failure is not expected at all. This is in contrast to a short-circuit fault taking place in the stator in which the faulty phase current can be relatively larger than that of the healthy phase. On the other hand, the more the number of the short-circuited rotor turns is, the closer the faulty phase current to that of the healthy phase is. In addition, the ripple produced by the short-circuit fault decreases upon increasing the number of faulty turns in the rotor.

By increasing the number of faulty turns, the induced voltage is improved proportional to the number of turns. The corresponding faulty phase resistance increases as well. Therefore, the current should approximately maintain a constant amount. In the rotor, the most dominant part of the impedance is of a resistance

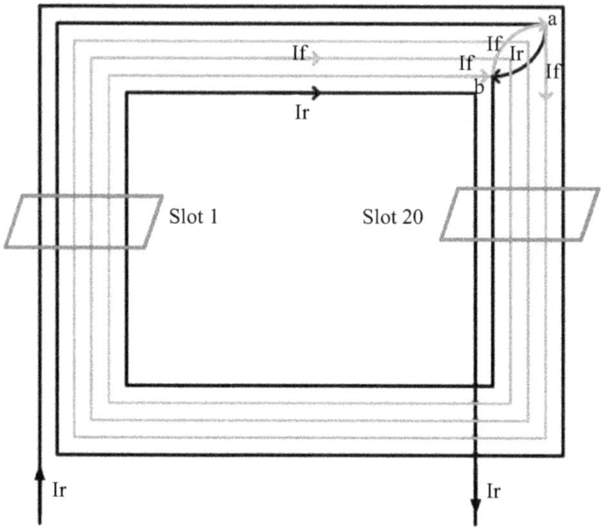


(a)

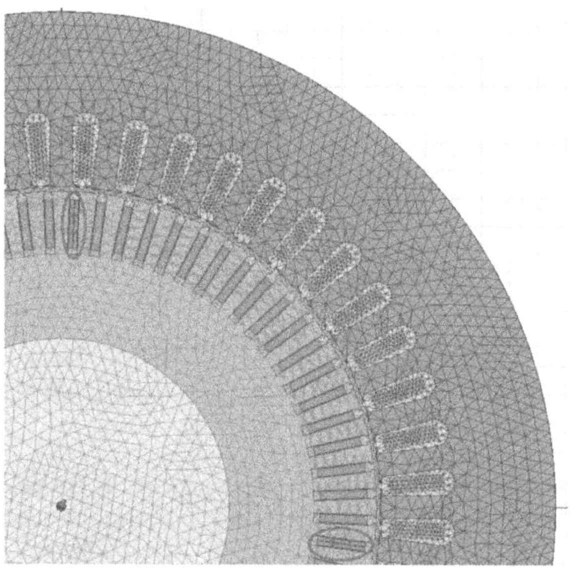

(b)

*Figure 6.33 (a) Three short-circuited turns in the rotor phase winding and (b) turn-wise model of the impacted rotor coil*

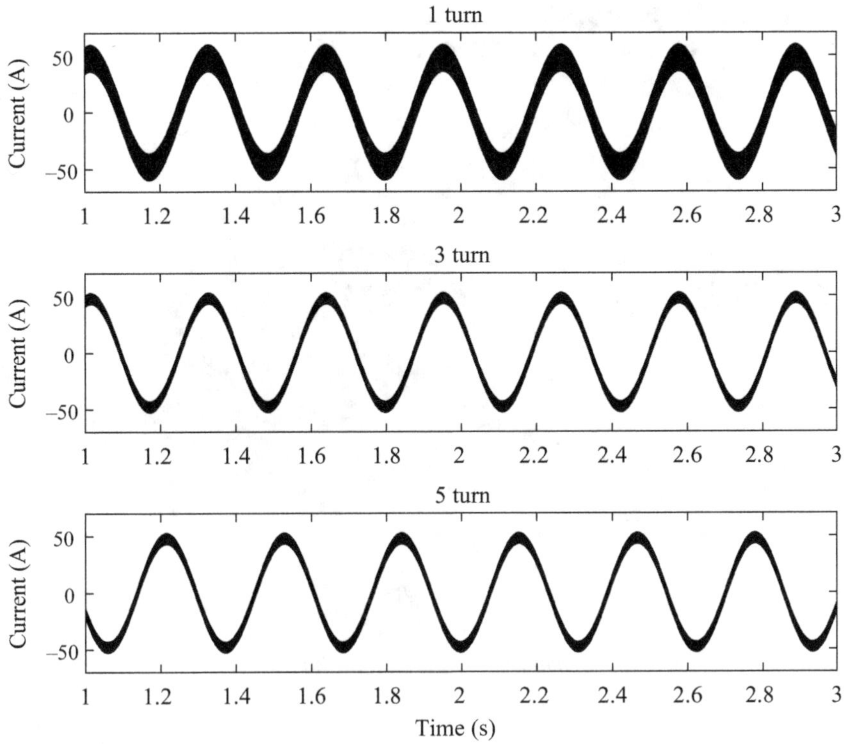


*Figure 6.34 Faulty phase current in a full-load condition*

nature as the frequency is very low. However, it is noteworthy that the self-inductance of the faulty turns increases as a square function of the number of turns. This has two significant results: first, the rate of increase in the induced voltage is less than the rate of increase in the faulty phase impedance; therefore, the current level should decrease by increasing the number of turns. Second, by increasing the number of faulty turns, the corresponding inductance increases as well. Hence, the series RL filter representing the faulty phase impedance becomes more inductive. The high-frequency ripples are filtered out from the signal. Investigating the motor signals including the torque, speed and flux reveals the fact that a short-circuit fault in the rotor side does not affect the motor behavior considerably. Unlike this situation, even a very weak short-circuit fault in the stator causes a lot of problem for the maintenance process. The difference between the stator and rotor and their contributions to the short-circuit fault is normally because of some technical design issues discussed below.

- The first influential factor is the motor slip. Assume that there is only one short-circuited turn in the rotor phase. The amount of the induced voltage is equal to (1/272) induced voltage in the stator regardless of the motor slip.

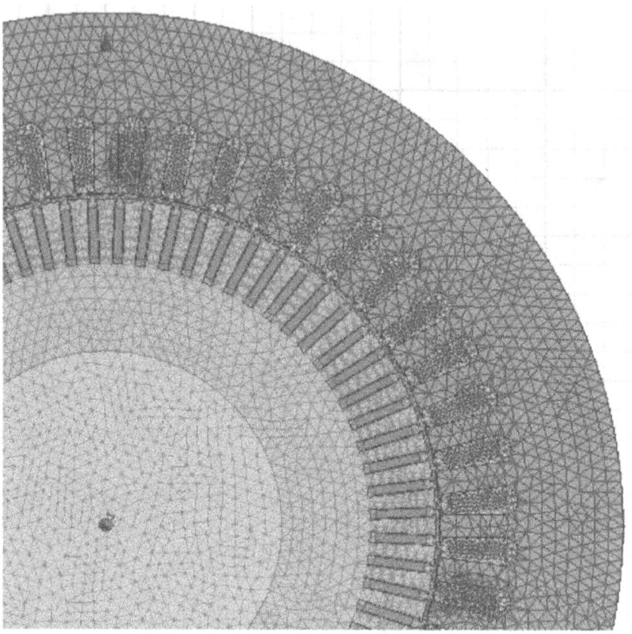

*Figure 6.35 Turn-wise model of the impacted stator coil*

272 is the number of the stator turns. Multiplying the induced voltage by the motor slip which is equal to 0.05 in the rated operation produced a negligible amount of voltage induced to the faulty turn. As a result, the corresponding current level is very small.

- The second factor is the source of induction phenomenon. In a short-circuit fault, whether in the stator or in the rotor, it is the stator coil that induces a voltage on the faulty turns. According to Lenz's law, the direction of the induced current is such that it contradicts the cause of induction phenomenon. In a stator short-circuit fault, the magnetic field produced by the induced current contradicts the main field, so the distortion is reflected to the torque and the speed. However, in a rotor short-circuit fault, the induced current produces a field in the same direction that the rotor main field rotates. Therefore, they are not conflicting fields at all.

#### 6.9.4.3 Modeling stator interturn fault

Stator short-circuit fault, particularly interturn fault, is implemented in the same fashion that the rotor fault is. The difference is that the fault current should have a direction opposite to that of the stator healthy phase. Faulty turns must be implemented turn-wise, and the rest of the phases might be modeled as one single group (see Figure 6.35).

In this case, the resistance of the short-circuited turns is calculated and no current-limiting resistance is connected to the fault connections. Having adjusted

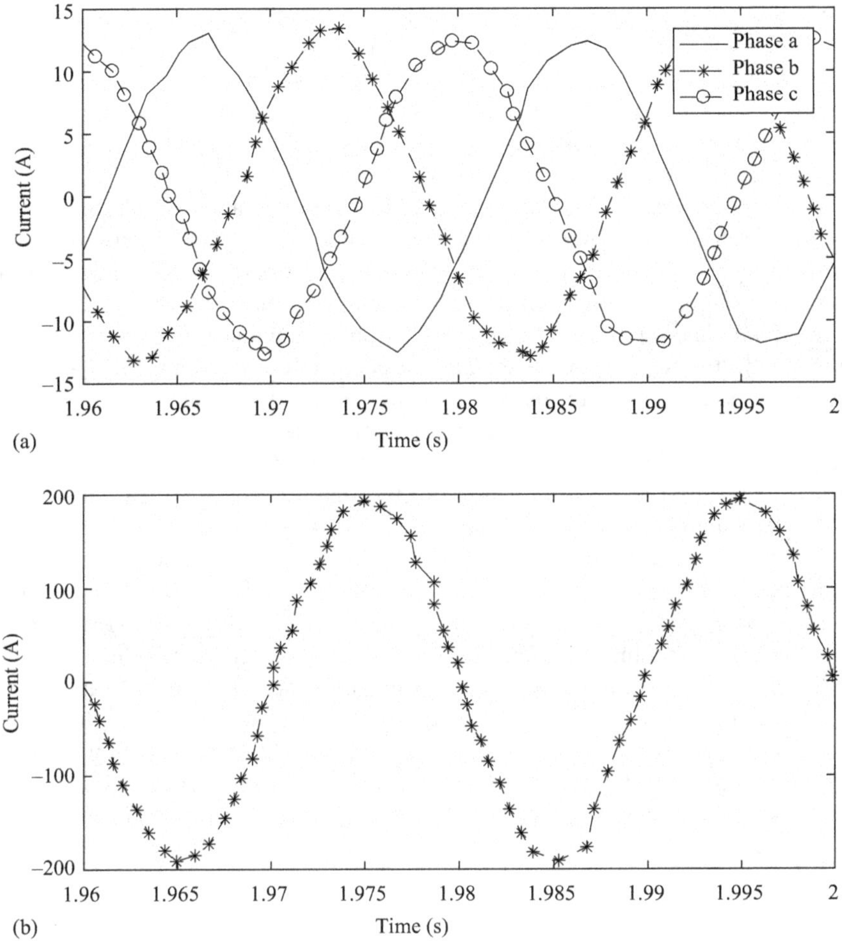


*Figure 6.36 (a) Healthy stator phase current and (b) faulty stator phase current for one interturn fault*

the simulation model and applied the full-load condition to it, a fault case in which one stator turn is short-circuited is simulated, and the corresponding current levels are shown in Figure 6.36.

The rms current value of healthy phases are 8.6 A while that of the faulty phase is equal to 9.07 A. There is a 6% increase with respect to the healthy case. On the other hand, the faulty phase has a considerably large amount of current, about 128 A.

What was discussed in this chapter is the most accurate way of dealing with fault-related simulations. Although a useful winding function-based motor model was previously proposed in Chapter 5, it suffers from lack of detailed magnetic analysis which is used for a comprehensive saturation and loss characterization of a motor. For example, the local saturation cannot be modeled and simulated by means of a winding function method. Moreover, the slotting effect, the skewing effect, the unbalanced magnetic pull, the stored energy and the iron losses are not easily obtainable using analytical methods. Even if an analytical solution exists, it is not guaranteed that it follows the same trends as experiments properly. Therefore, it is proposed to utilize an FEM package for the diagnosis purposes if a suitable computational power is available.

The signals measured by means of the FEM are pure faulty signals building the very initial steps of a diagnosis process. Significantly, no one has access to inside of motors to check if it is healthy or not. Therefore, the only way to understand and monitor the motors is the signals available outside the machine. Current and torque are two examples of this kind of signals. Of course, the magnetic signals including the magnetic flux density of losses are not accessible directly an FEM should be used to estimate faulty motor behavior. However, the signals such as motor current often reflect the unbalanced magnetic behavior caused by faults. Therefore, measuring and processing the available signals seem to be enough for the diagnosis purposes. Nevertheless, no one can deny the importance of a measurement tool such as search coils by means of which the air-gap flux is monitored.

Taking the mentioned points into consideration, "signal processing" is required to process measured signals and extract a knowledge based on which the diagnosis procedure is performed. The fundamentals of different types of faults, along with both the analytical and numerical approximations of motor signals in healthy and faulty cases, have been already proposed. Moreover, experimental implementations have also been discussed deeply with the goal of providing a basis for validating the quality and accuracy of the simulation processes. Now, it is time to go through processing the raw signals which might or might not include fault information. The processing step is held in three different domains including:

- time domain
- frequency domain
- time-frequency domain

each having its own advantages and disadvantages in terms of capability to diagnose faults accurately. So, it is essential to get used to different types of processors used for processing motor signals in the transient and steady-state modes. Some of the processors are only able to extract information in a steady-state mode, while others might be applied to both the transient and steady-state regimes.

## References

[1] Salon S. Finite element analysis of electrical machines, New York: Springer, 1995.

[2] Sayed-Ahmed A. and Demerdash N.A.O. Fault-tolerant operation of delta-connected scalar- and vector-controlled AC motor drives, *IEEE Transactions on Power Electronics*, 2012; **PE-27**: 3041–3049.

[3] Marcsa D. Induction motors simulation by finite element method and different potential formulations with motion voltage term, *B.Sc Thesis, Széchenyi István Universit, 2008.*

[4] Kathir I., Balakrishnan S. and Bevila R.J. Fault analysis of induction motor, *2011 International Conference on Emerging Trends in Electrical and Computer Technology*, Tamil Nadu, 2011: 476–479.

[5] Mini V.P. and Ushakumaric S. Rotor fault analysis of an induction motor using FEM, *2012 2nd International Conference on Power, Control and Embedded Systems*, Allahabad, 2012; 1–7.

[6] Praveen K.N., Isha T.B. and Balakrishnan P. Radial electro-magnetic field analysis of induction motor under faulty condition using FEM, *2016 Biennial International Conference on Power and Energy Systems: Towards Sustainable Energy (PESTSE)*, Bangalore, 2016; 1–6.

[7] Faiz J., Ebrahimi B.M. and Toliyat H.A. Effect of magnetic saturation on static and mixed eccentricity fault diagnosis in induction motor, *IEEE Transactions on Magnetics*, 2009; **MAG-45**: 3137–3144.

[8] Popescu M., Dorrell D.G., Alberti L., Bianchi N., Staton D.A. and Hawkins D. Thermal analysis of duplex 3-phase induction motor under fault operating conditions, *2012 XXth International Conference on Electrical Machines*, Marseille, 2012; 1875–1881.

[9] Faiz J., Ebrahimi B.M., Toliyat H.A. and Akin B. Diagnosis of a mixed eccentricity fault in a squirrel-cage three-phase induction motor using time stepping finite element technique, *2007 IEEE International Electric Machines & Drives Conference*, Antalya, 2007: 1446–1450.

[10] Lv Q., Fang Y., Cheng Z. and Bao X. Finite element analysis of a squirrel cage induction motor with an oval stator under eccentricity conditions, *2013 5th International Conference on Power Electronics Systems and Applications (PESA)*, Hong Kong, 2013; 1–5.

[11] Watson J.F. and Paterson N.C. The development of an accurate finite element model to investigate the factors which influence the fault-indicating components of current in 3-phase induction motors, *1999. Ninth International Conference on Electrical Machines and Drives (Conf. Publ. No. 468)*, Canterbury, 1999; 247–252.

[12] Weili L., Ying X, and Shukang C. Characterization analysis of effects in an induction motor on asymmetrical operation, *2008 IEEE Vehicle Power and Propulsion Conference*, Harbin, 2008; 1–5.

[13] Zaabi W., Bensalem Y. and Trabelsi H. Co-simulation of induction motor fed by PWM inverter under a broken bar fault, *2015 IEEE 12th International Multi-Conference on Systems, Signals & Devices (SSD15)*, Mahdia, 2015; 1–6.

[14] Mohammed O.A., Abed N.Y. and Ganu S. Modeling and characterization of induction motor internal faults using finite element and discrete wavelet transforms, *2007 IEEE Electric Ship Technologies Symposium*, Arlington, VA, 2007; 250–253.

[15] Immovilli F., Bianchini C., Cocconcelli M., Bellini A. and Rubini R. Bearing fault model for induction motor with externally induced vibration, *IEEE Transactions on Industrial Electronics*, 2013; **IE-60**: 3408–3418.

[16] Povinelli R.J., Johnson M.T., Bangura J.F. and Demerdash N.A.O. A comparison of phase space reconstruction and spectral coherence approaches for diagnostics of bar and end-ring connector breakage faults in poly-phase induction motors using current waveforms, *Conference Record of the 2002 IEEE Industry Applications Conference. 37th IAS Annual Meeting (Cat. No.02CH37344)*, Pittsburgh, PA, USA, 2002; 1541–1547.

[17] Xie Y. and Cao W. Influence of broken bar fault on the performance of squirrel-cage induction motor at standstill, *2011 International Conference on Electrical Machines and Systems*, Beijing, 2011; 1–5.

[18] Janda M., Makki Z. and Konicek P. Calculation of the vibration induction motor using the finite element method, *Proceedings of the 16th International Conference on Mechatronics - Mechatronika 2014*, Brno, 2014; 526–529.

[19] Bangura J.F. and Demerdash N.A.O. Effects of broken bars/end-ring connectors and airgap eccentricities on ohmic and core losses of induction motors in ASDs using a coupled finite element-state space method, *IEEE Transactions on Energy Conversion*, 2000; **EC-15**: 40–47.

[20] Belahcen A., Martinez J. and Vaimann T. Comprehensive computations of the response of faulty cage induction machines, *2014 International Conference on Electrical Machines (ICEM)*, Berlin, 2014; 1510–1515.

[21] Sudar V.A., Nagarajan S. and Reddy S.R. Detection and analysis of broken bar in three phase squirrel cage induction motor using FEM, *2012 International Conference on Computing, Electronics and Electrical Technologies (ICCEET)*, Kumaracoil, 2012; 40–50.

[22] Samonig M.A. and Wolbank T.M. Transient magnetic analysis of voltage pulse excited induction machines with respect to diagnosis and fault detection, *2013 15th European Conference on Power Electronics and Applications (EPE)*, Lille, 2013: 1–10.

[23] Sizov G.Y., Yeh C.C. and Demerdash N.A.O. Magnetic equivalent circuit modeling of induction machines under stator and rotor fault conditions, *2009 IEEE International Electric Machines and Drives Conference*, Miami, FL, 2009; 119–124.

[24] Karami M., Mariun N., Mehrjou M.R., Ab Kadir M.Z.A., Misron N. and Mohd Radzi M.A. Diagnosis of static eccentricity fault in line start permanent magnet synchronous motor, *2014 IEEE International Conference on Power and Energy (PECon)*, Kuching, 2014; 83–86.

[25] Ilamparithi T. and Nandi S. Comparison of results for eccentric cage induction motor using Finite Element method and Modified Winding Function Approach, *2010 Joint International Conference on Power Electronics, Drives and Energy Systems & 2010 Power India*, New Delhi, 2010; 1–7.

[26] Povinelli R.J., Bangura J.F., Demerdash N.A.O. and Brown R.H. Diagnostics of bar and end-ring connector breakage faults in poly-phase

induction motors through a novel dual track of time-series data mining and time-stepping coupled FE-state space modeling, *IEEE Power Engineering Review*, 2002; **PER-22**: 58–59.

[27] Aileen C.J., Nagarajan S. and Reddy S.R. Detection of broken bars in three phase squirrel cage induction motor using finite element method, *2011 International Conference on Emerging Trends in Electrical and Computer Technology*, Tamil Nadu, 2011: 249–254.

[28] Cusido J., Romeral L., Delgado M., Garcia A. and Ortega J.A. Induction machines fault simulation based on FEM modelling, *2007 European Conference on Power Electronics and Applications*, Aalborg, 2007; 1–8.

[29] Rojas C., Melero M.G., Cabanas M.F., Cano J.M., Orcajo G.A. and Pedrayes F. Finite element model for the study of inter-turn short circuits in induction motors, *2007 IEEE International Symposium on Diagnostics for Electric Machines, Power Electronics and Drives*, Cracow, 2007; 415–419.

[30] Barzegaran M., Mazloomzadeh A. and Mohammed O.A. Fault diagnosis of the asynchronous machines through magnetic signature analysis using finite-element method and neural networks, *IEEE Transactions on Energy Conversion*, 2013; **EC-28**: 1064–1071.

[31] Weili L., Ying X., Jiafeng S. and Yingli L. Finite-element analysis of field distribution and characteristic performance of squirrel-Cage induction motor with broken bars, *IEEE Transactions on Magnetics*, 2007; **MAG-43**: 1537–1540.

[32] Hanafy H.H., Abdo T.M. and Adly A.A. Analysis and force computations of single-phase permanent-capacitor induction motors using FEM at broken bar conditions, *7th IET International Conference on Power Electronics, Machines and Drives (PEMD 2014)*, Manchester, 2014; 1–6.

[33] Subash M., Nagarajan S. and Rama R.S. Detection and analysis of eccentricity in three phase squirrel cage induction motor using FEM, *2013 International Conference on Circuits, Power and Computing Technologies (ICCPCT)*, Nagercoil, 2013, 203–212.

[34] Rodriguez P.J., Belahcen A., Laiho A., Resina E. and Arkkio A. Signatures of electromechanical faults in stress distribution and vibration of induction motors, *IEEE International Symposium on Diagnostics for Electric Machines, Power Electronics and Drives*, Cracow, 2007; 67–172.

[35] Zouzou S.E., Khelif S., Halem N. and Sahraoui M. Analysis of induction motor with broken rotor bars using finite element method, *2011 2nd International Conference on Electric Power and Energy Conversion Systems (EPECS)*, Sharjah, 2011: 1–5.

[36] Holik P.J. and Holik S.M. A finite-element model for induction machines incorporating winding faults, *5th IET International Conference on Power Electronics, Machines and Drives (PEMD 2010)*, Brighton, UK, 2010; 1–3.

[37] Sprooten J. and Maun J.C. Influence of saturation level on the effect of broken bars in induction motors using fundamental electromagnetic laws and finite element simulations, *IEEE Transactions on Energy Conversion*, 2009; **EC-24**: 557–564.

[38] Barbour A. and Thomson W.T. Finite element study of rotor slot designs with respect to current monitoring for detecting static airgap eccentricity in squirrel-cage induction motors, *Industry Applications Conference, Thirty-Second IAS Annual Meeting, IAS '97., Conference Record of the 1997;* New Orleans, LA: 1997;112–119.

[39] Chang-Eob Kim, Yong-Bae Jung, Sang-Baeck Yoon and Dal-Ho Im, The fault diagnosis of rotor bars in squirrel cage induction motors by time-stepping finite element method, in *IEEE Transactions on Magnetics*, 1997; **MAG-33**: 2131–2134.

[40] Zaabi W., Bensalem Y. and Trabelsi H. Fault analysis of induction machine using finite element method (FEM), *2014 15th International Conference on Sciences and Techniques of Automatic Control and Computer Engineering (STA)*, Hammamet, 2014; 388–393.

[41] Martinez J., Belahcen A. and Arkkio A. Combined FE and two-dimensional spectral analysis of broken cage faults in induction motors, *IECON 2013 - 39th Annual Conference of the IEEE Industrial Electronics Society*, Vienna, 2013; 2674–2679.

[42] Barbour A. and Thomson W.T. Finite element analysis and on-line current monitoring to diagnose air gap eccentricity in 3-phase induction motors, *1997 Eighth International Conference on Electrical Machines and Drives (Conf. Publ. No. 444)*, Cambridge, 1997: 150–154.

[43] Mohammed O.A., Abed N.Y. and Ganu S. Modeling and characterization of induction motor internal faults using finite-element and discrete wavelet transforms, *IEEE Transactions on Magnetics*, 2006; MAG-42: 3434–3436.

[44] Watson J.F., Paterson N.C. and Dorrell D.G. The use of finite element methods to improve techniques for the early detection of faults in 3-phase induction motors, *1997 IEEE International Electric Machines and Drives Conference Record*, Milwaukee, WI, 1997; WB3/9.1–WB3/9.3.

[45] Fiser R. and Ferkolj S. Application of a finite element method to predict damaged induction motor performance, *IEEE Transactions on Magnetics*, 2001; MAG-37: 3635–3639.

# Chapter 7
# Signal-processing techniques utilized in fault diagnosis procedures

## 7.1 Introduction

Signal processing is a widely used technology that pervades every aspect of our lives which are actually unknown to many. This chapter surveys some basic concepts which drive the field.

1. What is a "Signal"? A signal describes how some physical quantity varies over time and/or space [1–3]. For example, velocity is a quantity which is measurable over time as it possesses values at different instants. A set of values which is obtained by measuring a quantity over time, frequency or space forms a signal. The quantities such as acceleration, temperature, humidity, magnetic flux density, torque, current and voltage are also measurable in time. A spatial distribution of the motor flux is another signal expressed in space. By sampling these quantities at different instants, a signal is generated. Technically, a signal is a representation of a real phenomenon which progresses over space or time.
2. What is "Signal processing"? Signal processing is the science of analyzing the measured or sampled signals [4–7]. The goal of analysis is to extract useful information on a subject which can be detection, determination or diagnosis of a machine fault. If the analysis is performed on a discretely sampled signal, i.e., a digital signal, the technique is called "digital signal processing," and the associated mathematical method used for formulating the processing procedure is called a signal processor.

The processing techniques are often used for measuring, filtering, compressing or extracting information from a signal. Normally, the first step is to digitalize an analog signal as the processors are implemented in a computer or a digital signal processor which works based on digital technology. Therefore, the focus henceforward will be on digital signal processing. The conversion from analog to digital is handled by an analogue-to-digital converter which converts an analog signal to series of data points forming analyzed signal. Due to the great advantages of digital signal-processing techniques such as the capability of compressing data, they are used widely nowadays.

Generally, there are the following three types of signal-processing techniques:

- time-domain techniques [8–10]
- frequency-domain techniques [11,12]
- time-frequency-domain techniques.

By means of the above-mentioned signal-processing techniques, both the transient and steady-state operations of motor signals can be analyzed in order to extract fault features. In fact, there are two general types of monitoring systems, those which are applied to transient operation and those which are applied to steady-state operation. In medium- and high-power applications in which there are many start–stop operations, the transient analysis of faulty motors becomes important as transient operations place much more stress on the motor components, especially the faulty ones; hence, monitoring and detecting faults in their incipient regime and more importantly when the motor just starts up is crucial.

The time-domain analysis of induction motors has been the subject of various studies as the very initial steps of diagnosis procedures. However, there are several influential factors affecting the time-domain variations and oscillations in an almost arbitrary fashion, not revealing a satisfactory fault feature. As a result, the trend has been switched to utilizing frequency and time-frequency transforms for processing signals in the hope that appropriate features will be extracted.

The Fourier transform (FT) is the fundamental tool building the foundation of most of the advanced transforms developed so far. The FT is widely used in condition monitoring and fault diagnosis of various applications including biomedical devices, aerospace apparatus, communication as well as electrical machines. This transform is normally used to process stationary signals having a repetitive or periodic nature with almost fixed amplitudes. An obvious example of such a signal is motor current signal sampled in steady-state operation. Although the FT is a very promising and powerful tool to extract potential fault-related components, it is only applied to stationary signals, and the corresponding output has nothing to do with the time-domain information which might be useful for the diagnosis process. Hence, the transforms such as the short-time FT (STFT) and the wavelet transform (WT) have been proposed in the literature.

The STFT and the WT are time-frequency transforms mapping signals on two apparently independent domains including time and frequency represented on the same plane. Thus, not only the frequency components are detectable, but also the time-related information of every single frequency is in hand. The significant drawback associated with the time-frequency transforms is the poorer resolution than that of the FT. The FT acts very well in satisfying resolution, while the STFT and WT always suffer from lack of resolution. However, the WT provides a better flexibility in determining the resolution trade-offs between the time and frequency components. Both the WT and the STFT are based on the FT.

Another time-frequency transform called the Hilbert–Huang transform (HHT) is also available. This transform was first used in aerospace applications and then moved toward the biomedical monitoring devices. Recently, some major work in the field of diagnosis of transient behavior of electrical machines has been handled by means of the HHT; therefore, it is becoming more popular. Moreover, this transform enables us to tackle the monitoring problems in an inverter-fed application. In fact, it provides a very straightforward tool to differentiate the effect of various controlling modes on the diagnosis process. The HHT is not actually based on the FT, and it is of a totally different nature requiring an adaptive and iterative implementation unlike the other transforms which only need a one-pass implementation.

The time-domain variations of the faulty motors have been already discussed in the previous chapters, and there is no specific implementation rule for this type of analysis. It is just based on observing faulty signals and extracting potential information. Nevertheless, there are some time-domain techniques such as time series data mining used for addressing some kinds of faults. They will be pointed out in the next chapters where necessary. For now, this chapter focuses on the other two types of analyses, i.e., the frequency- and time-frequency analyses. First, the mathematical concepts are explained; then MATLAB® implementation of transforms, along with intuitive examples, is provided in order to address the corresponding technical issues during the implementation and postprocessing analysis.

## 7.2 Fourier transform

This section covers the fundamentals of the FT and provides an intuitive way of thinking of the FT. So two equations are written below immediately without going through the historical discussions on the importance of the FT (see (7.1)). It might probably look a bit difficult to comprehend it first;

$$X(f) = \int_{-\infty}^{\infty} x(t)e^{-2\pi jft}dt \tag{7.1}$$

$$x(t) = \int_{-\infty}^{\infty} X(f)e^{+2\pi jft}df \tag{7.2}$$

The first (7.1) is called FT or the forward FT. The second equation is called the inverse. As it can be observed, (7.1) turns a function time, i.e., $x(t)$, into a function frequency, i.e., $X(f)$ and does the reverse. So the process of turning one into another is called transformation. The transform is actually a mapping between two different sets of data or domains. In this case, it changes information in the time domain into information in the frequency domain. Here, we are using the variable $f = \frac{\omega}{2\pi}$ [Hz] to represent the frequency in units of Hertz or cycle per second. The data in these two domains looks different and can even vary in length of the data, but it still represents exactly the same information.

Any continuous signal in the time domain can be represented by a sum of carefully chosen sinusoids, and as sinusoids can be completely described by their amplitude, their frequency and their phase, those three parameters are all the information we need to know. At that point, we can plot the amplitude and phase at every frequency or across the entire spectrum and call it frequency-domain representation of the signal. This information is extremely useful in fault diagnosis procedures used for condition monitoring of electrical machines such as induction motors. For example, imagine a situation that an induction motor operates with a broken rotor bar which introduces $2sf_s$ frequency components to the motor current containing a fundamental frequency of 50 Hz. Therefore, the current will have both 50 and $2sf_s$ sinusoids added to each other to form the entire signal. A very clear example of the FT applied to the mentioned faulty signal will be provided at the end of this chapter. Of course, most signals are not just comprised of a single sinusoid but multiple ones at different amplitudes and phases across the entire spectrum. Why is the signal decomposed into sinusoids? Why is it not decomposed into some other repeating signals like a square wave or sawtooth or into different periodic signals altogether? Actually, what FT tries to do is to transform signals into something easier to work with at least for our particular purpose which is tracking the trend of variations of sinusoidal or quasi-sinusoidal signals. The FT gives us the ability to move from one domain to another domain allowing us to have a better insight into fault-related components of motor quantities.

Here a very simple but intuitive example of signals and their corresponding FT is provided. Consider the following three signals in the time domain, $x_1$, $x_2$ and $x_3$. $x_3$ is the sum of $x_1$ and $x_2$.

$$x_1 = 2 \times \sin(2\pi t) \tag{7.3}$$

$$x_2 = 0.5 \times \sin(8\pi t) \tag{7.4}$$

$$x_3 = x_1 + x_2 = 2 \times \sin(2\pi t) + 0.5 \times \sin(8\pi t) \tag{7.5}$$

The time-domain variation of the signals is shown in Figure 7.1.

The amplitudes of the signals are 2 and 0.5 for the $x_1$ and $x_2$, respectively. We can add the two signals to build a new signal determined by $x_3$. Therefore, $x_3$ is a combination of two different frequencies, namely 1 and 4 Hz. Each of the two sinusoids $x_1$ and $x_2$ is the monotone signal representing a pure sinusoidal shape with a specific amplitude while $f_3$, although periodic, is not a pure sinusoid and oscillates in its envelope. So the first look at $x_3$ might not reveal the included frequencies, i.e., 1 and 4 Hz. The only deducible information is the periodicity of the signal. This is exactly where the FT comes into play. Figure 7.1(b) illustrates the frequency components of the investigated signals. The spectrum of each signal peaks at the previously specified frequencies. For example, $x_1$ has a jump at 1 Hz with an amplitude of 2 while the spectrum of $x_2$ peaks at 4 Hz with the amplitude of 0.5. $x_3$ is the sum of the mentioned terms including 1 and 4 Hz. All the three signals have been sampled by a sampling frequency of 50 Hz. The term "Sampling frequency" is different from the "frequency of signal." The sampling frequency is used to digitalize the signal by taking multiple points at different instants in order to model the signal in a computer. This means that we always deal with digital signals, not an analogue one. There are several approaches to the implementation of FT in discrete signals. One of them is called Fast FT (FFT) available in packages such as MATLAB and will be studied later in this chapter. Therefore, we are going to

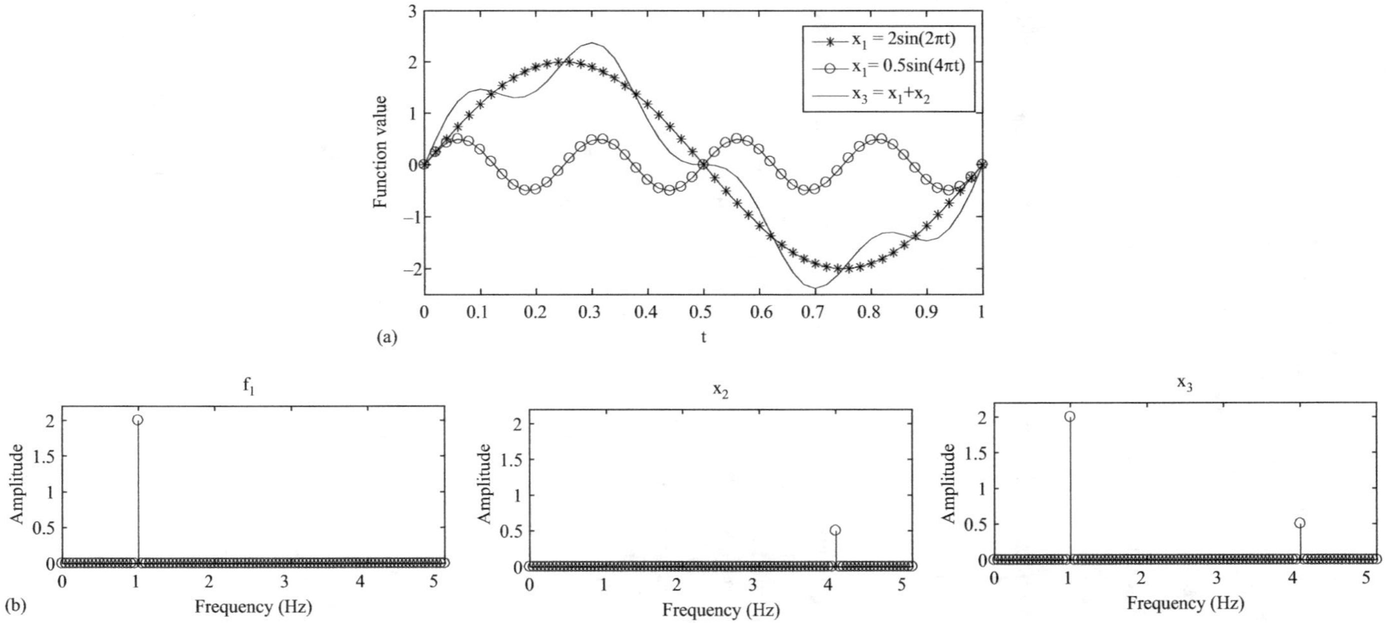


*Figure 7.1 (a) Time-domain signals and (b) frequency-domain signals*

provide the fundamentals of the FFT. Before that, some of the required prerequisites are discussed first.

The first important point is the concept of periodic sequences in a digital working environment. The sequence can be observed in time or any other domain. However, the focus here is on a time-dependent periodicity. A periodic sequence is usually determined as follows:

$$\tilde{x}(n) = \tilde{x}(n + kN) \tag{7.6}$$

where $k$ is an integer. For example, the signal W which is equal to

$$W_N^{kn} = e^{-j\frac{2\pi}{N}kn} \tag{7.7}$$

is a periodic, symmetric and orthogonal sequence leading to the following conclusions:

$$W_N^{kn} = W_N^{(k+N)n} = W_N^{k(n+N)} \tag{7.8}$$

$$W_N^{-kn} = \left(W_N^{kn}\right)^* = W_N^{(N-k)n} = W_N^{k(N-n)} \tag{7.9}$$

This is a periodic signal, so taking the samples of one single period is enough to rebuild the model in a computer or processor. In addition, the next periods of the signal can be obtained by repeating the signal for the next N points. W looks like the exponential function used in (7.1). This is just a simple example of a discrete periodic signal and further discussion on the implementation of the FT continues as follows. The remarkable point is that the length of the sampled signal is always finite. However, if the signal is periodic, there is always a relation between the whole sequence and the sampled finite signal. By applying the FT, of course in a discrete form which is handled by replacing the integral by a summation term over the period of existence of the signal, the discrete FT (DFT) is expressed as follows:

$$X[k] = \sum_{n=0}^{N-1} x[n]e^{-2\pi jnk/N} \tag{7.10}$$

where $k = 0, 1, \ldots, N-1$. In the same fashion, the inverse DFT is

$$x[k] = \frac{1}{N}\sum_{k=0}^{N-1} X[k]e^{2\pi jnk/N} \tag{7.11}$$

where $n = 0, \ldots, N-1$.

Equation (7.11) is the general form of a DFT used for calculating the FT of a digital signal. Several implementations have been already proposed, and the FFT which is the most efficient one will be briefly discussed here. The underlying idea of the FFT is to rewrite (7.10) as follows:

$$X[k] = \sum_{n=0}^{N-1} x[n]e^{-2\pi jnk/N} = \sum_{n=0}^{N-1} x[n]W_N^{kn} \tag{7.12}$$

$W_N^{kn}$ is calculated many times as the Fourier analysis is performed. On the other hand, the term $W_N^{kn}$ is symmetric. Therefore, (7.12) is decomposed into:

$$\begin{aligned} X[k] &= \sum_{n=0}^{N-1} x[n]W_N^{kn} = \sum_{n=0}^{N-1} x[n]W_N^{kn}|_{even\ n} + \sum_{n=0}^{N-1} x[n]W_N^{kn}|_{odd\ n} \\ &= \sum_{r=0}^{\frac{N}{2}-1} x[2r]W_N^{2kr} + \sum_{r=0}^{\frac{N}{2}-1} x[2r+1]W_N^{k(2r+1)} \\ &= \sum_{r=0}^{\frac{N}{2}-1} x_1[r]W_{\frac{N}{2}}^{kr} + W_N^k \sum_{r=0}^{\frac{N}{2}-1} x_2[r]W_{\frac{N}{2}}^{kr} = X_1(k) + W_N^k X_2(k) \end{aligned} \tag{7.13}$$

According to (7.13), the DFT of an *N*-point signal is obtained by calculating two *N*/2-point transforms applied to the even and odd input data points, separately. Two types of implementations in a programming environment, i.e., the recursive and iterative types, are available (Tables 7.1 and 7.2).

### *7.2.1 Recursive FFT algorithm in MATLAB*

The recursive FFT can be coded in MATLAB as follow:

*Table 7.1 Recursive FFT algorithm*

```
function y = fft_rec(x)
n = length(x);
if n == 1
     y = x;
else
     m = n/2;
     y_top = fft_rec(x(1:2:(n-1)));
     y_bottom = fft_rec(x(2:2:n));
     d = exp(-2 * pi * i / n) .^ (0:m-1);
     z = d .* y_bottom;
     y = [ y_top + z , y_top - z ];
end
```

### *7.2.2 Iterative FFT algorithm in MATLAB*

In the case of the proposed MATLAB-based algorithms, the number of points in the frequency domain will be equal to the number of points of the samples signal. However, the MATLAB built-in function is able to change the number of points in the frequency domain independently. So, the MATLAB FFT is used as the reference to the program explained below.

### *7.2.3 Example*

In this section, a couple of examples is provided to make the FFT process more clear for readers. Moreover, the major technical issues which must be considered while

*Table 7.2 Iterative FFT algorithm*

```
function y = fft_it(x)
n = length(x);
x = x(bitrevorder(1:n));
q = round(log(n)/log(2));
for j = 1:q
    m = 2^(j-1);
    d = exp(-2 *pi * i /m).^(0:m-1);
    for k = 1:2^(q-j)
        s = (k-1)*2*m+1; % start-index
        e = k*2*m; % end-index
        r = s + (e-s+1)/2; % middle-index
        y_top = x(s:(r-1));
        y_bottom = x(r:e);
        z = d .* y_bottom;
        y = [y_top + z, y_top - z];
        x(s:e) = y;
    end
end
```

using the FFT are discussed. To continue, a sinusoid with the fundamental frequency of 5 Hz and the amplitude of 13 is considered, and the FFT is applied to it during a 1-s sampling period (see Figure 7.2(a)). Then, the FFT is applied to the signal, and the effect of variation of two important parameters, namely the sampling frequency and the number of data points in the frequency domain is investigated for a given period of signals presence, which is equal to 1 s (Table 7.3).

Some significant aspects of the demonstrated variations are as follows:

- For a given nfft, increasing the sampling frequency does apply a considerable variation to the amplitude of the fundamental frequency component. The similar trend is observed for a given sampling frequency and variable nfft.
- Increasing the sampling frequency to very large values, eliminates the improper effect of numerical analysis and imperfection of the FFT, and the amplitude gets closer to the real value exponentially.
- Increasing the nfft has a similar effect as that of increasing the sampling frequency.
- The initial parts of Figure 7.2(b) and c reveal the fact that the FFT returns an approximate half of the amplitude in its output. There is a reason for this claim. In general, the output of the FFT should be normalized by nfft to get closer to the amplitude of the signal. Doing this divides the power between the positive and negative sides of the FFT which is a symmetric transform. So, in order to get the expected value, the result must be multiplied by two.[1] This problem is resolved by increasing the sampling frequency or the nfft up to very large values. However, this is not practically applicable due to the existing

[1]https://www.mathworks.com/matlabcentral/answers/162846-amplitude-of-signal-after-fft-operation (access date: October 2016).

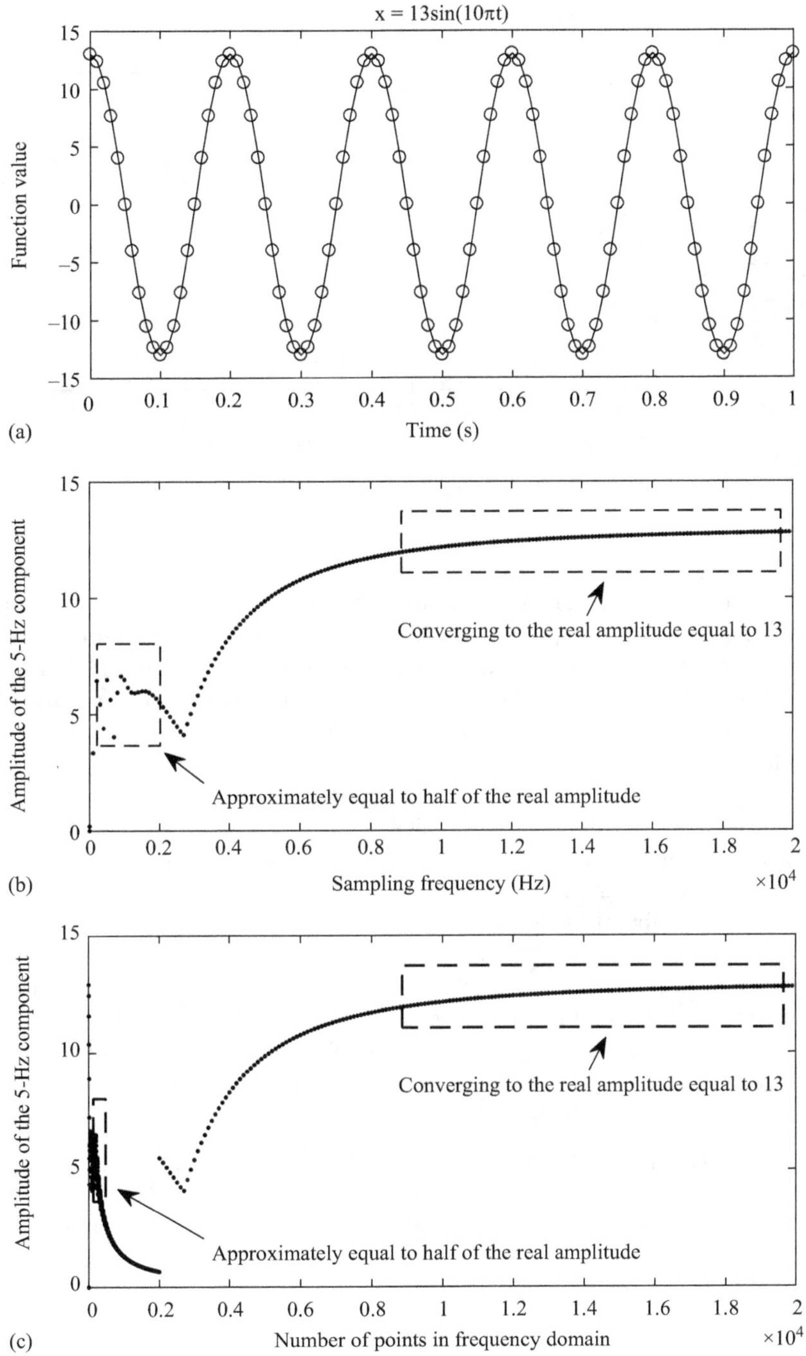


*Figure 7.2 (a) Input signal, amplitude of the fundamental component versus, (b) sampling frequency, (c) number of points in frequency domain and (d) fixed ration of the number of points to the sampling frequency*

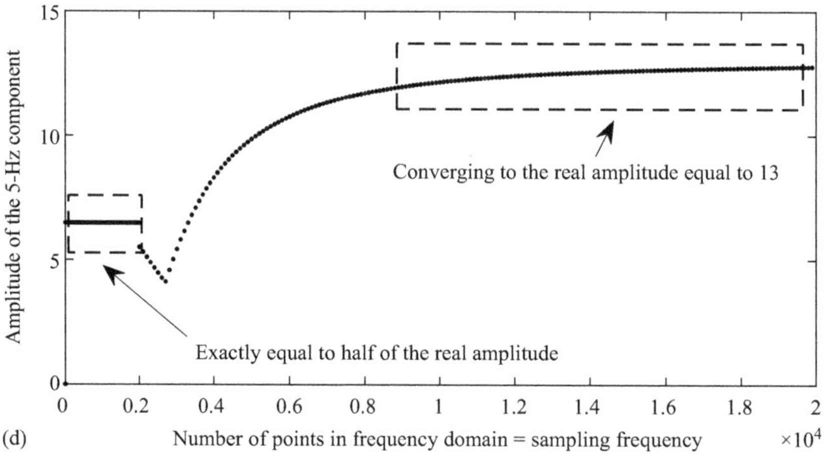


(d)

*Figure 7.2 (Continued)*

shortcoming in hardware and computational power. To deal with this problem, both the sampling frequency and the nfft must be increased simultaneously with the same rate. Moreover, it is proposed to keep the nfft as a factor of sampling frequency (Figure 7.2(d)). As a result, even smaller values of the sampling frequency and the nfft generate an amplitude which is exactly equal to the half of the expected value.

- Increasing the nfft, of course, increases the spectral resolution, although it requires a larger number of samples in the time domain to guarantee the accuracy of the extracted amplitudes. So, there should be a trade-off between the expected resolution and the computational power.
- As a rule of thumb, the Nyquist's law necessitates the sampling frequency being at least equal to twice the minimum frequency we are targeting in the spectrum. In practice, the sampling frequency should be very larger than this measure. Otherwise, the quality of the spectrum will not be high.
- Although keeping the nfft equal to the sampling frequency leads to an accurate outcome, increasing the period of the presence of the signal requires increasing both the sampling frequency and nfft.

It is noteworthy that the FT is normally applied to a time-domain signal while some applications to a spatial domain have also been reported in the literature. The underlying idea of the FT is to assume that the signal has a periodic extension up to infinity so that it does not matter at which instant the investigated frequency is present. What really matters is the power of the targeted frequency component of the signal regardless of its position in the time frame. This indicates a critical disadvantage of the FT and consequently the FFT in detecting the time-domain variation of different transient frequency components which might not be present at all-time instances with the same amplitude. These kinds of signals have a

*Table 7.3 MATLAB FFT applied to* $x = 13\sin(10\pi t)$

```
%%
% fs: Fundamental frequency
% Fs: Sampling frequency
% t: time
% x: signal
% nfft: Number of data points in the frequency domain
% X: Fourier transform of x
% f: frequency
% amplitude: Amplitude of the fundamental frequency component of x
%%
clc
clear
close all
%%
fs = 5;
Fs = 100;
t = 0:1/Fs:1;
x = 13*cos(2*pi*t*fs);
plot(t,x,'-o','Color',[0,0,0])
xlabel('Time(s)')
ylabel('Function value')
%% Variable Fs
fs = 5;
nfft = 200;
for i = 2:100:20000
Fs = i;
t = 0:1/Fs:1;
x = 13*cos(2*pi*t*fs);
X = fft(x,nfft);
X = X(1:nfft/2);
mx = abs(X);
f = (0:nfft/2-1)*Fs/nfft;
amplitude(i) = max(mx/nfft);
samplingFrequency(i) = i;
clear t Fs x
end
amplitude = amplitude';
figure
plot(samplingFrequency,amplitude,'.','Color',[0,0,0])
xlabel('Sampling frequency(Hz)')
ylabel('Amplitude of the 5Hz component')

%% Variable nfft
fs = 5;
Fs = 200;
for i = 2:2:2000
nfft = i;
t = 0:1/Fs:1;
x = 13*cos(2*pi*t*fs);
```

*(Continues)*

*Table 7.3 (Continued)*

```
X = fft(x,nfft);
X = X(1:nfft/2);
mx = abs(X);
f = (0:nfft/2-1)*Fs/nfft;
amplitude(i) = max(mx/nfft);
samplingFrequency(i) = i;
clear t nfft x
end
amplitude = amplitude';
figure
plot(samplingFrequency,amplitude,'.','Color',[0,0,0])
xlabel('Number of points in frequency domain')
ylabel('Amplitude of the 5Hz component')

%% Variable nfft and Fs
fs = 5;
for i = 2:2:2000
nfft = i;
Fs = i;
t = 0:1/Fs:1;
x = 13*cos(2*pi*t*fs);
X = fft(x,nfft);
X = X(1:nfft/2);
mx = abs(X);
f = (0:nfft/2-1)*Fs/nfft;
amplitude(i) = max(mx/nfft);
samplingFrequency(i) = i;
clear t nfft Fs x
end
amplitude = amplitude';
figure
plot(samplingFrequency,amplitude,'.','Color',[0,0,0])
xlabel('Number of points in frequency domain=Sampling frequency')
ylabel('Amplitude of the 5Hz component')
```

nonstationary nature requiring a more sophisticated frequency analysis technique to deal with. Considering the points, we can conclude that the FT only answers to the equation that "does the frequency $f$ exist in a signal?" It does not provide any information in the period that the expected frequency exists. So, focusing on the nature of the signal, either transient or stationary, is an essential task. In order to become more familiar with the concept of the FT and its shortcoming in defining the exact location of frequencies in a signal, consider to the following example.

Assume that we have a combination of four sinusoids with the frequencies of 5, 10, 20 and 50 Hz present at all instants. The second signal is again a combination of the mentioned frequencies while the different frequencies are present at different instants (see Figure 7.3).

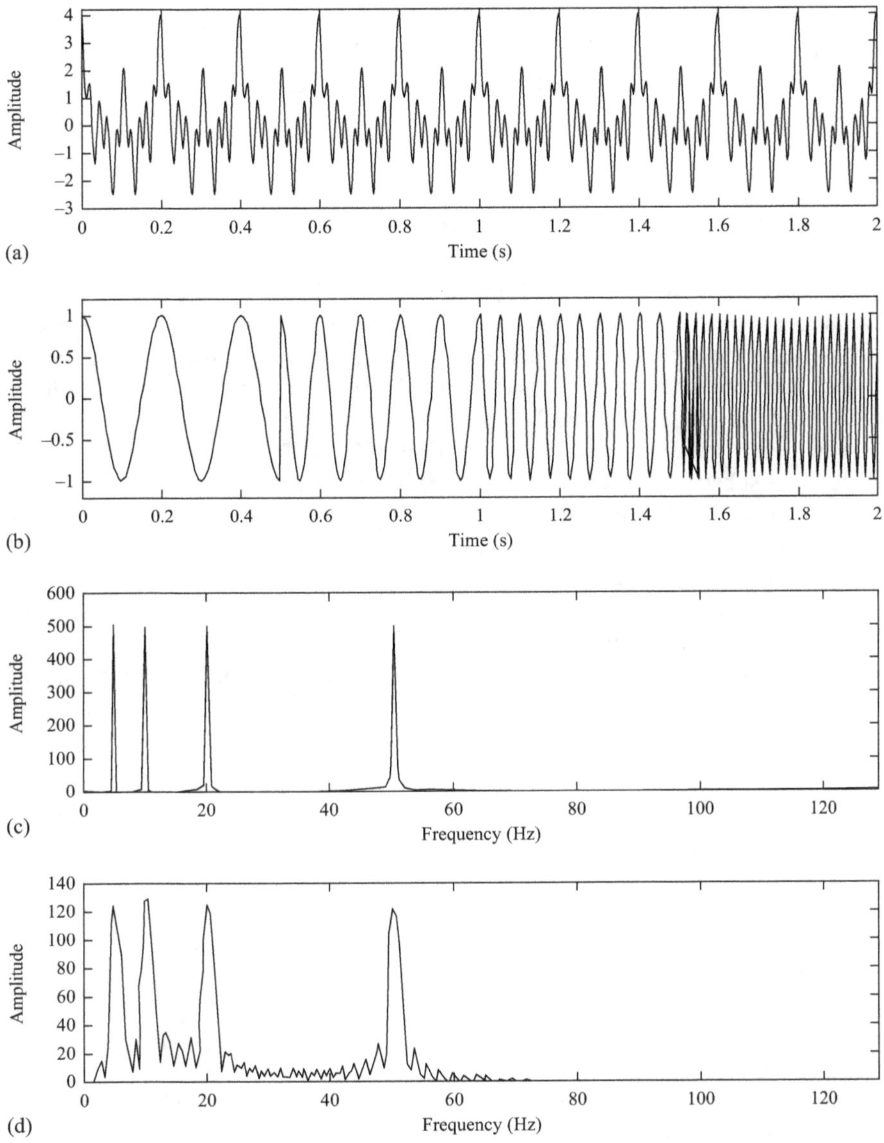


*Figure 7.3 (a) The first signal containing the frequencies 5, 10, 20 and 50 Hz at all instants, (b) the second signal containing the frequencies 5, 10, 20 and 50 Hz at specific instants, (c) FFT of the first signal and (d) FFT of the second signal*

As it can be seen in Figure 7.3, both FFTs have four peaks at 5, 10, 20 and 50 Hz. Nevertheless, the FFT of the second signal reveals some minor peaks around the mentioned ones. This somehow shows the difference between the nature of two signals while it is possible to actually differentiate them when only the spectrum is analyzed.

A time-domain demonstration of the signals is certainly needed for a clear understanding of the problem. Thus, it is concluded that the FT is not a good tool to analyze the transient signals and to realize how they are affected by various frequency components. The first effort to deal with this issue is to use the STFT whose MATLAB code is called STFT.

## 7.3 Short-time Fourier transform

In the previous section, it was observed that the FT is not capable of correctly revealing a nonstationary or transient signal behavior. The simplest way of tackling this problem is to take one small period of the signal and apply the FT to that part as a nonstationary signal is locally stationary during short periods of time unless a highly time-variant signal is to be analyzed. For example, Figure 7.3(b) shows a signal which is stationary within every period of 0.5 s. Therefore, the signal can be divided into substationary shorter signals by applying windows to it. It is crucial to choose a proper window length so that the subsignals are certainly stationary within the window length.

Considering the above-mentioned points, it is seen that there is no much difference between the FFT and STFT. The main difference is that in the STFT, the signal splits into several stationary subsignals. For this purpose, a window function called $w$ whose length in time is equal to the minimum length required for having the stationary signal assumption is used. Doing so, the STFT is expressed as follows:

$$STFT_x^W(\tau,f) = \int_{-\infty}^{+\infty} x(t)W^*(t-\tau)e^{-j2\pi ft}dt \tag{7.14}$$

where $x(t)$, $w$, $f$ and $\tau$ are the input signal, the window, the frequency and the temporal variable, respectively. In fact, the STFT is the same as the FFT of the $x(t)$ to which the window has been applied. The window function is multiplied by the input signal, and the corresponding FFT is calculated. Then, the window is shifted toward positive instants by $\tau$, and the FFT is again applied to the multiplication of the new part of $x(t)$ and the window. Therefore, for every pair of f and $\tau$, the STFT is calculated. The approach is demonstrated in Figure 7.4.

Focusing on the STFT formulation, it is obvious that the STFT is a kind of time-frequency transform as there are two independent outputs, i.e., the time and the frequency. Thus, the STFT can be shown in a 3D coordinate whose $x$ and $y$ axes are the time and the frequency, respectively, and the amplitude of the components are assigned to the $z$-axis.

It is worth noting that the resolution in frequency domain obtained by a simple FFT is usually high enough. In the similar fashion, the time-domain values of the input signal have been already known. Therefore, the time-domain resolution is also fine. Moreover, we should know that the term producing an acceptable frequency-domain resolution of the FFT is the exponential term $e^{-j2\pi ft}$ which is expanded from minus infinity to plus infinity. However, in an STFT, the window

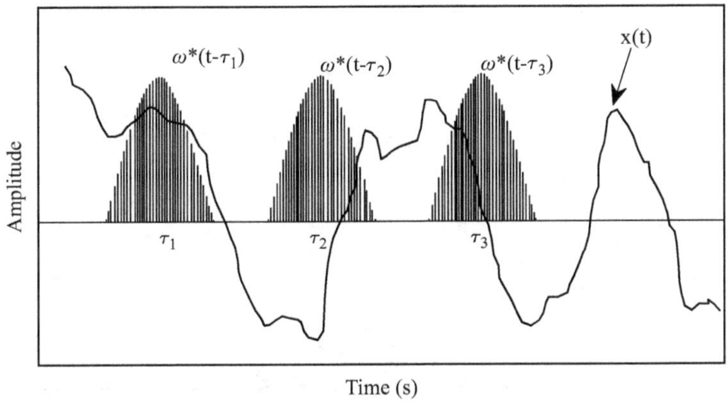


*Figure 7.4 Demonstration of windowing approach applied to a nonstationary signal*

length is finite which degrades the resolution. Hence, we might not end up with a clear understanding of the frequency components. This means that it is not usually possible to extract the exact values of the frequencies present in the signal. However, only one frequency range is available. This causes a lower quality of the time-frequency resolution of the STFT compared to the FFT. The larger the window length is, the more the STFT looks like to the FFT. In other words, increasing the window length increases the frequency resolution of the STFT output while a smaller window improves the time resolution of the STFT output. The window length is always fixed in an STFT; hence, there should be a trade-off between the time and frequency resolutions. Improving both is not possible at the same time.

In order to become more familiar with the STFT, two examples, i.e., the STFT of a chirp and a saw-tooth signals are provided (Figure 7.5).

The corresponding MATLAB code which utilizes the function "spectrogram" to calculate the STFT is explained in Table 7.4.

## 7.4 Multiresolution analysis

The problem of a fixed resolution of the STFT can be justified by the Heisenberg uncertainty principle. According to this principle, it is not possible to have an exact time-frequency demonstration of a signal. However, the only understandable thing is the approximate frequency band of the signal. This directly affects the frequency resolution. Although the issues associated with the time and frequency resolutions are caused by a physical phenomenon and are independent of the type of transform used, an alternative analysis which is called (multiresolution analysis) might be utilized. This concept helps develop a new type of transform called the WT.

Multiresolution analysis means that the signal is analyzed using different levels of resolution implemented by applying various window lengths. Therefore,

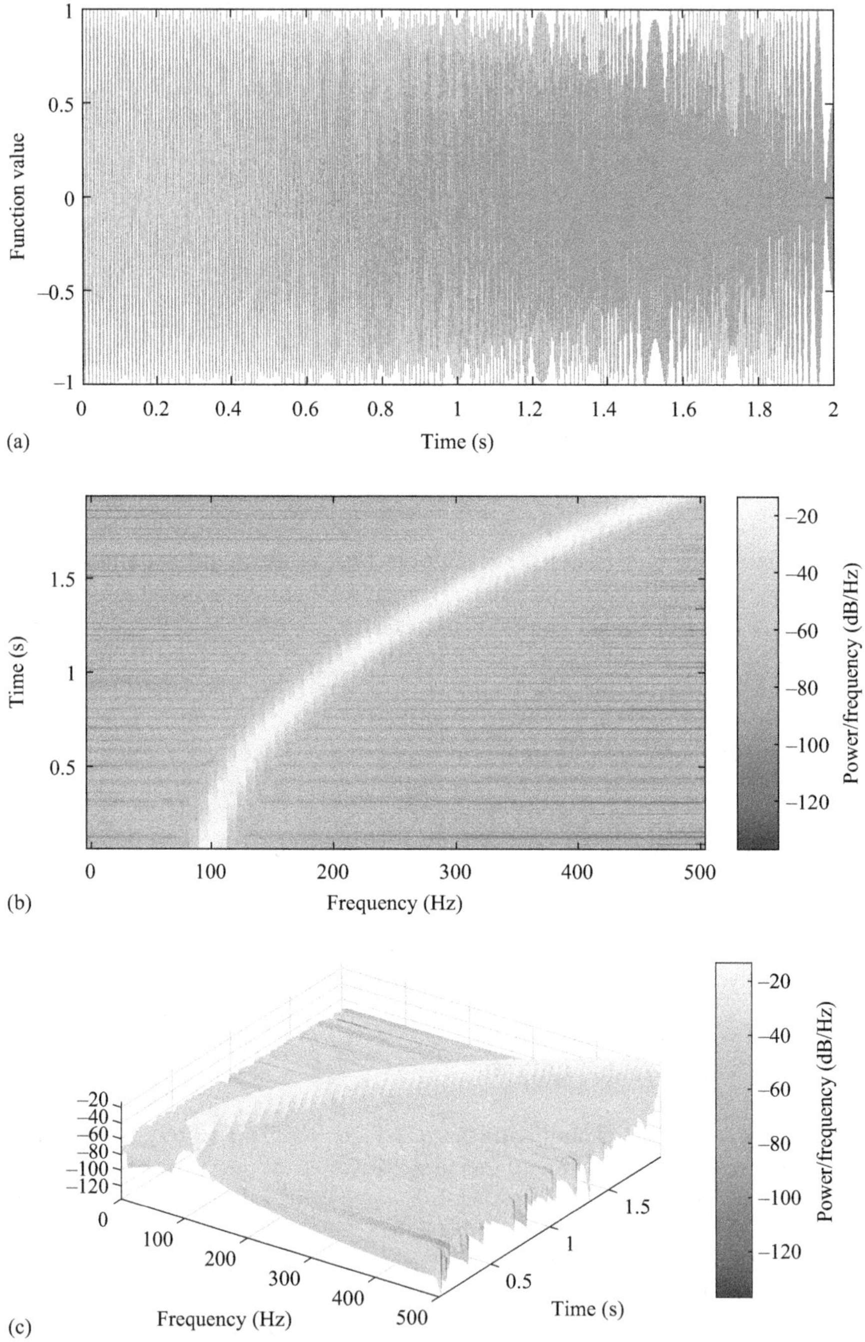


*Figure 7.5 (a) Time-domain variation of a chirp signal, (b) 2D STFT of chirp signal, (c) 3D STFT of chirp signal, (d) time-domain variation of a sawtooth signal (e), 2D STFT of sawtooth signal and (f) 3D STFT f sawtooth signal*

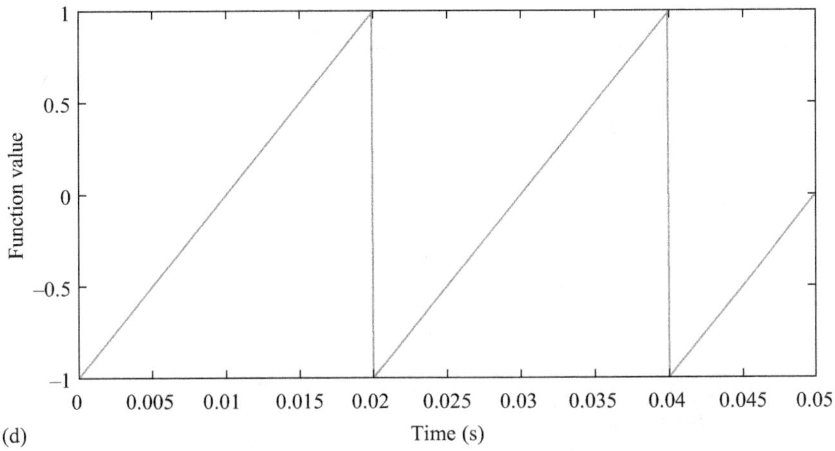


(d)

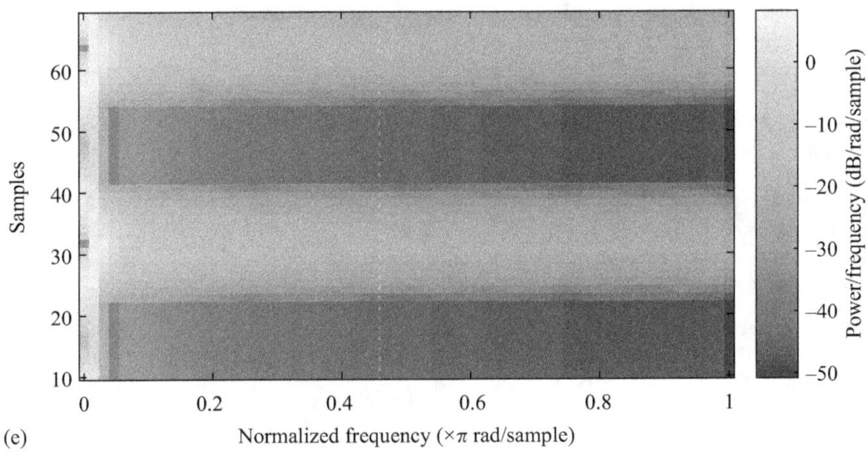


(e)

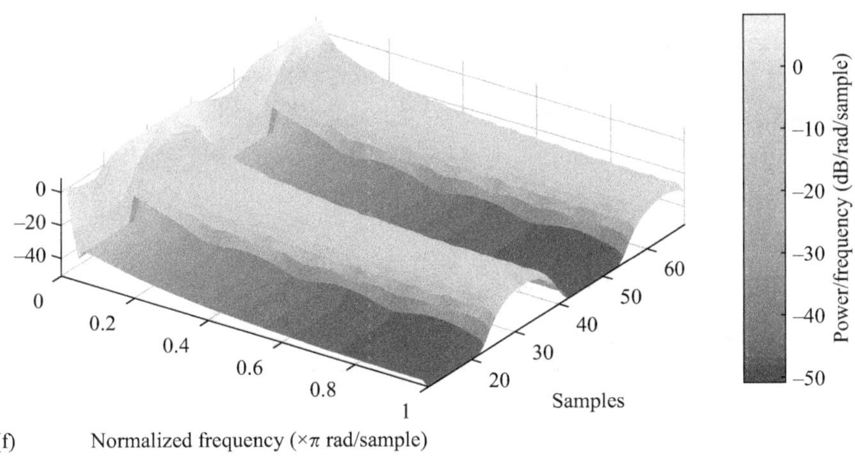


(f)

*Figure 7.5 (Continued)*

*Table 7.4 STFT code*[2]

```
%% Chirp signal
t = 0:0.001:2;
x = chirp(t,100,1,200,'quadratic');
plot(t,x)
xlabel('Time(s)')
ylabel('Function value')
fig = gcf;
set(0,'DefaultFigureWindowStyle','normal')
set(fig,'Position', [500 300 600 300]);
box on
ax = gca;
ax.LineWidth = 2;
ax.FontSize = 10;
ax.FontAngle = 'normal';
ax.FontWeight = 'Bold';
figure
spectrogram(x,128,120,128,1e3)
fig = gcf;
set(0,'DefaultFigureWindowStyle','normal')
set(fig,'Position', [500 300 600 300]);
box on
ax = gca;
ax.LineWidth = 2;
ax.FontSize = 10;
ax.FontAngle = 'normal';
ax.FontWeight = 'Bold';

%% Saw-tooth signal
t = 0:.0001:0.05;
x = sawtooth(2*pi*50*t);
plot(t,x)
xlabel('Time(s)')
ylabel('Function value')
fig = gcf;
set(0,'DefaultFigureWindowStyle','normal')
set(fig,'Position', [500 300 600 300]);
box on
ax = gca;
ax.LineWidth = 2;
ax.FontSize = 10;
ax.FontAngle = 'normal';
ax.FontWeight = 'Bold';
figure
spectrogram(x,128,120,128)
fig = gcf;
set(0,'DefaultFigureWindowStyle','normal')
set(fig,'Position', [500 300 600 300]);
box on
ax = gca;
ax.LineWidth = 2;
ax.FontSize = 10;
ax.FontAngle = 'normal';
ax.FontWeight = 'Bold';
```

[2]http://www.mathworks.com/help/signal/ref/spectrogram.html (access date: October 2016).

unlike the STFT, different frequency components will be treated differently. The goal is to have a good time-resolution and rather proper frequency-resolution assigned to higher frequencies and a rather good time-resolution and good frequency-resolution assigned to lower frequencies.

## 7.5 Wavelet transform

The continuous WT (CWT) is an alternative to the STFT and its goal is to overcome the shortcomings associated with the STFT. In fact, the poor quality of the STFT resolution is dealt with by means of the concept of the CWT. In the wavelet analysis, the signal is multiplied by a function (wavelet) which plays the role of a window discussed in the previous section. Likewise, the wavelet is applied to different subsignals which are obtained by splitting the main signal into many. However, there are two major differences between the STFT and the CWT as follows:

- In the CWT, the FT is not applied to the wavelet; hence, the individual peaks corresponding to one sinusoid are not calculated.
- In the CWT, the variation of the window length is the same as that of the frequency components. This is the main feature of the WT.

On this basis, the WT is formulated as follows:

$$CWT_x^{\psi}(\tau, s) = \Psi_x^{\psi}(\tau, s) = \frac{1}{\sqrt{|s|}} \int_{-\infty}^{+\infty} x(t)\psi^*\left(\frac{t-\tau}{s}\right)dt \tag{7.15}$$

where $\tau$ and $s$ are the translation and scaling parameters, respectively. The translation concept is the same as the time-domain shifting process in the STFT and defines the amount of displacement of the window. Obviously, the translation consists of the signal information in the time-domain. However, unlike the STFT, there is no frequency parameter in the CWT. Instead, there is a new parameter called $s$ which is inverse proportional to the frequency. In other words, $s = 1/f \cdot \psi$ is the window function and called the mother wavelet. The reason for using the term "wavelet" in this transform is the short length of the window function. The mother wavelet is a prototype used for producing other windows.

Scaling, as a mathematical operator, shrinks or expands the signal. Thus, for larger values of scaling parameter, the details are more visible. It should be noted that the parameter $s$ is the denominator of the CWT which leads to the signal expansion if $s > 1$ and vice versa.

### *7.5.1 Resolution in time-frequency map*

This section provides a more comprehensive insight into the time-frequency resolution of a CWT. Figure 7.6 demonstrates the concept of resolution in time, frequency and time-frequency maps. In a time-frequency map, every rectangular map has a nonzero area, conveying the idea that an accurate time or frequency resolution

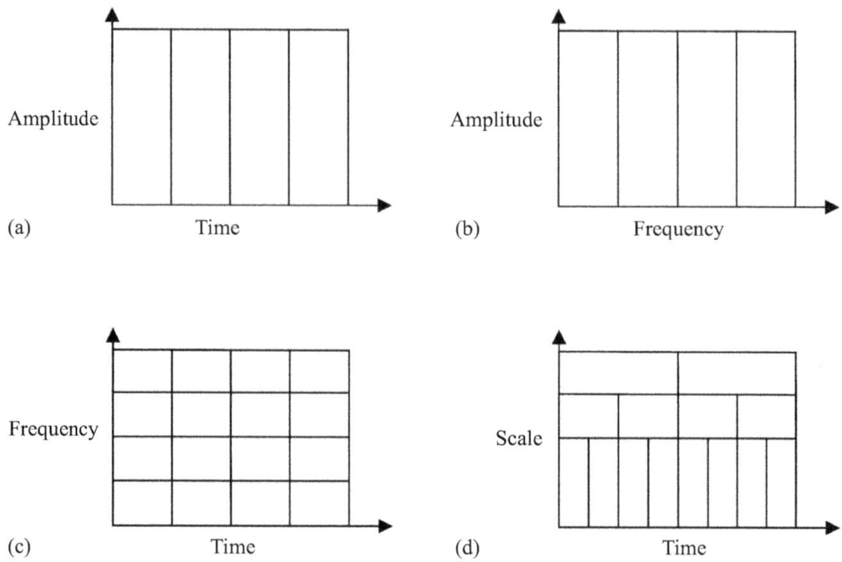


*Figure 7.6 Resolution in (a) time, (b) frequency, (c) time-frequency for STFT and (d) time-frequency for WT*

is not available. Reducing the width of the boxes to a value close to zero increases the resolution while in an STFT, it is somehow impossible to have a high resolution in time and frequency at the same time. All the points located inside one box are dealt with the same wavelet or window.

According to Figure 7.6, as the size of the boxes are fixed and the same in the STFT, the whole time-frequency map has a uniform resolution. However, in the WT, the width and height of the boxes change, while the area is fixed. Therefore, every box represents a fixed area of the time-frequency map. In lower frequencies, the height of the boxes is smaller than that of the region corresponding to higher frequencies. As a result, the time resolution gets worse. It is noteworthy that the area of the boxes depends on the type of mother wavelet. It can be mathematically proved that the minimum value of the area should be equal to $\pi/4$ regardless of the type of mother wavelet. In this section, the main idea of the WT is provided. However, some mathematical developments required for preparing the bases are discussed first.

Assume that we have a space vector $v$ containing linearly independent vectors, i.e., $v$. The vectors $v$ are such that they can be formulated as a linear combination of the other vectors in the space $V$. So the demonstration of each vector, i.e., $v$, is as the following:

$$v = \sum_{k=1}^{N} a_k b_k \tag{7.16}$$

where $b_k$ is the vector in the space $V$, $a_k$ is the coefficient of the linear combination and $N$ is the dimension of the space. Similarly, the concept can be extended to a function version by replacing the vectors $b_k$ by some function forms. As a result, $f(t)$ is expressed as follows:

$$f(t) = \sum_{k=1}^{N} a_k \phi_k(t) \tag{7.17}$$

As it was noticed before, complex exponential functions, are the bases for computing the FT. In addition, these functions are orthogonal; hence, it is possible rebuilt the original time-domain signal by means of the components in the frequency domain. If we assume that $f(t)$ and $g(t)$ are two orthogonal functions, the dot product of these two functions is equal to

$$< f(t), g(t) >= \int f(t) g^*(t) dt \tag{7.18}$$

Accordingly, the CWT can be reformulated as a dot product of the signal $x(t)$ and a function:

$$CWT_x^{\psi}(\tau, s) = \Psi_x^{\psi}(\tau, s) = \frac{1}{\sqrt{|s|}} \int_{-\infty}^{+\infty} x(t) \psi^* \left( \frac{t - \tau}{s} \right) dt = \langle x(t), \psi_{\tau,s}(t) \rangle \tag{7.19}$$

in which

$$\psi_{\tau,s}(t) = \frac{1}{\sqrt{|s|}} \psi \left( \frac{t - \tau}{s} \right) \tag{7.20}$$

Using the definition provided above, the WT is, in fact, like measuring the similarity of the signal $x(t)$ and mother wavelet. By the similarity, we mean the common frequency components. In other words, the coefficients of the WT are an indication of how close the signal $x(t)$ is to the wavelets. Therefore, if $x(t)$ includes a frequency component similar to one of the wavelets used for calculating the WT, the corresponding wavelet will be the main factor in extracting the expected frequency, so the coefficient of that wavelet will be larger. The $a_k$ corresponding to the orthogonal functions are:

$$a_k = \langle f(t), \phi_k(t) \rangle = \int f(t) \phi_k^*(t) dt \tag{7.21}$$

Having calculated the coefficients, the function $f(t)$ can be rebuilt as follows:

$$f(t) = \sum_{k=1}^{N} a_k \phi_k(t) = \sum_{k=1}^{N} \langle f(t), \phi_k(t) \rangle \phi_k(t) \tag{7.22}$$

The following are two types of very well-known mother wavelets

Mexican Hat wavelet

$$\psi(t) = \frac{d^2}{dt^2}\left(\frac{1}{\sqrt{2\pi}\sigma}e^{\frac{-t^2}{2\sigma^2}}\right) = \frac{1}{\sqrt{2\pi}\sigma^3}\left(e^{\frac{-t^2}{2\sigma^2}}\left(\frac{-t^2}{\sigma^2} - 1\right)\right) \tag{7.23}$$

And

Morlet wavelet

$$\psi(t) = e^{\frac{jt}{a}} \cdot e^{\frac{-t^2}{2\sigma^2}} \tag{7.24}$$

## 7.6 Inverse wavelet transform

In this section, the inverse WT is briefly explained, and the requirements of an invertible WT are discussed mathematically. The WT is invertible if:

$$\int \psi(t)dt = 0 \tag{7.25}$$

In order to satisfy (7.25), the mother wavelet must have an oscillating nature. Most of the time, this rule is easily satisfied using ordinary mother wavelets regardless of the wavelets being orthogonal or not. In this case, the inverse WT is as the following:

$$x(t) = \frac{1}{c_\psi^2}\iint \Psi_x^\psi(\tau, s)\frac{1}{s^2}\psi\left(\frac{t-\tau}{s}\right)d\tau ds \tag{7.26}$$

where $c_\psi$ is a constant and depends on the wavelet. The process of rebuilding the signal highly depends on this term. Generally, this constant is called "Admissibility constant" and is equal to

$$c_\psi = \left(2\pi\int_{-\infty}^{+\infty}\frac{|\hat{\psi}(\xi)|^2}{|\xi|}d\xi\right)^{\frac{1}{2}} < \infty \tag{7.27}$$

in which, $\hat{\psi}$ is the FT of the mother wavelet.

## 7.7 Discrete of wavelet transform

All the materials discussed so far in terms of the WT contribute to the continuous form of the transform, and it is possible to apply them to a discrete signal which is the one available in a computer. Therefore, it is necessary to implement a discrete version of the WT. The concept is like the DFT; hence, the methodology is to sample time-frequency map with the purpose of obtaining data distributed on the map discretely. Equally important, a uniform sampling process is the simplest and the most straightforward way of dealing with this issue. Of course, in a WT, we can have trade-off between the sampling rate and the scaling factor. In higher frequencies, the sampling rate can be reduced regarding the Nyquist's law. Therefore, assuming that the sampling rate is equal to $N_1$ for the scaling factor $s_1$, the sampling rate $N_2$ related to a scaling factor $s_2 > s_1$ should be smaller than $N_1$.

$$N_2 = \frac{s_1}{s_2} N_1 = \frac{f_1}{f_2} N_1 \tag{7.28}$$

Moreover, the Nyquist's law can even be violated if rebuilding the original signal is not the aim. For building a discrete WT, the parameter $s$ should be discretized using a logarithmic grading system. Then, the parameter of time is discretized regarding the parameter $s$ such that a different sampling rate is chosen for different scaling factors. Figure 7.7 shows the discretization method applied to a WT.

Accordingly, the mother wavelet is discretized as follows:

$$\psi_{j,k}(t) = s_0^{-j/2} \psi\left(s_0^{-j} t - k\tau_0\right) \tag{7.29}$$

Using (7.29), the discrete form of the WT becomes equal to

$$\Psi_x^{\psi_{j,k}} = \int x(t) \psi_{j,k}^*(t) dt \tag{7.30}$$

Likewise, the inverse discrete WT is expressed as follows:

$$x(t) = c_\psi \sum_j \sum_k \Psi_x^{\psi_{j,k}} \psi_{j,k}(t) \tag{7.31}$$

Although the mentioned discretized version of the WT can be handled by computers and digital processors, it is not basically a discrete transform. In fact, the discrete

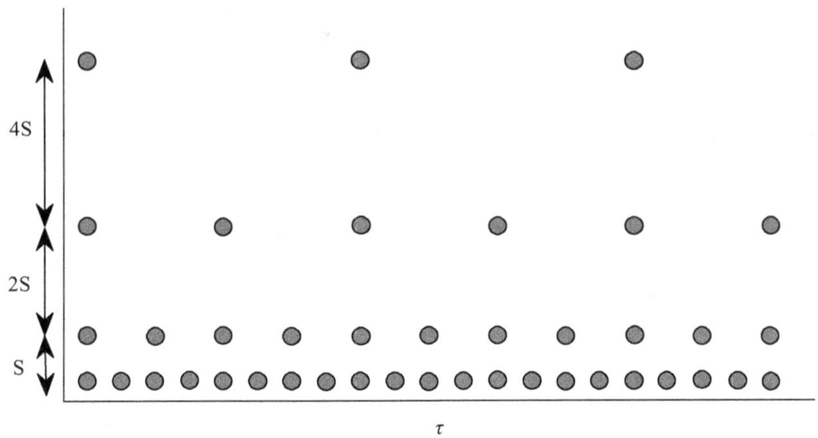


*Figure 7.7 Locations of the wavelets in a Dyadic grading process*

version is the sum of fundamental terms inspired by the CWT. This usually introduces extra computational burdens and should be modified in a more efficient way. An alternative which is based on the "subband coding" is proposed. The underlying idea somehow looks like the continuous transform in which a specific type of time-frequency representation is implemented using digital filters. It should be remembered that the WT is the output of the convolution of frequency contents of the input signal and wavelets. In the discrete version, filters with different cut-off frequencies are applied to different scales. Then, by passing the signal through low- and high-pass filters, different frequency contents of the input signal are analyzed. The resolution is controlled by filters operations, and the scaling factor is controlled using the down-sampling or up-sampling process. Usually, the change of variation rate of samples is handled by a Dyadic network, using $s_0 = 2$ and $\tau_0 = 1$.

The process is initiated in the following trend. First, the input signal passes through a low-pass digital filter whose impulse response is equal to $h[n]$. Therefore, the output of the filter is equal to the convolution of the input and the impulse response of the filter. As a result, all frequency components of the signal which are larger than half of the maximum frequency present in the signal are removed. As the maximum available frequency in the output of the filter is equal to $\frac{\pi}{2}$ rad, half of the samples are removable. Thus, a proper output is achieved by removing every two samples without losing a lot of information. A similar trend is applied to a high-pass filter with the impulse response of $g[n]$. Therefore, two versions of the filtered signal are available as follows:

$$y_{high}[k] = \sum_{n} x[n] \cdot g[2k - n] \tag{7.32}$$

$$y_{low}[k] = \sum_{n} x[n] \cdot h[2k - n] \tag{7.33}$$

Doing so, the time-resolution becomes half while the frequency resolution is doubled. The process can be repeated using the low-pass part of the signal by applying the low- and high-pass filters to it. In each step, the time-frequency decreases while the frequency resolution increases. Figure 7.8 shows a three-level implementation of the mentioned process. The output coefficients of the low-pass filters follow the initial signal curve. So the corresponding coefficients are called "Approximation." The coefficients of the high-pass filters demonstrate the high-frequency details, so they are called "Details."

The WT and its different implementations are available in MATLAB. So we refer readers to the signal-processing toolbox of MATLAB for further information. The following are some of the examples implemented in MATLAB.

Table 7.5 and Figure 7.9 show an example of the WT applied to a signal with and without noise. Actually, Figure 7.9(a) and (b) illustrates a noise-free signal along with the corresponding WT. In addition, Figure 7.9(a) and (b) shows the same signal with a powerful noise level applied to it. The "rand" function is used to have specific level of noise. The noise is present at all instants and it normally affects higher frequency levels. So the WT of the noisy signal clearly proves this

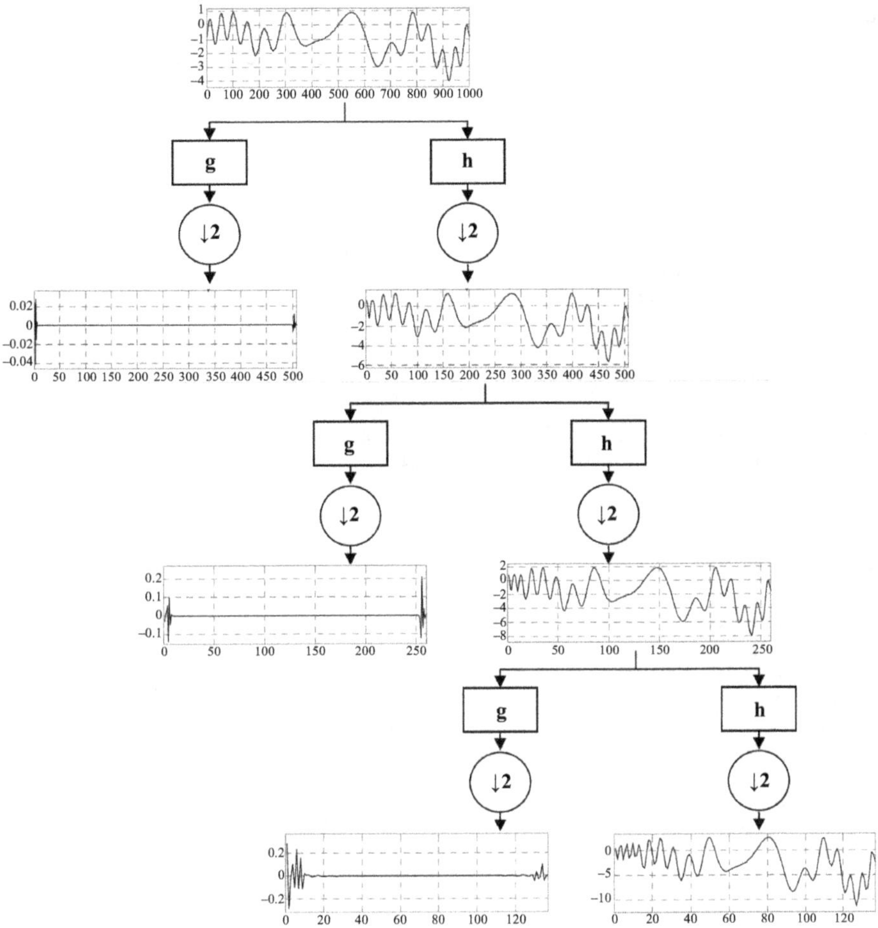


*Figure 7.8 Decomposition of signal using wavelet*

claim as the amplitudes of the wavelet coefficients are close to zero for the lower-frequency band. Figure 7.9(b) shows a very good correlation between the WT and the general behavior of the time-domain signal. The first pure sinusoidal part of the signal which has a frequency of 32 Hz is of a moderate amplitude around 1 shown by the corresponding yellow color in Figure 7.9(b) between 0.1 and 0.3 s. The largest amplitude is related to the frequency of approximately 20 Hz (Figure 7.9(b)). Note that the word "approximately" should be used as if the original time-domain signal is not available, we have only a vague approximate of the location of the expected frequency component in the time-frequency map. The WT is not able to have high-frequency resolution like what is obtainable using FFT. However, the frequencies can be detected clearly if a noise-free signal is in hand. Otherwise, depending on the noise level, frequency components might or might not

*Table 7.5 An example of wavelet transform*

```
clc
clear
close all
%%
Fs = 1e3;
t = 0:1/Fs:2;
x = cos(2*pi*32*t).*(t>=0.1 & t<0.3) + exp(-t.^4).*sin(2*pi*64*t).
      *(t>=0.7 & t<1)+...
   4*exp(-t).*sin(2*pi*20*t).*(t>=0.9);
pplot(t,x,'color',[0 0 0])
xlabel('Time(s)')
ylabel('Function value')
fig = gcf;
set(0,'DefaultFigureWindowStyle','normal')
set(fig,'Position', [500 300 600 300]);
box on
ax = gca;
ax.LineWidth = 2;
ax.FontSize = 10;
ax.FontAngle = 'normal';
ax.FontWeight = 'Bold';
figure
cwt(x,Fs)
fig = gcf;
set(0,'DefaultFigureWindowStyle','normal')
set(fig,'Position', [500 300 600 300]);
box on
ax = gca;
ax.LineWidth = 2;
ax.FontSize = 10;
ax.FontAngle = 'normal';
ax.FontWeight = 'Bold';

%%
figure;
Fs = 1e3;
t = 0:1/Fs:2;
x = cos(2*pi*32*t).*(t>=0.1 & t<0.3) + exp(-t.^4).*sin(2*pi*64*t).
      *(t>=0.7 & t<1)+...
   4*exp(-t).*sin(2*pi*20*t).*(t>=0.9);
wgnNoise = .4*randn(size(t));
x = x + wgnNoise;
plot(t,x,'color',[0 0 0])
xlabel('Time(s)')
ylabel('Function value')
fig = gcf;
set(0,'DefaultFigureWindowStyle','normal')
set(fig,'Position', [500 300 600 300]);
box on
ax = gca;
ax.LineWidth = 2;
```

*(Continues)*

*Table 7.5* (*Continued*)

```
ax.FontSize = 10;
ax.FontAngle = 'normal';
ax.FontWeight = 'Bold';
figure
cwt(x,Fs)
fig = gcf;
set(0,'DefaultFigureWindowStyle','normal')
set(fig,'Position', [500 300 600 300]);
box on
ax = gca;
ax.LineWidth = 2;
ax.FontSize = 10;
ax.FontAngle = 'normal';
ax.FontWeight = 'Bold';
```

be masked. If they are masked by any kind of noise (Figure 7.9(a)), it is likely that at least some frequencies are not detectable (Figure 7.9(d), the second term of the signal oscillating at the frequency of 64 Hz). Nevertheless, the applied noise is very powerful.

The STFT, together with WT, provides highly promising tools to analyze a signal in the time-frequency domain. They each have their own advantages and disadvantages in terms of the complexity of the implementation as well as the quality of their outputs. However, they are both used to return a time-frequency map in which the time-domain variations of different frequency components of a signal are investigable. Although the frequency resolution is not as high as that of the FT, we will end up with an acceptable discrimination between the frequency components if a well-tuned time-frequency process is used. In general, there is no way of having a 100% FT-like time-frequency transform which has the same resolution as that of the FT. Even the other type of the time-frequency transforms such as the HHT does not reveal a precisely detectable frequency components. Nevertheless, there is still a lot of fields of study, especially those related to the monitoring of things, which utilizes the existing approaches with no trouble as they are somehow the only techniques available so far. We brought up the name of a new time-frequency transform which has proved to be very useful in medical application. This transform is a combination of a conventional mathematical map called Hilbert transform and a recently developed empirical mode decomposition (EMD) approach. It is called HHT whose basis is to decompose a signal into various energy levels including specific frequency band/range. It is noteworthy that the mentioned transform, or processor, has widely been used in the diagnosis of electrical machines since the past decade. Therefore, we think it should be investigated in this book as well as some of the significant works dealing with fault diagnosis of inverter-fed machines rely on this type of transform. Therefore, it is

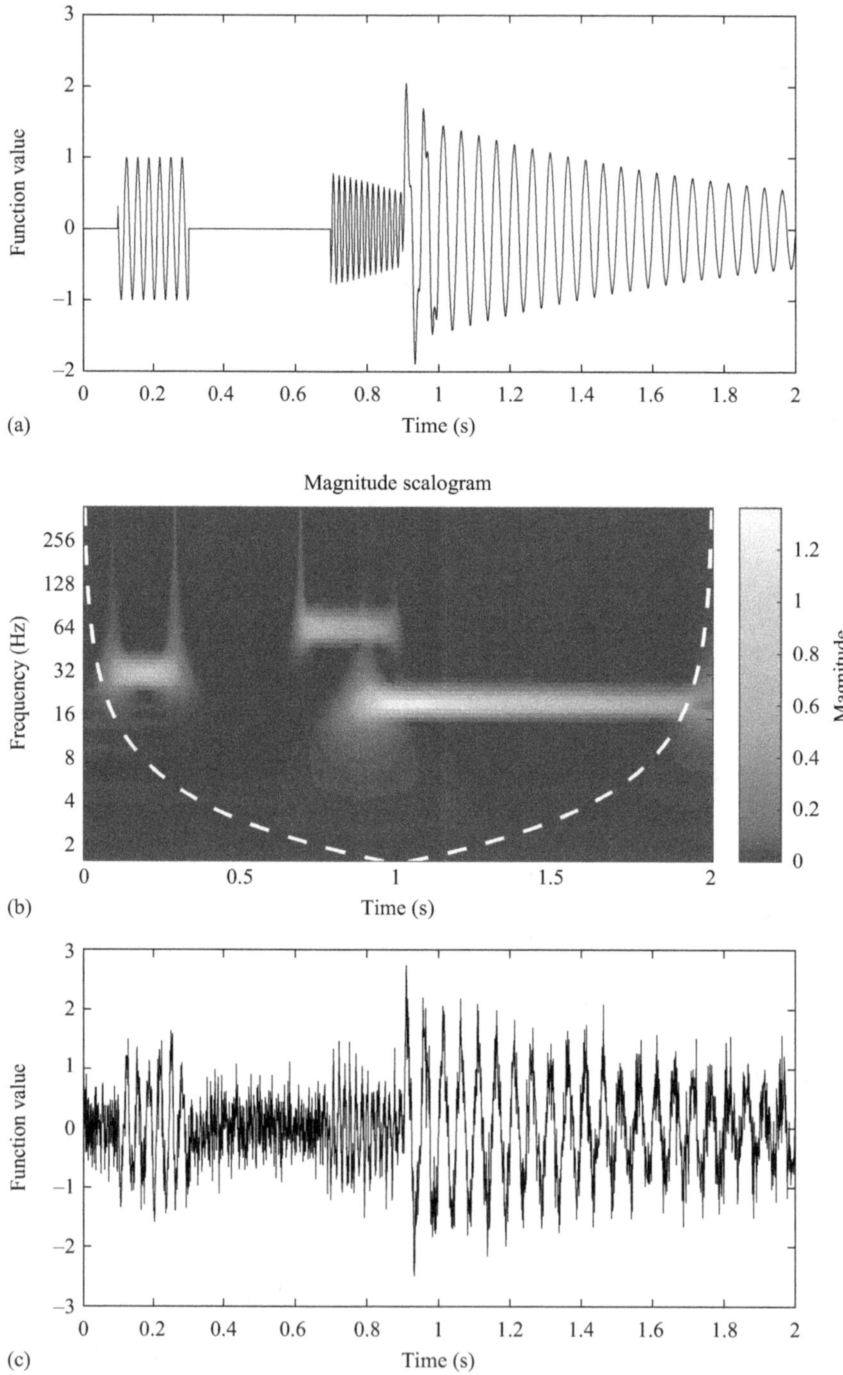


*Figure 7.9* *(a) Noise-free signal, (b) WT of noise-free signal, (c) noisy signal and (d) WT of noisy signal*

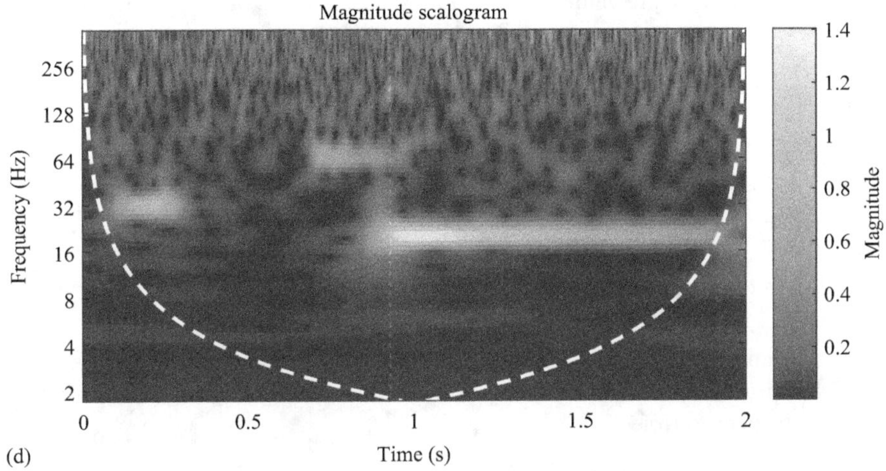


(d)

*Figure 7.9 (Continued)*

highly proposed to check out the following section. The underlying concept is easy to understand while comprehending the outputs is not as easy as that of the previously discussed techniques. Thus, we try to cover the corresponding technical issue as much as possible.

## 7.8 Hilbert–Huang transform

Real signals, either those which are man-made or natural signals, usually represent a nonstationary behavior at least for a specific period. Different motor faults are not exceptions, and occurrence of any type of fault produces an incipient variation of the motor signals, and then, although they might have a stationary or, in other words, periodic steady-state operation, they are not the same as that of healthy motors. Therefore, there is always a transition from healthy to faulty signals. This kind of variation is known as "harmonic distortion" which might have either stationary or nonstationary trend. In the case of fault diagnosis procedure, this distortion has certainly a physical and mathematical meaning. Furthermore, the processed signals in time, frequency or even time-frequency domains obtained by the previously mentioned transforms have a mathematical meaning. The physical meaning of any distortion of a physical signal is hidden in the instantaneous frequencies of the signal. "Instantaneous frequency" is defined as the time-domain variation of frequency components of a signal. The HHT directly deals with the instantaneous frequencies of a signal revealing the variation of specific frequency bands during specific period. The basis of the HHT is to use the Hilbert transform, which is discussed below, to extract the instantaneous frequencies first. This is handled by calculating the complex conjugate of the signal $x(t)$, called $y(t)$.

$$\mathrm{H}[x(t)] = y(t) = \frac{1}{\pi} PV \int_{-\infty}^{+\infty} \frac{x(\tau)}{t-\tau} d\tau \tag{7.34}$$

where PV is the principle value of a singular integral. Then, the signal $z(t)$ is formed as follows:

$$z(t) = x(t) + jy(t) = a(t)e^{j\theta(t)} \tag{7.35}$$

in which

$$a(t) = \sqrt{x^2 + y^2} \tag{7.36}$$

$$\theta(t) = arctan\left(\frac{y}{x}\right) \tag{7.37}$$

where $a(t)$ and $\theta(t)$ are the instantaneous amplitude and phase functions, respectively. Simply, the instantaneous frequency is calculated by taking the derivate of $\theta(t)$ with respect to time.

$$\omega = \frac{d\theta}{dt} \tag{7.38}$$

Focusing on the Hilbert transform, we will find out that (7.34) is nothing but the convolution of $x(t)$ with the function "$1/t$." Although Hilbert transform is useful on its own, it is not capable of extracting a proper frequency information. Therefore, it is normally combined with the EMD with the purpose of increasing the frequency resolution. This really matters when a nonstationary signal is targeted. Depending on the nature of the input signal and also some of the major adjustments, the output might vary widely. Moreover, the type of decomposition used in this method relies on some specific assumptions as follows:

- The input signal includes intrinsic mode functions (IMF) oscillating at specific frequency.
- Each IMF has the same number of extrema and zero-crossings.
- The oscillation should be symmetric around the mean value of the IMF.

The mentioned oscillatory modes, i.e., IMFs, follow the following rules:

- The number of zero-crossings and the number of extrema must be equal or differ at most by one.
- The mean value of the point of the positive and negative envelopes formed by the maxima and minima, respectively, must by zero.

The above definitions indicate a function which looks like a harmonic function, i.e., a sinusoid, in the time domain. The difference is that unlike a simple harmonic function, an IMF generally includes all functions with a possibly variable amplitude or frequency.

The basis of the EMD is the calculation of the time variations of a phenomenon or signal and presentation of its variations over different frequencies. The input signal is then decomposed into IMFs calculated using a sifting process shown in Figure 7.10.

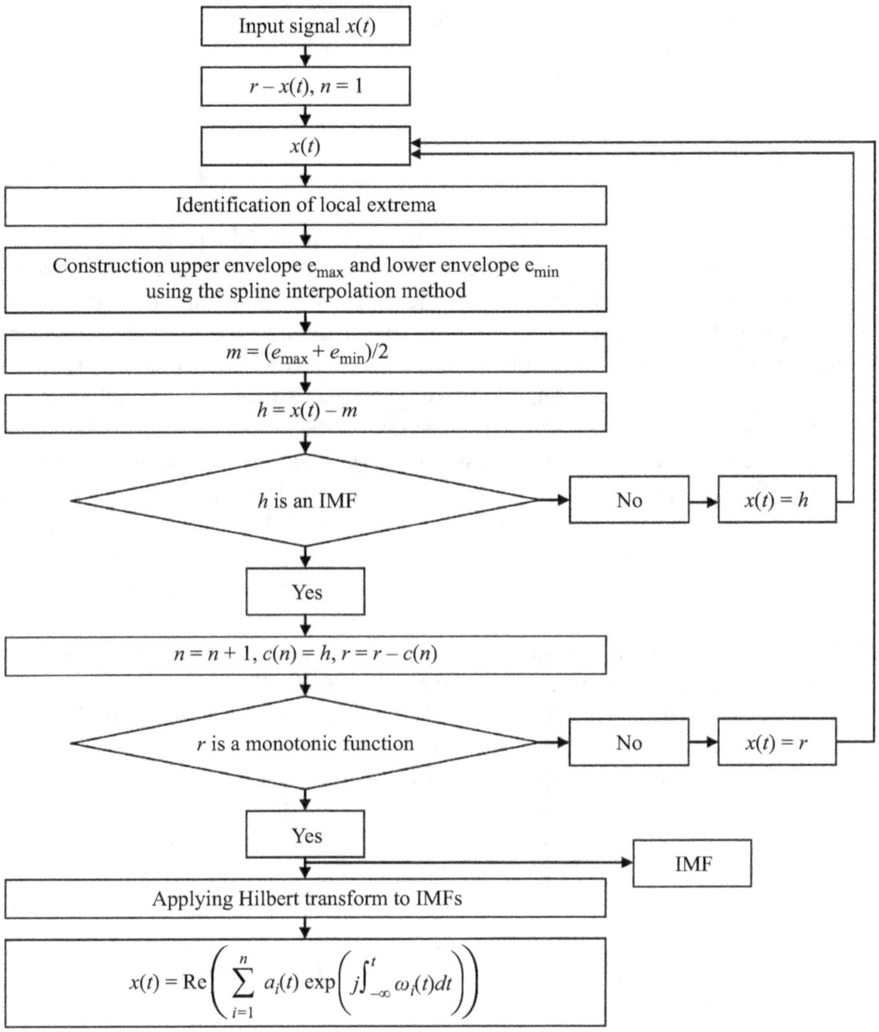


*Figure 7.10 Empirical mode decomposition*

The positive and negative envelopes extracted from the extrema points are determined by the spline interpolation. $x(t)$, $r$, $m$, $h$, $n$ and $c$ are the input signal or the stator single-phase current, residual function showing the last calculated IMF, negative and positive envelops of the signal, difference between the input signal in every stage with $m$, number of IMF and final IMF in every stage, respectively.

The mean value of the two envelops are subtracted from the fundamental signal, and the existence of the IMF is investigated. If this condition is not satisfied, the signal $x(t)$ is substituted by h, and this process continues until the condition ($h$ is an IMF) is satisfied. As the input signal has been already discretized, calculations are done numerically. Therefore, satisfying both IMF conditions is not possible, and a criterion is considered to stop the loop of condition holding (Equation (6)). The following standard deviation (SD):

$$SD = \sum_{t=0}^{T} \left[ \frac{\left|(h_{(n-1)}(t) - h_n(t))\right|^2}{h_{(n-1)}^2(t)} \right] \tag{7.39}$$

must be relatively small, where $n$ is the iteration number and $c$ is the IMF obtained at $n$th step. The value of $SD$ depends on the quality of the required frequency separation. As the dominant frequency of the supply is not high and fault harmonics have separable difference with the fundamental harmonic, SD can be chosen large. Hence, iterations number and consequently IMFs evaluation time are reduced. IMFs are calculated up to point in which $r$ or $s$ residual signal becomes a monotonic function. Consequently, the fundamental signal is decomposed into different frequencies. The first IMF has the highest and the last one has the lowest frequency band. The fundamental harmonic component can be decomposed by $IMF_1$, and only the components triggered by the fault are preserved. It is noted that the number of sampled points must be large enough. The resolution of the –HHT highly depends on this number. The sampling frequency is determined such that the frequency band of the IMF related to the fundamental frequency is limited to 0.5–1 Hz. SD must be chosen such that the $IMF_1$ frequency band consisting of the fundamental component of the current is large. So the EMD analysis finishes. Then, the Hilbert transform is applied to every IMF and the last step where the instantaneous frequencies are extracted. Instead of applying the Hilbert transform to the entire $x(t)$, it is applied to the decompositions of $x(t)$. Therefore, a more resalable frequency resolution is obtained as the frequency bands are already separated by means of the EMD. In the case of applying Hilbert transform to $x(t)$, only one single frequency band is obtained. So it might not be useful when the transients or oscillations caused by faults are relatively smaller amplitudes compared to the fundamental frequency of the input signal.

Finally, $x(t)$ can be rebuilt by combining the real parts of the Hilbert transform of the IMFs expressed as the following:

$$x(t) = \mathrm{Re}\left( \sum_{i=1}^{n} a_i(t) \exp\left( i \int_{-\infty}^{t} \omega_i(t)dt \right) \right) \tag{7.40}$$

Note that the last IMF obtained when the stopping criterion is met, is called the "residual." The residual consists of the lowest frequencies of the signal while the first IMF includes the highest frequencies in it. It is normally a monotonic function of time.

There are two influential factors in determining the quality of the IMFs;

- the SD: increasing this value leads to a faster calculation process while the sifting process may not be accurate enough.
- the number of IMFs: increasing the number of IMFs means that a better decomposition is performed.

One of the disadvantages of the HHT is the marginal effect observable in the output of the process. If the sampling period is smaller than one cycle of the expected frequency component, the marginal effect might totally deform the corresponding IMF shape, producing an ambiguous signal. On the other hand, if there are different types of faults whose related frequencies are close to each other, the HHT will not be able to decompose them unless the corresponding amplitudes differ a lot. The HHT works very well in the case of nonstationary signals; however, its application to stationary signals is still valid.

Table 7.6 shows the MATLAB code for the HHT. Using this code, an example of a transient signal consisting of one pure sinusoid and one exponentially decaying sinusoid with the frequencies of 30 and 60 Hz, respectively, are processed with the purpose of demonstrating the application of the HHT to a transient signal. The code shown in Table 7.6 is used for extracting the corresponding IMFs of the signal coded in Table 7.7.

Figure 7.11 shows the six IMFs obtained by applying the HHT to the signal described in Table 7.7. Note the following features extracted from this figure:

- There is an exponentially vanishing part at the very initial instants. This is the transient part of the signal, and it vanishes when the amplitude of the exponential term becomes relatively small compared to the other term. This happens at around $t = 1.2$s. There is a significant change in the $IMF_1$, $IMF_2$ and $IMF_3$.
- The $IMF_1$ consists of the highest frequency components of the signal which have the dominant amplitudes.
- The $IMF_2$ consists of the smaller frequencies compared to the $IMF_1$. Up to $t =$ 1.2s, the $IMF_2$ has a considerable amplitude while its amplitude gets close to zero as there is no more components within the frequency band which is included in the $IMF_2$. This is the nature of the HHT. Every IMF contains a specific frequency band sorted from high to low regarding the order of IMFs, 1 to 6 in this study. The upper and lower limits of the frequency bands highly depend on the nature of the input signal as well as the number of IMFs before starting the process. The number of IMFs is a factor controlled by the user, so for example, if it is equal to one, the $IMF_1$ will be so similar to the input signal, and it contains all the frequency components of the signal. In other words, no decomposition is done. On the other hand, increasing the number of IMFs might give a better resolution depending on the signal and the sampling rate.
- Even if the number of IMFs is increased to a value beyond the validity of the condition of being an IMF, the maximum number of IMFs will be equal to what satisfies the nonmonotonicity of the IMF. For example, in the case of the investigated signal, the number of IMFs equal to 6 is the maximum available value as the corresponding residual is a monotonic signal; hence, no more decomposition will be done. This is handled automatically by the algorithm.
- A relatively large jump is observed in the IMFs, especially the $IMF_1$, $IMF_2$ and $IMF_3$ at the starting and ending instants. This is what we call marginal effect of the HHT. So these jumps are of no interest at all as they are just numerical errors and unreliable parts of the HHT output.

*Table 7.6 Hilbert–Huang Transform*[3]

```
%%%%%%%%%%%%%%%%%%%%%%%% IMF calculations %%%%%%%%%%%%%%%%%%%%%%%%%%%%%%%
% x: Input signal
% t: Time vector
% imf: IMF of the input signal
% amplitude: Vectors of the Hilbert-transform amplitudes
% theta: Instantaneous positions
% w: Instantaneous frequencies
% t2: Time vectors of instantaneous frequencies
function [imf plot_array amplitude theta t2 w] = vHHT(t,x,imf_number)
close all
clc
main=x; % storing the main data
n_imf=1; % number of intrinsic mode functions
while(n_imf<=imf_number)
    h1=x;
    SD=Inf; % SD: primary value of stoppage criteria
iter=1;
    while (SD > 0.5) || ~IMF(h1)
       positive_peak=vfindpeaks(t,h1); % local maxima envelope production
       negative_peak=-vfindpeaks(t,-h1); % local minima envelope production
       m=(positive_peak+negative_peak)/2; % mean value of local maxima and minima envelopes
       h11=h1-m; % IMF component calculation
       SD=sum((h1-h11).^2)/sum(h1.^2); % SD calculation
       h1=h11;
       if(iter>1000)
           break;
       end
       iter=iter+1;
    end
```

```
    imf(n_imf,:)=h1; %IMF component
    x=x-h1; % residue component
    n_imf=n_imf+1;

    if(monotonic(x)==1)
       break;
    end
end
k=floor((n_imf-1)/4)+1;
k1=1;
plot_array(1,:)=main;
plot_array((2:n_imf),:)=imf(:,:);
plot_array(end+1,:)=x; %final residual
[amplitude theta t2 a2 w]=HHT_plot(t,plot_array);
Hspectrum(t2,a2,w);
end

% %%%%%%%%%%%%%%%%%%% Check if the function is an IMF %%%%%%%%%%%%%%%%%%% function tf = IMF(signal)

d=diff(signal);
ppeak=zeros(1,length(d));
j=1;
for i=1:length(d)-1
    if(d(i)>0)&&(d(i+1)<0)
        ppeak(j)=signal(i+1);
        j=j+1;
    elseif(d(i)==0)&&(d(i+1)<0)
        ppeak(j)=signal(i+1);
        j=j+1;
    end
end
```

(Continues)

*Table 7.6 (Continued)*

```
ppeak=ppeak(1:j-1);
%d=zeros(1,length(signal));
d=diff(-signal);
j=1;
npeak=zeros(1,length(d));
for i=1:length(d)-1
    if(d(i)>0)&&(d(i+1)<0)
        npeak(j)=signal(i+1);
        j=j+1;
    elseif(d(i)==0)&&(d(i+1)<0)
        npeak(j)=signal(i+1);
        j=j+1;
    end
end
npeak=npeak(1:j-1);
N=length(signal);
n1=length(ppeak)+length(ppeak);
n2=sum(signal(1:N-1).*signal(2:N) < 0);
if abs(n1-n2) > 1
    tf = 0;
else
    tf = 1;
end
end
%%%%%%%%%%%%%%%%%%%%%%%%%%%% find envelopes %%%%%%%%%%%%%%%%%%%%%%%%%%%%%%%
function xx = vfindpeaks(t,x)
d=zeros(1,length(x)-1);
d=diff(x);
peak=zeros(2,length(x));
```

```
peak(1,1)=t(1);
peak(2,1)=x(1);
j=2;
for i=1:length(d)-1
    if(d(i)>0)&&(d(i+1)<0)
        peak(1,j)=t(i+1); %peak(1,:)= time array
        peak(2,j)=x(i+1); %peak(2,:)= value array
        j=j+1;
    elseif(d(i)==0)&&(d(i+1)<0)
        peak(1,j)=t(i+1); %peak(1,:)= time array
        peak(2,j)=x(i+1); %peak(2,:)= value array
        j=j+1;
    end
end
peak(1,j)=t(end);
peak(2,j)=x(end);
%xx=spline(peak(1,:),peak(2,:),t);
xx = spline(peak(1,1:j),peak(2,1:j),t);
end
%%%%%%%%%%%%%%%%%%%%%%%% if the function is monotonic %%%%%%%%%%%%%%%%%%%
function u = monotonic(x)
u1 = length(findpeaks(x))*length(findpeaks(-x));
if u1>0
    u=0;
else
    u=1;
end
end
%%%%%%%%%%%%%%%%%%%HHT plot%%%%%%%%%%%%%%%%%%%%%%%%%%%%%%%%%%%%%%%%%%%%%
function [amplitude theta t2 a2 w2] = HHT_plot(t,x)
[a b]=size(x);
if(floor(a/5)==(a/5))
    fn=(a/5);
```

*(Continues)*

*Table 7.6 (Continued)*

```
else
    fn=floor(a/5)+1;
end
k=1;
for i=1:fn
    figure
    for j=1:5
        if(j+k-1>a)
            break;
        else
        subplot(5,1,j),plot(t,x(j+k-1,:));
        end
    end
    k=k+5;
end
figure,subplot(2,1,1),plot(t,x(1,:)),title('Original signal')
subplot(2,1,2),plot(t,sum(x(2:end,:)));title('Sum of IMFs and final residual')
amplitude=zeros(a-2,length(x));
theta=zeros(a-2,length(x));
w=zeros(a-2,length(x)-1);
for i=2:(a-1)
    amplitude(i-1,:)=abs(hilbert(x(i,:)));
    theta(i-1,:)=angle(hilbert((x(i,:))));
    w(i-1,:)=(diff(theta(i-1,:))/(t(2)-t(1)));
    k1=find(w(i-1,:)>0);
    %plot(t(1,1:(length(t)-1)),w(i-1,:),'.','MarkerSize',3)
    %hold all
end
```

```
% ylim([0 100000])
t2=t;
a2=amplitude;
w2=w;
end

%%
function Hspectrum(t,amplitude,w)
figure
t=t(1:length(t)-1);
w=w/2/pi;
a=abs(amplitude(:,1:length(t)));
s=pcolor(t,w,a);
set(s,'FaceColor','none','EdgeAlpha',[0],'Marker','.','MarkerSize',[4],'EdgeColor','flat');
%xlim([0 1]);
ylim([0 1000]);
xlabel('Time(s)');
ylabel('Frequency(Hz)');
colorbar;
h = gca;
set(h,'Color',[1 1 1]);
title('HilbertHuang Spectrum');
end
```

[3] https://www.mathworks.com/matlabcentral/fileexchange/19681-hilbert-huang-transform (access date: October 2016).

*Table 7.7 An example of the HHT of a transient signal*

```
clc
clear
close all
%%
t = 0:.0001:5;
x = 5*sin(2*pi*30*t)+8*exp(-t).*sin(2*pi*60*t);
[imf plot_array amplitude theta t2 w]=vHHT(t,x,6);
```

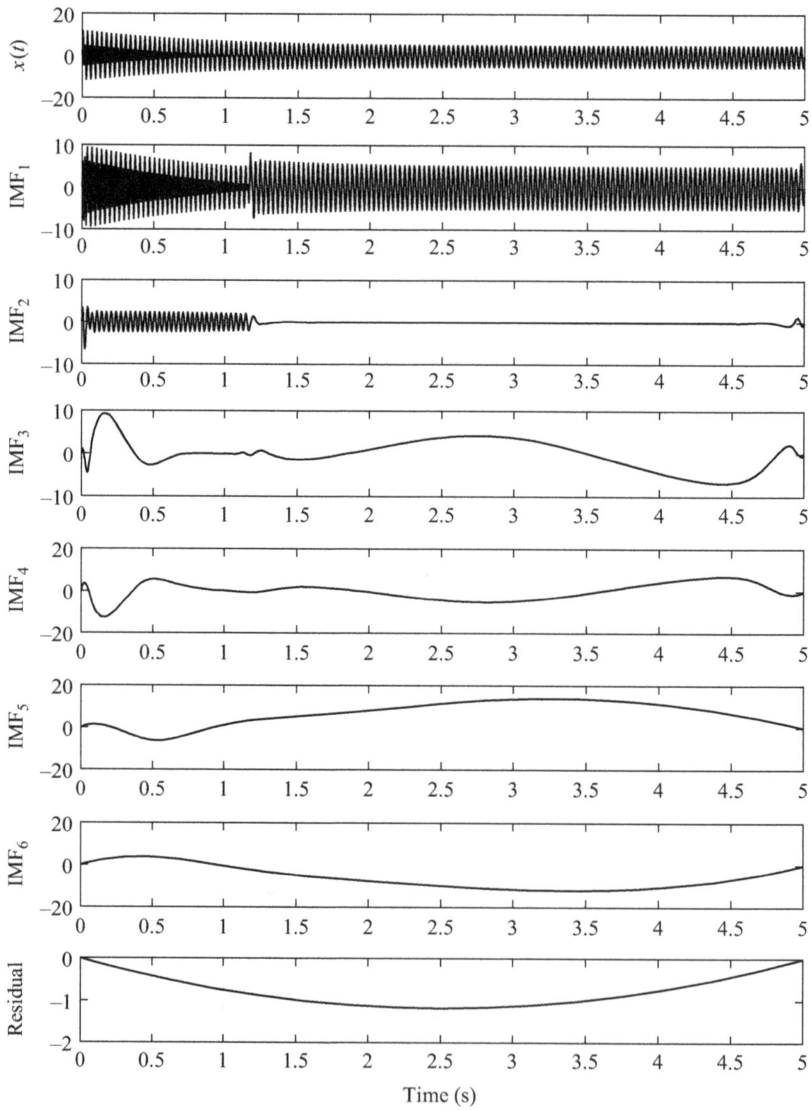


*Figure 7.11 IMFs of $x(t) = 5\, sin(2\pi \times 30 \times t) + 8e^{-t} \times sin(2\pi \times 60t)$ considering six IMFs*

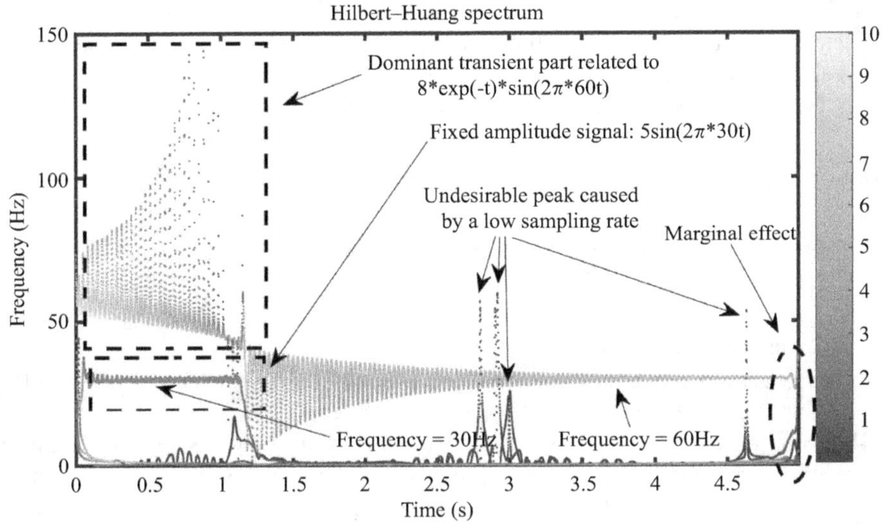


*Figure 7.12 Hilbert–Huang spectrum of* $x(t) = 5\ sin(2\pi \times 30 \times t) + 8e^{-t} \times sin(2\pi \times 60t)$

Now, the Hilbert–Huang spectrum, which is the demonstration of the instantaneous frequency over the sampled period, is illustrated in Figure 7.12. The noticeable point in this figure is that there are some undesirable peaks caused by a low sampling rate. This means if an appropriately large sampling rate/frequency is used, the peaks will be eliminated. The correct value of the sampling rate depends on the largest frequency existing in the signal.

In this chapter, four significant signal-processing tools were discussed in detail with the purpose of providing a solid background for the next chapters in which the mentioned tools are going to be used as mediums for extracting fault features. In particular, we focused on the FT, the STFT, the WT as well as the HHT.

The FT is used to process the stationary signals representing motors behavior in a steady-state operation. The FT has a very precise frequency resolution, although it has nothing to do with the time-domain information of signals. What we will obtain from the FT is frequency-wise implication of the motor signals regardless of the time-dependent variations of various frequency components. Equivalently important, almost all the analytical formulations in terms of the fault features investigated so far were based on the steady-state operation of motors. This is the most important practice in fault diagnosis efforts as extracting features in a nonstationary or transient operation is often based on what we have already inferred from the steady-state operation analysis.

On the other hand, the other three transforms, namely the STFT, the WT and the HHT are all useful to monitor signals variations in a time-frequency map leading to a detailed understanding of not only potential fault-related frequency components but also the correct instants at which fault occurs and evolves. This is a major issue in medium- and high-power motors operating in start–stop applications such as hybrid vehicles. The significant point is that these transforms require a relatively large period of motor transient signals to be present during the process of the signal. Otherwise, the time resolution will not be appropriate. This is why a high-inertia load is usually connected to the motor shaft where it is tried to experiment the transient faulty behavior. This problem will be further discussed in the next chapters.

## References


[1] http://www.mathworks.com/help/pdf_doc/signal/signal_tb.pdf (Access date: October 2016).

[2] Lyons R.G. Understanding digital signal processing, 2012; Upper Saddle River, NJ: Prentice Hall.

[3] Oppenheim A.V. and Schafer R.W. Discrete-time signal processing, 2009; Upper Saddle River, NJ: Prentice Hall.

[4] Mehrjou M.R., Mariun N., Marhaban M.H. and Misron N. Fourier analysis of motor current for diagnosis of broken bars in squirrel-cage induction machines, *2010 IEEE Student Conference on Research and Development (SCOReD)*, Putrajaya, 2010; 475–478.

[5] Climente-Alarcon V., Antonino-Daviu J.A., Riera-Guasp M. and Vlcek M. Induction motor diagnosis by advanced notch FIR filters and the Wigner–Ville distribution, *IEEE Transactions on Industrial Electronics*, 2014; **IE-61**: 4217–4227.

[6] Kia S. H., Henao H. and Capolino G.A., Fault index statistical study for gear fault detection using stator current space vector analysis, *IEEE Transactions on Industry Applications*, 2016; **IA-52**: 4781–4788.

[7] Yin S., Hu N., Chen L. and Hu L. Teager energy operator and its application in the study of induction motor rotor broken bars fault, *Prognostics and System Health Management Conference (PHM)*, Beijing, 2015; 1–5.

[8] Kral C., Habetler T.G., Harley R.G. *et al.* A comparison of rotor fault detection techniques with respect to the assessment of fault severity, *4th IEEE International Symposium on Diagnostics for Electric Machines, Power Electronics and Drives, 2003*. SDEMPED 2003, 2003; 265–270.

[9] Korkua S., Jain H., Lee W.J. and Kwan C. Wireless health monitoring system for vibration detection of induction motors, *2010 IEEE Industrial and Commercial Power Systems Technical Conference – Conference Record*, Tallahassee, FL, USA, 2010; 1–6.

[10] Jevaanand. S and Mathew A.T. Condition monitoring of induction motors using Wavelet based analysis of vibration signals, *2008 Second International Conference on Future Generation Communication and Networking Symposia*, Sanya, 2008; 75–80.

[11] Zhaoxia W., Fen L., Shujuan Y. and Bin W. Motor fault diagnosis based on the vibration signal testing and analysis, *2009 Third International Symposium on Intelligent Information Technology Application*, Nanchang, 2009; 433–436.

[12] Deekshit Kompella K.C., Rao M.V.G., Rao R.S. and R.N. Estimation of nascent stage bearing faults of induction motor by stator current signature using adaptive signal processing, *2013 Annual IEEE India Conference (INDICON)*, Mumbai, 2013; 1–5.

# Chapter 8
# Diagnosis of broken bars fault in induction motors

## 8.1 Introduction

Having discussed the fundamentals of induction motors with broken bars fault and explained how the breakage affects the motor electromagnetic and mechanical properties, the diagnosis approaches and the fault indices along with the corresponding specifications including:

- implementation method (offline or online) [1–15]
- the ability to be applied to line-start and inverter-fed motors [16–33]
- the nature of the fault (invasive or noninvasive) [34–50]
- being fault-location-sensitive location-sensitive or not [51–67]
- the ability to diagnose the fault over a wide range of motor load [68–81]
- being load-sensitive or not [82–95]
- being speed-sensitive or not [96–105]
- being bandwidth-sensitive or not [106–118]

are going to be discussed in detail with the purpose of introducing a comprehensive diagnosis process. There are also other specifications of the diagnosis approaches which will be introduced and explained in this chapter.

On the other hand, different types of motor supplies play significant role in the fault diagnosis process. These supplies can be divided into two main categories: (1) line-start and (2) inverter-fed mode. The second category uses both open-loop and closed-loop control strategies. In line-start mode, a pure sinusoidal supply voltage is applied to the motor while in inverter-fed mode, the supply voltage contains various harmonics. In the first case, the motor variables directly depend on the load level, but in the inverter-fed mode, electrical and mechanical variables are controllable; therefore, fault detection is more complicated. The reason is that the number of effective factors in fault diagnosis process increases considerably. The major influential factor is the high-frequency switching supply voltage, which amplifies the core saturation level and causes more distortion in the air gap flux density. In an inverter-fed motor in which the motor speed is controllable regardless of the motor slip, the characteristics of the rotor broken bar indices, which depend on the motor slip, is different from that of the line-start mode. Moreover, the electromagnetic torque is adjusted independent of the speed, using the capability of changing amplitude and frequency in inverter-fed mode. The saturation profile again changes upon the change of amplitude of the supply and frequency. On the basis of the above discussions, selecting a proper signal, which merely consists of the fault-related information, is inevitable in precise fault diagnosis process. On the other hand, a proper processor must be used for diagnosis. Although all the motor variables reflect the fault effects, some of them such as air gap flux density are not easily measurable, or some such as torque may have less sensitivity to the fault. For magnetic flux, the use of a search coil is the most reliable method. This coil is inserted in the stator slot and therefore it is an invasive technique. As the complete motor behavior depends on the air gap magnetic flux, the flux method can be considered as the most reliable technique. As the search coil must be placed inside the motor frame, sampling the flux is not as easy as the motor current, voltage, torque and speed.

Taking the above-mentioned specifications into account, one can conclude that not all the diagnosis approaches and indices introduced so far in the literature are able to cover all the practical situations encountered in an industrial environment. For instance, the amplitude of the sideband components which is the direct outcome of the broken bars fault is not able to return the fault features in a light-load level if a pure Fast Fourier Transform (FFT) is used. This conveys the idea that although the fault-related components do exist in the motor quantities as soon as the fault occurs, they may not be detected in some certain conditions. Therefore, sometimes it is proposed to analyze a signal other than the motor current, for example, the flux density which is more sensitive to the fault than the motor current. However, measuring the flux density requires an invasive process leading to a more complex diagnosis scheme.

On the other hand, some quantities such as the torque and flux of the motor are smoothly controlled where a closed-loop application is used as there is a direct closed flux or torque loop, whether sensor-less or with sensor. Therefore, it is going to be demonstrated that the fault-related components are also damped by means of the closed loops. Therefore, the fault features might not be properly reflected into the motor torque or flux signals, leading to a decision based on which other signals such as the motor current should be targeted. Moreover, the presence of a drive affects the fault diagnosis procedure, depending on the drive adjustments including the PI controller and its bandwidth. It is also shown that the proportional and integral gains of the PI controller apply changes to the amplitude of both sideband components of the motor.

Considering the mentioned points, the fault diagnosis procedure be a rather sophisticated task depending on different operating conditions. The bottom line of this procedure is to investigate various possible conditions and check for their effects on the detection, determination and diagnosis process and then devise a plan to introduce a comprehensive procedure. Certainly, the procedure cannot be a single routine which allows to diagnose the fault, using one specific motor signal, fault feature and also diagnosis process. What is needed is a multilevel procedure taking advantage of different motor signals, the drive signals, the signal processing techniques, the fault indices and the fault features. This is applied to all kinds of faults, and the corresponding knowledge is not achievable unless variety of operating conditions are explored and investigated. Therefore, this is the focus of this chapter as well as the next two chapters focusing on the eccentricity and short-circuit faults.

A promising way of diagnosing the fault is the frequency-domain analysis of the motor variables. This analysis is more qualified than the time-domain analysis.

The reason is the presence of harmonic components of the different faults at different frequency patterns which is not the case in time-domain analysis. Nevertheless, the indices, neither in time domain nor in frequency domain, are addressed to be fully useful under every operating condition. So being familiar with and knowing the situations in which an index could be used, is very important in practice. On the other hand, combination of the time and frequency analysis provides an easy and reliable method to diagnose the fault even in the worst operating conditions including the transient periods.

Although all motor variables are affected by the fault, it is necessary to choose a proper method enabling accurate detection, determination and diagnosis of the fault degree upon the alternation of various influential quantities such as load and speed levels. Besides, not all the diagnosing methods are easily applicable in any operating environment. For example, in a noisy environment, an acoustic signal for condition monitoring is not recommended. The other restriction is the type of sensor used for sampling a specific signal. The sensor could be used in invasive or noninvasive manner. For invasive method, considering a very pure way of connecting the sensor to the motor is crucial because the sensor itself could be a reason of producing error in diagnosing procedure.

As discussed above, there are a lot of influential factors affecting fault detection process. Without considering the above factors, any effort is definitely useless because the output is probably not reliable. In addition, the literature has mostly focused on a specific index under a particular operating condition which simplifies the fault diagnosis process. So a practitioner who needs to know the details will be obviously mislead through the diagnosis procedure. So therefore, providing a general basis for the users is necessary in terms of various indices and their capabilities in different operating conditions. In this chapter, different fault indices are investigated to address the above-mentioned issues.

This chapter aims at providing a qualitative diagnosis approach to the broken bars fault by illustrating the time, frequency and time-frequency representations of the motor signals for some specific fault levels at different operating conditions. Using this, the way that the indices should be extracted from the representations are thoroughly discussed; then at some points, the quantitative analysis is provided by means of numerical values of the index amplitudes in order to show how the index values must be followed in a practical apparatus for having a precise diagnosis process. Finally, we will move toward the next chapter which concerns the eccentricity fault, and a more quantitative approach rather than a qualitative one is provided.

## 8.2 Motor current signature analysis (MCSA)

Assuming that a line-start motor is studied herein, the idea of supply voltage and current signals can be formulated as follows:

$$V = V_m \sin(\omega_s t) \tag{8.1}$$

$$I_s = I_m \sin(\omega_s t + \alpha) \tag{8.2}$$

The broken bars fault introduces an improved sideband component with the frequency pattern $2sf_s$ where $s$ is the slip, $f_s$ is the fundamental frequency and $k$ is an integer.

However, the supply voltage is fixed and not changed as a function of the fault. Therefore, the motor current is expanded as the following [25,28,75]:

$$\begin{aligned} I_s = {} & I_m \sin(\omega_s t + \alpha) + I_{l1} \sin((1-2s)\omega_s t) + \alpha_{l1}) + I_{r1} \sin((1+2s)\omega_s t) + \alpha_{r1}) \\ & + I_{l2} \sin((1-4s)\omega_s t) + \alpha_{l2}) + I_{r2} \sin((1+4s)\omega_s t) + \alpha_{r2}) \\ & + I_{l3} \sin((1-6s)\omega_s t) + \alpha_{l3}) + I_{r3} \sin((1+6s)\omega_s t) + \alpha_{r3}) + \cdots \end{aligned} \tag{8.3}$$

where $I_{lk}$, $I_{rk}$, $\alpha_{lk}$, $\alpha_{rk}$ ($k = 1, 2, 3, \ldots$) are the amplitude of the left sideband components, amplitude of the right sideband components, phase current angle of left sideband components, phase current angle of right sideband components, respectively. Extracting the components with the angular position of $\omega_s t$ and gathering them up in one group returns:

$$\begin{aligned} I_s = {} & I_m \cos(\alpha) \sin(\omega_s t) + I_m \sin(\alpha) \cos(\omega_s t) + I_{l1} \begin{bmatrix} \sin(\omega_s t) \cos(2s\,\omega_s t - \alpha_{l1}) - \\ \cos(\omega_s t) \sin(2s\,\omega_s t - \alpha_{l1}) \end{bmatrix} \\ & + I_{r1} [\sin(\omega_s t) \cos(2s\,\omega_s t + \alpha_{r1}) + \cos(\omega_s t) \sin(2s\,\omega_s t + \alpha_{r1})] \\ & + I_{l2} [\sin(\omega_s t) \cos(4s\,\omega_s t - \alpha_{l2}) - \cos(\omega_s t) \sin(4s\,\omega_s t - \alpha_{l2})] \\ & + I_{r2} [\sin(\omega_s t) \cos(4s\,\omega_s t + \alpha_{r2}) + \cos(\omega_s t) \sin(4s\,\omega_s t + \alpha_{r2})] \\ & + I_{l3} [\sin(\omega_s t) \cos(6s\,\omega_s t - \alpha_{l3}) - \cos(\omega_s t) \sin(6s\,\omega_s t - \alpha_{l3})] \\ & + I_{r3} [\sin(\omega_s t) \cos(6s\,\omega_s t + \alpha_{r3}) + \cos(\omega_s t) \sin(6s\,\omega_s t + \alpha_{r3})] + \ldots \end{aligned} \tag{8.4}$$

Reformulating (8.4) leads to

$$\begin{aligned} I_s = {} & \sin(\omega_s t) \begin{bmatrix} I_m \cos(\alpha) + I_{l1} \cos(2s\omega_s t - \alpha_{l1}) + I_{r1} \cos(2s\omega_s t + \alpha_{r1}) \\ + I_{l2} \cos(4s\omega_s t - \alpha_{l2}) + I_{r2} \cos(4s\omega_s t + \alpha_{r2}) \\ + I_{l3} \cos(6s\omega_s t - \alpha_{l3}) + I_{r3} \cos(6s\omega_s t + \alpha_{r3}) + \ldots. \end{bmatrix} \\ & + \cos(\omega_s t) \begin{bmatrix} I_m \sin(\alpha) - I_{l1} \sin(2s\omega_s t - \alpha_{l1}) + I_{r1} \sin(2s\omega_s t + \alpha_{r1}) \\ - I_{l2} \sin(4s\omega_s t - \alpha_{l2}) + I_{r2} \sin(4s\omega_s t + \alpha_{r2}) \\ - I_{l3} \sin(6s\omega_s t - \alpha_{l3}) + I_{r3} \sin(6s\omega_s t + \alpha_{r3}) - \ldots. \end{bmatrix} \\ = {} & \sin(\omega_s t) \left[ \underbrace{I_m \cos(\alpha) + \sum_{k=1} I_{lk} \cos(2ks\omega_s t - \alpha_{lk}) + I_{rk} \cos(2ks\omega_s t + \alpha_{rk})}_{A(t)} \right] \\ & + \cos(\omega_s t) \left[ \underbrace{I_m \sin(\alpha) + \sum_{k=1} (-I_{lk}) \sin(2ks\omega_s t - \alpha_{lk}) + I_{rk} \sin(2ks\omega_s t + \alpha_{rk})}_{B(t)} \right] \\ = {} & A(t) \sin(\omega_s t) + B(t) \cos(\omega_s t) \end{aligned} \tag{8.5}$$

Therefore, the motor current signal is reformulated as follows:

$$I_s = A \sin(\omega_s t) + B \cos(\omega_s t) = \sqrt{A^2 + B^2} \sin\left(\omega_s t + \tan^{-1}\left(\frac{A}{B}\right)\right) \tag{8.6}$$

According to (8.6), the broken bars fault is reflected as a periodic oscillation into the motor current envelope in the time domain. The frequency of the oscillations is equal to $2ksf_s$, meaning that not only are the first-order sideband components, which have the largest amplitudes, detected in the motor current, the higher order components are also influential. Normally, the amplitude of the sideband components increases by increasing the fault level in one location. Therefore, it is expected to have a larger envelope oscillation in the time-domain variation of the current signal. Figure 8.1 illustrates the stator current of an induction motor in a steady-state

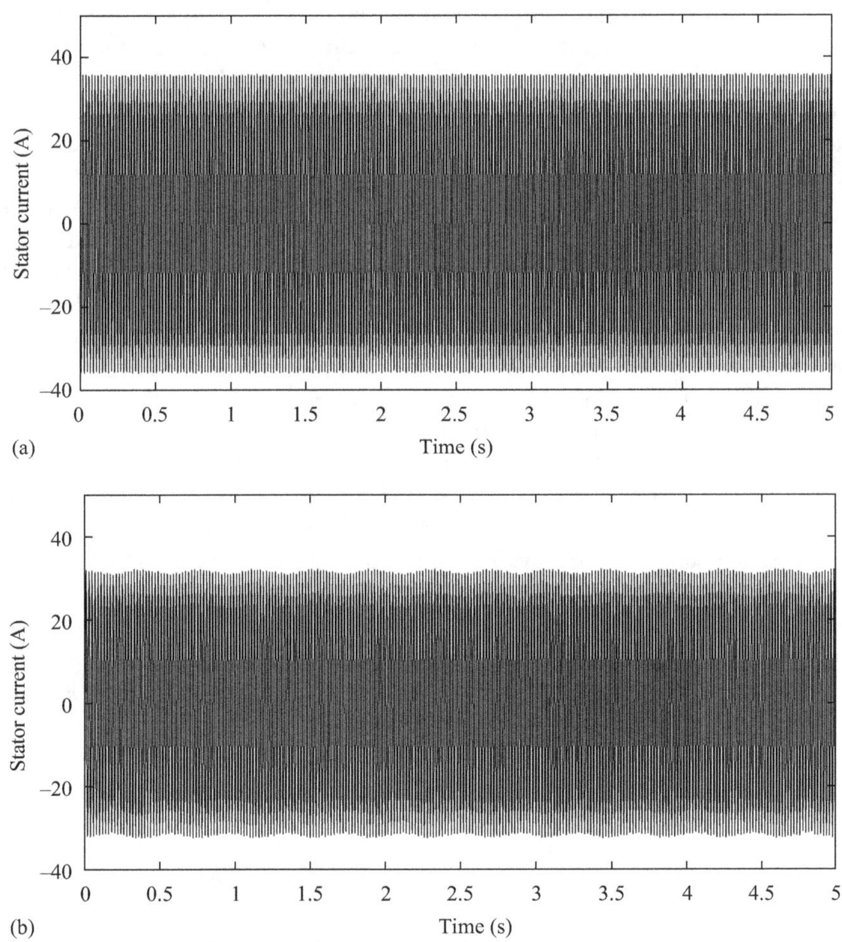


*Figure 8.1 Stator current of an induction motor (a) healthy and (b) one broken bar*

operation. Figure 8.1(a) demonstrates a healthy motor and Figure 8.1(b) shows the motor with one broken bar. The periodic oscillations of the current envelope are clearly observed in Figure 8.1(b). As the term $2ksf_s$ is a function of the motor slip, it is expected to have a larger frequency of the oscillations if the motor load increases. Besides, it will be shown that increasing the load level while fixing the supply frequency leads to a considerable increase in the fault features.

The discussed time-domain variation of the motor current is the primary feature of the broken bars fault. Other types of faults also have their own specific variations depending on the fault-related components. Investigating the motor current, either in the time or frequency domains, is called the motor current signature analysis (MCSA) known as one of the best, not necessarily the most reliable, technique in the fault diagnosis procedures. The reason for the widespread usage of this technique might be the easiness of the corresponding measurements as it only requires non-invasive and rather cost-effective sensors such as the current transducers. Moreover, if a reliable shielded cable is used, the monitoring unit where technicians are located can be far away from the motor location. This indeed facilitates the condition monitoring process. More importantly, while the motor current signal is measured, the fault-related data can be transferred using a well-tuned wireless transmission equipment. Therefore, a wireless implementation is also possible. In terms of the motor current, unlike the motor voltage, measuring the current is not subjected to a considerable human health risk. This gives rise to the popularity of the MCSA.

The MCSA is not limited to the time-domain analysis of the motor current signature, and it is indeed a more comprehensive concept including time, frequency and time-frequency domains and the corresponding analysis. However, most of the time, it is the sideband components and their amplitudes which are dealt with during the MCSA. In the case of other types of faults, different harmonic components are investigated. The MCSA also covers the inverter-fed applications regardless of the type of the inverter and the control strategy. To clarify this issue, look at Figure 8.2 illustrating the time-domain variation of the envelope of the current for an inverter-fed motor in different fault levels including 0, 1 and 4 broken bars and two different speeds including 1500 and 700 rpm. It is noteworthy that the frequency of oscillations is a function of the reference speed which in turn determines the supply frequency. The larger the reference speed is, the larger the frequency of oscillations for the same slip level is. Moreover, increasing the fault level greatly affects the amplitude of the oscillations in a way that it increases as the number of the broken bars increases. Therefore, it is implied that the amplitude of the sideband components expressed by the frequency pattern of $(1 \pm 2ks)f_s$ should increase. Figure 8.2 is the demonstration of the output of a voltage-source inverter in which the controlled signal is the motor voltage, and it has nothing to do with the control of the motor current. On the other hand, there is another type of inverters called "current-source inverter" which applies some strict conditions to the motor current in order to keep it as smooth as possible. This is usually held by some sort of hysteresis controllers with a specific bandwidth. Therefore, depending on the bandwidth of the current controllers, the fault-related components, not only in the case of broken bars fault but also in other cases, might or might not be filtered out

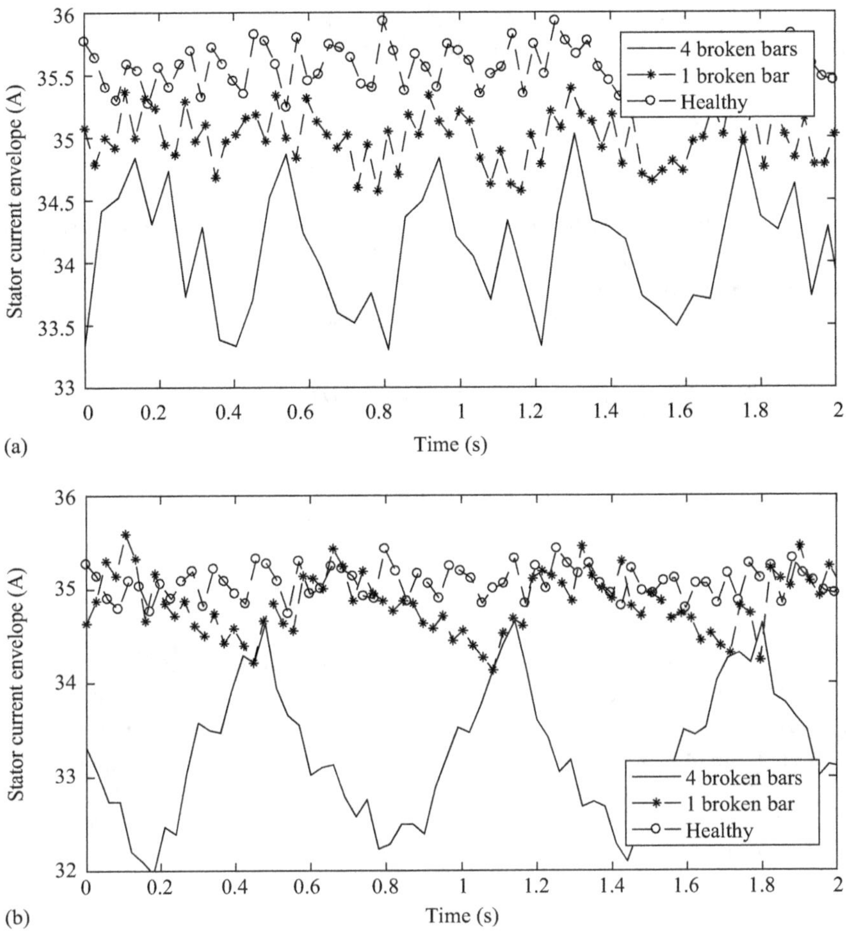


*Figure 8.2 Time variations of current envelope in inverter-fed motor: (a) 1500 rpm and (b) 700 rpm*

from the motor-drive circuit. This issue has not been tackled yet and is a very open area of research nowadays. However, a very useful approach will be proposed in this chapter to address at least one part of the remaining issues.

The other point regarding Figure 8.2 is the average values of the oscillations getting smaller by increasing the fault. This naturally happens by increasing rotor resistance, making the total motor resistance become larger. Thus, the motor current is expected to decrease. This trend is observed regardless of the supply mode, either the line-start or the inverter-fed motor, and the only difference is between the absolute average and amplitude values.

The time-domain analysis of the healthy motor current does not reveal any oscillation or deficiency. This does not necessarily mean that there is not a sideband component and it only implies that the time-domain analysis is unable to detect the sideband components which indeed exist even in the case of a healthy motor. This issue will be further discussed using the FFT of the motor current.

## 8.3 Pendulous oscillation

Due to the negligible inherent asymmetry of induction motors, diagnosis of the time-dependent oscillations caused by the inherent sideband components is somehow impossible using the current profile. However, it can be addressed by means of a new fault index called the "Pendulous oscillation" defined as the phase shift between the voltage and current space vectors formulated by (8.10) as follows [40,41]:

$$\vec{i}_s(t) = \frac{2}{3}\left((i_a - i_b) + \alpha(i_b - i_c) + \alpha^2(i_c - i_a)\right) \tag{8.7}$$

$$\vec{v}_s(t) = \frac{2}{3}\left(v_{ab} + \alpha v_{bc} + \alpha^2 v_{ca}\right) \tag{8.8}$$

$$\alpha = \exp\left(\frac{j2\pi}{3}\right) \tag{8.9}$$

$$\delta(t) = \angle\vec{i}_s(t) - \angle\vec{v}_s(t) \tag{8.10}$$

The pendulous oscillation should be measured through a low-pass filter in order to eliminate the ripples caused by the high-frequency switching phenomenon which might affect the correct measurement process. Therefore, both the voltage and current signals should be passed through a low-pass filter and then sampled into the processor. It is interesting to know that the pendulous oscillation is applicable to both the line-start and inverter-fed motors. Doing so, an example of the corresponding variations as the function of the number of the broken bars is shown in Figure 8.3.

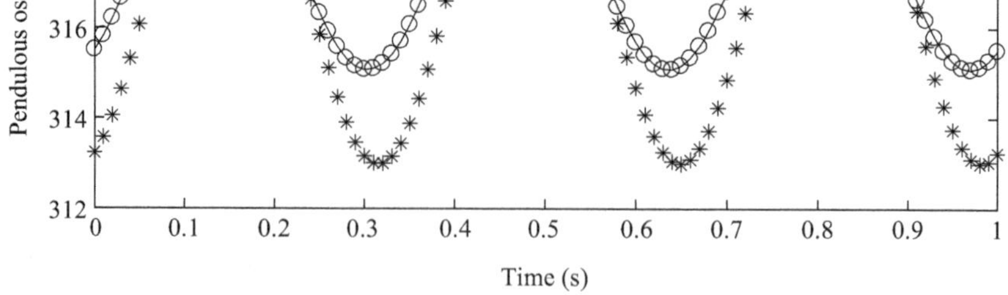


*Figure 8.3 Pendulous oscillation applied to a PWM excited motor*

The amplitude of the pendulous oscillation increases by increasing the fault and also load level. Remarkably, the amplitude of the pendulous oscillation for the healthy motor is not zero, proving the concept in terms of the existence of the sideband components. The trend of variations, not the peak-to-peak amplitude of the oscillations, is independent of the supply mode.

It is worth noting that more investigates are required to accept this index for a precise diagnosis and condition monitoring tool as different phenomena other than the broken bars fault, those producing a backward field in the air gap, might introduce the same behavior leading to an improper decision. However, the capability of the pendulous oscillation to diagnose a pure broken bars fault has been already approved in the literature. The major assumption is that there is no other type of fault producing a backward field. The pendulous oscillation is a time-domain index enabling technicians to diagnose the broken bars fault in different fault and load levels. Moreover, an accurate implementation of this technique leads to a precise discrimination of various supply modes including the line-start, open-loop and closed-loop strategies. So, it can be considered as one of the most useful diagnosis techniques applied in the time domain. However, the capability of this index to detect the location of the fault has not been studied yet.

## 8.4 Virtual current technique

As one of the significant topics of the condition monitoring techniques, the inverter-fed applications and their effect on the diagnosis procedure have been already studied widely. The virtual current technique (VCT) is certainly one of the most reliable techniques applied to the closed-loop field oriented control (FOC) control strategy. The VCT is based on measuring the $2sf_s$ components of the $d$-axis rotor flux and relating it to the knowledge of motor parameters. On the basis of the VCT, the number of adjacent broken bars is approximately equal to:

$$n_F \approx \frac{\Delta i_{ds}}{\bar{i}_{qs}} N_b \tag{8.11}$$

where $n_F$ is the number of broken bars, $\Delta i_{ds}$ is the amplitude of the variations of the $d$-axis current flowing through the closed flux loop, $\bar{i}_{qs}$ is the average value of the $q$-axis current component flowing through the closed torque loop and $N_b$ is the number of the rotor bars. This method is easy to apply where a FOC controlled induction motor is under operation. Otherwise, extracting the $dq$-axis components requires additional processors which increase the complexity of the diagnosis procedure. Here, some measurement examples for a motor with 28 rotor bars are provided in order to give the insight into how the quantities related to the VCT vary by the load, fault, speed and bandwidth level. These are the most influential factors giving change to the fault indices in almost any considerable situations. In fact, no fault index which is prominently independent of all the mentioned factors has been proposed. This necessitates investigating the corresponding effects during any diagnosis procedure.

The considerable advantage of the VCT is its capability in being implemented online without interrupting the motor operation. The same is possible for the MCSA as well as the pendulous oscillation. This provides a promising tool to diagnose the fault immediately after it takes place. However, it should be first proved that the index is capable of extracting fault information during a transient regime in which the motor quantities change from one state to another. Mostly, the indices presented in the time domain cannot handle this issue. Nevertheless, the VCT seems to be very responsive to the broken bars fault in the steady-state operation. The only remaining issue is that it only applies to adjacent broken bars. In addition, the existence of the $2sf_s$ components in the *d*-axis components highly depends on the bandwidth of the flux loop. If the bandwidth becomes relatively small, it is likely the $2sf_s$ component vanishes. However, this is not the case in high-power induction motors.

## 8.5 Air gap flux density

Measuring the air gap flux density requires an invasive equipment called "search coil" inserted into the motor body. It returns the flux linkage not the flux density itself. In order to have the flux density values and its spatial distribution, all the stator teeth should be equipped with search coils. Therefore, using this technique as a diagnosis medium is not recommended as it is practically impossible in industrial environment. However, it provides a very intuitive insight into the motor behavior. Assuming that there is such kind of measurement setup, Figure 8.4 shows the flux density over half the mechanical cycle. The third harmonic component of the flux density is a very good reason for an improved local saturation in one part of the motor core. This is another justification of the presence of a locally saturated region

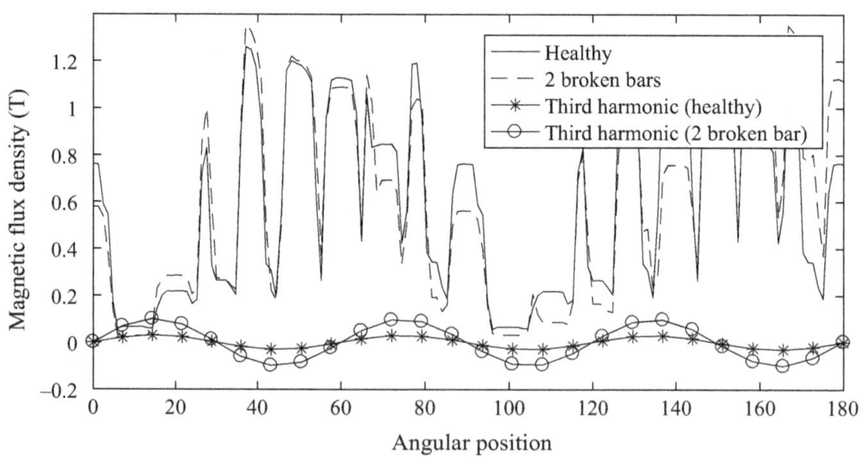


*Figure 8.4 Spatial distribution of magnetic flux density at air gap level along with corresponding third harmonic component*

caused by the broken bars fault. If a time-domain distribution of the magnetic flux is acquired by means of a search coil, a very promising signal will be obtained to analyze any kind of faults. Nevertheless, the significance of the motor flux signal does not affect the decision of not using it as it requires a very invasive approach.

In addition, the mentioned approach can be applied to both the line-start and inverter-fed motors regardless of the control strategy. It is naturally an online method and capable of addressing the effect of variations in load and speed. Besides, if a set of search coils is installed on all stator teeth, one can easily decide on the location of the broken bar fault by investigating the undesirable peaks present in the spatial distribution of the flux density.

The defected motor flux is reflected in the electro-motive force (EMF) signals. Therefore, the stator winding senses an unbalanced back EMF from the rotor side. So the undesirable fluctuations are added to the induced voltage. This concept is the same in the line-start and inverter-fed modes and has been used to introduce a new criterion for the detection of rotor broken bars fault. In the proposed off-line technique, the motor is temporarily disconnected from the supply for a very short period in order to measure the back EMFs. Then, the zero-sequence component is calculated as the fault index. This is a very good example of offline techniques. Its applicability has not been investigated at higher degree of the fault and load variations; hence, there are many questions to be answered. Moreover, the rotor broken bars fault usually takes place in industrial zones where it is impossible to disconnect the motor from the supply. This index is a time-dependent one and therefore not reliable enough as other types of faults, especially the short-circuit fault, affects the zero-sequence component.

## 8.6 Speed fluctuations

Most of the time, induction motor faults take place during the motor operation; hence, there is a transition from the healthy to faulty condition leading to some sort of undesirable transient fluctuations which stabilize after a couple of cycles up to the steady-state operation. Diagnosing the fault at its very initial instances is very important especially when the motor operates in start–stop operations as the mechanical transients caused by abrupt change of the motor speed increase the probability of fault occurrence. Therefore, having an online monitoring system is recommended in order to acquire the transient behavior of the motors caused by not only the start–stop operations but also the fault. The motor speed is one of the promising time-domain signals providing the opportunity to detect the incipient broken bars fault. This arises from the fact that the broken bars fault introduces a very clear $2sf_s$ harmonic component to the motor speed, and the corresponding effect is improved by increasing the fault level. Figure 8.5 shows the speed fluctuations for a line-start induction motor under three different fault levels including 1, 2 and 3 broken bars. The healthy motor operation is also shown to make the comparisons of the results. An incipient broken bars fault clearly introduces the undesirable speed fluctuations whose amplitude gets larger by increasing the fault level. The transient part of the signals which takes place at $t = 0.3$ s is of a great

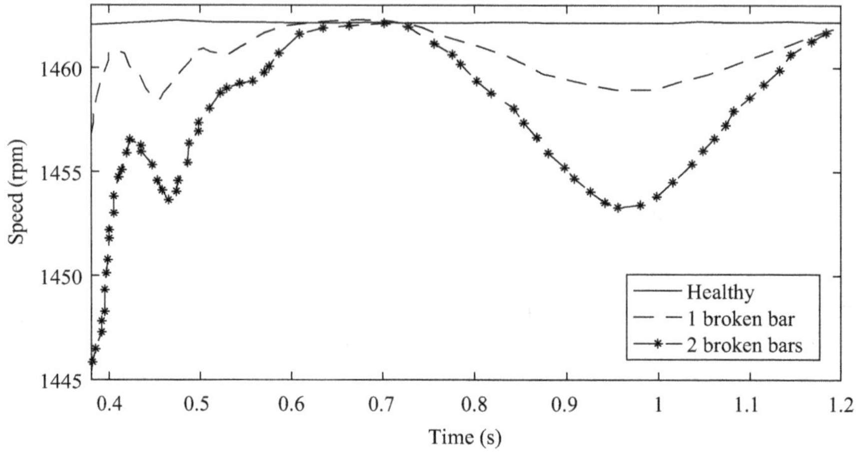


*Figure 8.5 Speed fluctuations caused by rotor broken bar fault in line-start mode*

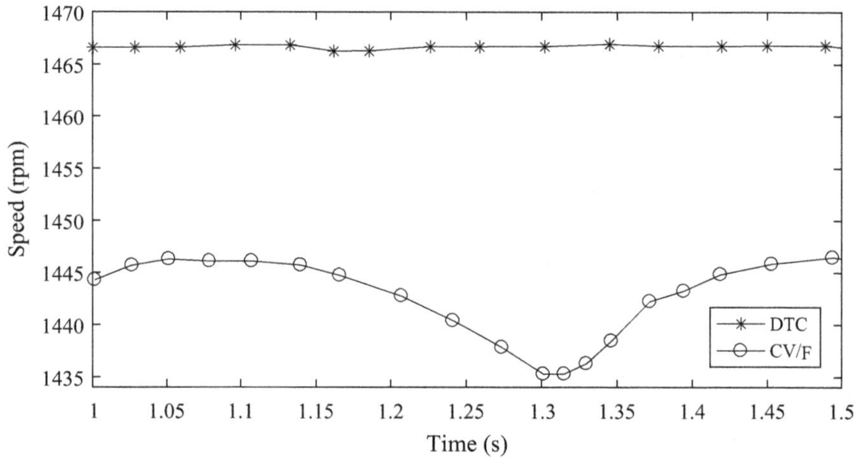


*Figure 8.6 Speed fluctuations caused by rotor broken bar fault in inverter-fed modes (2 broken bars)*

amplitude making the motor tolerate an improper mechanical stress which in turn is a factor of defecting other motor components. The transient part of the signals vanishes as time goes on up to $t = 0.6$ s. In addition, the speed signals come to their steady-state operations which are totally different from that of the healthy motor operation. The $2sf_s$ oscillations are obvious in Figure 8.5.

Figure 8.5 is an illustration of the motor signals in a line-start mode while Figure 8.6 represents the speed oscillations for the direct torque control (DTC) and constant voltage per frequency (CV/F) control modes considering two broken bars and the steady-state operation. This figure clearly proves the claim that incorporating various control modes can greatly affect the diagnosis procedure as amplitude of the oscillations is approximately equal to zero for the DTC strategy while the one related to the CV/F strategy reveals a relatively large amplitude. As there is a closed speed loop in a DTC drive, the corresponding mechanical speed is always regulated by the bandwidth of the PI controller; therefore, the amplitude of the oscillations might be lessened to zero if the bandwidth is so small.

## 8.7 Gyration radius

One of the most referable indices in this area is the Gyration Radius (*GR*) applied to the time-domain oscillations caused by the fault [73]. This method is applied to the signals with rather high variation rates. The results show that the index increases upon the higher fault degree. However, when the broken bar fault spreads such that one complete pole is eliminated, the mentioned trend does not continue anymore. Note that if the amplitude of the oscillations are not much enough to be detected with naked eyes, the GR will not probably be useful for the diagnosis purposes. Therefore, applying it to the closed-loop applications where there is a direct feedback applied to the motor quantities is not usually recommended.

The underlying idea of the GR has been applied to the current envelope oscillations which are similar to the torque or speed variations, but with different amplitudes. The corresponding phase space is build by calculating the points of the current envelop in time domain with the following order:

- The $i(k-10)$th array in horizontal axis during signal existing period
- The $i(k)$th array in vertical axis during signal existing period

where $k$ is an integer limited to the number of measured samples. The noticeable point of this index is that there is a considerable difference between no-load and on-load motors. It shows that the index can determine the faulty motor behavior at light-load conditions. As the local saturation level at no-load condition does not largely change by the fault degree increase, diagnosing the fault is usually impossible unless a proper method is introduced. In fact, the amplitudes of $2sf_s$ components are so small that they are masked by the fundamental component. The *GR* of the current envelope modulates the current envelope which only consists of fault-related components.

Tables 8.1 and 8.2 demonstrate the *GR* values for different operations including various load, and speed and fault levels in the line-start and inverter-fed modes.

*Table 8.1 GR in line-start mode*

| Load (%) | Healthy | 1 Broken bar | 2 Broken bars | 3 Broken bars | 4 Broken bars |
|---|---|---|---|---|---|
| 25 | 0.0045 | 0.0048 | 0.0050 | 0.0053 | 0.0068 |
| 50 | 0.0064 | 0.0064 | 0.007 | 0.0082 | 0.0129 |
| 75 | 0.0086 | 0.0091 | 0.0114 | 0.0145 | 0.0248 |
| 100 | 0.0099 | 0.0110 | 0.0153 | 0.0168 | 0.0331 |

*Table 8.2 GR for closed-loop inverter-fed mode ($n_r$: drive reference speed)*

| | Load (%) | 25 | 50 | 75 | 100 |
|---|---|---|---|---|---|
| Healthy | $n_r = 700$ | 0.0639 | 0.0547 | 0.0807 | 0.092 |
| | $n_r = 1{,}100$ | 0.0319 | 0.0557 | 0.0687 | 0.0962 |
| | $n_r = 1{,}500$ | 0.048 | 0.0647 | 0.0677 | 0.0852 |
| 1 Broken bar | $n_r = 700$ | 0.0705 | 0.0799 | 0.0842 | 0.1008 |
| | $n_r = 1{,}100$ | 0.0618 | 0.0760 | 0.0805 | 0.0998 |
| | $n_r = 1{,}500$ | 0.0605 | 0.0759 | 0.792 | 0.0858 |
| 2 Broken bars | $n_r = 700$ | 0.0816 | 0.080 | 0.1013 | 0.1197 |
| | $n_r = 1{,}100$ | 0.0082 | 0.084 | 0.081 | 0.1097 |
| | $n_r = 1{,}500$ | 0.081 | 0.094 | 0.0783 | 0.0997 |
| 3 Broken bars | $n_r = 700$ | 0.1030 | 0.1111 | 0.1254 | 0.1253 |
| | $n_r = 1{,}100$ | 0.1012 | 0.0991 | 0.113 | 0.1353 |
| | $n_r = 1{,}500$ | 0.1000 | 0.121 | 0.0990 | 0.1353 |
| 4 Broken bars | $n_r = 700$ | 0.1014 | 0.1500 | 0.1767 | 0.1834 |
| | $n_r = 1{,}100$ | 0.1024 | 0.1509 | 0.1547 | 0.1736 |
| | $n_r = 1{,}500$ | 0.1336 | 0.1449 | 0.1601 | 0.1636 |

By increasing the fault severity and load level, the index increases while it decreases upon the increase in speed level. The *GR* for the inverter-fed mode is generally larger than that of the line-start mode. This clarifies that the injection of fault components into the control loops disturbs the supply voltage waveform; therefore, a higher total harmonic distortion (THD) is expected. The GR is a noninvasive type of indices and can be applied to the on-line applications. The major problem with the *GR* is that it is not only a function of the broken bars fault, and the values vary even if other faults such as the eccentricity exists. Thus, the results are reliable only if it is possible to make sure that there is only one type of fault in the motor. Besides, the values provided in Tables 8.1 and 8.2 are only valid for the tested motor. The motor structure is the same as the one with 28 rotor bars investigated in Chapter 6. Equally important, the effect of the light-load level has been tested by means of precise experimental instruments which might not be available everywhere.

The point is that almost all fault indices must have a common basis for the assessments. In fact, the faulty signals should be compared to a common basis which is usually the index values in the healthy condition. Therefore, depending on the motor specifications, not only does the index value in the healthy condition vary, but the corresponding faulty operations will also produce different index values. As a result of this ambiguity caused by the motor structure and specifications, what is intended in the diagnosis procedures is the trend and rate of variations of indices as functions of the fault, load and speed level. The absolute value of the indices is often useless unless a reasonable normalized index value is introduced. However, as a trend in the fault diagnosis procedures, index values must be predetermined for the healthy operation. This is applied to both time- and frequency-domain indices.

Having mentioned the concept of a normalized index, it is noteworthy to say that one of the most important indices of this kind is the normalized amplitude of the sideband components of the motor current in the frequency domain. This issue will be further discussed in this chapter. Although the mentioned normalized index somehow addresses the previously explained issue in terms of different index values for different motors, it is still a challenge in the field, and no one has proposed a comprehensive index which covers a wide range of classes of induction motors.

## 8.8 Time-domain analysis of nonadjacent broken bars fault

During the diagnosis procedure, four concepts, namely, "detection," "determination," "location" and the "diagnosis" must be differentiated. These terms refer to apparently different aspects and are clarified in this chapter as the following:

- Detection is the process of realizing that the fault has taken place or not.
- Determination is the process of finding the fault level, for example, the number of broken bars
- Location is the process of finding the location of the fault. In the case of broken bar fault, different bars might be broken in different nonadjacent locations.
- Diagnosis is the process of gathering all the information obtained from the previous steps and exporting a value or measure which tells us about the motor behavior correctly.

The first aspect, detection, is probably the easiest part and can be dealt with using both the time- and frequency-domain analysis. However, without hiring the frequency domain, the detection process likely fails to detect the fault type unless a clear oscillation such as $2sf_s$ is observed in the time-domain variations of motor signals. The second aspect, determination, is indeed challenging, and it has been the most important focus of the research works done so far. It requires investigating motors behavior in different operating conditions including various load, speed, fault and temperature in order to check for the possible variation of the motor signals and the corresponding features which are used to extract fault indices. The location process has been targeted in works dealing with the broken bars fault. This is a significant aspect of the diagnosis of broken bars fault as determining the correct number of broken bars is the function of the location of the broken bars. As a matter of fact, as all origins of our perception are the variations of motor signals which might be affected by the fault level and the fault location. Understanding the correct motor condition requires investigating all possible influential factors. Thus, we try to address the issue by considering three critical fault locations and their effects on the motor time-domain signals (Figure 8.7). $\text{Case}_1$ shown in Figure 8.7(a) demonstrates two adjacent broken bars. $\text{Case}_2$ shows two broken bars with the distance of half of the pole pitch, and finally $\text{case}_3$ illustrates two broken bars located one pole pitch away from each other.

Figure 8.8(a) shows the current envelope for the mentioned three fault locations. On the basis of this figure, $\text{case}_1$, which is an illustration of two consecutive broken bars in one pole, has the largest peak–peak amplitude of oscillation while the corresponding average value is less than the others. This obviously proves that adjacent broken bars have the greatest effect on the motor behavior while the

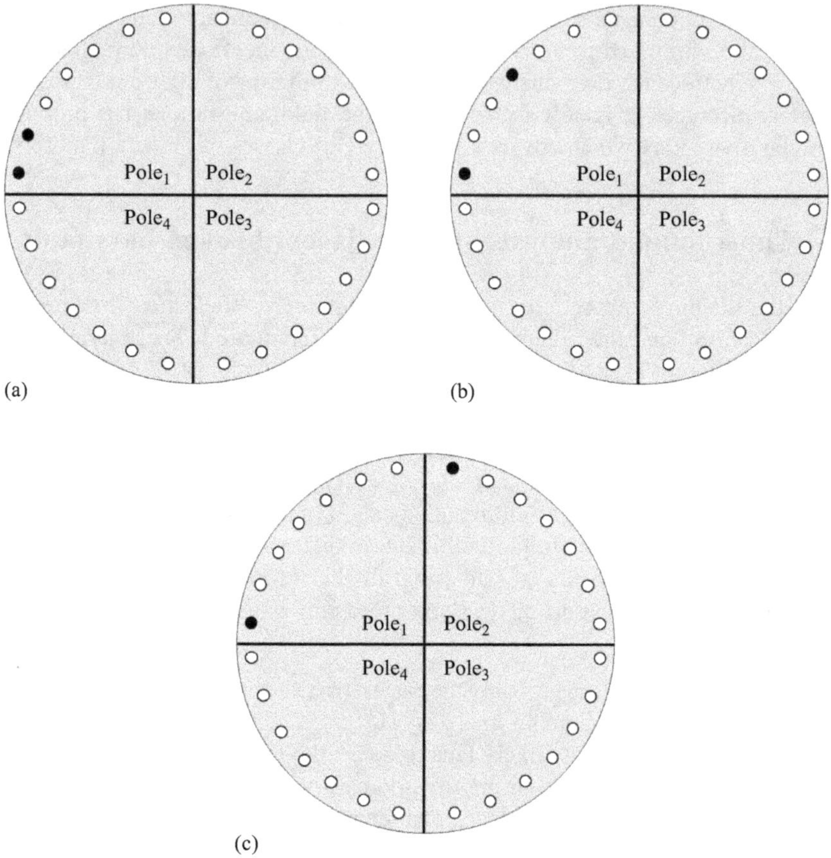


*Figure 8.7 Rotor broken bar (bb) locations (a) 2 adjacent bbs (case$_1$), (b) 2 bbs at half pole-pitch distance (case$_2$) and (c) 2 bbs at one pole-pitch distance (case$_3$), "bb" stands for "broken bar"*

broken bars located at a half-pole-pitch distance have the smallest effect. The broken bars located at one-full-pitch away from each other exhibit a medium influence compared to the other two cases. The important outcome of this investigation is that the two broken bars, if not adjacent, might be mistaken by a lower number of broken bars if the location of the fault is not taken into account. A similar trend is observed in the motor speed profile shown in Figure 8.8(b).

On the basis of Figure 8.8(b),

- The order of rise-time of the motor speed is as the following:

  $$\text{case}_3 > \text{case}_2 > \text{case}_1$$

  Rise-time means the first time that the speed reaches its average value in steady-state operation

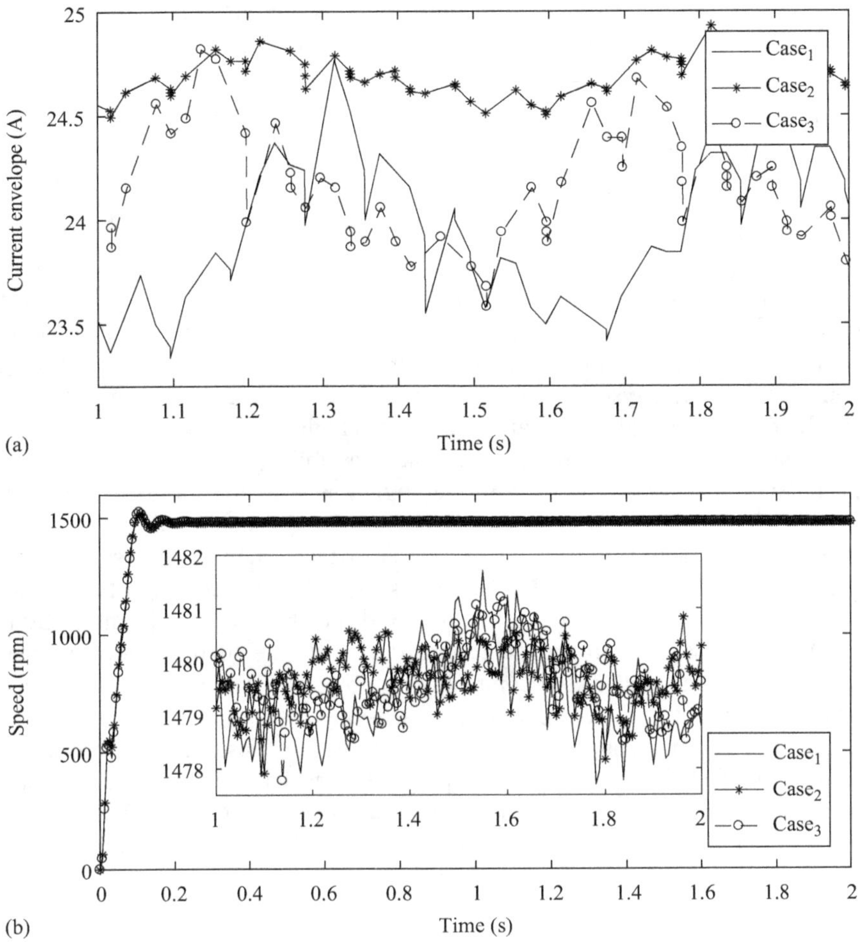


*Figure 8.8 Variations of motor variables for different fault locations (a) motor current profile, (b) motor speed profile and (c) motor speed in steady-state operation*

- The order of value of overshoot is as follows: (Figure 8.6(b)):

$$\text{case}_1 > \text{case}_3 > \text{case}_2$$

- The order of amplitude of oscillations in steady-state operation is as the following:

$$\text{case}_1 > \text{case}_3 > \text{case}_2$$

- The more severe the fault is, the more the average of the speed (steady-state) decreases.

So far in this chapter, the time-domain variation of the motor quantities including the current, the speed and the flux has been explained through illustrations in different operating conditions as follows:

- different load levels
- different fault levels
- different speed levels
- different fault locations.

On the basis of the results, one can generally conclude that increasing the fault and load level increases the value of the indices. On the other hand, depending on the motor speed or the fault location, various amplitudes of oscillations of the motor quantities are obtained. Most of the time-domain features can be extracted through an online monitoring system. Therefore, an incipient fault detection might also be possible if the processed signals and features are fault-sensitive enough. Moreover, the introduced concepts can differentiate the open-loop control strategies from the line-start one in the case where a proper signal such as the motor speed is processed. However, diagnosing the broken bars fault and generally all types of faults in a closed-loop application using the motor speed is almost impossible as the motor speed is directly controlled by means of a closed speed loop leading to a smooth speed operation without an explicitly detectable oscillation caused by the fault. When we say that it is possible to differentiate the supply modes, it means that the broken-bar-related $2sf_s$ component is obviously detected in an open-loop application while it does not apply to a closed-loop case.

The discussion mentioned above necessitates the presence of a process for selecting a proper signal for the diagnosis purposes. Not all the motor signals are able to reflect the fault in the same detectable fashion. The motor speed and torque signals are the best examples which do not satisfy the diagnosis goals in a closed-loop application. Selecting a proper signal requires consideration of some major factors including:

- It should be sensitive to the fault.
- A proper index can be extracted from the signal in the time, frequency or time-frequency domains.
- The extracted index should be more sensitive to the fault than the other factors such as the load and speed levels.
- The measurement process should preferably be noninvasive.
- An online implementation should be possible.

Normally, the time-domain signals are not satisfactory enough in terms of fulfilling the factors. Although some processors such as the GR make it possible to extract useful fault information, the results have been obtained in ideal operating conditions, using the precise laboratory instruments and also detailed simulation models. The industrial environment can be totally different, and there is a lot of factors harmfully affecting the measurements and the diagnosis procedure. In addition, if other types of defects affect the motor at the same time that the broken bar exists, the pure time-domain analysis is not recommended at all. Instead, a frequency domain or a combination of the time and frequency domains helps the diagnosis procedure better. Having discussed the issues associated with the time-domain signals, we proceed further through the frequency-domain analysis of the faulty motor signals. It should be noted that the general idea of applying the time, frequency or time-frequency analysis to the motor signals is not unique to the broken bars fault, and all types of faults such as eccentricity and short-circuit faults benefit from the same fashion of analysis. The difference is the selection of features or indices used for the diagnosis of a specific type of fault. Otherwise, the whole concept is the same. Therefore, the features and the indices related to the broken bars fault are comprehensively studied in this chapter with the purpose of explaining how the features can be explored and found in the time- and frequency domains. After getting used to the corresponding process, the next chapters which cover the eccentricity and short-circuit faults address the main indices without probably going through the depth of how they are extracted.

## 8.9 Spectrum of motor current

The motor current and its signature are the most commonly used signals in the diagnosis procedures of electrical machines as its measurement is very straightforward, and also it contains most of the required information for the fault diagnosis [80]. In addition, the corresponding measurement process is noninvasive. Although it is not the most sensitive signal for the diagnosis, it is generally appreciated in industry. Figure 8.9 shows an interesting FFT of the motor current, called the current spectrum, whose $y$-axis indicates the normalized amplitudes of the harmonic components of the motor current in the line-start mode. The spectrum has been normalized with respect to the amplitude of the fundamental frequency component which is equal to 50 Hz for the tested motor. This motor has 36 stator slots and 28 rotor bars. The rated current is equal to 24 A, and the corresponding voltage is 380V. As the entire spectrum is normalized, it is expected to have a fundamental frequency with the amplitude of zero. Note that a logarithmic scale is used for the calculations. Moreover, the factors of the fundamental harmonic components including 100, 150, 200 and 250 Hz can be seen in this figure. These are the main harmonic components of the motor which represent greater amplitudes compared to other slot/bar-related components located around them. The even harmonics are of a lower amplitude, and as they are close to the noise level, i.e. −80 dB, they can be neglected while the odd harmonics exhibit higher amplitudes. Therefore, they certainly affect the motor behavior. Normally, by increasing the frequency, the amplitude of the harmonic orders decreases unless a specific type of motor or defect is under the test.

Let us look back at the fundamental frequency, 50 Hz shown in Figure 8.9(b). The left and right sideband components are clearly observed in this figure. The frequency pattern of these components, as mentioned before, is $(1 \pm 2ks)f_s$. However, in Figure 8.9(b), the highlighted components are not related to different $k$ [the integer in $(1 \pm 2ks)f_s$]. They correspond to the first-order right and left sideband components obtained by setting k equal to 1 and varying the load level. The different load level must be distinctly discussed as it is one of the major factors affecting the diagnosis process. The "load level" is defined by slip, not the electromagnetic

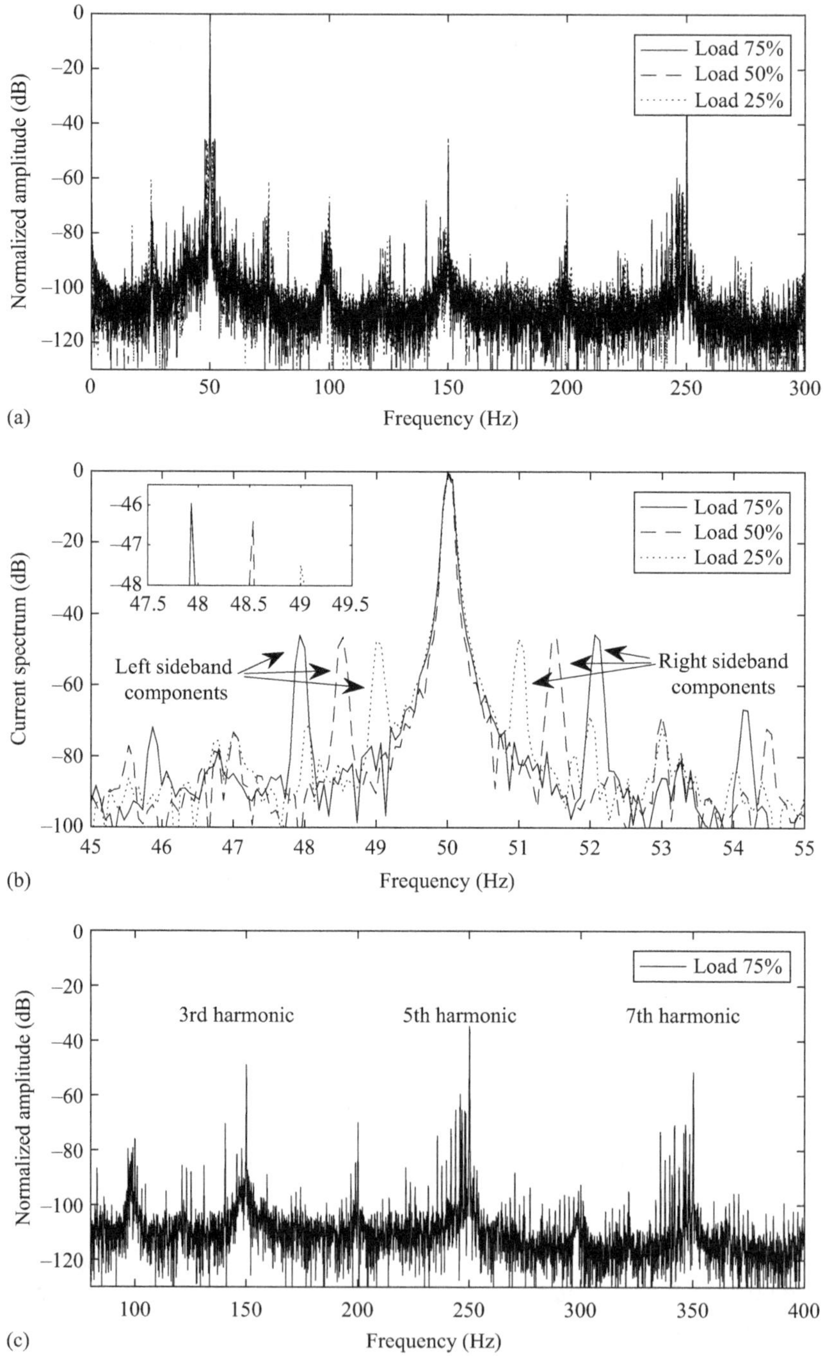


*Figure 8.9 Motor current spectrum in different frequency ranges (operating condition: one broken bar, three different load levels, line-start mode and steady-state operation) (a) 0–300 Hz, (b) 45–55 Hz and (c) 80–400 Hz*

torque level. Nevertheless, the percentage of electromagnetic torque will be equal to the slip expressed in percentage only if the motor supply frequency and the voltage are equal to the rated values. Otherwise, the only measure of the motor load to be used in the diagnosis procedure is the slip, not the torque level.

Taking the above-mentioned points into account and by increasing the motor slip level or the load level, the amplitude of the sideband components increases, while the components take longer distance from the fundament component. The left sideband component moves toward zero, and the right one moves toward higher frequencies. This trend is always fixed for induction motors, and the only difference is in the amplitude and the rate of change in the amplitude of sideband components. It is worth mentioning that the FFTs shown in Figure 8.9 have been obtained by applying hamming window to the FFT. If not so, you will not probably be able to observe the sideband components, especially in light-load levels. Moreover, the sampling time is larger than 5 s for a sampling frequency of 12 kHz. Although this is more than enough for the fault diagnosis of the line-start motors, we prefer to maintain the same adjustments to have proper basis for comparisons between the line-start and inverter-fed modes. On the other hand, as the variation of sideband components is not much as the load level increases, reducing the resolution of the frequency-domain might lead to false results. In Figure 8.9, the amplitude of the right sideband component increases by the increase in the load level. This is a very well-appreciated trend which does not usually happen in practice. The right component highly depends on the load and speed fluctuations such that any small disturbance in the speed causes a large change of amplitude in the right sideband components. Therefore, it is not guaranteed that by increasing the load, an increase in right sideband amplitude occurs.

The following features are extractable from Figure 8.10.

- In no-load case, the sideband component is eliminated from the current spectrum, no matter how much the fault level is (Figure 8.10(a)). This is called the masking effect of the fundamental harmonic component and will be dealt with later in this chapter.
- At a fixed load level, increasing the fault severity leads to an increase in the amplitude of the sideband components (Figure 8.10(b)). The rate of increase is usually larger than that of increase in the load level.
- Increasing the fault level makes the higher order sideband components visible as their amplitudes become larger than that of the noise level which exists in the entire frequency range of the spectrum. So this is called the white noise (Figure 8.10(c)). The amplitude of the orders of the sideband components decreases as the order increases.

As mentioned earlier, the no-load condition is very challenging due to the presence of the masking effect of the fundamental components. This also matters in the time-domain analysis of the motor current. However, a very wise mathematical transform, the envelope formulation, can somehow eliminate the masking effect if a high-resolution FFT is used and applied to the motor current envelope (Figure 8.11). Even in the no-load condition, the $2ksf_s$ components of the current envelop are detectable in the spectrum. This means that the masking effect is by-passed by

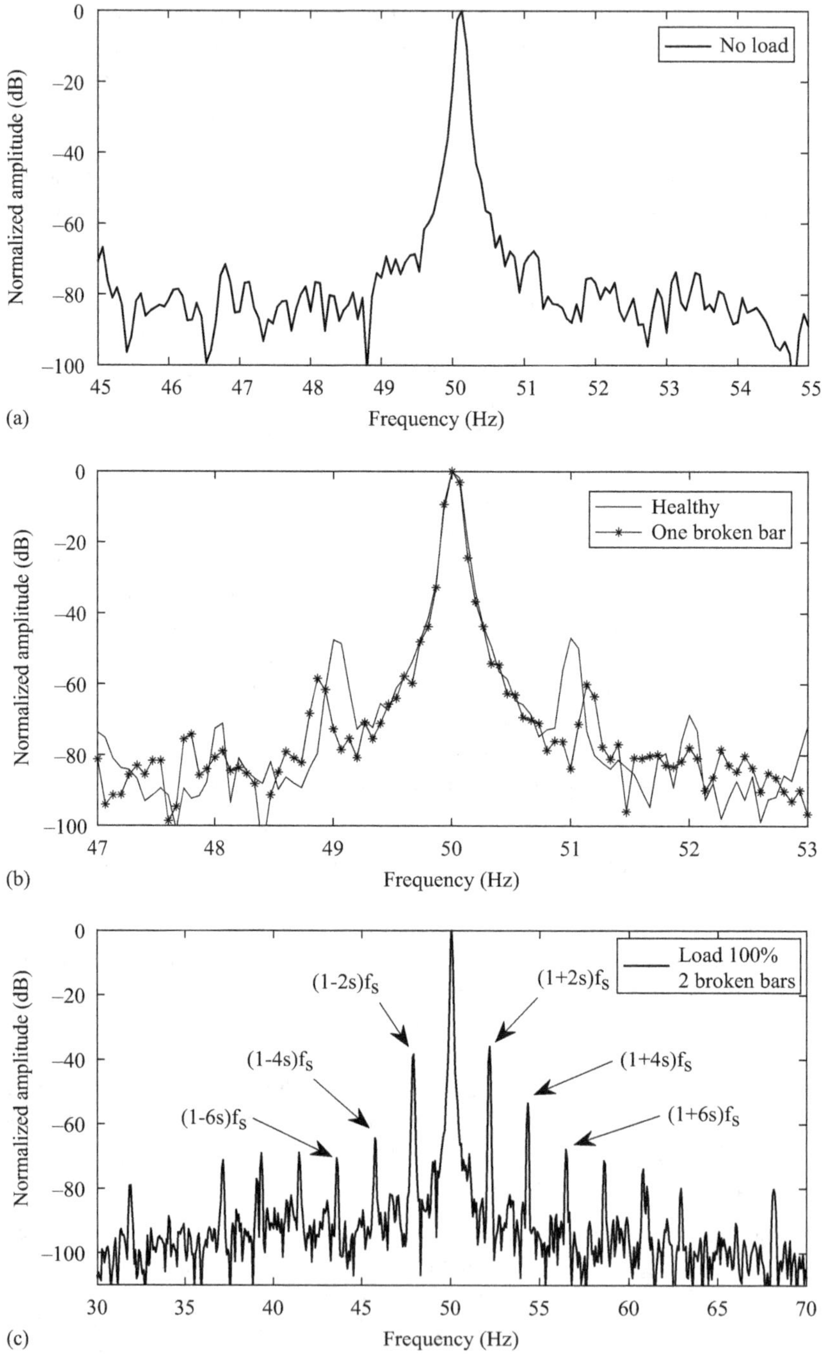


*Figure 8.10 Motor current spectrum in line-start mode (a) one broken and no-load condition, (b) healthy and one broken bar rotor and partial load (25%) condition and (c) two broken bars and full-load condition*

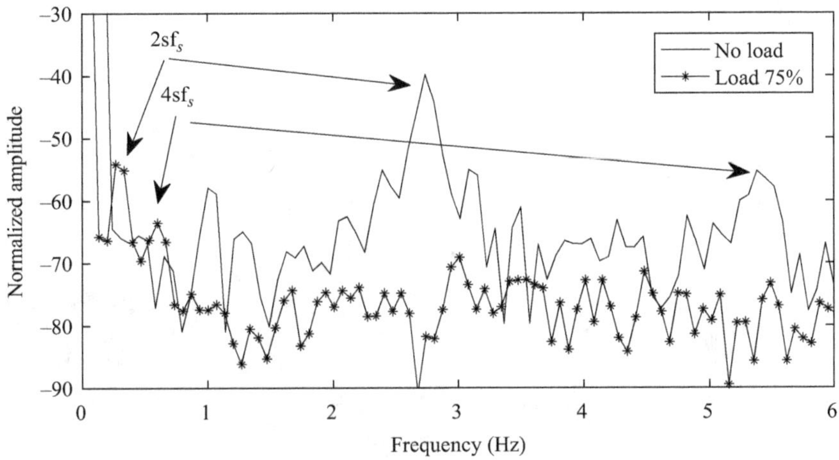


*Figure 8.11 Spectrum of motor current envelope in line-start mode*

modulating the sideband components to the low-frequency components. This clever idea is the foundation for the HHT-based diagnosis procedure of induction motors with the broken bars fault and will be further discussed later in this chapter. Increasing the load level has the same effect on the fault-related components of the current envelope as it has on the sideband components of the current itself.

The current spectrum is definitely the most cited approach to the diagnosis of the broken bars fault in induction motors. Therefore, many attempts have been already made to improve its applicability to different motor applications including the line-start and inverter-fed modes. Figure 8.12 shows the current spectrum in the open-loop CV/F and DTC modes for a motor with two broken bars. The tested motor is the same as the one used for obtaining the previous figures in this chapter. The features of the motor current spectrum in the inverter-fed modes are:

- The sideband components also exist in the inverter-fed mode regardless of the control strategy.
- The amplitude of the sideband components differs in open- and closed-loop applications.
- The positions of the sideband components as well as the fundamental component in the $x$-axis of the spectrum are different regarding the control mode. That of the CV/F is somehow similar to the line-start application in which changing the load level does not change the position of the fundamental frequency (50 Hz). However, that of the DTC mode moves toward larger frequencies as the motor load increases. This comes from the fact that induction motors slip increases by increasing the load level. Besides, the closed-loop drive always tries to keep the track of the reference frequency and applies it to the rotor mechanical speed. Therefore, in the DTC mode, the rotor mechanical speed is equal to 1500 rpm as this is a four-pole machine, and the supply frequency should be larger than 50 Hz in order to compensate for the slip.

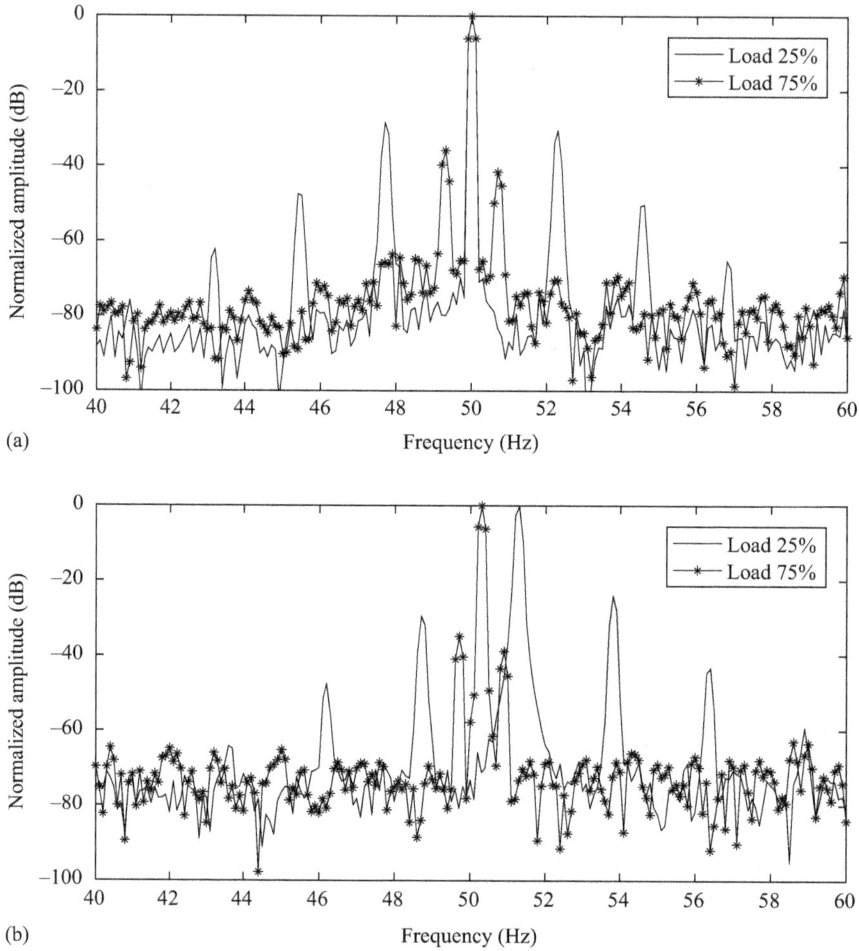


*Figure 8.12 Motor current spectrum (2 broken bars and reference frequency of 50 Hz) (a) open-loop CV/F mode and (b) closed-loop DTC mode*

Using the current spectrum or its derivatives requires a noninvasive technique in which the current transducers are used to measure the current. Then, a processor is required to apply the FFT to the measured signal through an online process. This signal, the motor current spectrum, is very sensitive to both the fault and load level. It is also a function of the supply frequency. This is almost the same in all the fault indicators used for detecting the broken bars, eccentricity and short-circuit faults. Generally, there is no fault index/indicator which is sensitive to fault, not the other factors. This conveys the idea that all the factors should be studied in order to have a reliable fault diagnosis procedure. The main disadvantage of a pure spectral analysis of the motor current is that it only addresses the steady-state operation and has nothing to do with the transient mode of the motor operation. Nevertheless, it can resolve some of the issues related to the inverter-fed applications. Tables 8.3–8.5

*Table 8.3 Normalized amplitude of first-order sideband components in line-start mode (unit: dB, supply frequency: 50 Hz)*

| | **$(1-2s)f_s$** | | | **$(1+2s)f_s$** | | |
|---|---|---|---|---|---|---|
| **Load level (%)** | **25** | **50** | **75** | **25** | **50** | **75** |
| **Healthy** | −58.44 | −56.7 | −55.97 | −60.04 | −57.78 | −55.84 |
| **1 Broken bar** | −47.49 | −46.4 | −45.96 | −47.81 | −46.67 | −45.99 |
| **2 Broken bars** | −38.55 | −38.4 | −38.32 | −38.61 | −36.34 | −35.9 |
| **3 Broken bars** | −34.71 | −33.51 | −33.1 | −34.48 | −34.02 | −32.06 |
| **4 Broken bars** | −31.2 | −30.02 | −29.57 | −30.58 | −29.83 | −29.21 |

*Table 8.4 Normalized amplitude of first-order sideband components in CV/F mode (unit: dB, reference frequency: 50 Hz)*

| | **$(1-2s)f_s$** | | | **$(1+2s)f_s$** | | |
|---|---|---|---|---|---|---|
| **Load level (%)** | **25** | **50** | **75** | **25** | **50** | **75** |
| **Healthy** | −56.87 | −55.24 | −55.01 | −57.43 | −55.71 | −55.5 |
| **1 Broken bar** | −47.81 | −47.1 | −46.2 | −48.6 | −47.14 | −45.88 |
| **2 Broken bars** | −39.44 | −37.83 | −37.45 | −40.69 | −35.84 | −33.8 |
| **3 Broken bars** | −36.46 | −33.13 | −32.97 | −36.34 | −32.5 | −30.73 |
| **4 Broken bars** | −31.99 | −29.91 | −29.26 | −33.1 | −30.08 | −28.26 |

*Table 8.5 Normalized amplitude of first-order sideband components in DTC mode (unit: dB, reference frequency: 50 and 30 Hz)*

| | | **$(1-2s)f_s$** | | | **$(1+2s)f_s$** | | |
|---|---|---|---|---|---|---|---|
| **Load level (%)** | | **25** | **50** | **75** | **25** | **50** | **75** |
| **Healthy** | $f_s = 50$ Hz | −55.18 | −54.05 | −53.8 | −58.56 | −54.21 | −53.98 |
| | $f_s = 30$ Hz | −65.59 | −59.58 | −59.41 | −69.75 | −60.78 | −57.09 |
| **1 Broken bar** | $f_s = 50$ Hz | −51.09 | −46.45 | −46.12 | −51.23 | −46.79 | −44.66 |
| | $f_s = 30$ Hz | −48.24 | −48.25 | −45.55 | −50.39 | −48 | −46.3 |
| **2 Broken bars** | $f_s = 50$ Hz | −40.5 | −38 | −37.37 | −41.68 | −36.52 | −35.17 |
| | $f_s = 30$ Hz | −41.15 | −39.04 | −38.59 | −45.21 | −40.08 | −37.73 |
| **3 Broken bars** | $f_s = 50$ Hz | −34.98 | −34 | −33.57 | −35.49 | −31.47 | −31.13 |
| | $f_s = 30$ Hz | −35.17 | −34.86 | −34.53 | −40.04 | −35.68 | −34.97 |
| **4 Broken bars** | $f_s = 50$ Hz | −32.19 | −31.27 | −30.84 | −31.68 | −28.94 | −28.46 |
| | $f_s = 30$ Hz | −32.34 | −30.52 | −31.07 | −37.25 | −32.16 | −31.25 |

give the amplitude of the first-order left and right sideband components in the line-start, CV/F and DTC modes. The line-start and CV/F motor are only tested at the supply frequency of 50 Hz, while the DTC one contains two different frequencies, namely 30 and 50 Hz in order to consider the effect of different speed levels.

The noticeable points are as follows:

- The rate of increase in the index amplitude as the function of the fault level is greater than the rate as the function of the motor load level. This is the case for all the three investigated modes.
- The rate of increase is reduced as the number of the broken bars increases. In fact, there is a trend like saturation.
- Decreasing the motor speed in the DTC mode which is handled by changing the supply frequency leads to a general increase in the index values.
- In general, the CV/F mode represents a more disturbed current signal as the corresponding sideband components have larger amplitudes. The DTC and line-start modes exhibit almost the same values. However, it will be proved that the inverter-fed mode, either open- or closed-loop, increases the amplitude of the indices. The Hilbert–Huang transform (HHT) will be used to discriminate various control strategies and differentiate them from the line-start mode.
- The data provided in the tables above are machine-specific which means that the values, although normalized, are subjected to change if a new motor with a different structure is used. It is noteworthy that the normalized values will be almost the same for two motors with the same structure but different power ratings.

Having discussed the inverter-fed applications and their effect on the diagnosis process, we can see that the amplitudes of the sideband components are different compared to the line-start mode due to the fact that the supply voltage of the inverter-fed applications is not a pure sinusoid, and it consists of higher order harmonic components (discussed in Chapter 2). Particularly, the sideband components are detectable in the motor voltage in the case of closed-loop applications. This issue is demonstrated in Figure 8.13. The simulated machine is a three-phase,

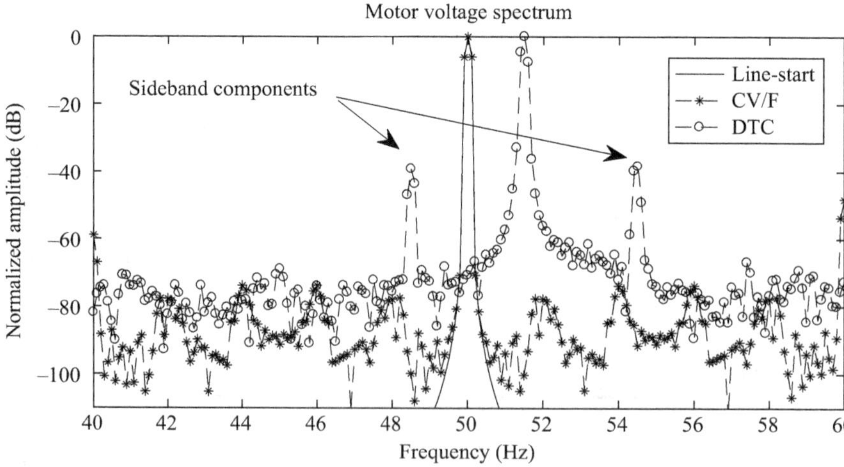


*Figure 8.13 Motor voltage spectrum (operating condition: full load, one broken bar)*

11 kW induction motor. To arrange for a common basis between the supply modes, the reference speed of the CV/F and DTC modes are set to 1,500 rpm which means the reference frequency is equal to 50 Hz. This is exactly equal to that of the line-start mode, the network frequency.

The most significant features deducible from Figure 8.13 are as follows:

- The fundamental frequency component of the DTC mode is almost 1.5 Hz away from that of the line-start and CV/F modes. This arises from the fact that the reference frequency has been set to 50 Hz. As the DTC mode is equipped with a speed closed loop applied to the rotor, the rotor speed should come to 1,500 rpm or 50 Hz. In effect, the stator synchronous frequency should be larger than 50 Hz to compensate for the slip frequency which is almost equal to 1.5 Hz in full-load condition.
- Surprisingly, the fault-related components, the sideband components with the pattern of $(1 \pm 2ks)f_s$, are only observed in the voltage spectrum of the DTC mode. This literally proves the previous claim in terms of the possibility of flowing the fault-related components through the closed loops.
- The line-start and CV/F modes reveal the same supply frequency without any sideband components. This is a compelling reason for the claim that the voltage signal is not recommended for the broken bar diagnosis in any supply mode with an open-loop nature as there is no fault information inside.
- If the goal is to detect the sideband components regardless of their positions in the spectrum (Figure 8.13, the DTC mode), a method which is independent of the motor supply frequency and also the motor load is preferred.

## 8.10 Effect of closed loops on faulty motor signals

There are two scenarios for qualifying the existence of the voltage sideband components. One is the presence of the closed speed loop, and the other is the closed flux loop utilized in a closed-loop drive such as the DTC [105]. Each loop has its own contribution to the motor harmonic component, depending on the bandwidth of the loop. Therefore, investigating the bandwidth of the loops will be probably useful for determining which one contributes to the fault-related voltage components. To this end, the closed speed loop is analyzed first as it is possible to formulate the corresponding transfer function analytically. Let us look at Figure 8.14.

The motor is modeled by its first-order transfer function $G_m(S) = \frac{1}{(J_m S + B)}$ where $J_m$, $S$ and $B$ are the motor inertia, Laplace argument and the viscos friction coefficient, respectively. The transfer function of the speed controller including the Proportional ($P$) and Integral ($I$) gains is expressed as follows:

$$G_{sc}(S) = k_p + \frac{k_I}{S} \tag{8.12}$$

where $k_p$ and $k_I$ are the proportional and integral coefficients, respectively. As the motor current and accordingly the motor torque are always restricted to a specific

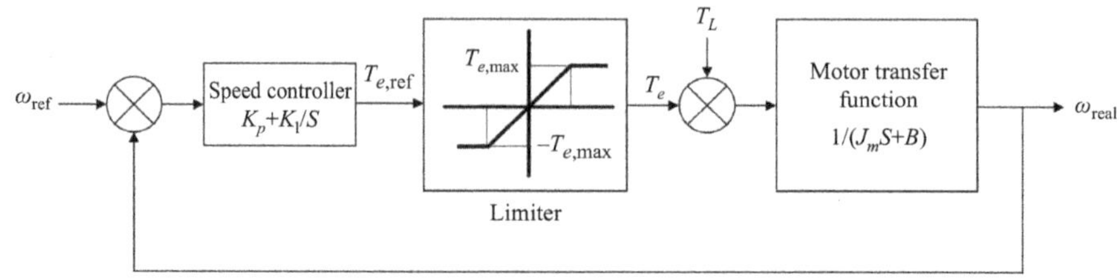


*Figure 8.14 The block diagram of the closed speed loop*

value demonstrating the maximum available current and torque quantities by means of a limiter, the second block from the left in Figure 8.14 is combined and called the "torque-limiter" which applies upper and lower bounds to the motor torque. The corresponding transfer function is as the following:

$$T_e = \begin{cases} \dfrac{T_{e,\max}}{T_{e,\text{ref}}}, & T_{e,\text{ref}} \geq T_{e,\max} \\ 1, & -T_{e,\max} < T_{e,\text{ref}} < T_{e,\max} \\ -\dfrac{T_{e,\max}}{T_{e,\text{ref}}}, & T_{e,\text{ref}} \leq T_{e,\max} \end{cases} \tag{8.13}$$

where $T_{e,\max}$, $T_{e,\text{ref}}$ and $T_e$ are the maximum available torque, the reference torque and the achievable torque, respectively. As seen, the limiter block is independent of the frequency and also the Laplace argument. Therefore, it does not produce any phase shift and only affects the amplitude of signal. That is why it can be easily modeled as a constant expressed by $G_{\text{lim}}(S) = C$. Having calculated the block-wise transfer functions, the total transfer function of the closed loop is calculated as follows:

$$G(S) = \frac{G_{sc}(S)G_{lim}(S)G_m(S)}{1 + G_{sc}(S)G_{lim}(S)G_m(S)} = \frac{\left(K_p + \dfrac{K_I}{S}\right)(C)\left(\dfrac{1}{(J_mS + B)}\right)}{1 + \left(K_p + \dfrac{K_I}{S}\right)(C)\left(\dfrac{1}{(J_mS + B)}\right)}$$
$$= \frac{(K_pS + K_I)}{(J_m/C)S^2 + ((B/C) + K_p)S + K_I} \tag{8.14}$$

The matching Bode diagram corresponding to (8.14) is shown in Figure 8.15 for the parameters listed in Table 8.6.

The amount of $J_m$ reported in Table 8.6 embraces only the shaft inertia while the load inertia should also be added to this value if a more realistic analysis is targeted. Therefore, it is assumed that the total inertia is equal to twice that of the single rotor to which no load is connected. The cut-off frequency of the speed control loop is almost equal to 1 Hz beyond which the amplitude of the frequency components is weakened depending on the filter gain, located on the negative slope of the "magnitude plot" (Figure 8.15, the top subplot). Therefore, any frequency

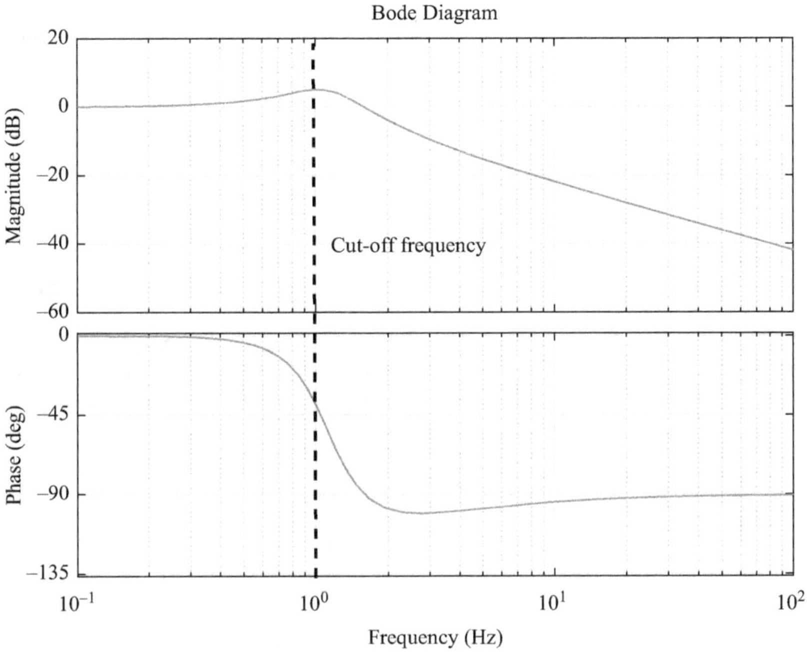


*Figure 8.15 Frequency response of transfer function of speed closed loop*

*Table 8.6 Parameters of motor*

| | |
|---|---|
| $K_p/K_I$ | 10 |
| $T_{e,\max}$ | 70 N m |
| $C$ | 1 |
| $J_m$ | 0.0625 (kg/m$^2$) |
| $B$ | 0.0001 (kg/s m) |

component larger than 1 Hz will be practically filtered out which means that the $(1 \pm 2ks)f_s$ pattern will not be detectable throughout this loop, and the $2ksf_s$ pattern hardly passes through this loop as the corresponding frequencies are typically located above 2 or 3 Hz depending on the motor slip (Figure 8.11). Hence, it is expected to have very smooth time- and frequency-domain variations of the motor speed in the closed-loop applications (Figure 8.6).

In order to explain the issue further, the time- and frequency-domain analysis of the motor speed for the line-start, CV/F and DTC operations are provided in Figure 8.16. On the basis of this figure, the low-frequency $2ksf_s$ components do not even exist in the DTC mode, neither in the time domain nor in the frequency domain.

This proves that the speed/torque control loop does not contribute to the production of the voltage sideband components in the DTC mode as the fault-related

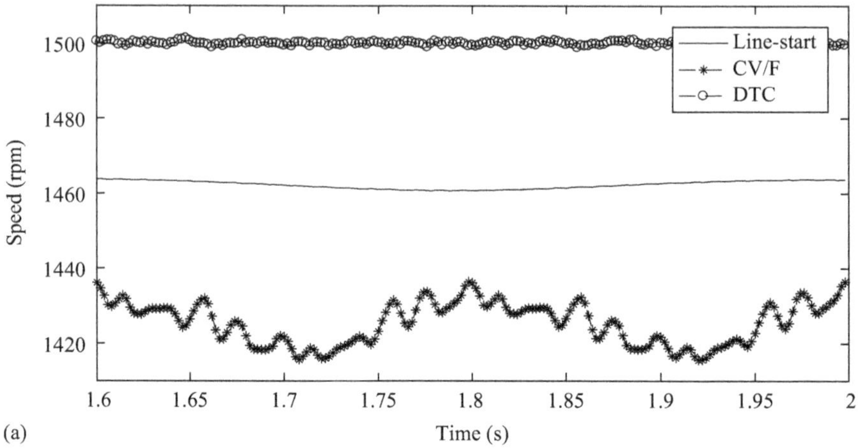


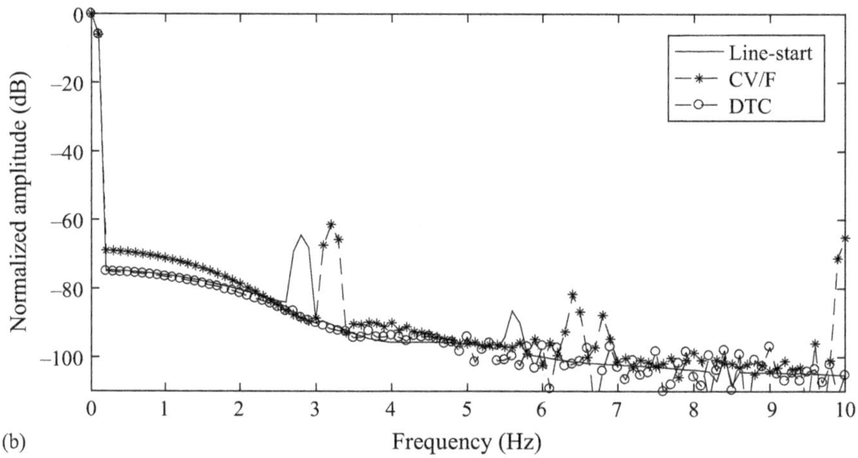


*Figure 8.16 (a) Time-domain variation and (b) frequency-domain variation of motor speed (operating condition: full load and one broken bar)*

components are filtered out by the transfer function of the loop. However, $2ksf_s$ pattern is clearly observed in speed spectrum of the line-start and CV/F modes. Remarkably, the amplitudes of the CV/F components are larger than that of the line-start because higher order harmonic components are always applied to the motor in the inverter-fed modes while the line-start mode has a pure sinusoidal supply. The DTC speed is completely free of the fault-related components at least in terms of the speed signals. Reaffirming that the motor speed/torque control loop cannot be considered as a sideband-producing loop, the remaining loop is the flux control loop. If the motor flux consists of the fault-related components, it would be the only reason to produce the sideband components of the motor voltage. Figure 8.17 shows the flux spectrum of the faulty motor for the DTC mode.

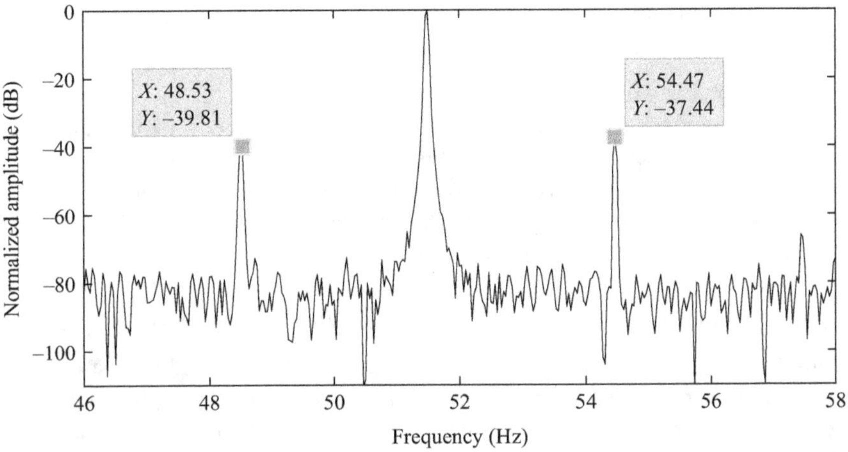


*Figure 8.17 Stator flux spectrum for DTC mode (operating condition: full load and one broken bar)*

Obviously, the sideband components with the amplitudes very close to that of the voltage spectrum are visible in the stator flux spectrum (Figure 8.17). One can easily conclude that it is the flux control loop which is responsible for flowing the fault-related components into the motor supply voltage in the DTC mode. Although there is also a closed loop applied to the motor flux, it functions in the fundamental component, not the fault-related components. Besides, there is no low band-pass PI controller in the flux control loop; hence, unlike the speed control loop in which the fault components are filtered out, the fault components can freely flow through the flux loop. Grounded on the information provided so far, the following guidelines are recommended:

- The supply voltage reveals the fault information only in the DTC mode. In fact, unlike the line-start and CV/F voltages which only contain the fundamental harmonic components of the network or the switching supply, the DTC supply voltage possess a more complex pattern in the presence of the broken bar fault. Equation (8.15) represents the corresponding voltage formulation:

$$V_{DTC} = \sum_{n} \left( V_{mn} \sin(n\omega_s t) + \sum_{k} \begin{pmatrix} V_{l_{nk}} \sin((1-2ks)n\omega_s t - \alpha_{vl_{nk}}) \\ + V_{r_{nk}} \sin((1+2ks)n\omega_s t - \alpha_{vr_{nk}}) \end{pmatrix} \right) \tag{8.15}$$

where $m$, $n$, $V_{mn}$, $\omega_s$, $t$, $k$, $V_{l_{nk}}$, $\alpha_{vl_{nk}}$, $V_{r_{nk}}$ and $\alpha_{vr_{nk}}$ are the order of the voltage harmonic component, the amplitude of the $n$th voltage harmonic component, the synchronous frequency in rad/s, time, and integer indicating the $k$th sideband around the voltage components, the $k$th left sideband of the $n$th voltage component, the phase angle of the $k$th left sideband of the $n$th voltage component, the $k$th right sideband of the $n$th voltage component and the phase angle of the $k$th right sideband of the $n$th voltage component, respectively.

- The motor speed discloses the fault information only in the line-start and CV/F modes unless the cut-off frequency of the speed loop is set to a value larger than that of the $2ksf_s$ pattern. In this case, the declared components will be changed into the supply voltage through the torque control loop.
- The speed oscillations triggered by the broken bar fault are well damped by the speed closed loop in the DTC mode.
- The source of producing a fault-related component in the motor voltage is generally the flux control loop, not the speed/torque control loop.

It seems that the CV/F and line-start modes have the highest and lowest effects on the sideband components, respectively while the DTC mode applies a moderate effect inferred from matching left sideband amplitude. The right sideband component is not reliable as it is mostly a function of the motor speed and load fluctuations, not the fault. The illustrated trend in Figure 8.7 might be violated depending on the topology of the inverter. A voltage source inverter has been utilized in this study while a current source inverter defines a different behavior due to the presence of current controllers. What is more, not only the frequency-domain, but also the time-domain characteristics of the motor current are affected by the fault regardless of which operating modes are dealt with. That is exactly why the motor current and its various modulations have been greatly utilized in the literature.

## 8.11 Analytical analysis of the effect of speed variation on sideband components

As mentioned before, increasing the motor speed leads to a decrease in the amplitude of the sideband components. This is an ideal case in which there is no oscillation produced by the load. To justify the discussed trend, the analytical form of the left sideband amplitude is considered as (8.16). This equation is obtained by considering a linear magnetization profile of the silicon steel material and can be found in the literature:

$$i_l = \sqrt{3}I_l' e^{j[(1-2s)\omega_s t - a_l]}\left[1 - \frac{3p\Psi^2}{24s^2\omega_s Z_s J}\right], \quad Z_s = \sqrt{R_s^2 + X_s^2} = \sqrt{R_s^2 + (\omega_s L)^2} \tag{8.16}$$

where $i_l$, $\psi$, $I_l'$, $\omega_s$, $a_l$, $s$, $t$, $Z_s$, $J$, $R_s$ and $X_s$ are the current component related to the left sideband component, the motor flux linkage, the amplitude of the left sideband component, the synchronous frequency, the phase angle of the left sideband component, the slip, the time, the stator impedance, the rotor inertia, the stator resistance, the stator inductance, respectively. Often, a constant flux ($\psi$) control strategy is used to control the motor speed below the rated speed. On the other hand, the maximum available torque is constant in the constant flux region (see Figure 8.18). Therefore, $s\omega_s$ is fixed upon decreasing the motor speed level unless the load level changes.

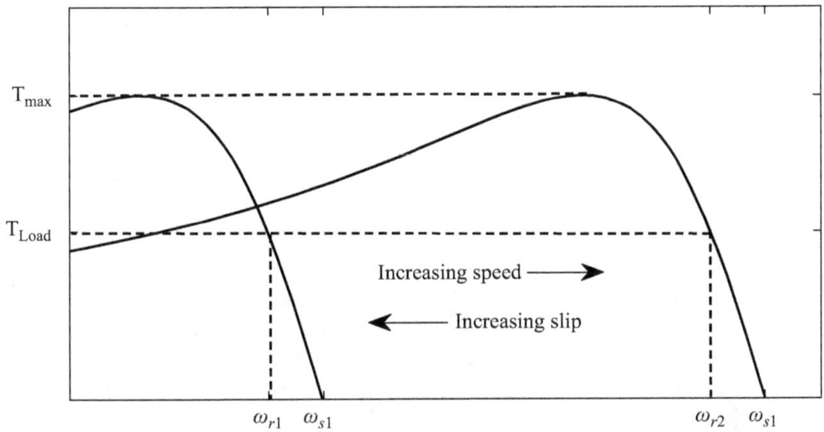


*Figure 8.18 Torque–speed curve at different speed*

Note that the slop of the stable operation region of the torque–speed profile does not change in the constant flux region. Considering all the mentioned points, we have:

$$s = \frac{\omega_s - \omega_r}{\omega_s} \Rightarrow s\omega_s = \omega_s - \omega_r \quad \underset{\omega_s - \omega_r = \text{fixed}}{\overset{\text{Constant flux region}}{\Rightarrow}} \quad s\omega_s = \text{fixed} \tag{8.17}$$

The amplitude of the left sideband component of the motor current is obtained as follows:

$$\begin{aligned} \text{Amplitude of left sideband} &= \sqrt{3}I_l' \left[ 1 - \frac{3P\Psi^2}{24J \times (s\omega_s) \times s \times \sqrt{R_s^2 + (\omega_s L)^2}} \right] \\ &= \left[ \sqrt{3}I_l' - \frac{k}{s\sqrt{R_s^2 + (\omega_s L)^2}} \right], \quad k = \frac{3\sqrt{3}I_l' P\Psi^2}{24J(s\omega_s)} \end{aligned} \tag{8.18}$$

The rotor reference speed rises by increasing $\omega_s$ and hence this leads to the increase in the amplitude of the sideband component. Moreover, by increasing $\omega_s$, the slip is reduced because the term $(\omega_s - \omega_r)$ is fixed at a constant load level. On the other hand, the slip reduction, which is caused by increasing the supply frequency/speed, lessens the amplitude of the sideband component. Therefore, increasing the speed while decreasing the slip reveals counter effects. So investigating the sensitivity of the amplitude of the sideband component with respect to the motor speed and slip levels will be helpful to see which one has greater effect so that we can decide if the amplitude should decrease in general or increase. The provided discussion is only valid for the regions where s is nonzero. On the other hand, due to the inductive slippery nature of induction motors, there is always a difference between $\omega_r$ and $\omega_s$.

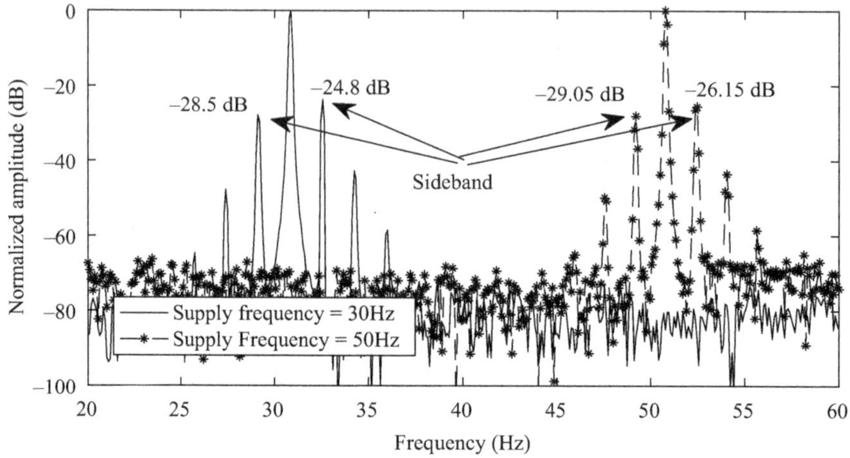


*Figure 8.19 Motor current spectrum for different supply frequencies (operating condition: full load and one broken bar)*

so the assumption of a nonzero slip is always true. Even if the drive reference speed is adjusted at synchronous frequency, the supply frequency would be slightly greater than that of the reference value in order to compensate for the slip losses. This is needed to ensure the proper operation of a closed-loop inverter-fed induction motor. The sensitivity analysis can be carried out as follows:

$$S_s^{\text{amplitude}} = \left.\frac{\partial \text{amplitude}}{\partial s}\right|_{\omega_s = cte} = \frac{k}{s^2\sqrt{R_s^2 + (\omega_s L)^2}} \tag{8.19}$$

$$S_{\omega_s}^{\text{amplitude}} = \left.\frac{\partial \text{amplitude}}{\partial \omega_s}\right|_{s = cte} = \frac{kL^2\omega_s}{s\left(R_s^2 + (\omega_s L)^2\right)^{3/2}} \tag{8.20}$$

To realize the factor with a larger effect, the ratio of two sensitivities is calculated as the following:

$$\frac{S_s^{\text{amplitude}}}{S_{\omega_s}^{\text{amplitude}}} = \frac{R_s^2 + (\omega_s L)^2}{sL^2\omega_s} > 1 \tag{8.21}$$

As the motor slip is normally very smaller than one, this ratio is always larger than one. Therefore, the slip variation has a greater effect on the amplitude of the sideband component compared to the speed variation. The typical ratio of the tested induction motor at the rated load and the angular frequency of 157 rad/s is equal to 189.32. This amount shows that by increasing the speed and consequently decreasing the slip, the amplitude of the sideband should be reduced in the constant flux-operating region.

## 8.12 Motor power spectrum

The motor power spectrum is another signal which reflects the broken bars fault clearly. In order to deal with the power spectrum and provide justifications for the corresponding fault-related components, an interesting analytical development is provided. As the motor power is nothing but the product of the current and voltage signals, the voltage and current formulations under the fault condition are derived as (8.22) and (8.23). This is a general derivation which consists of all possible harmonic components caused by the fault in both the current and voltage signals. Therefore, any kind of power supply mode including the line-start, open- and closed-loop inverter-fed one can be investigated by applying the related conditions to the signals. For example, if a line-start motor is targeted, the motor voltage equation should contain $n = 1$ and $k = 0$ as the voltage should be a pure sinusoid, and the higher order voltage components produced by the PWM technique and also those related to the fault must be removed. Note that $n$ is the order of the main harmonic components, and $k$ is the order of the sideband components. On the other hand, if a closed-loop application is investigated, $n$ depends on the switching topology while $k$ is an integer.

$$V = \sum_{n} \left( V_{mn} \sin(n\omega_s t) + \sum_{k} (V_{l_{n,k}} \sin((1-2ks)n\omega_s t - \alpha_{vl_{n,k}}) \right. \\ \left. + V_{r_{n,k}} \sin((1+2ks)n\omega_s t - \alpha_{vr_{n,k}})) \right) \tag{8.22}$$

$$I = \sum_{n} \left( I_{m,n} \sin(n\omega_s t - \alpha_{I_n}) + \sum_{k} (I_{l_{n,k}} \sin((1-2ks)n\omega_s t - \alpha_{Il_{nk}}) \right. \\ \left. + I_{r_{n,k}} \sin((1+2ks)n\omega_s t - \alpha_{Ir_{n,k}})) \right) \tag{8.23}$$

The above equations convey the idea of the presence of sideband components around the first, second, third... harmonic components depending on the supply mode. Therefore, the motor power signal is affected by all components of the motor voltage and current, and it can be calculated analytically by multiplying the voltage and current signals. However, to simplify the process, the first harmonic of the current and voltage of the motor with their first-order sideband components are considered, and products are evaluated. Thus,

$$V = \underbrace{V_{m1} \sin(\omega_s t)}_{v_1} + \underbrace{V_{l_{1,1}} \sin((1-2s)\omega_s t - \alpha_{vl_{1,1}})}_{v_2} + \underbrace{V_{r_{1,1}} \sin((1+2s)\omega_s t - \alpha_{vr_{1,1}})}_{v_3} \tag{8.24}$$

$$I = \underbrace{I_{m1} \sin(\omega_s t - \alpha_{I_1})}_{i_1} + \underbrace{I_{l_{1,1}} \sin((1-2s)\omega_s t - \alpha_{Il_{1,1}})}_{i_2} + \underbrace{I_{r_{1,1}} \sin((1+2s)\omega_s t - \alpha_{Ir_{1,1}})}_{i_3} \tag{8.25}$$

where $V_{m1}$, $V_{l_{1,1}}$, $V_{r_{1,1}}$, $\alpha_{vl_{1,1}}$, $\alpha_{vr_{1,1}}$, $I_{m1}$, $I_{l_{1,1}}$, $I_{r_{1,1}}$, $\alpha_{Il_{1,1}}$, $\alpha_{Ir_{1,1}}$ and $\alpha_{I_1}$ are the amplitude of voltage fundamental component, the amplitude of the left sideband component of the voltage fundamental component, the amplitude of the right sideband component of the voltage fundamental component, the phase angle of the left sideband component of the voltage fundamental component, the phase angle of the right sideband component of the voltage fundamental component, the amplitude of current fundamental component, the amplitude of the left sideband component of the current fundamental component, the amplitude of the right sideband component of the current fundamental component, the phase angle of the left sideband component of the current fundamental component, the phase angle of the right sideband component of the current fundamental component and the phase angle of the current fundamental component, respectively. The instantaneous power is as follows:

$$
\begin{aligned}
P &= 3vi = 3\left[\underbrace{v_1(i_1+i_2+i_3)}_{P_1}+\underbrace{v_2(i_1+i_2+i_3)}_{P_2}+\underbrace{v_3(i_1+i_2+i_3)}_{P_3}\right] \\
P_1 &= \frac{V_{m1}I_{m1}}{2}\left[\cos(\alpha_{I_1})-\cos(2\omega_s t-\alpha_{I_1})\right] \\
&\quad+\frac{V_{m1}I_{l_{11}}}{2}\left[\cos(2s\omega_s t+\alpha_{Il_{11}})-\cos((1-s)2\omega_s t-\alpha_{Il_{11}})\right] \\
&\quad+\frac{V_{m1}I_{r_{11}}}{2}\left[\cos(-2s\omega_s t+\alpha_{Ir_{11}})-\cos((1+s)2\omega_s t-\alpha_{Ir_{11}})\right] \\
P_2 &= \frac{V_{l_{11}}I_{m1}}{2}\left[\cos(-2s\omega_s t-\alpha_{vl_{11}}+\alpha_{I1})-\cos((1-s)2\omega_s t-\alpha_{vl_{11}}-\alpha_{I1})\right] \\
&\quad+\frac{V_{l_{11}}I_{l_{11}}}{2}\left[\cos(\alpha_{I1_{11}}-\alpha_{vl_{11}})-\cos((1-2s)2\omega_s t-\alpha_{vl_{11}}-\alpha_{I1_{11}})\right] \\
&\quad+\frac{V_{l_{11}}I_{r_{11}}}{2}\left[\cos(-4s\omega_s t+\alpha_{Ir_{11}}-\alpha_{vl_{11}})-\cos(2\omega_s t-\alpha_{vl_{11}}-\alpha_{Ir_{11}})\right] \\
P_3 &= \frac{V_{r_{11}}I_{m1}}{2}\left[\cos(2s\omega_s t-\alpha_{vr_{11}}+\alpha_{I1})-\cos((1+s)2\omega_s t-\alpha_{vr_{11}}-\alpha_{I1})\right] \\
&\quad+\frac{V_{r_{11}}I_{l_{11}}}{2}\left[\cos(4s\omega_s t+\alpha_{I1_{11}}-\alpha_{vr_{11}})-\cos(2\omega_s t-\alpha_{vr_{11}}-\alpha_{I1_{11}})\right] \\
&\quad+\frac{V_{r_{11}}I_{r_{11}}}{2}\left[\cos(\alpha_{Ir_{11}}-\alpha_{vr_{11}})-\cos((1+2s)2\omega_s t-\alpha_{vr_{11}}-\alpha_{Ir_{11}})\right]
\end{aligned}
\tag{8.26}
$$

Figure 8.20 illustrates a practical example of the motor power spectrum under the broken bar fault condition. This figure shows the power spectrum within a 250-Hz frequency range divided into two subplots in order to have a better focus on the frequency components.

Considering Figure 8.20, there is a direct current (DC) component with the amplitude and zero frequency. This means that the DC component which produces the average motor power is the most powerful component, and the other

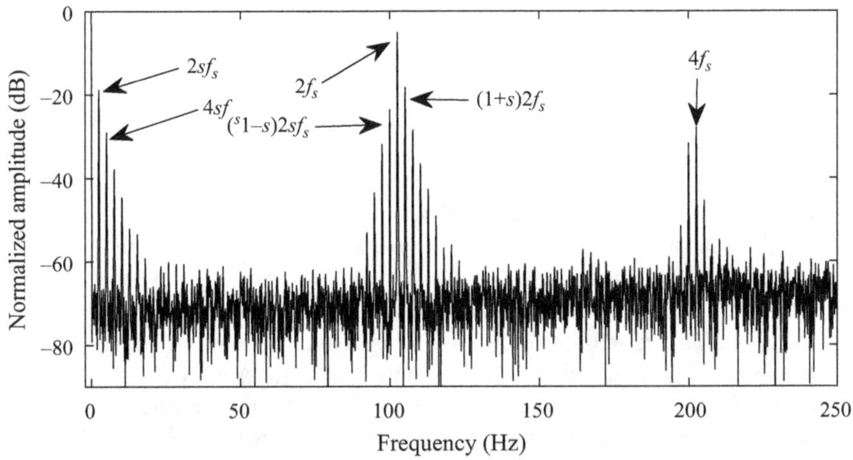


*Figure 8.20 Motor power spectrum*

*Table 8.7 Frequency patterns included in power spectrum of an induction motor with broken bars fault*

| DC | $2kf_s$ ($k = 1, 2, 3, \ldots$) | $2ksf_s$ ($k = 1, 2, 3, ..$) | $(1 \pm ks)2nf_s \begin{cases} k = 1,2,3,... \\ n = 1,2,3,... \end{cases}$ |
|---|---|---|---|

components related to the higher order voltage and current signals as well as the fault-related components have a modest value depending on their order. On the basis of Figure 8.20 and also (8.26), the power spectrum of a broken bar motor consists of the frequency patterns listed in Table 8.7.

Measuring the motor power spectrum requires voltage and current transducers; hence, the process of obtaining the corresponding indices is noninvasive. Besides, it can be applied online and used for the incipient fault diagnosis procedure.

Sometimes, loads connected to the motor shaft possess an oscillating rotation with a specific frequency. If the corresponding frequency is equal to the $2sf_s$ pattern produced by the broken bars fault, the amplitude of the sideband components becomes larger even if there is no rotor defect. This is a source of uncertainty in diagnosing the broken bars fault. However, the use of active and reactive power spectrum of the motor somehow resolves the issue. The similar technique might be used to discriminate the load oscillation from the eccentricity fault. The underlying idea of using the motor power spectrum for discriminating the fault from the load oscillations is to follow the variation rate of the fault indices which in this case are the amplitude of the fault-related components of the motor power. Although this method has not been widely used in the diagnosis procedures, it might be used as a complement to the MCSA with the goal of introducing a more reliable diagnosis procedure.

## 8.13 Additional frequency-domain fault indices

Vibration is one of the conventional signals which have been used for the fault diagnosis purposes. Although, its major application is to detect the eccentricity fault in large generators, there are some broken bars-related frequency components in its spectrum which make it possible to monitor the induction motor behavior. Usually, accelerometers are used to measure the vibration signal, so depending on the location in which the sensors are installed, various vibration signals are obtained. In other words, the measurement process is indeed invasive regarding the location of the vibration sensor. Sometime, the vibration signal is bolted to the stator body inside the motor housing. This is another definition of invasive methods in which the motor structure is somehow changed in order to mount the sensor. Vibration signal has been the matter of various studies from different points of view. Some of them have focused on the sideband components of the vibration signal. The others have concentrated on the factors of the fundamental component of the vibration signal which is equal to the angular mechanical speed. Similar to the MCSA, the amplitude of the sideband components of the vibration signal decreases by decreasing the load level. As the trend looks like that of the motor current, it is expected to have a troublesome diagnosis process in light-load levels. Moreover, the information of the healthy motor is necessary for the fault diagnosis. Unfortunately, the effect of the influential factors such as the drive type and the reference speed, in terms of the vibration signal, has not been studied yet. Therefore, the results are limited to the line-start mode.

Combining the motor current and vibration spectrum and then calculating their correlation provides a very interesting feature useful for the diagnosis of the half-broken bar fault. However, it is not guaranteed that this technique works well in any circumstances because the quality of the output directly depends on the location of accelerometers mount on the motor body; therefore, this technique is not recommended. Instead, a global index, which enables us to detect the partial broken bars, is available. The index is obtained by summing and subtracting the left and right sideband components of the motor current. This method has proved to be helpful in a wide range of motor load levels.

Most of the previous studies have used a band-pass filter to eliminate the masking effect of fundamental component from current spectrum. Nevertheless, filtering needs an extra processor; besides, the accuracy of the technique is reduced when the supply frequency increases. The Hilbert transform (HT) is another technique for precise extraction of the sideband components at light loads. The applicability of this index has not been studied for detecting different locations of the fault in both direct-drive and inverter-fed induction motors. Therefore, further investigations are required to make sure if the technique is general enough. Multiple signal classification (MUSIC) transform is another tool to diagnose the fault at no- or light-load conditions. Both the HT and the MUSIC use the normalized current spectrum and transform it into a new frequency scale in order to detect the broken bars fault. So it is acceptable that the general behavior of the regular current spectrum is observed in the HT and MUSIC. The basis of these techniques is to allocate a higher weight to the fault-related components and a lower weight to the fundamental component. The problem with these methods is that the accurate measurement of the motor slip is essential in order to detect the $(1 \pm 2ks)f_s$ pattern. Both techniques can be applied online.

## 8.14 Motor transient operation

Motor behavior in the transient mode partly differs from that of the steady-state mode. In fact, the motor quantities consist of transient signals with varying frequencies. Therefore, extracting the fault information has been a big challenge recently. Although sometimes the frequency-domain indices have been introduced to deal with the transient signals, they are case-specific techniques which are only applicable in certain situations. When the pure time-based information is used to diagnose the fault in the transient mode, varying frequency characteristics are ignored. Both time and frequency information is useful and essential for precise diagnosis of the fault. While there are both time- and frequency-related variations in transient signals, the previous works tend to use the transforms which are capable of extracting the information in time-frequency domain simultaneously. There are usually two types of processing tools not only for the rotor broken bars fault, but also for the other types of faults in transient modes:

- the wavelet transform (WT): decomposing the signal into its frequency bands within the time period that the signal exists via a mathematical equation
- the HHT: decomposing the signal into its frequency bands within the period that the signal exists via a recursive and repetitive algorithm.

Both transforms calculate the energy level of different frequency bands of the induction motor signals including the current, torque, vibration and speed in different periods. The quality of the fault diagnosis is related to the signal selected for processing. The frequency band limits are determinable in the WT while the corresponding bands in the HHT are the output of a predefined algorithm discussed in Chapter 7. Both WT and HHT are the powerful tools for analyzing the systems with a transient nature. Nevertheless, they are also used in steady-state analysis.

Typically, there are two major categories for the transient modes as follows:

- The transient start-ups where the motor speed is zero at $t = 0$ and then speeds up to the steady-state operation as the time goes on.
- The transients over two different steady-state operations in which the motor operates with a certain load and speed level and suddenly the load or speed level is changed.

The literature indicates that the rotor broken bars fault can be categorized into the above-mentioned groups while most of the works out there focus on the first group of the transient operations. Induction motors accelerate from zero speed to the steady-state speed which depends on the load level in the line-start mode and the

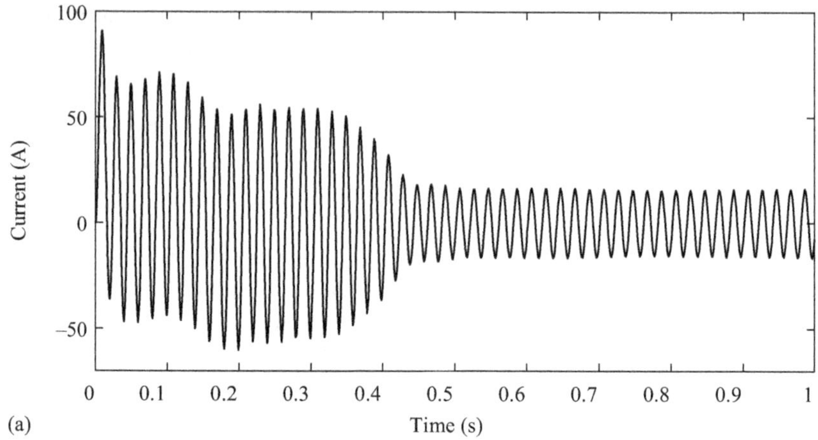

(a)

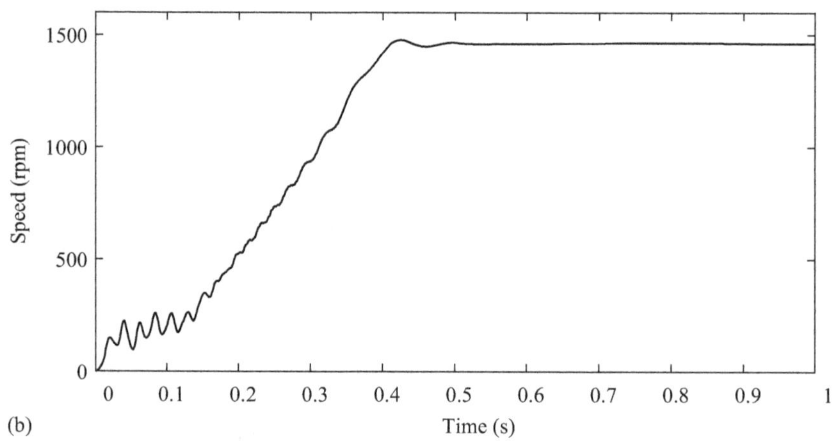

(b)

*Figure 8.21 (a) IM transient current and (b) IM transient speed*

reference speed in the inverter-fed mode. In the line-start mode, the transient period is directly forced by the motor structure, the load and the input voltage. The major issue in the line-start mode is that the transient period is not long enough to be useful for extracting the fault information. For example, the transient period for an 11-kW industrial induction motor is about 0.5 s which is very short. The suitable corresponding value must be at least equal or greater than 1 s so that the fault information and features become detectable (Figure 8.21).

This major issue is usually dealt with using a high inertia load on the shaft where a line-start mode is under the test. In the case of inverter-fed applications, the motor speed acceleration time can be easily controlled by the drive. Therefore, there is no need to a high-inertia load. The high-inertia load is used just to mimic the behavior of high-power induction motors whose acceleration time is naturally longer than 1 s. Thus, in real applications, it is not required to use a high-inertia load for testing.

Figure 8.21 illustrates the current and speed profiles of a faulty motor during the start-up. The supply frequency is fixed in the line-start mode, and the motor speed increases from zero to its steady-state value. As a result, the motor slip has a decreasing trend shown in Figure 8.22. In this case, not only the value of $(1 \pm 2s)f_s$, but also the amplitude of the pattern changes. Figure 8.22(b) and (c) shows the frequency and amplitude variations of the left sideband component, respectively, during the start-up. This concept is exactly what was mentioned before regarding time-frequency variation of the fault-related components. This trend is usually used to diagnose the rotor broken bars fault using WT. It is known that the motor current includes different inconsistent frequency components some of which decay as time goes on, but others, such as the fundamental component and also the sideband components, are consistent.

In addition to the transient operation, the steady-state operation can also be studied by means of time-frequency transform for enhancing the diagnosis quality and addressing some of the challenges which exist in the time or frequency domains. For example, the masking effect of the fundamental component of the motor current can be somehow resolved using the HHT and its energy representation. Although the time-frequency transforms suffer from a low-frequency resolution compared to the FFT, their application to enhance the capability of knowledge extraction in light-load levels is appreciated. Therefore, an example of both the transient and steady-state implementation of the time-frequency transforms is discussed in this chapter.

## 8.15 Application of wavelet transform to the diagnosis of broken bars fault

The WT is often applied to the motor transient signals, for example the current, in order to extract fault information during the transient operation. One of the significant applications is to apply the WT to the motor current whose fundamental frequency is filtered out by subtracting a pure sinusoidal signal from the motor transient current. The approach for filtering out the fundamental frequency component has been thoroughly discussed in [110]. Therefore, we assume it has been done; hence, we move toward calculating the wavelet coefficients extracted from the filtered current.

Remembering different decomposition levels of the WT explained in Chapter 7, there is an "approximation" as well as a "detail" signal at every single level of the decomposition which is handled by low- and high-pass filters. Herein, the detail signals will be used to show the differences between the healthy and faulty motors. Examining the first six detailed signals does not reveal a difference while the detail signals 7 to 10 clearly exhibit interesting differences which can be essentially used for the diagnosis of the broken bars fault. However, it is shown that the detail signal at level 9 consists of the best fault measure. To conduct the comparisons, have a look at [110] in which healthy and broken bar motors are investigated in the full-load condition. Note that the motor operates in the line-start mode.

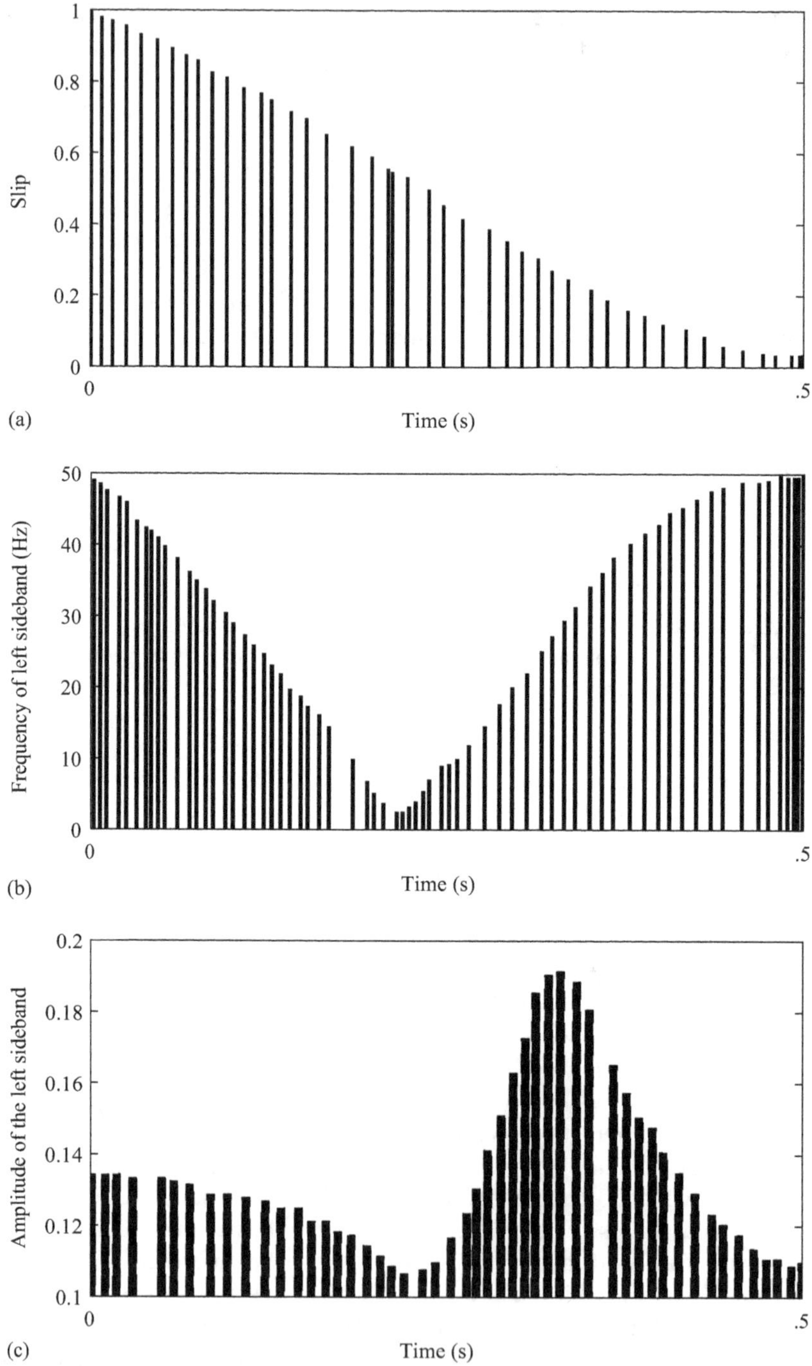


*Figure 8.22 (a) Slip of IM during start-up, (b) instantaneous frequency of left sideband component during start-up and (c) instantaneous amplitude of left sideband component during start-up*

Three different load levels including 100%, 60% and 30% are illustrated. If an inverter-fed motor is used, the process of filtering out the fundamental frequency of the motor current becomes really complicated as a variable frequency acceleration is usually applied to the motor during the start-up in order to have soft start-up.

## 8.16 Application of Hilbert–Huang transform to the diagnosis of broken bars fault

The HHT, as a time-frequency-based transform, has been widely used in biomedical and aerospace applications. However, no one can deny its competence in the fault diagnosis procedures of electrical machines. The HHT is a fixed-routine algorithm consisting of just one input which is the faulty signal such as the current and a set of outputs expressing energy levels of various frequency bands possessed by the input signal. The margins of the frequency bands are determinable in the WT while it is not so in the HHT. This might be one of the drawbacks of the HHT. However, the HHT has been utilized in discovering fault information throughout an iterative algorithm discussed in Chapter 7.

The HHT decomposes the signal into some frequency bands called the intrinsic mode functions (IMF). The number of IMFs is predetermined as input of the process and their resolutions highly depend on this number as well as the sifting process discussed in Chapter 7. Figure 8.23 shows the IMFs of a typical faulty and healthy motor current. In this figure, the 1st, 2nd, 3th, 4th and 5th are the input signals (faulty motor current), $IMF_1$, $IMF_2$, $IMF_3$ residual function, respectively. $IMF_1$ includes the highest frequency band of the signal. The frequency limitations of the other bands are reduced by the order of IMF. It means that the components of $IMF_3$ are present at lower frequencies. The residual function is the difference between the input signal and sum of the IMFs. As seen in Figure 8.23(b), a mono-frequency fluctuation is observable in $IMF_2$ when the fault occurs. This is the exact frequency band that must be investigated. The reason for selecting this frequency band is illustrated in Figure 8.23.

Using the mentioned concept as well as a new one which demonstrates the instantaneous energy (IE) level of the HHT, a very promising tool is introduced for diagnosing the broken bars fault. This method will be applied to the motor current as it is proved that the MCSA is certainly the most reliable diagnosis approach. Recalling the HHT formulation from Chapter 7, we have:

$$H(\omega,t) = Re\left(\sum_{i=1}^{m} a_i(t)exp\left(j\int_{-\infty}^{t} \omega_i(t)dt\right)\right) \tag{8.27}$$

The IE level is defined as follows:

$$IE(t) = \int H^2(\omega,t)d\omega = \left[IMF_1^2 + IMF_2^2 + IMF_3^2 + \ldots\right]_{\text{at fixed time instances}} \tag{8.28}$$

where $m$ denotes the number of energy levels, in other words the IMFs, into which the input signal is decomposed. $a$ expresses the signal amplitude at a specific time $t$

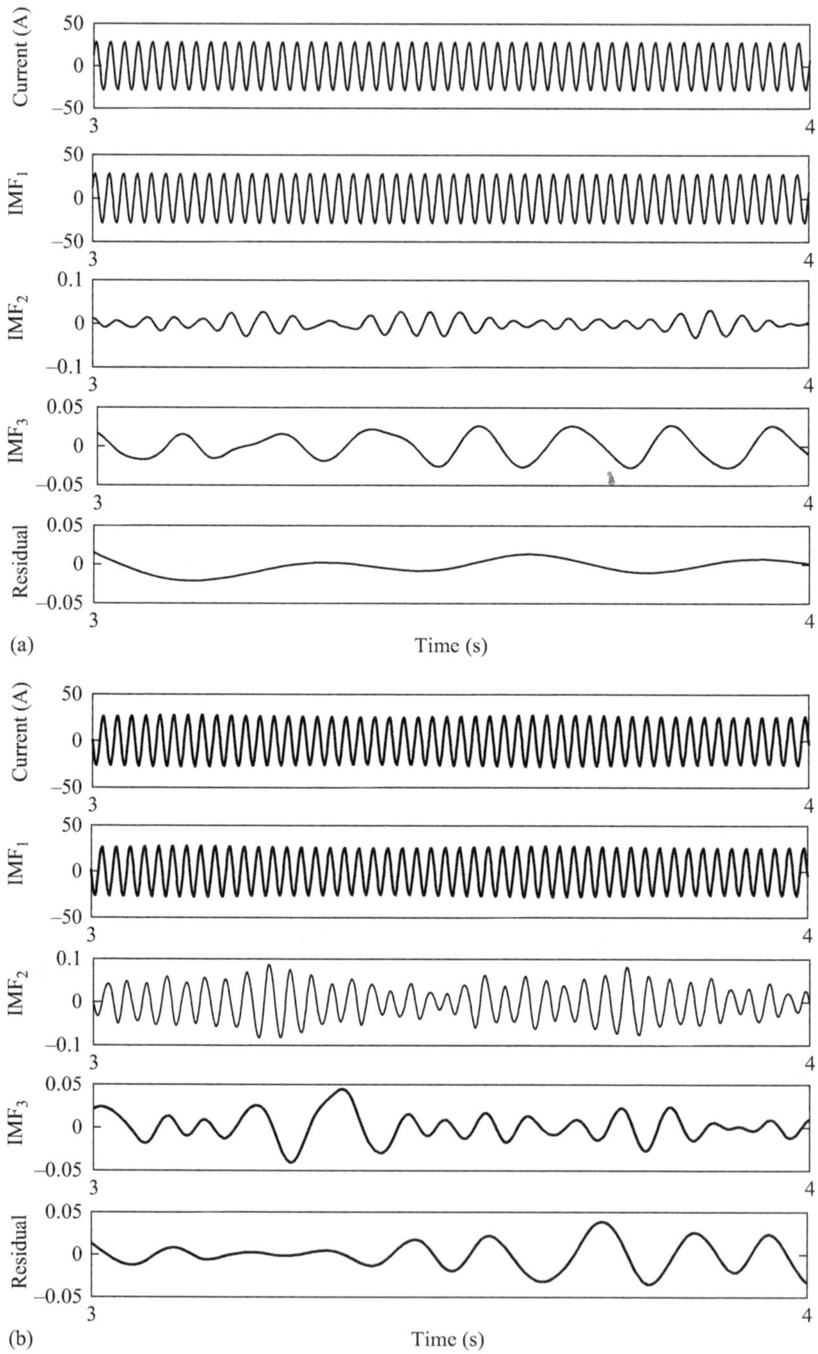


*Figure 8.23 IMFs of the motor current for a fixed load (a) healthy and (b) one broken bar*

and the instantaneous frequency $\omega$. $a$ and $\omega$ are obtained from the HT of the input signal for different frequency components. The HHT is based on calculation of the time variations of a phenomenon and presentation of its variations over different frequency bands. As it is an almost well-known algorithm found in the literature, for the sake of briefness, we skip detailed discussion of the algorithm. The energy levels are called IMF. The frequency content of the IMFs decreases by increasing the IMF order. In other words, the last energy level consists of the lowest frequency range. This last IMF is usually called "residual." Figure 8.23 shows the experimental result of the HHT of the motor current for one broken bar and line-start mode. On the basis of this figure, the most sensitive and revealing IMF with respect to the broken bar fault is the $IMF_2$ which oscillates with the same frequency as that of the speed or the current envelope. The corresponding energy level of the inverter-fed motor is the $IMF_3$. In this case, the $IMF_1$ is dedicated to the high-frequency components produced by the switching topology, and the $IMF_2$, $IMF_3$ ... would represent the same frequency bands as $IMF_1$, $IMF_2$ ... in the line-start mode. This discussion holds only if similar adjustments and parameters are maintained for the measurements in both the line-start and inverter-fed modes. Otherwise, the IMFs might contain the frequency components different from what was just mentioned. Actually, depending on the number of IMFs, various frequency bands, which might be totally different from what has been reported in the literature, are extracted. Along with the marginal effects, this is the most important shortcoming of the HHT. As a matter of fact, there is no 100-percent control on the frequency components of the IMFs, and it is a matter of trial and error unless an intelligent algorithm manages the problem. On the other hand, the sideband components are normally expanded far away from the fundamental frequency (Figure 8.10), namely within the range 30–70 Hz or even further if a proper window such as Hamming is applied to the FFT. This point, along with the previous comments on the possible large variations of the IMF components, brings up the idea of dealing with the total energy of the HHT, not the energy level of one single IMF, in order to cover all frequency bands which include the fault-related components. Therefore, (8.29) is introduced and examined over various fault cases to show its applicability through the experiments.

$$\textit{Amplitude of oscillations of } [IE(t) - \textit{average } (IE(t))] \tag{8.29}$$

The index signifies the amplitude of oscillations of IE (AOIE) regardless of its average value. Figure 8.24 illustrates the proposed measure for the line-start, CV/F and DTC supply modes in different operating conditions including various load and fault levels.

In the inverter-fed mode, as the higher order voltage components, other than the fundamental ones, are also applied to the motor, they produce undesirable backward magnetic torque which acts against the fundamental torque component if the switching topology contains a high level of total harmonic distortion. Therefore, in order to provide a proper basis for the comparison of line-start and inverter-fed modes in terms of the load levels experienced by the motor, it is better to take the motor slip as the load-level indicator, not the mechanical torque expressed in Newton-meter

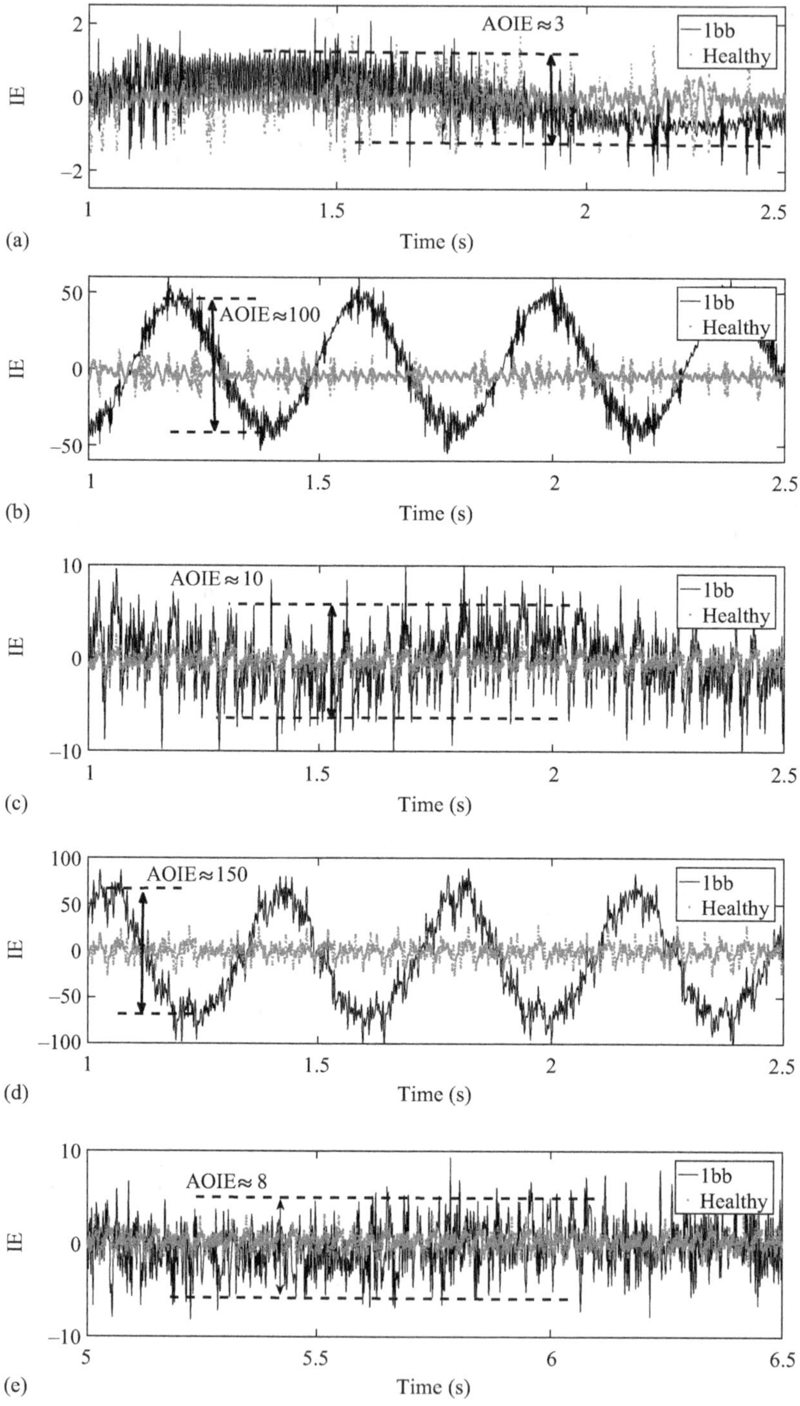


*Figure 8.24 AOIE for supply frequency of 50Hz (a) line-start mode and 25% of full load, (b) line-start mode and 100% of full load, (c) CV/F mode and 25% of full load, (d) CV/F mode and 100% of full load, (e) DTC mode and 25% of full load and (f) DTC mode and 100% of full load*

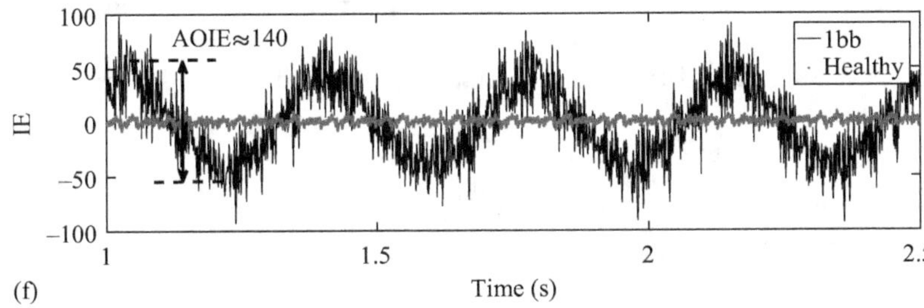


*Figure 8.24 (Continued)*

*Table 8.8 AOIE for different operating conditions*

| **Supply frequency** | **Number of broken bars** | **Mode** | **Load level (%)** | | | | |
|---|---|---|---|---|---|---|---|
| | | | **0** | **25** | **50** | **75** | **100** |
| **30** | **0** | CV/F | 0 | 0 | 0 | 0 | 0 |
| | | DTC | 0 | 0 | 0 | 0 | 0 |
| | **1** | CV/F | 0 | 11 | 50 | 102 | 159 |
| | | DTC | 0 | 8 | 47 | 91 | 144 |
| | **2** | CV/F | 0 | 14 | 71 | 143 | 243 |
| | | DTC | 0 | 13 | 67 | 138 | 207 |
| **50** | **0** | CV/F | 0 | 0 | 0 | 0 | 0 |
| | | DTC | 0 | 0 | 0 | 0 | 0 |
| | | Line-start | 0 | 0 | 0 | 0 | 0 |
| | **1** | CV/F | 0 | 10 | 48 | 95 | 150 |
| | | DTC | 0 | 8 | 44 | 88 | 140 |
| | | Line-start | 0 | 3 | 29 | 67 | 100 |
| | **2** | CV/F | 0 | 13 | 68 | 137 | 228 |
| | | DTC | 0 | 13 | 63 | 130 | 203 |
| | | Line-start | 0 | 5 | 42 | 103 | 147 |

connected to the shaft. Thus, one can make sure that the motor senses the same load level if the motor slip is fixed for different supply modes. Besides, the experiments show that the sampling frequency should be large enough, for example 25.8 kHz, in order to sweep a smooth IE. Otherwise, undesirable outsized spikes would be detected in the IE. On the basis of Figure 8.24, the AOIE is equal to zero for the healthy motor regardless of the supply mode and the load level. The high-frequency ripples carried by the IE can be easily eliminated by means of a low-pass filter with a cut-off frequency smaller than 10 Hz. However, to show the real measured quantity, it is ignored in this study. The line-start mode reveals a smaller AOIE compared to the CV/F and DTC modes for the same load and fault level. Moreover, the corresponding AOIE of the CV/F and DTC modes are very close to each other. Nevertheless, the CV/F mode holds a relatively larger AOIE. In order to quantify the AOIE for different operating conditions, the corresponding values are listed in Table 8.8. The operating conditions have different fault levels including 0, 1 and 2 broken bars, different load levels including 0%, 25%, 50%, 75% and 100% of full load, two operating supply frequencies, i.e. 30 and 50 Hz and also various supply modes including the line-start, CV/F and DTC modes.

The following diagnosis guidelines are extractable:

- The line-start mode reveals the smallest fault measures while the inverter-fed modes are of the largest values.
- The larger the motor load is, the larger the AOIE is.
- The larger the fault level is, the larger the AOIE is.
- The AOIE is exactly equal to zero in both the healthy motor and the zero-load level.
- The CV/F mode possesses the largest AOIE values while the DTC mode is of a moderate value in a fixed load and fault levels.
- The AOIE depends more on the motor load level than the fault level.
- Increasing the supply frequency slightly decreases the AOIE. This claim might change depending on the way the supply frequency is controlled. The reported values are valid for a linear voltage over speed adjustments. However, the voltage/speed profile can be easily changed as a user-defined function so that the motor flux varies so much. This means the fault-dependent components are subjected to various amplitude saturation levels and consequently the amplitudes changes.
- The line-start mode has been only discussed for the supply frequency of 50 Hz as the network frequency is unchangeable unless an interconnection is used. However, this contradicts the nature of the line-start mode.
- The difference between the line-start and inverter-fed modes becomes larger for higher load levels.

On the basis of the experimental results, the fundamental frequency of the IE is equal to $2sf_s$. An analytical proof of this is provided in the next section.

For the sake of simplicity, only two right sideband components, along with two left sideband components of the fundamental component of the motor current, are considered in the following calculations. The $IMF_1$ includes the fundamental and right sideband components. On the other hand, the $IMF_2$ contains the left sideband components. The component-wise description of the IE is calculated using the following equation:

$$\begin{aligned} IE(t) &= IMF_1^2 + IMF_2^2 + \ldots \\ &= \left( [I_m \sin(\omega_s t) + I_{r1} \sin((1+2s)\omega_s t) + I_{r2} \sin((1+4s)\omega_s t)]_{IMF_1} \right)^2 \\ &\quad + \left( [I_{l1} \sin((1-2s)\omega_s t) + I_{l2} \sin((1-4s)\omega_s t)]_{IMF_2} \right)^2 \end{aligned} \tag{8.30}$$

where $I_m$, $I_{r1}$, $I_{r2}$, $I_{l1}$ and $I_{l2}$ are the amplitude of the fundamental, first right sideband, second right sideband, first left sideband and second left sideband components, respectively. Simplifying (8.30) to obtain pure cosine functions yields:

$$
\begin{aligned}
IE(t) = {} & \left(\frac{I_m^2}{2} + \frac{I_{r1}^2}{2} + \frac{I_{r2}^2}{2} + \frac{I_{l1}^2}{2} + \frac{I_{l2}^2}{2}\right) + \frac{I_m^2}{2}\cos(2\omega_s t) \\
& + (I_m I_{r1} + I_{l1} I_{l2} + I_{r1} I_{r2})\cos(2s\omega_s t) + (I_m I_{r2})\cos(4s\omega_s t) \\
& - (I_m I_{r1})\cos((1+s)2\omega_s t) - \left(\frac{I_{r1}^2}{2} + I_m I_{r2}\right)\cos((1+2s)2\omega_s t) \\
& - (I_{r1} I_{r2})\cos((1+3s)2\omega_s t) - \left(\frac{I_{r2}^2}{2}\right)\cos((1+4s)2\omega_s t) \\
& - \left(\frac{I_{l1}^2}{2}\right)\cos((1-2s)2\omega_s t) - (I_{l1} I_{l2})\cos((1-3s)2\omega_s t) \\
& - \left(\frac{I_{l2}^2}{2}\right)\cos((1-4s)2\omega_s t)
\end{aligned}
\tag{8.31}
$$

The first term on the right side of (8.31) inside the parentheses represents the average (DC) value of the other frequency components listed in Table 8.9 as well. Note that the higher order components of the $2ksf_s$ and $(1 \pm ks)2f_s$ patterns obtained by increasing $k$ can also be calculated and added to (8.31) if the 3rd, 4th, 5th and higher order right and left sideband components are incorporated in (8.30). For example, if the $(1-4s)f_s$ and $(1-6s)f_s$ components are taken into account in (8.30), the $(1-s)f_s$ components of the IE would also be present in (8.31). The goal of this analytical calculation is to find the frequency component of the IE, not the exact value of their amplitudes because an accurate calculation of the amplitudes requires a more comprehensive formulation considering the effect of material saturation as well as the higher order harmonic components. Table 8.9 is a very worthy prescription of the existing harmonic components. The proper insight into the correct amplitudes is obtained through the experiments (see Figure 8.25).

Table 8.9 illustrates the IE spectrum for the faulty motor in the line-start mode. The fundamental frequency of the network, 50 Hz, is greatly degraded, almost eliminated from the spectrum, and it is generally modulated to twice the fundamental frequency, 100 Hz. Together with the mathematical proof, i.e. (8.30) and (8.31), Figure 8.25 is a clear justification of the dominant $2sf_s$ oscillations of the IE. So it is truly expected to observe the IE oscillations with the frequency of $2sf_s$. Of course, it is a function of the motor slip; hence, the frequency of oscillations increases upon increasing the motor load level.

*Table 8.9 Frequency components of the IE*

| DC | $2ksf_s$<br>$k$: integer | $(1 \pm ks)2f_s$<br>$k$: integer | $2f_s$ |
|---|---|---|---|

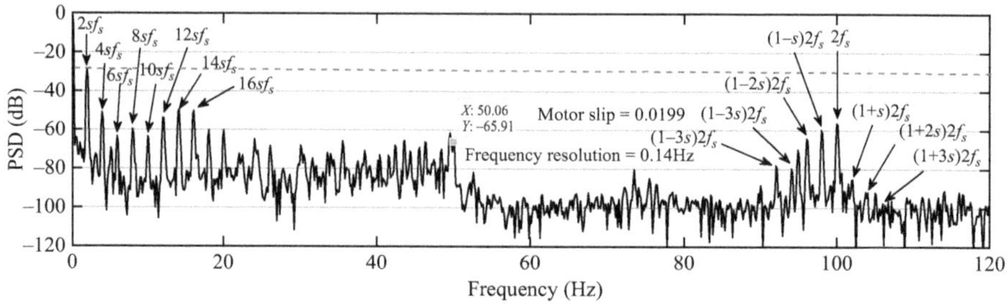


*Figure 8.25 IE spectrum*

## 8.17 Loss characterization of induction motors with broken bars fault

One of the important parameters which catches the attention during the monitoring and maintenance process of induction motors is the losses, the significant one of which is the loss in silicon steel materials. In induction motors, losses are categorized into two groups, namely the fixed and variable losses. The fixed losses consist of the magnetic core losses and the friction losses. The variable losses generally consist of the ohmic loss in the stator and rotor circuits. There are some other types of losses whose origins are somehow unknown and are called the stray losses. What we are going to focus on is generally the losses associated with silicon steel material; then, we will move toward illustrating how the broken bars fault affects the loss distribution inside the motor. This term is usually affected not only by means of the fault, but also by incorporating various supply modes which apply different harmonic orders which produce different operating points in the BH curve of the silicon steel material.

As discussed in Chapter 6, there are two types of magnetic losses, namely the hysterics and eddy-current losses. These are the main sources of the iron loss (the loss produced in silicon steel materials) and their summation builds up the total iron losses. In order to address the loss variations inside the motor in the healthy and faulty conditions, an finite element method (FEM) package is used to compute the hysteresis and eddy-current components over time or space. Then, the results at different motor components are reported. Note that, all FEM-based simulations are held for a full-load case in a line-start motor. A 28-bar, 36-slot, 4-pole and 11 kW motor, which is already simulated in Chapter 6, is used here for investigating the losses. The supply voltage is equal to 380 V applied through a delta connection, and the rated current of the motor is equal to 24 A. The FEM-based reported data is the time average spatial distribution of the losses including the Ohmic, hysteresis and eddy-current ones during a 500-ms simulation run time starting from the zero speed to the steady-state operation of the motor at the torque level of 70 N m. The losses are calculated at the stator teeth level (radius = 76 mm) and the rotor teeth radius (radius = 74 mm with respect to the rotor center). Figure 8.26 shows the corresponding losses distribution for every stator and rotor slots or teeth.

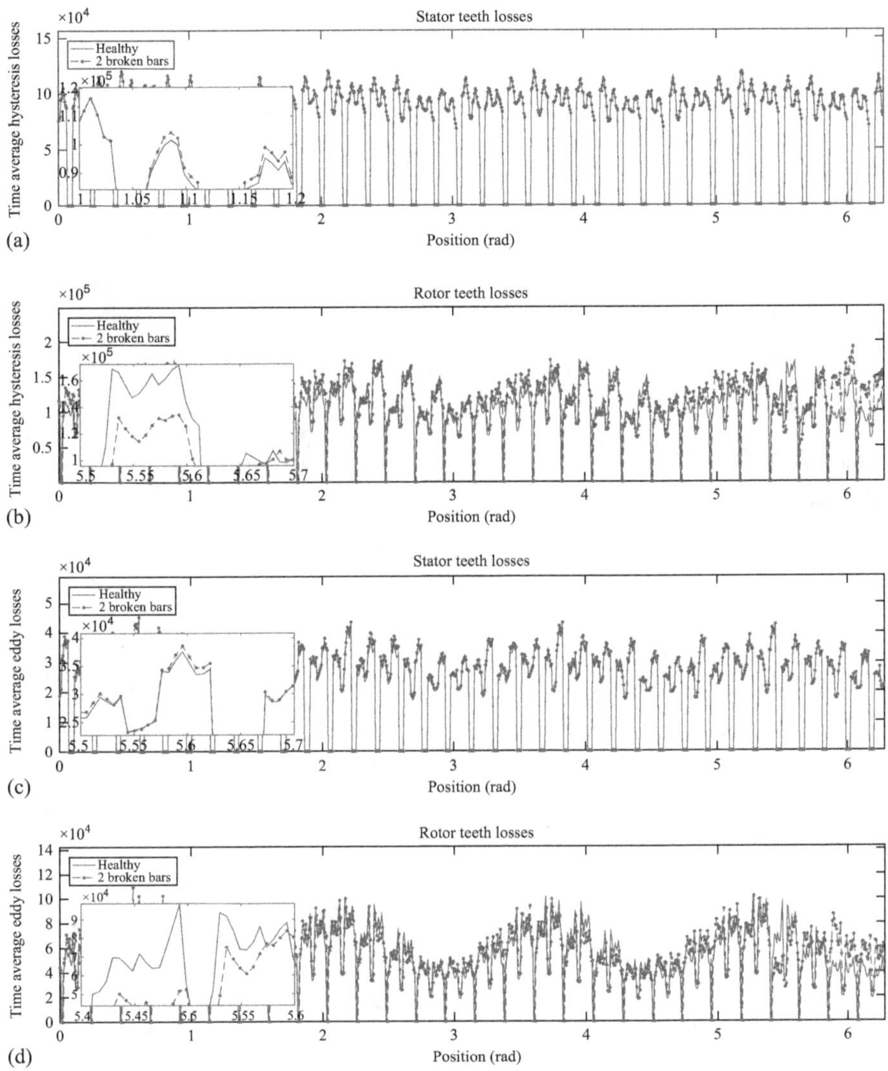


*Figure 8.26 Motor losses in different components (full load, line-start, healthy and 2 adjacent broken bars, simulation run time is equal to 500 ms) (a) stator teeth hysteresis losses, (b) rotor teeth hysteresis losses, (c) stator teeth eddy losses, (d) rotor teeth eddy losses and (e) stator slot ohmic losses, (f) rotor slot ohmic losses*

Figure 8.26 Motor losses in different components (full load, line-start, healthy and 2 adjacent broken bars, simulation run time is equal to 500 ms) (a) stator teeth hysteresis losses, (b) rotor teeth hysteresis losses, (c) stator teeth eddy losses, (d) rotor teeth eddy losses, (e) stator slot ohmic losses and (f) rotor slot ohmic losses

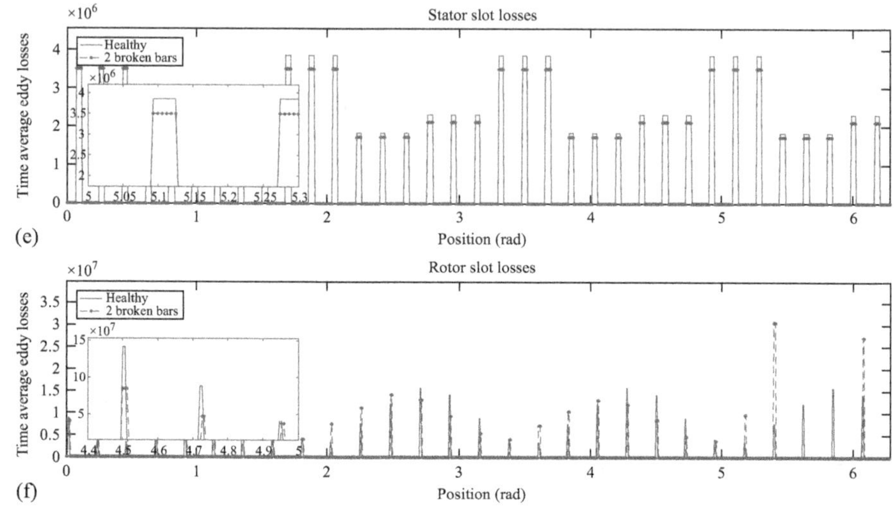


*Figure 8.26 (Continued)*

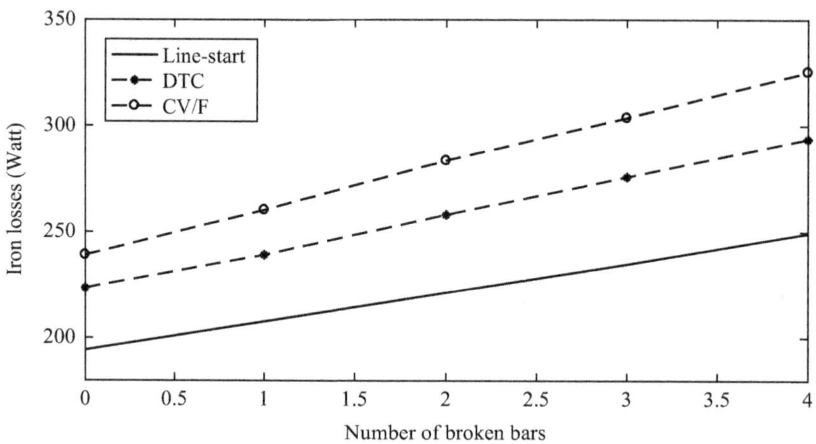


*Figure 8.27 Motor iron losses in different supply modes*

As a useful complement to this subsection which demonstrates the losses inside the induction motor upon the occurrence of broken bars fault, the experimental results in terms of the motor losses in different supply modes are provided to see how the motor quality is affected as a function of the supply modes (Figure 8.27). The experiments are related to the same motor simulated in an FEM package. Note that the motor iron losses are not normally a function of the motor load level. Therefore, it is not required to report the load dependency of this parameter.

## 8.18 Conclusion

Three different procedures based on the time, frequency and time-frequency domains were presented in this chapter through a progressive development of the concepts of the rotor broken bar fault and the corresponding motor behavior. The discussions and results are the outputs of the various visions of the fault which are extracted from the analytical, simulation-based and experimental efforts. The broken bars fault has a direct effect on both the time and frequency characteristics of the IM and may be detected by processing the changes and alternations of the motor quantities including the flux, current, voltage, torque and speed.

In order to achieve a precise fault diagnosis, selecting a proper signal to be processed, signal processing technique and finally extracting a fault-sensitive feature are the hierarchy of the diagnosis procedure. The progressive steps of fault diagnosis can be divided into the following items:

1. selecting a proper domain (time, frequency, time-frequency) in which the measured signals are processed
2. selecting an appropriate signal
3. applying an appropriate signal processing technique (GR, FFT, WT, HHT ...)
4. extracting a fault-sensitive index or feature which enables us to address the concepts of detection, determination, location and diagnosis thoroughly
5. The index should be preferably less sensitive to the factors other than the fault. These factors include the supply frequency, the load level, the drive bandwidth and any other factor changing the value of the indices.

The mentioned points are not broken bar fault-specific, and they should be followed for any other types of faults. Therefore, in this chapter, we tried to explain all the possible ways to deal with the time, frequency and time-frequency investigations of the motor signals by providing an example of the broken bar induction motor. Moreover, the ways to extract the fault-sensitive indices were introduced. So in the next chapter in which the eccentricity fault is dealt with, we will not go through how the indices are extracted as it can be overwhelming.

In the case of broken bars fault, the following points provide an appropriate guideline for the diagnosis approaches of the rotor broken bar fault:

- Although the time-domain variations reflect the changes caused by the rotor broken bar fault, they depend on other mechanical, magnetic, thermal and electric asymmetries of the induction motors (IMs). For example, both short-circuit and eccentricity faults disturb the current, voltage and torque profiles. So it is proposed not to use time-domain analysis, if possible. In order to use time-domain analysis, one needs to make sure whether other faults do not exist.
- Frequency spectrum of the motor quantities, especially current, consists of the rotor broken bar fault-related components which are almost independent of other types of faults. Therefore, using the frequency analysis as much as possible is recommended.

- The major issue with the conventional time and frequency analysis is that the corresponding techniques are usually applicable to only the steady-state mode while sometimes it is necessary to detect the fault during the transient mode, specifically the motor start-up. This is an important thing in high-power induction motors. In these situations, using time-frequency analysis is recommended.
- The mostly used motor signal is the motor current which is easily measurable using the current transducers. Moreover, the current is sensitive enough to the rotor broken bar fault. It can be measured safely, and electromagnetic interference problems can be avoided if an appropriate shielding is considered. This issue is one of the significant aspects when dealing with inverter-fed motors.
- In addition to the motor current analysis, motor flux density analysis has been the subject of very important investigations in the field of fault diagnosis. As the magnetic flux is the most sensitive and reliable signal for reflecting the motor behavior, it is crucial to get a proper knowledge of its variation. It will be great if a noninvasive flux measurement tool is developed while such a tool is not available today, at least easily. Therefore, using the flux or flux density analysis for the diagnosis process is not recommended.
- The other remarkable point is that most of the proposed techniques are healthy condition-dependent which means that it is necessary to have the index value in the healthy condition for conducting comparisons in the faulty cases.
- The MCSA has proved to be working in different operating conditions including various faults, load and speed levels. However, more investigations are required to diagnose precisely the fault locations and the effect of the bandwidth on inverter-fed motors. These issues are applied to time, frequency and time-frequency analysis.
- Fault diagnosis of inverter-fed motor is a novel field of study, and not many reliable sources are available in this case, especially when it comes to the analysis of the motor behavior in transient regime.

## References

[1] Nemec M., Ambrozic V., Nedeljkovic D., Fiser R. and Drobnic K. Detection of broken bars in induction motor using voltage pattern analysis, *IEEE International Symposium on Diagnostics for Electric Machines, Power Electronics and Drives, SDEMPED 2009*, 2009; 1–6.

[2] Nemec M., Drobnic K., Nedeljkovic D., Fiser R. and Ambrozic V. Detection of broken bars in induction motor through the analysis of Supply voltage modulation, *IEEE Transactions on Industrial Electronics*, 2010; **IE-57**: 2879–2888.

[3] Pineda-Sanchez M., Riera-Guasp M., Roger-Folch J., Antonino-Daviu J.A., Perez-Cruz J. and Puche-Panadero R. Diagnosis of induction motor faults in time-varying conditions using the polynomial-phase transform of the current, *IEEE Transactions on Industrial Electronics*, 2011; **IE-58**: 1428–1439.

[4] Faiz J. and Ebrahimi B.M. Locating rotor broken bars in induction motors using finite element method, *International Journal of Energy Conversion and Management*, 2009; **JECM-50**: 125–131.

[5] Ben Sasi A.Y., GU F., Li Y. and Ball A.D. A validated model for the prediction of rotor bar failure in squirrel-cage motors using instantaneous angular speed, *Journal of Mechanical Systems and Signal Processing*, 2006; **MSSP-20**: 1572–1589.

[6] Faiz J., Tabatabaie I. and Sharfi-ghazvini E. A Precise electromagnetic modeling and performance analysis of a three-phase squirrel-cage induction motor under mixed eccentricity condition, *International Journal of Electromagnetics*, 2004; **IJE-24**: 471–489.

[7] Abu-Elhaija W.S., Faiz J. and Ebrahimi B.M. Analytical prediction of instantaneous torque and speed for induction motors with mixed-eccentricity fault using magnetic-field equations, *International Journal of Electromagnetics*, 2010; **IJE-30**: 525–540.

[8] Faiz J., Ghorbanian V. and Ebrahimi B.M. Modification of winding function method for broken bars fault diagnosis in induction motors, *Pacific Symposium on Applied Electromagnetics & Mechanics*, Ho Shi Min City, Vietnam, 2012.

[9] Bellini A., Filippetti F., Franceschini G., Tassoni C. and Kliman G.B. Quantitative evaluation of induction motor broken bars by means of electrical signature analysis, *IEEE Transactions on Industry Applications*, 2001; **IA-37**: 1248–1255.

[10] Ying X. Characteristic performance analysis of squirrel cage induction motor with broken bars, *IEEE Transactions on Magnetics*, 2009; **MAG-45**: 759–766.

[11] Wu T.Y., Chen J.C. and Wang C.C. Characterization of gear faults in variable rotating speed using Hilbert-Huang transform and instantaneous dimensionless frequency normalization, *Mechanical Systems and Signal Processing*, 2012; **MSSP-30**: 103–122.

[12] Faiz J., Ebrahimi B.M. and Sharifian M.B.B. Finite element transient analysis of an on-load three-phase squirrel-cage induction motor with static eccentricity, *International Journal of Electromagnetics*, 2007; **IJE-27**: 207–227.

[13] Bangura J.F. and Demerdash N.A.O. Effects of broken bars/end-ring connectors and airgap eccentricities on ohmic and core losses of induction motors in ASDs using a coupled finite element-state space method, *International Conference on Electric Machines and Drives, IEMD '99,* 1999; 339–341.

[14] Faiz, J., Ghorbanian V. and Ebrahimi B.M. Locating broken bars in line-start and inverter-fed induction motors using modified winding function method, *International Journal of Electromagnetics*, 2012; **IJE-32**: 173–192.

[15] Cruz S.M.A., Stefani A., Filippetti F. and Marques Cardoso A.J. A new model-based technique for the diagnosis of rotor faults in RFOC induction motor drives, *IEEE Transactions on Industrial Electronics*, 2008; **IE-55**: 4218–4228.

[16] Watson J.F., Paterson N.C. and Dorrell D.G. The use of finite element methods to improve techniques for the early detection of faults in 3-phase

induction motors, *1997 IEEE International Conference on Electric Machines and Drives*, 1997; WB3/9.1–WB3/9.3.

[17] da Silva A.M., Povinelli R.J. and Demerdash N.A.O. Induction machine broken bar and stator short-circuit fault diagnostics based on three-phase stator current envelopes, *IEEE Transactions on Industrial Electronics*, 2008; **IE-55**: 1310–1318.

[18] Abu-Elhaija W.S., Ghorbanian V., Faiz J. and Ebrahimi B.M. Impact of closed-loop control on behavior of inverter-fed induction motors with rotor broken-bars fault, 2012 *IEEE International Conference on Power Electronics, Drives and Energy Systems (PEDES)*, Bagalura, India, 2012; 1–4.

[19] Yukun L., Liwei G., Qixiang W., Guoqing A., Ming G. and Hao L. Application to induction motor faults diagnosis of the amplitude recovery method combined with FFT, *Journal of Mechanical Systems and Signal Processing*, 2010; **MSSP-24**: 2961–2971.

[20] Salehi Arashloo R. and Jalilian A. Design, implementation and comparison of two wavelet based methods for the detection of broken rotor bars in three phase induction motors, *2010 1st Power Electronic & Drive Systems & Technologies Conference (PEDSTC)*, Tehran, Iran, 2010, pp. 345–350.

[21] Guillaume B. Oblique projection pre-processing and TLS application for diagnosing rotor bar defects by improving power spectrum estimation, *Journal of Mechanical Systems and Signal Processing*, 2013; **MSSP-41**: 301–312.

[22] Ceban A., Pusca R. and Romary R. Study of rotor faults in induction motors using external magnetic field analysis, *IEEE Transactions on Industrial Electronics*, 2012; **IE-59**: 2082–20932.

[23] Cabanas M.F., Pedrayes F., Melero M.G., *et al.* Unambiguous detection of broken bars in asynchronous motors by means of a flux measurement-based procedure, *IEEE Transactions on Instrumentation and Measurement*, 2011; **IM-60**: 891–899.

[24] Henao H., Demian C. and Capolino G.A. A frequency-domain detection of stator winding faults in induction machines using an external flux sensor, *IEEE Transactions on Industrial Electronics*, 2003; **IE-39**: 1272–1279.

[25] Bossio G.R., Angelo C.H.De, Garcia G.O., Solsona J.A. and Valla M.I. Effects of rotor bar and end-ring faults over the signals of a position estimation strategy for induction motors, *IEEE Transactions on Industry Applications*, 2005; **IA-41**: 1005, 1012.

[26] Joáo F.M., Vitor F.P. and Amaral T. Induction motor fault detection and diagnosis using a current state space pattern recognition, *Pattern Recognition Letters*, 2011; **PRL-32**: 321–328.

[27] Faiz J., Ghorbanian V. and Ebrahimi B.M. FEM-based analysis of induction motors with adjacent and non-adjacent broken rotor bars, *International Electromotion Symposium*, Gluj, Romania, 2013.

[28] Ebrahimi B.M., Takbash A.M. and Faiz J. "Losses Calculation in Line-Start and Inverter-Fed Induction Motors Under Broken Bar Fault," *IEEE Transactions on Instrumentation and Measurement*, Jan. 2013; **vol. 62**, no. 1: 140–152.

[29] Takbash A.M., Faiz J. and Ebrahimi B.M. Losses characterization in voltage-fed PWM inverter induction motor drives under rotor broken bars fault, *IEEE Transactions on Magnetics*, 2013; **MAG-49**: 1516–1525.

[30] Bangura J.F. and Demerdash N.A.O. Diagnosis and characterization of effects of broken bars and connectors in squirrel-cage induction motors by a time-stepping coupled finite element-state space modeling approach, *IEEE Transactions on Energy Conversion*, 1999; **EC-14**: 1167–1176.

[31] Byunghwan K., Kwanghwan L., Jinkyu Y., Sang-Bin L., Wiedenbrug E.J. and Shah M.R. Automated detection of rotor faults for inverter-fed induction machines under standstill conditions, *IEEE Transactions on Industry Applications*, 2011; **IA-47**: 55–64.

[32] Soufi Y., Bahi T., Harkat M. and Merabet H. Detection of broken bars in squirrel cage induction motor, *2009 4th International Design and Test Workshop (IDT)*, 2009; **1**: 5.

[33] Akin B., Orguner U., Toliyat H.A. and Rayner M. Phase-sensitive detection of motor fault signatures in the presence of noise, *IEEE Transactions on Industrial Electronics*, 2008; **IE-55**: 2539–2550.

[34] Haji M. and Toliyat H.A. Pattern recognition-A technique for induction machines rotor broken bar detection, *IEEE Power Engineering Review*, 2001; **PER-21**: 67–67.

[35] Faiz J., Ghorbanian V. and Ehya H. Broken bar fault diagnosis in industrial induction motors using the current envelope modulation via TSDM, *15th International Conference on Applied Electromagnetics, 2013, NIS, Siberia*.

[36] Luis A. Escudero G., Duque-Perez O., Morinigo-Sotelo D. and Perez-Alonso M. Robust condition monitoring for early detection of broken rotor bars in induction motors, *Expert Systems with Applications*, 2011; **ESA-38**: 2653–2660.

[37] Wang J., Liu S., Gao R.X. and Yan R. Current envelope analysis for defect identification and diagnosis in induction motors, *Journal of Manufacturing Systems*, 2011; **JMS-31**: 380–387.

[38] Kim C.E., Jung Y.B., Yoon S.B. and Im D.H. The fault diagnosis of rotor bars in squirrel cage induction motors by time-stepping finite element method, 1997; **MAG-33**: 2131–2134.

[39] Said M.S.N., Benbouzid M.E.H. and Benchaib A. Detection of broken bars in induction motors using an extended Kalman filter for rotor resistance sensorless estimation, *IEEE Transactions on Energy Conversion*, 2000; **EC-15**: 66–70.

[40] Mirafzal B. and Demerdash N.A.O. Effects of load magnitude on diagnosing broken bar faults in induction motors using the pendulous oscillation of the rotor magnetic field orientation, *IEEE Transactions on Industry Applications*, 2005; **IA-41**: 771–783.

[41] Mirafzal B. and Demerdash N.A.O. On innovative methods of induction motor inter-turn and broken-bar fault diagnostics, *IEEE Transactions on Industry Applications*, 2006; **IA-42**: 405–414.

[42] Mirafzal B. and Demerdash N.A.O. Induction machine broken-bar fault diagnosis using the rotor magnetic field space vector orientation, *Conference*

*Record of 38th IAS Annual Meeting, Industry Applications Conference*, 2003; 1847–1854.

[43] Yeh C.C., Sizov G.Y., Sayed-Ahmed A., *et al.* A reconfigurable motor for experimental emulation of stator winding interturn and broken bar faults in poly-phase induction machines, *IEEE Transactions on Energy Conversion*, 2008; **EC-23**: 1005–1014.

[44] Eltabach M., Antoni J., Shanina G., Sieg-Zieba S. and Carniel X. Broken rotor bars detection by a new non-invasive diagnostic procedure, *Mechanical Systems and Signal Processing*, 2009; **MSSP-23**: 1398–1412.

[45] Xie Y. Performance evaluation and thermal fields analysis of induction motor with broken rotor bars located at different relative positions, *IEEE Transactions on Magnetics*, 2010; **MAG-46**: 1243–1250.

[46] Acosta G.G., Verucchi C.J. and Gelso E.R. A current monitoring system for diagnosing electrical failures in induction motors, *Mechanical Systems and Signal Processing*, 2006; **MSSP-20**: 953–965.

[47] Henry D., Zolghadri A., Monsion M. and Cazaurang F. Fault diagnosis in induction machines using the generalized structured singular value, *Control Engineering Practice*, 2002; **CEP-10**: 587–598.

[48] Li W., Xie Y., Sheng J. and Luo Y. Finite-element analysis of field distribution and characteristic performance of squirrel-cage induction motor with broken bars, *IEEE Transactions on Magnetics*, 2007; **MAG-43**: 1537,1540.

[49] Faiz J. and Ebrahimi B.M. Determination of number of broken rotor bars and static eccentricity degree in induction motor under mixed fault *International Journal of Electromagnetics*, 2008; **IJE-28**: 433–449.

[50] Bangura J.F., Povinelli R.J., Demerdash N.A.O. and Brown R.H. Diagnostics of eccentricities and bar/end-ring connector breakages in poly-phase induction motors through a combination of time-series data mining and time-stepping coupled FE-state-space techniques, *IEEE Transactions on Industry Applications*, 2003; **IA-39**: 1005–1013.

[51] Povinelli R.J., Bangura J.F., Demerdash N.A.O. and Brown R.H. Diagnostics of bar and end-ring connector breakage faults in poly-phase induction motors through a novel dual track of time-series data mining and time-stepping coupled FE-state space modeling, *IEEE Power Engineering Review*, 2002; **PER-22**: 58–59.

[52] Riera-Guasp M., Cabanas M.F., Antonino-Daviu J.A., Pineda-Sanchez M. and Garcia C.H.R. Influence of nonconsecutive bar breakages in motor current signature analysis for the diagnosis of rotor faults in induction motors, *IEEE Transactions on Energy Conversion*, 2010; **EC-25**: 80–89.

[53] Ye Z., Sadeghian A. and Wu B. Mechanical fault diagnostics for induction motor with variable speed drives using Adaptive Neuro-fuzzy Inference System, *Electric Power Systems Research*, 2006; **EPSR-76**: 742–752.

[54] Dobrodeyev P.N., Volokhov S.A., Kildishev A.V. and Nyenhuis J.A. Method for detection of broken bars in induction motors, *IEEE Transactions on Magnetics*, 2000; **MAG-36**: 3608–3610.

[55] Kral C., Pirker F. and Pascoli G. Model-based detection of rotor faults without rotor position sensor-the sensorless Vienna monitoring method, *IEEE Transactions on Industry Applications*, 2005; **IA-41**: 784–789.

[56] Kral C., Wieser R.S., Pirker F. and Schagginger M. Sequences of field-oriented control for the detection of faulty rotor bars in induction machines-the Vienna monitoring method, *IEEE Transactions on Industrial Electronics*, 2000; **IE-47**: 1042–1050.

[57] Kathir I., Balakrishnan S. and Bevila R.J. Fault analysis of induction motor, *International Conference on Emerging Trends in Electrical and Computer Technology* (ICETECT), 2011; **476**, 479.

[58] Romary R., Pusca R., Lecointe J.P. and Brudny J.F. Electrical machines fault diagnosis by stray flux analysis, *IEEE Workshop on Electrical Machines Design Control and Diagnosis (WEMDCD)*, 2013; 247–256.

[59] Razik H., de Rossiter Correa M.B. and da Silva E.R.C. A novel monitoring of load level and broken bar fault severity applied to squirrel-cage induction motors using a genetic algorithm, *IEEE Transactions on Industrial Electronics*, 2009; **IE-56**: 4615–4626.

[60] Faiz J. and Ebrahimi B.M. Locating rotor broken bars in induction motors using finite element method, *Energy Conversion and Management*, 2009; **ECM-50**: 125–131.

[61] Benbouzid M.E.H. and Kliman G.B. What stator current processing-based technique to use for induction motor rotor faults diagnosis? *IEEE Transactions on Energy Conversion*, 2003; **EC-18**: 238–244.

[62] Chen S. and Živanović R. Estimation of frequency components in stator current for the detection of broken rotor bars in induction machines, *Measurement*, 2010; **M-43**: 887–900.

[63] Elkasabgy N.M., Eastham A.R. and Dawson G.E. Detection of broken bars in the cage rotor on an induction machine, *IEEE Transactions on Industry Applications*, 1992; **IA-28**: 165–171.

[64] Benbouzid M.E.H., Vieira M. and Theys C. Induction motors' faults detection and localization using stator current advanced signal processing techniques, *IEEE Transactions on Power Electronics*, 1999; **PE-14**: 14–22.

[65] Kerszenbaum I. and Landy C.F. The existence of large inter-bar currents in three phase squirrel cage motors with rotor-bar and/or end-ring faults, *IEEE Transactions on Power Apparatus and Systems*, 1984; **PAS-103**: 1854–1862.

[66] Walliser R.F. and Landy C.F. Determination of inter-bar current effects in the detection of broken rotor bars in squirrel cage induction motors, *IEEE Transactions on Energy Conversion*, 1994; **EC-9**: 152–158.

[67] Rodríguez P.V.J., Negrea M. and Arkkio A. A simplified scheme for induction motor condition monitoring *Mechanical Systems and Signal Processing*, 2008; **MSSP-22**: 1216–1236.

[68] Bacha K., Ben-Salem S. and Chaari A. An improved combination of Hilbert and Park transforms for fault detection and identification in three-phase induction motors, *International Journal of Electrical Power & Energy Systems*, 2012; **JEPES-43**: 1006–1016.

[69] Khezzar A., Kaikaa M.Y., El Kamel O.M., Boucherma M. and Razik H. On the use of slot harmonics as a potential indicator of rotor bar breakage in the induction machine, *IEEE Transactions on Industrial Electronics*, 2009; *IE-56*: 4592–4605.

[70] Aydin I., Karakose M. and Akin E. A new method for early fault detection and diagnosis of broken rotor bars, *Energy Conversion and Management*, 2011; **ECM-52**: 1790–1799.

[71] Didier G., Ternisien E., Caspary O. and Razik H. Fault detection of broken rotor bars in induction motor using a global fault index, *IEEE Transactions on Industry Applications*, 2006; **IA-42**: 79–88.

[72] Costa F.F., De Almeida L.A.L, Naidu S.R. and Braga-Filho E.R. Improving the signal data acquisition in condition monitoring of electrical machines, *IEEE Transactions on Instrumentation and Measurement*, 2004; **IM-53**: 1015–1019.

[73] Sun L. Detection of rotor bar breaking fault in induction motors based on Hilbert modulus gyration radius of filtered stator current signal, *2008 International Conference on Electrical Machines and Systems (ICEMS 2008)*, 2008; 877–881.

[74] Ben Salem S., Bacha K. and Gossa M. Induction motor fault diagnosis using an improved combination of Hilbert and Park transforms, *2012 16th IEEE Mediterranean Electrotechnical Conference (MELECON)*, 2012; 1141–1146.

[75] Xu B., Sun L. and Hui R. A new criterion for the quantification of broken rotor bars in induction motors, *IEEE Transactions on Energy Conversion*, 2010; **EC-25**: 100–106.

[76] Puche-Panadero R., Pineda-Sanchez M., Riera-Guasp M., Roger-Folch J. Hurtado-Perez E. and Perez-Cruz J. Improved resolution of the MCSA method via Hilbert transform, enabling the diagnosis of rotor asymmetries at very low slip, *IEEE Transactions on Energy Conversion*, 2009; **EC-24**: 52–59.

[77] Hu. N.Q., Xia L.R., ShouGu F. and Qin G.J. A novel transform demodulation algorithm for motor incipient fault detection, *IEEE Transactions on Instrumentation and Measurement*, 2011; **IM-60**: 480–487.

[78] Kia S.H., Henao H. and Capolino G.A. A high-resolution frequency estimation method for three-phase induction machine fault detection, *IEEE Transactions on Industrial Electronics*, **IE-54**: 2305–2314.

[79] Cupertino F., De Vanna E., Salvatore L. and Stasi S. Analysis techniques for detection of IM broken rotor bars after supply disconnection, *IEEE Transactions on Industry Applications*, 2004; **IA-40**: 526–533.

[80] Milimonfared J., Kelk H.M., Nandi S., Minassians A.D. and Toliyat H.A. A novel approach for broken-rotor-bar detection in cage induction motors, *IEEE Transactions on Industry Applications*, 1999; **IA-35**: 1000–1006.

[81] Akin B., Orguner U., Toliyat H.A. and Rayner M. Low order PWM inverter harmonics contributions to the inverter-fed induction machine fault diagnosis, *IEEE Transactions on Industrial Electronics*, 2008; **IE-55**: 610–619.

[82] Concari C., Franceschini G. and Tassoni C. Rotor fault detection in closed loop induction motors drives by electric signal analysis, *2008. 18th International Conference on Electrical Machines (ICEM 2008),* 2008; 1–6.

[83] Bellini A., Filippetti F., Franceschini G. and Tassoni C. Closed-loop control impact on the diagnosis of induction motors faults, *IEEE Transactions on Industry Applications*, 2000; **IA-36**: 1318–1329.

[84] Liu Z., Yin X., Zhang Z., Chen D. and Chen W. Online rotor mixed fault diagnosis way based on spectrum analysis of instantaneous power in squirrel cage induction motors, *IEEE Transactions on Energy Conversion*, 2004; **EC-19**: 485–490.

[85] Dulce F.P., Fernáo-Pires V., Martins J.F. and Pires A.J. Rotor cage fault diagnosis in three-phase induction motors based on a current and virtual flux approach, *Energy Conversion and Management*, 2009; **ECM-50**: 1026–1032.

[86] Liang B., Iwnicki S.D. and Zhao Y. Application of power spectrum, spectrum, higher order spectrum and neural network analyses for induction motor fault diagnosis, *Mechanical Systems and Signal Processing*, 2013; **MSSP-39**: 342–360.

[87] Faiz J., Ghorbanian V. and Ebrahimi B.M. EMD-based analysis of industrial induction motors with broken rotor bar for identification of operating point at different supply modes, *IEEE Transactions on Industrial Informatics*, 2014; **TIE-10**: 957–966.

[88] Filho P.C.M., Brito J.N., Silva V.A.D. and Pederiva R. Detection of electrical faults in induction motors using vibration analysis, *Journal of Quality in Maintenance Engineering*, 2013; **QME-19**: 364–380.

[89] Rangel-Magdaleno J., Romero-Troncoso R.M., Osornio-Rios R.A., Cabal-Yepez E. and Contreras-Medina L.M. Novel methodology for online half-broken-bar detection on induction motors, *IEEE Transactions on Instrumentation and Measurement*, 2009; **IM-58**: 1690–1698.

[90] Rodriguez P.J., Belahcen A. and Arkkio A. Signatures of electrical faults in the force distribution and vibration pattern of induction motors, *IEE Proceedings, Electric Power Applications*, 2006; **EPA-153**: 523–529.

[91] Medina L.M.C., de Jesus Romero-Troncoso R., Cabal-Yepez E., de Jesus Rangel-Magdaleno J. and Millan-Almaraz J.R. FPGA-based multiple-channel vibration analyzer for industrial applications in induction motor failure detection, *IEEE Transactions on Instrumentation and Measurement*, 2010; **IM-59**: 63–72.

[92] Gritli Y., Di Tommaso A.O., Filippetti F., Miceli R., Rossi C. and Chatti A. Investigation of motor current signature and vibration analysis for diagnosing rotor broken bars in double cage induction motors, *2012 International Symposium on Power Electronics, Electrical Drives, Automation and Motion (SPEEDAM)*, 2012; 1360–1365.

[93] Muller G.H. and Landy C.F. A novel method to detect broken rotor bars in squirrel cage induction motors when interbar currents are present, *IEEE Power Engineering Review*, 2002; **PER-22**: 58–58.

[94] Medina L.M.C., de Jesus Romero-Troncoso R., Cabal-Yepez E., de Jesus Rangel-Magdaleno J. and Millan-Almaraz J.R. FPGA-based multiple-channel vibration analyzer for industrial applications in induction motor failure detection, *IEEE Transactions on Instrumentation and Measurement*, 2010; **IM-59**: 63–72.

[95] Tandon A. and Choudhury A. Review of vibration and acoustic measurement methods for the detection of defects in rolling element bearings, *International Journal of Tribology*, 1999; **JT-32**: 469–480.

[96] Menacer A., Moreau S., Champenois G., Said M.S.N. and Benakcha A. Experimental detection of rotor failures of induction machines by stator current spectrum analysis in function of the broken rotor bars position and the load, *The International Conference on Computer as a Tool*, 2007; 1752–1758.

[97] Sizov G.Y., Sayed-Ahmed A., Chia-Chou Y. and Demerdash N.A.O. Analysis and diagnostics of adjacent and nonadjacent broken-rotor-bar faults in squirrel-cage induction machines, *IEEE Transactions on Industrial Electronics*, 2009; **IE-56**: 4627–4641.

[98] Briz F., Degner M.W., Diez A.B. and Guerrero J.M. Online diagnostics in inverter-fed induction machines using high-frequency signal injection, *IEEE Transactions on Industry Applications*, 2004; **IA-40**: 1153–1161.

[99] Faiz J. and Ebrahimi B.M. A new pattern for detecting broken rotor bars in induction motors during start-up, *IEEE Transactions on Magnetics*, 2008; **MAG-44**: 4673–4683.

[100] Pineda-Sanchez M., Riera-Guasp M., Antonino-Daviu J.A., Roger-Folch J., Perez-Cruz J. and Puche-Panadero R. Diagnosis of induction motor faults in the fractional Fourier domain, *IEEE Transactions on Instrumentation and Measurement*, 2010; **IM-59**: 2065–2075.

[101] Yu Y., Yigang H., Junsheng C. and Dejie Y. A gear fault diagnosis using Hilbert spectrum based on MODWPT and a comparison with EMD approach, *Measurement*, 2009; **M-42**: 542–551.

[102] Climente-Alarcon V., Antonino-Daviu J., Riera-Guasp M., Pons-Llinares J., Roger-Folch J., Jover-Rodriguez P. and Arkkio A. Transient tracking of low and high-order eccentricity-related components in induction motors via TFD tools, *Mechanical Systems and Signal Processing*, 2011; **MSSP-25**: 667–679.

[103] Espinosa A.G., Rosero J.A., Cusido J., Romeral L. and Ortega J.A. Fault detection by means of Hilbert–Huang transform of the stator current in a PMSM with demagnetization, *IEEE Transactions on Energy Conversion*, 2010; **EC-25**: 312–318.

[104] Ebrahimi B.M., Faiz J., Lotfi-fard S. and Pillay P. Novel indices for broken rotor bars fault diagnosis in induction motors using wavelet transform, *Mechanical Systems and Signal Processing*, 2012; **MSSP-30**: 131–145.

[105] Faiz J., Ghorbanian V. and Ebrahimi B.M. A new criterion for rotor broken bar fault diagnosis in line-start and inverter-fed induction motors using Hilbert-Huang transform, *IEEE International Conference on Power Electronics, Drives and Energy Systems (PEDES)*, 2012; 1–6.

[106] Shi P., Chen Z., Vagapov Y. and Zouaoui Z. A new diagnosis of broken rotor bar fault extent in three phase squirrel cage induction motor, *Mechanical Systems and Signal Processing*, 2013; **MSSP**; 388–403.

[107] Cusido J., Romeral L., Ortega J.A., Rosero J.A. and Garcia Espinosa A. Fault detection in induction machines using power spectral density in

Wavelet decomposition, *IEEE Transactions on Industrial Electronics*, 2008; **IE-55**: 633–643.

[108] Pineda-Sanchez M., Riera-Guasp M., Antonino-Daviu J.A., Roger-Folch J., Perez-Cruz J. and Puche-Panadero, R. Instantaneous frequency of the left sideband harmonic during the start-up transient: A new method for diagnosis of broken bars, *IEEE Transactions on Industrial Electronics*, 2009; **IE-56**: 4557–4570.

[109] Pons-Llinares J., Antonino-Daviu J.A., Riera-Guasp M., Pineda-Sanchez M. and Climente-Alarcon V. Induction motor diagnosis based on a transient current analytic Wavelet transform via frequency B-splines, *IEEE Transactions on Industrial Electronics*, 2011; **IE-58**: 1530–1544.

[110] Douglas H., Pillay P. and Ziarani A.K. Broken rotor bar detection in induction machines with transient operating speeds, *IEEE Transactions on Energy Conversion*, 2005; **EC-20**: 135–141.

[111] Georgakopoulos I.P., Mitronikas E.D. and Safacas A.N. Detection of induction motor faults in inverter drives using inverter input current analysis, *IEEE Transactions on Industrial Electronics*, 2011; **IE-58**: 4365–4373.

[112] Antonino-Daviu J., Riera-Guasp M., Roger-Folch J., Martínez-Giménez F. and Peris A. Application and optimization of the discrete wavelet transform for the detection of broken rotor bars in induction machines, *Applied and Computational Harmonic Analysis*, 2006; **ACHA-21**: 268–279.

[113] Briz F., Degner M.W., Garcia P. and Bragado D. Broken rotor bar detection in line-fed induction machines using complex Wavelet analysis of startup transients, 2008; *IEEE Transactions on Industry Applications*, 2008; **IE-44**: 760–768.

[114] Zhang Z., Ren Z. and Huang W. A novel detection method of motor broken rotor bars based on wavelet ridge, *IEEE Transactions on Energy Conversion*, 2003; **EC-18**: 417–423.

[115] Ruqiang Y. and Gao R.X. Hilbert–Huang transform-based vibration signal analysis for machine health monitoring, *IEEE Transactions on Instrumentation and Measurement*, 2006; **IM-55**: 2320–2329.

[116] Antonino-Daviu J.A. and Riera-Guasp M. Pineda-Sanchez M. and Perez, R.B. A critical comparison between DWT and Hilbert–Huang-based methods for the diagnosis of rotor bar failures in induction machines, *IEEE Transactions on Industry Applications* , 2009; **IA-45**: 1794–1803.

[117] Antonino-Daviu J.A., Riera-Guasp M., Pons-Llinares J., Roger-Folch J., Perez R.B. and Charlton-Perez C. Toward condition monitoring of damper windings in synchronous motors via EMD analysis, *IEEE Transactions on Energy Conversion*, 2012; **EC**; 432–439.

[118] Rezazadeh Mehrjou M, Mariun N.H., Marhaban M. and Misron N. Rotor fault condition monitoring techniques for squirrel-cage induction machine—A review, *Mechanical Systems and Signal Processing*, 2011; **MSSP-25**: 2827–2848.

*Chapter 9*

# Diagnosis of eccentricity fault in induction motors

## 9.1 Introduction

The eccentricity fault and the corresponding diagnostic techniques are focused in the current chapter. Like the trend previously used in Chapter 8, the time, frequency and time-frequency domain indices, along with the investigation of the effects of various influential factors on the indices, are explained. Unlike the previous chapter in which the way to extract the indices were carefully discussed, the aim here is to explain the direct application of indices to the diagnosis of eccentricity fault in different operating points without overwhelming readers with details of how the indices are extracted. Even though readers prefer such knowledge, they might refer to the previous chapter since the governing rules are almost the same. The only difference is in the time or frequency pattern to be looked for. Nevertheless, the knowledge provided in this chapter in terms of operating conditions and their effects on the eccentricity-related indices can be examined for the broken bars fault in order to possibly introduce a competent diagnosis approach. Actually, Chapter 8 was sorted in a qualitative fashion while the current chapter aims at providing a quantitative analysis by means of which numerical index values and the corresponding variations are discussed in detail. Therefore, combining Chapters 8–10 provides a very interesting guideline for researchers to obtain a comprehensive knowledge of the field from scratch. Taking this into account, the following aspects will be comprehensively discussed in present chapter:

- to introduce the eccentricity-related indices in time, frequency and time-frequency domains commonly used in condition monitoring of induction motors [1–15]
- to investigate the effect of motor loading [16–23]
- to investigate the effect of various supply modes [24–30]
- to investigate various fault severities and types [31–38]
- to characterize faulty motor losses
- to provide more technical aspects of the diagnosis process of eccentricity fault which might affect the diagnosis process [39–52].

As the underlying concepts of the fault diagnosis have been already discussed in previous chapters, we directly go through the diagnosis procedure of the eccentricity fault without wasting time.

## 9.2 Effect of mixed eccentricity fault on time-domain variation of speed and torque signals

In Chapters 3 and 6, it was clearly shown that any type of the eccentricity fault including the static and dynamic ones produces variations in the time domain. Hence, all the motor quantities are affected. However, unlike the broken bars fault, the time-domain variation of the motor current might not be useful enough in the case of eccentricity fault. On the other hand, the motor speed profile exhibits a better distinction between healthy and faulty motors. Therefore, the motor speed is illustrated in three different cases including (Figure 9.1):

- healthy motor
- static eccentricity: 10% and dynamic eccentricity: 20%
- static eccentricity: 20% and dynamic eccentricity: 10%.

The following features are extracted from Figure 9.1:

- The healthy motor reveals the smoothest operation among three cases.
- Surprisingly, the occurrence of eccentricity fault reduces the motor rise time which is defined as the time instance at which the motor speed becomes equal

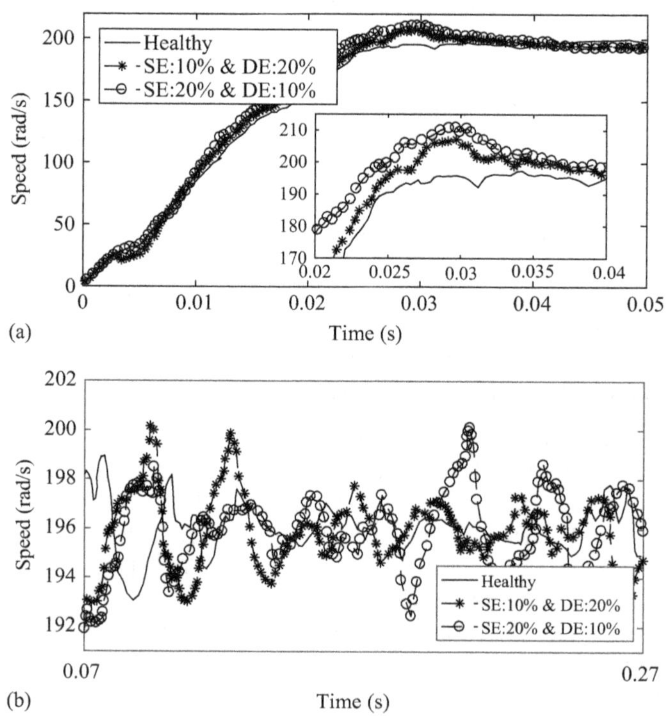


*Figure 9.1 Time variation of speed of a healthy and faulty full-load induction motor (a) transient operation and (b) steady-state operation*

to its average steady-state value. This is a proof of the claim that the fundamental harmonic component of the motor torque corresponding to the faulty case has somehow more power than that of the healthy motor. Nevertheless, no one can deny the improperly large overshoot of the motor speed in faulty cases. This is a source of mechanical tension and might lead to a higher level of not only the eccentricity fault but also the broken bars fault.

- If the definition of the motor rise time is changed to a time instance at which the motor speed comes to its stable steady-state operation, the healthy motor still bits the faulty cases.
- The larger the static eccentricity severity is, the larger the motor speed overshoots will be.
- The larger the static eccentricity severity is, the larger the speed fluctuations in the steady-state operation are.
- The dynamic eccentricity normally presents a smoother motor signal fluctuation.
- The time-domain variation of motor speed does not reflect a clearly observable harmonic component related to the eccentricity fault, and what exists is a fluctuation representing a nonconsistent variation trend. The same deduction is usually applied to the time-domain analysis of other motor signals including the current and torque.

Figure 9.2 illustrated the steady-state operation of the motor torque in three distinct operations, namely:

- healthy
- 50% static eccentricity
- 50% dynamic eccentricity.

In this case, two types of eccentricity faults have been applied to the motor separately to check if the separation yields an understandable time-domain difference between the healthy and faulty cases. The answer is no, and a more detailed analysis obtained with the aim of the frequency-domain signals is required in order to start detecting the fault. Then, we can move forward with different aspects of a comprehensive diagnosis and condition monitoring procedure. Before going through the details of a frequency-domain diagnosis procedure, we will refer readers to a time-domain approach discussed in the next subsection.

## 9.3 Normalized splitting severity factor

The normalized splitting severity factor (NSSF) is a time-domain index of the eccentricity fault. This index is examined here by addressing its variations corresponding to a 4-pole, 28-rotor bar and 36-stator slot motor modeled and simulated in Chapter 6. In a four-pole induction motor, every mechanical cycle of magnetic field is equal to two cycles of the stator current. The static eccentricity produces an unbalanced and nonuniformly distributed flux density which is a function of the air gap length as well as the magnetic reluctance. This leads to an unbalanced stator current during the rotation of the rotor. The Park's transform of

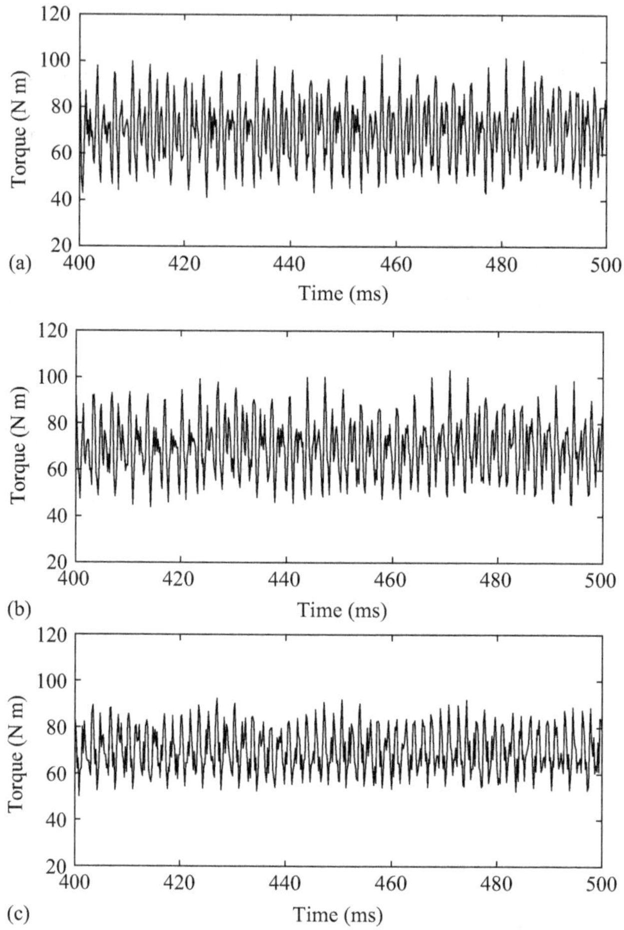


*Figure 9.2 Time-domain variation of motor torque (a) healthy, (b) 50% static eccentricity and (c) 50% dynamic eccentricity*

the stator current can clearly illustrate the difference between the healthy and faulty cases. The NSSF is a simple measure for calculating the difference between two consecutive cycles of the stator current. If $i_{dk}$ and $i_{qk}$ are the measured samples of the current components for two consecutive electric cycles using $N$ samples, then the NSSF is calculated as follows:

$$(\Delta\rho)_N = \frac{(\Delta\rho)_{av}}{\rho_{av}} \tag{9.1}$$

The denominator is the average length of the Park's vector of the stator current expressed by

$$\rho_{av} = \frac{\sum_{k=1}^{N} \sqrt{i_{dk}^2 + i_{qk}^2}}{N} \tag{9.2}$$

and the numerator is equal to the average difference between two cycles of the stator current expressed as

$$(\Delta\rho)_{av} = \frac{\sum_{k=1}^{N/2} \left| \sqrt{i_{dk}^2 + i_{qk}^2} - \sqrt{i_{dN/2+k}^2 + i_{qN/2+k}^2} \right|}{N/2} \tag{9.3}$$

The NSSF as a function of the static eccentricity severity has an increasing trend. Basically, diagnosing the static eccentricity fault below 60% is difficult and somehow impossible in practice since the variation rate of the index is negligible for the fault severity between 0% and 60%. The easiness of the corresponding measurements and the robustness of the index against noise as well as the number of rotor bars make it an interesting way of detection. However, the results might be compromised for low levels of the fault. Moreover, its applicability to the diagnosis of a mixed eccentricity fault is still questionable. The effect of load variation has not been addressed yet.

## 9.4 Ratio of area enclosed by stator current in two consecutive cycles to average area of two stator current cycles

In a four-pole induction motor, if the magnitude of the stator current, i.e., $\sqrt{i_q^2 + i_d^2}$, in two consecutive electrical cycles is plotted versus the rotor position, two completely different waves are obtained if an eccentricity fault exists. The introduced index is the ratio of the area enclosed by the stator current in two consecutive cycles to average area of the stator current cycles. This index and the NSSF are actually two definitions of the same phenomenon; hence, they reveal the same characteristics in terms of capability and incapability. If the number of measured points of the stator current is fairly large, the NSSF and the index introduced lately lead to the same value. This claim can be proved mathematically. First, the second index is formulated as follows:

$$\text{Second index value} = \frac{\Delta A}{A_{av}} \tag{9.4}$$

where $\Delta A$ is the area enclosed by the stator current in two consecutive cycles, and $A_{av}$ is the average area of the two curves, the current for two consecutive cycles. If the number of samples in two cycles is equal to $N$ and if the time step of the measurements is fixed such that the first $N/2$ samples are assigned to the first electric cycle, the second $N/2$ samples are associated with the second electric cycle. The numerator and denominator of (9.4) are calculated as the following, using the trapezoidal integration:

$$\Delta A = \frac{2\pi}{N/2} \left( \sum_{k=1}^{N/2} \left( \left| i_k - i_{N/2+k} \right| \right) - \frac{1}{2} \left| i_1 - i_{N/2+1} \right| - \frac{1}{2} \left| i_{N/2} - i_N \right| \right) \tag{9.5}$$

$$A_{av} = \frac{2\pi}{N/2}\left(\frac{1}{2}\sum_{k=1}^{N/2}(i_k + i_{N/2+k}) - \frac{1}{4}(i_1 + i_{N/2+1}) - \frac{1}{4}(i_{N/2} + i_N)\right)$$
$$= \frac{2\pi}{N/2}\left(\frac{1}{2}\sum_{k=1}^{N}(i_k) - \frac{1}{4}(i_1 + i_{N/2+1}) - \frac{1}{4}(i_{N/2} + i_N)\right) \tag{9.6}$$

in which $i_k$ is the value of the $k$th samples of the stator current used for calculating $d$ and $q$ components of the stator current.

$$i_k = \sqrt{i_{dk}^2 + i_{qk}^2} \tag{9.7}$$

Considering the above equations, (9.5) and (9.6) can be reformulated as the functions of NSSF parameters.

$$\Delta A = 2\pi\left((\Delta\rho)_{av} - \frac{1}{N}\left|i_1 - i_{N/2+1}\right| - \frac{1}{N}\left|i_{N/2} - i_N\right|\right) \tag{9.8}$$

$$A_{av} = 2\pi\left(\rho_{av} - \frac{1}{2N}(i_1 + i_{N/2+1}) - \frac{1}{2N}(i_{N/2} + i_N)\right) \tag{9.9}$$

By increasing the number of samples, i.e., $N$, not only is the accuracy of the integration process improved, but also the second and third terms on the right-hand side of (9.8) and (9.9) converge at zero, and the values of two indices become equal. Hence, considering the same measurement setup, along with the same operating conditions, both approaches should theoretically lead to the same values as the fault index. If a two-pole induction motor is under the test, the absolute value of the two half cycles of the stator current should be utilized for diagnosing the fault.

In this section, the behavior of the NSSF in the line-start, closed-loop DTC and open-loop CV/F modes is investigated. Earlier in this chapter, it was noted that the index is very sensitive to the static eccentricity fault if the fault level is above 60%. We aim at investigating different levels of mixed eccentricity fault and the corresponding effect on the index value in different operating conditions including various load levels and different supply modes. Figure 9.3 shows the corresponding variations in detail. The static and dynamic eccentricity levels are determined by $\rho_s$ and $\rho_d$, respectively and are identified in Chapter 3.

As seen in Figure 9.3, in the line-start motor, the index reveals an almost zero-value in light-load operation while the value increases up to 2% by increasing the load level. As the mixed eccentricity fault ($\rho_s = 0.115$ and $\rho_d = 0.115$) takes place, the index value slightly increases for the light load and decreases for the heavy loads. This is the case for a low-severity fault while increasing the fault severity to the level ($\rho_s = 0.462$ and $\rho_d = 0.254$) causes more disturbance and a clear difference is observable in Figure 9.3(a). Increasing the load leads to a decrease in the index value. Unlike the broken bars fault where the values of indices are generally proportional to the load level, the eccentricity fault is inverse-proportional to the load level. This arises from the fact that the eccentricity fault is of a mechanical

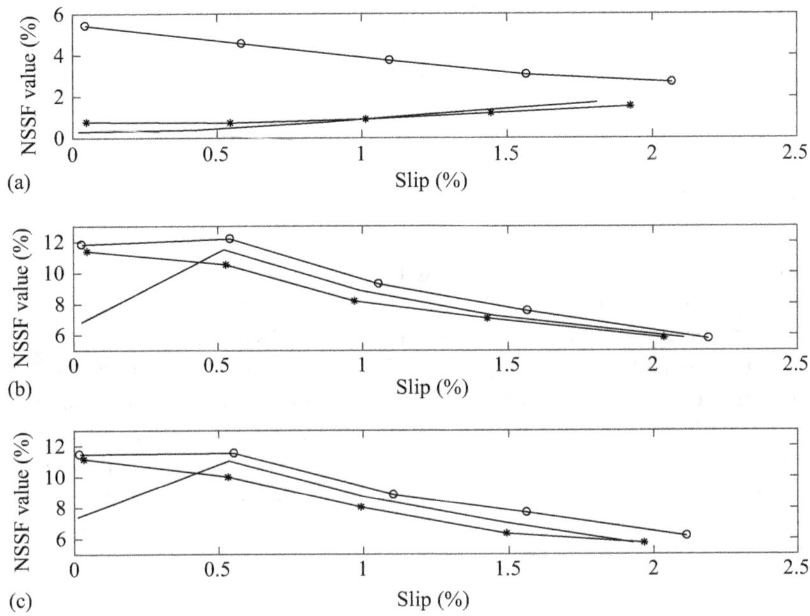


*Figure 9.3 Variation of NSSF versus motor load (slip) level in healthy motor (–), eccentric motor with $\rho_s = 0.115$ and $\rho_d = 0.115$ (–*) and eccentric motor with $\rho_s = 0.462$ and $\rho_d = 0.254$ (–o) (a) line-start motor, (b) closed-loop DTC motor and (c) open-loop CV/F motor*

nature, and increasing the load acts as a damper for the mechanical oscillations of the motor. Therefore, it is expected to observe lower fluctuations so that for the higher load levels, for example, slip = 2%, the index is getting closer to that of the healthy motor.

While supplying the motor using the DTC mode, the index value is very larger than that of the line-start mode. In this case, increasing the load gives rise to the index value initially; then the index value decreases for the slips higher than 0.5%. In a wide range of slip levels ranging from 0.5% to 2.5%, the low fault level ($\rho_s = 0.115$ and $\rho_d = 0.115$) brings the index below that of the healthy motor. However, increasing the fault level ($\rho_s = 0.462$ and $\rho_d = 0.254$) increases the index significantly. Again, the index values in healthy and faulty cases converge at the same point by increasing the slip/load level. The CV/F mode exhibits a similar trend as that of the DTC mode. The main reason for increasing the index in inverter-fed modes is the nonconformal switching angles in two different electric cycles.

The above-mentioned characteristics of the NSSF are the best claims that the index is improper for a precise diagnosis procedure since there is no fixed trend of variations with respect to the motor load or fault level. Therefore, a more reliable approach is required.

## 9.5 High-frequency components of stator current

It is analytically proved that the eccentric induction motor reveals some high-frequency components of the stator current expressed by:

$$f_{\text{high}} = f_s\left[(kR \pm n_d)\frac{1-s}{P} \pm \nu\right] \tag{9.10}$$

where $n_d$ is an integer defining the order of the eccentricity. $n_d = 0$ denotes the static eccentricity, and higher orders are related to the dynamic eccentricity fault. $f_s$ is the synchronous frequency supplied by the network or the drive, $R$ is the number of rotor slots/bars, $s$ is the slip, $P$ is the number of pole-pairs, $k$ is an arbitrary integer and $\nu$ is the order of time harmonics of the motor supply. The mentioned high-frequency harmonic components are/will be present only if their frequency is not equal to the multiples of three of $f_s$. Their frequency is not equal to the multiples of three of $f_s$. Moreover, a specific relation between the number of stator and rotor slots must be held for their presence. This issue has been already discussed in Chapter 4. In addition, for $k = 1$, $n_d = 0$ and $\nu = 1$, the corresponding harmonic components are detectable in the frequency domain of a healthy motor as well, and these harmonic components are called the principle slot harmonics (PSH). Hence, they are inherent harmonic components of induction motors and are not originated by the eccentricity fault. The eccentricity fault only changes the amplitude of the harmonic components. Since the amplitude of these harmonic components are very smaller than the fundamental harmonic component, the $f_s$ component should be first filtered out from the signal, or a (dB) scale is used to demonstrate the frequency-domain signals. Otherwise, the high-frequency components will not be detectable. An example of the high-frequency components of the stator current is illustrated in Figure 9.4.

In Figure 9.4, the current spectrum for the healthy and mixed-eccentric motor in which the static eccentricity severity dominates the dynamic eccentricity severity is shown. The motor has 28 rotor slots and the slip is 0.28% . The most powerful harmonic component of the spectrum is the PSH which is located around 780 Hz. The analytical frequency pattern of PSHs is fairly accurate and is proved by FEM and experimental results. This is true in terms of the frequency of the pattern, not the amplitude which highly depends on the inclusiveness of the simulation model. The possibility of discriminating the static and dynamic types of eccentricity is another advantage of the proposed frequency pattern. On the other hand, there are some disadvantages associated with this index as follows:

- The small amplitude of the index compared to that of the fundamental frequency challenges the measurement process in a way that some accurate filtering process is required before taking the FFT of the signal.
- There is a possibility of interference with the frequency components evoked by load oscillations.

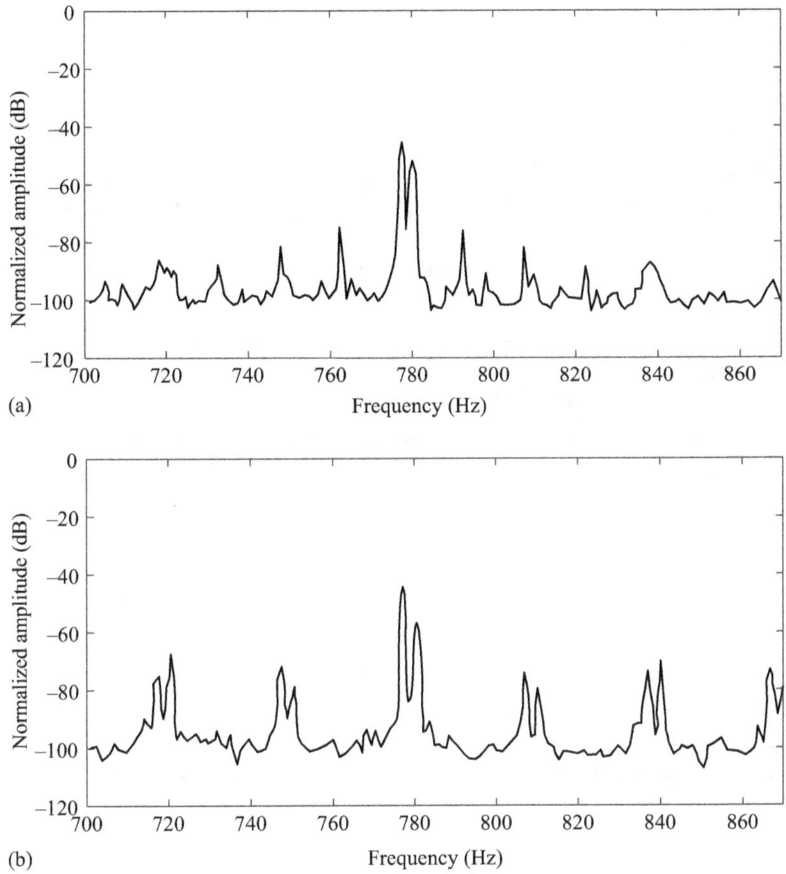


*Figure 9.4 Frequency spectrum of stator current of a 4-pole and 28-rotor bar induction motor (operating slip = 0.28% ) (a) healthy motor and (b) eccentric motor (41.37% static eccentricity and 20.69% dynamic eccentricity)*

- The presence of the index is a direct outcome of the proper combination of the stator and rotor slot numbers. For some specific combinations, the index vanishes completely.
- The rotor slot shape, the skewing effect, the inherent rotor asymmetry and the magnetic profile of the silicon steel material are some influential factors in the amplitude of the index.
- The time harmonics of the supply voltage affect the amplitude of the index regardless of the fault level.
- In a mixed eccentricity fault, the amplitude of the harmonic components corresponding to the dynamic eccentricity is affected by the static eccentricity.
- The amplitude of the index is changed by the motor load level.

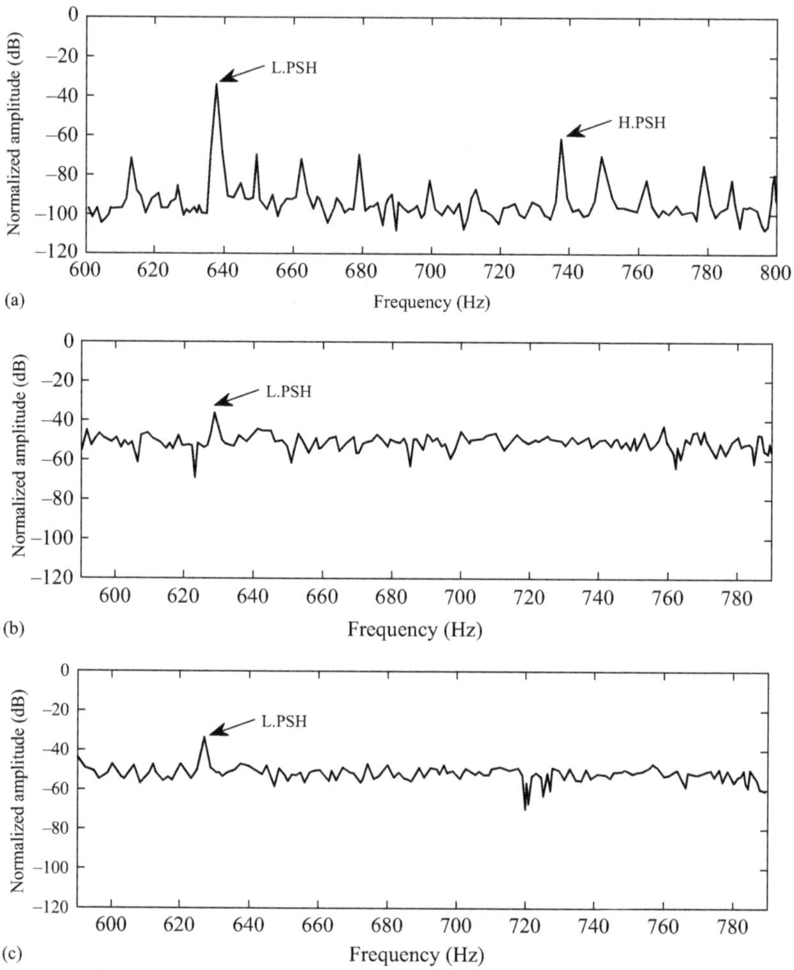


*Figure 9.5 Normalized frequency spectrum of stator current (static eccentricity = 46.2% and operating slip = 1.7%) (a) line-start motor, (b) closed-loop DTC motor and (c) open-loop CV/F motor*

As mentioned before, the static eccentricity amplifies the amplitude of the PSHs whose frequency pattern is given by (9.10). Figure 9.5 shows the frequency spectrum of the stator current at high frequencies for the static eccentricity severity of 46.2% and operating slip of 1.7%. As seen, both the lower and upper PSHs are detectable in the case of a line-start motor supplied by a pure sinusoidal network. The PSHs are shown by arrows in Figure 9.5. However, the closed-loop DTC and open-loop CV/F drives make the upper PSH vanish. It does not mean that the upper PSH does not exist at all. The reason is the presence of a higher noise level of inverter-fed applications caused by the switching phenomenon. The corresponding noise level is larger than the amplitude of the upper PSH; hence, the upper PSH is

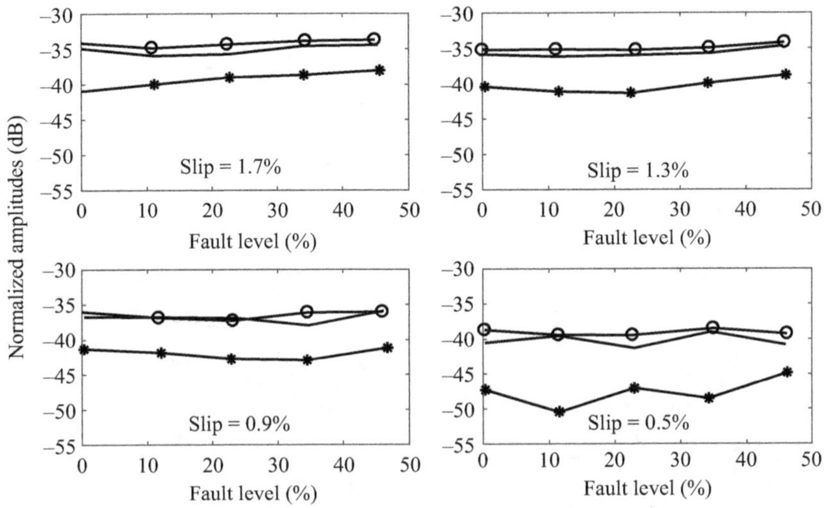


*Figure 9.6 Variation of the amplitude of lower PSH versus static eccentricity level and motor slip. Line-start motor (–), closed-loop DTC motor (–*), open-loop CV/F motor (–o)*

vanished from the spectrum. The noise level of the line-start motor is around −100 dB, while the corresponding value of the inverter-fed motors is equal to −50 dB. The amplitude of the PSH is equal to −61.47 dB. In addition, it is observed in Figure 9.5 that the DTC operation lessens the amplitude of the lower PSH component. This is much clearer in Figure 9.6.

Figure 9.6 illustrates the variation of the amplitude of lower PSH versus the static eccentricity fault and slip levels in three different modes of operations. According to this figure, the amplitude of the lower PSH is greatly reduced in the closed-loop DTC mode. The reduction is more visible in light loads demonstrated by lower slip levels. The corresponding values are smaller than that of the line-start motor. However, the CV/F mode leads to an increase in the index so that they become larger than that of the line-start motor. Increasing the load is a reason for an increase in the amplitude of the index. Obviously, the sensitivity of the index to the motor load level is greater that its sensitivity to the fault severity which is not a good behavior of a fault index. Therefore, the lower PSH is not a proper index for the diagnosis of the static eccentricity fault. This is addressed better using Figure 9.7.

The resultant frequencies of the dynamic eccentricity fault detected in the stator current spectrum are obtained using (9.10) for $n_d > 0$. In Figure 9.8, some dynamic eccentricity-related high-frequency components are highlighted by nontext arrows for a 49.6% dynamic eccentricity severity. The corresponding frequency patterns are clearly detectable in the spectrum of the line-start motor shown by the arrows. However, their amplitudes, like that of the upper PSH, are below the noise level in the inverter-fed applications. So they are expected to

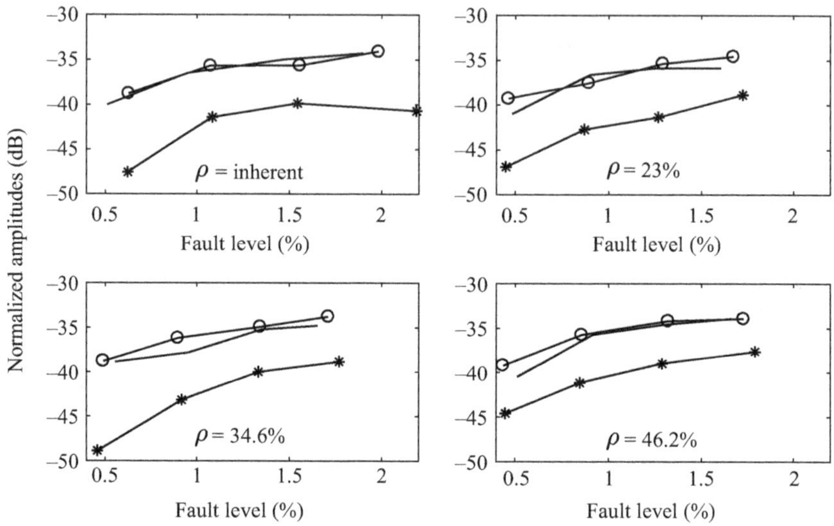


*Figure 9.7 Variation of the amplitude of lower PSH versus motor slip and static eccentricity. Line-start motor (–), closed-loop DTC motor (–*), open-loop CV/F motor (–o)*

vanish in the spectrum of the CV/F and DTC modes. Here again, the amplitude of the lower PSH decreases in the DTC mode. Taking the mentioned points in terms of the high-frequency pattern of the stator current into consideration, we should say that these are not reliable eccentricity fault indices, especially in the case of inverter-fed motors.

## 9.6 Low-frequency components of stator current

There are so many factors affecting the existence of the high-frequency components of the stator current expressed by (9.10). Besides, the corresponding amplitudes are highly affected by the factors other than the fault. On the other hand, due to their high-frequency nature, the interference with noise is somehow inevitable, and the effective spectrum is affected by the noise in inverter-fed applications. Therefore, additional processing tools are probably required in order to reveal more the indices in the frequency spectrum. Thus, it is proposed to find an index located in a lower frequency range and less sensitive to the noise level. It is analytically proved that in the case of a mixed-eccentric rotor, a low-frequency pattern expressed as follows:

$$f_{\text{low}} = |f_s \pm kf_r| \tag{9.11}$$

is visible in the motor spectrum. In (9.11), $f_r$ is the rotational frequency of the rotor, $k$ is an integer and $f_s$ is the fundamental frequency of the supply voltage. For $k = 1$,

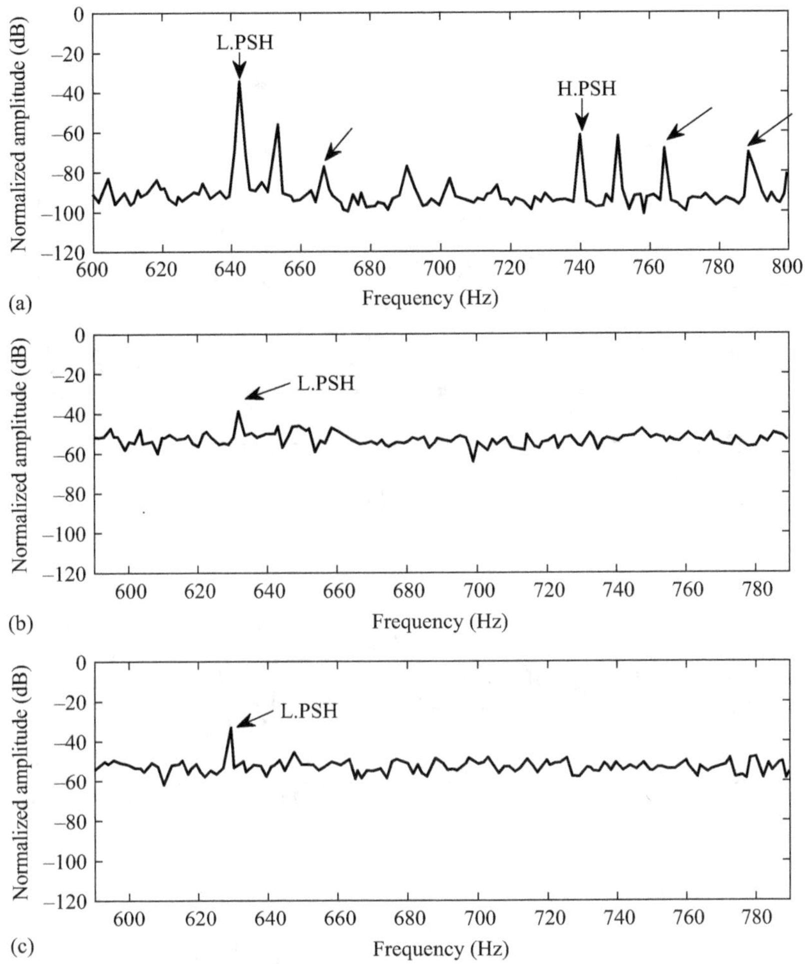


*Figure 9.8 Normalized frequency spectrum of stator current (dynamic eccentricity = 49.6% and operating slip = 1.6%) (a) line-start motor, (b) closed-loop DTC motor and (c) open-loop CV/F motor*

two harmonic components (eccentricity-related sideband components) located at the same distance around the fundamental frequency are detectable. Normally, these low-frequency sideband components are of greater amplitudes than those of the high-frequency patterns, so they are less sensitive to noise. Neither the combination of the stator and rotor slot numbers, nor the number of poles can challenge the existence of the low-frequency pattern. Theoretically, its existence only depends on the existence of the eccentricity fault. However, they are detectable in a spectrum of a healthy motor as well due to the fact that there is always an inherent level of eccentricity fault in any manufactured motor. Obviously, the sideband

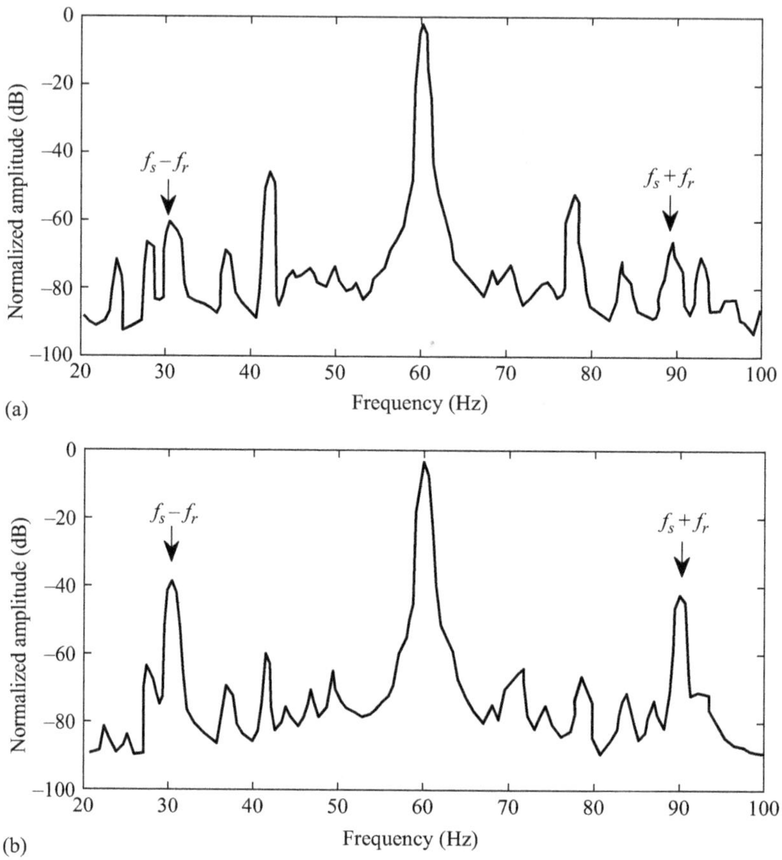


*Figure 9.9 Low-frequency pattern of stator current spectrum for a motor with 28 rotor bars and 4 poles (slip = 2.2%) (a) healthy motor and (b) 41.37% static eccentricity and 20.69% dynamic eccentricity*

components are visible even in the healthy motor, and the eccentricity fault causes the amplitudes of the components to get greatly higher than the healthy case. The easiness of the detection in the spectrum and the independence of this index from noise and motor structure make it a promising tool for diagnosis of the eccentricity fault, although the type of eccentricity is not discriminable using this index. The major disadvantage of the low-frequency pattern is its dependence on the voltage harmonic components (Figure 9.9).

Figure 9.10 shows the current spectrum for a low-frequency range containing the fundamental frequency (50 Hz) as well as the sideband components under a mixed eccentricity fault. Three different supply modes are investigated, and the related spectra are shown. The sideband components are completely visible in the line-start motor while they vanish in the inverter-fed mode. The reason is that

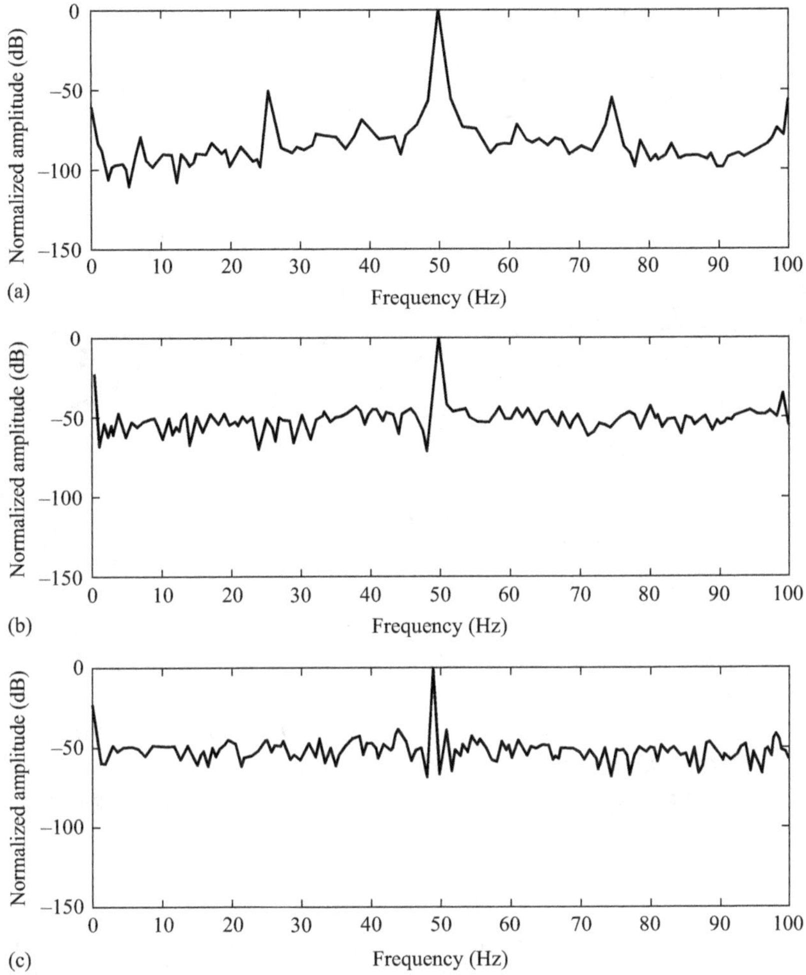


*Figure 9.10 Normalized frequency spectrum of stator current (dynamic eccentricity = 11.5%, static eccentricity = 11.5% and operating slip = 1.6%) (a) line-start motor, (b) closed-loop DTC motor and (c) open-loop CV/F motor*

the eccentricity severity is very small and close to the inherent asymmetry of the motor. However, increasing the fault severity will certainly make the sideband components detectable.

Figure 9.11 illustrates the variation of the amplitude of right (higher) and left (lower) sideband components (H.SB and L.SB, respectively) as well as the lower PSH in the rated speed and two fault cases. Three different supply modes are also investigated. The left column of Figure 9.11 demonstrates a situation where the

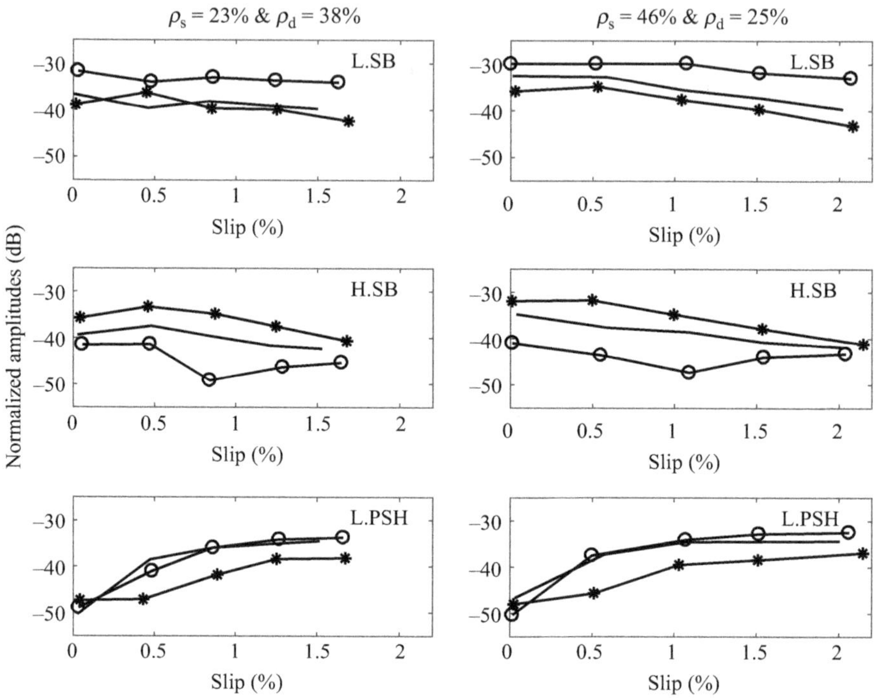


*Figure 9.11 Variation of the amplitude of sideband components (L.SB and H.SB) as well as lower PSH versus motor slip at two fault levels. Line-start motor (–), closed-loop DTC motor (–*), open-loop CV/F motor (–o)*

dynamic eccentricity is more powerful than the static one. On the other hand, the static eccentricity dominates the dynamic eccentricity in the right column. In general, the latter case is more sever in terms of the fault severity. Therefore, the subplots of the right column of Figure 9.11 show the amplitude of the indices which are shifted toward larger positive values compared to the subplots of the left column. In both fault cases, applying the DTC mode to the induction motor decreases the amplitude of the left sideband component (L.SB) compared to the line-start motor with only one exception which might be the result of experimental circumstances. On the other hand, the amplitude of the right sideband component (H.SB) increases compared to the line-start mode. The results of the CV/F mode reveal an opposite trend leading to the following discussion. Applying the CV/F mode to the motor increases the amplitude of the L.SB, while it decreases the amplitude of the H.SB compared to the line-start mode.

Figure 9.12 shows the variation of the amplitude of L.SB, H.SB and lower PSH (L.PSH) as a function of motor slip and fault severity in two different speeds, namely the rated and 55% rated ones. According to this figure, decreasing the

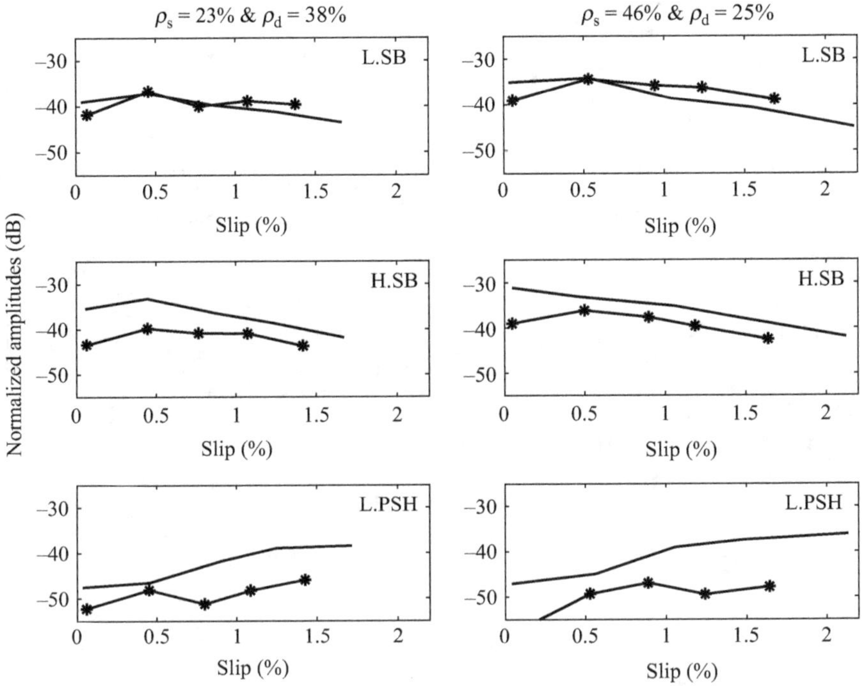


*Figure 9.12 Variation of the amplitude of sideband components (L.SB and H.SB) as well as lower PSH versus motor slip at two fault levels and DTC control mode. Rated speed (–), 55% of rated speed (–*)*

motor speed leads to a considerable decrease in H.SB and L.PSH, while the amplitude of the L.SB slightly increases.

Figure 9.13 shows the variation of indices for the same operating conditions as that of Figure 9.12, but for the CV/F control mode. Accordingly, the L.SB does not change greatly as the motor speed changes. However, the change in speed increases the H.SB and decreases the L.PSH. Figure 9.14 demonstrates the variation of the discussed indices versus the motor speed ranging from 800 rpm to close to 1,500 rpm. The values are compared to each other in two fixed slips, namely 1.5% and 1.8%. The L.SB corresponding to the DTC mode slightly decreases as the speed increases. Unlike the L.SB, the H.SB increases. The L.PSH shows a great increasing rate as the motor speed increases. The variation of the motor speed in the CV/F mode does not have a considerable effect on the L.SB. However, the H.SB and the L.PSH vary a lot.

Having considered the variation trend of the low- and high-frequency indices already discussed, the L.SB component has a lower sensitivity to the motor slip and speed variation in the CV/F mode. Therefore, the L.SB might be considered as a reliable index for the CV/F mode.

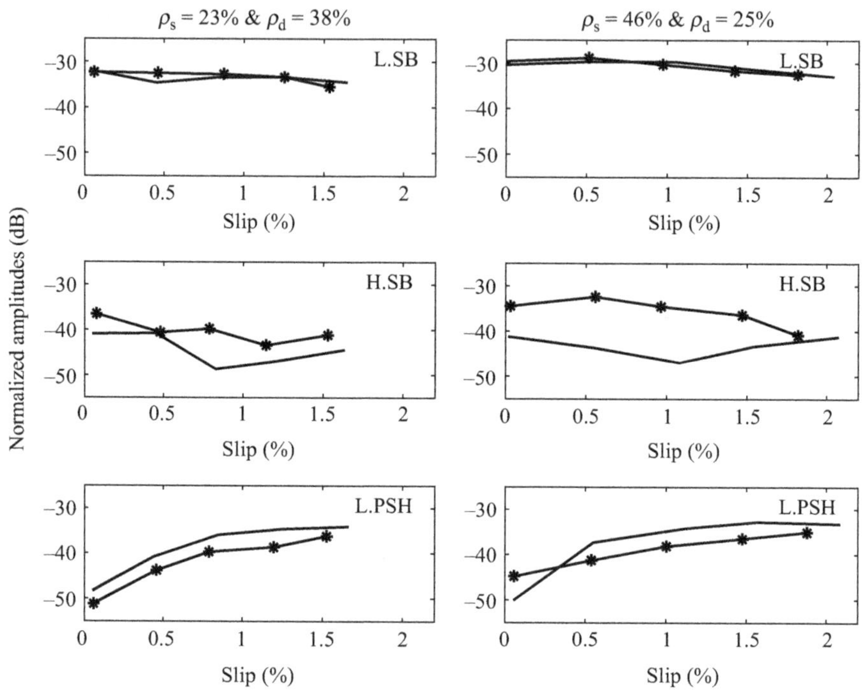


*Figure 9.13 Variation of the amplitude of sideband components (L.SB and H.SB) as well as lower PSH versus motor slip at two fault levels and CV/F control mode. Rated speed (–), 55% of rated speed (–*)*

In the DTC mode, the L.SB is still the best candidate because although the sensitivity to the load and speed is higher than that of the CV/F mode, it is still less than that of the H.SB and L.PSH.

## 9.7 Joint analysis of low- and high-frequency patterns of stator current

The low-frequency fault patterns of the stator current determined as the sideband components of the fundamental frequency, together with the static and dynamic eccentricity faults, affect the high-frequency patterns known as PSHs. Thus, it is very useful to simultaneously analyze the low- and high-frequency indices for the diagnosing and condition monitoring purposes of the eccentricity fault. The literature clearly reveals the weakening effect of the increasing motor load on the fault indices, especially those of low-frequency pattern. This really matters when the eccentricity severity, either static or dynamic, is relatively small. Therefore, the diagnosis procedure might face a big problem since the fault-related harmonic

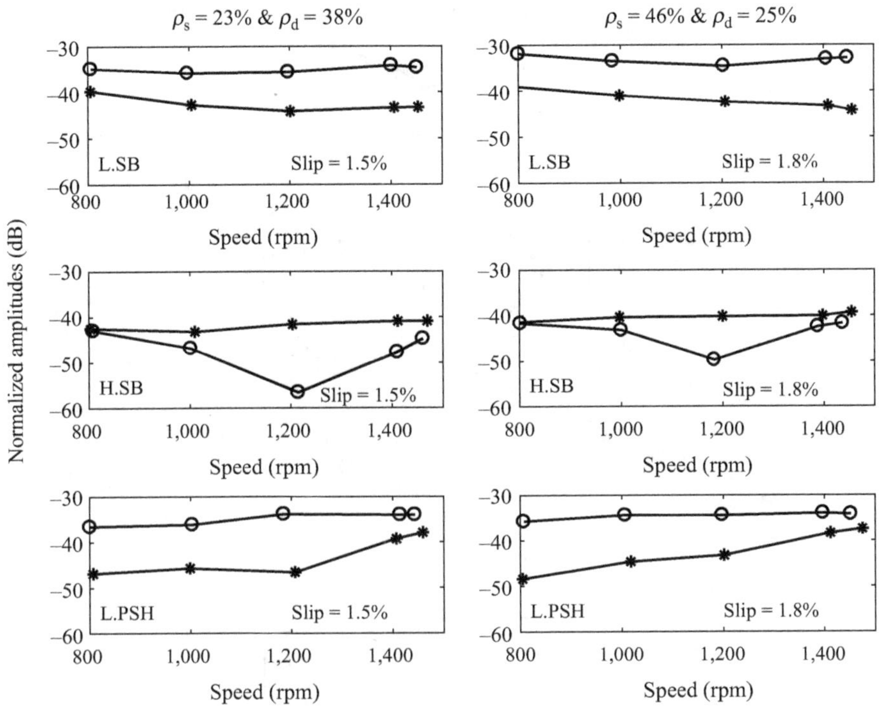


*Figure 9.14 Variation of the amplitude of sideband components (L.SB and H.SB) as well as lower PSH versus motor speed at two slip and fault levels. DTC (–*), CV/F (–o)*

components might not be detectable easily. Figure 9.15 shows how the indices are affected by load and fault levels.

## 9.8 Low-frequency components of voltage space vector and high-frequency components of current space vector

In an inverter-fed case, the motor voltage consists of higher-order time harmonics which in turn affect the motor behavior. In addition, the harmonic components of the torque and speed profile produced by the eccentricity fault might flow into the closed loops embedded in the drive circuit and hence affect the supply frequency as well as the motor current. Besides, the harmonics of the motor current pass into the circuit through the current controllers and make the diagnosis procedure trickier since it is not only the fault affecting the motor, and there are several factors other than the fault which define the motor behavior in relation to the fault. Considering the space vectors of the stator current and voltage signals in a synchronous reference frame, the fundamental components change into a DC component. Therefore, there is no need to filter them out of the signal in order to eliminate their masking effect.

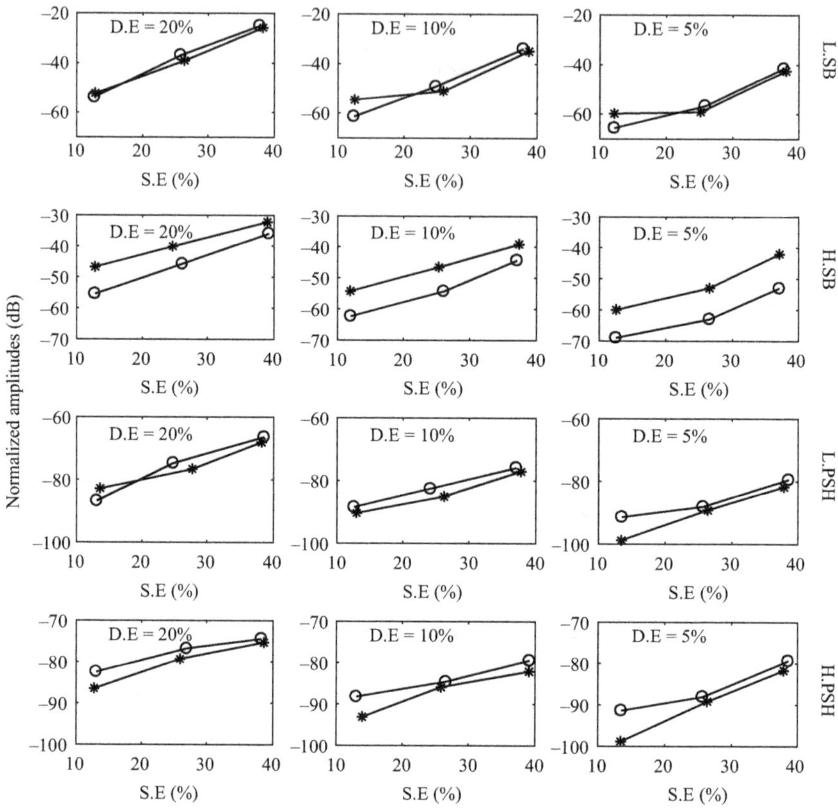


*Figure 9.15 Variation of the amplitude of sideband components (L.SB and H.SB) as well as higher and lower PSH versus motor slip and different fault levels in line-start mode. No-load operation (–*), loaded operation (–o), S.E and D.E stand for static eccentricity and dynamic eccentricity, respectively*

Taking this into account, the following frequency patterns are detectable in the space vectors of the motor voltage and current signals.

$$f_{\text{low}} = f_s \frac{1-s}{P} \tag{9.12}$$

$$f_{\text{high}} = f_s(R \pm 1)\frac{1-s}{P} \tag{9.13}$$

The investigations show that for a typical cutting frequency of the controllers, the low-frequency pattern of the voltage space vector, along with the high-frequency pattern of the current space vector, exhibits larger values. Thus, a combination of these two indices might be interesting for the diagnosis purposes where a mixed fault takes place in an inverter-fed induction motor. One of the main advantages of the mentioned approach is its less sensitivity to the load variations. Nevertheless, the effect of the saturation profile, the number of rotor bars, the number of poles and noise level on the indices have not been addressed yet.

According to the literature, the controllers embedded in the drive circuit normally reduce the amplitude of the fault-related components in the motor current, while the corresponding components of the motor voltage are amplified instead. Therefore, it is concluded that the motor current and voltage signals in a closed-loop application are somehow complementary factors and should be considered simultaneously if a precise diagnosis is required. On other hand, the space vectors of the motor voltage and current are easily accessible in an inverter-fed motor. This suggests the idea of adding the index values of the voltage and current up in order to introduce a competent measure of the eccentricity fault. As the quantities are already available in a drive, no more extension of the experimental equipment is needed.

To become familiar with the variation trend of the indices discussed above in the DTC and CV/F modes, the frequency spectrum of the stator current, the d-component of the stator current, the q-component of the stator current, the current space vector and the voltage space vector are shown in Figure 9.16–9.19 for a given fault and load (slip). The sideband components of the fundamental frequency are shown in Figure 9.16 and Figure 9.18 which demonstrate the frequency spectrum for a low-frequency range. The sideband components are detectable in the spectrum of the dq components of the stator current as well. In the spectrum of the current space vector, the sideband components change into other frequency patterns. The combination of the fundamental frequency and its sidebands produces a new pattern with the frequency as that of the rotor frequency shown in Figure 9.16 and Figure 9.18. Moreover, the PSH-related harmonic components are visible in the spectrum of the dq and space vector of the current signals. The corresponding amplitude is close to that of the phase current illustrated in Figure 9.17 and Figure 9.19. The interesting point is that almost none of the frequency patterns are detectable in the frequency spectrum of the motor voltage components. This is true for both the DTC and CV/F modes.

## 9.9 The ratio of sum of right (high) and left (low) sideband components to no-load current

The low-frequency sideband components of the motor current are one of the most common indices of the eccentricity fault, and they enable us to diagnose a mixed fault in induction motors. The most powerful sideband components are obtained by assigning $k = 1$ in (9.11) which returns the first-order sideband components located in the same distance from the fundamental component. These are the ones used for introducing a new index of the eccentricity fault as follows:

$$\text{fault index} = \frac{I_{\text{low}} + I_{\text{high}}}{I_{\text{no load}}} \tag{9.14}$$

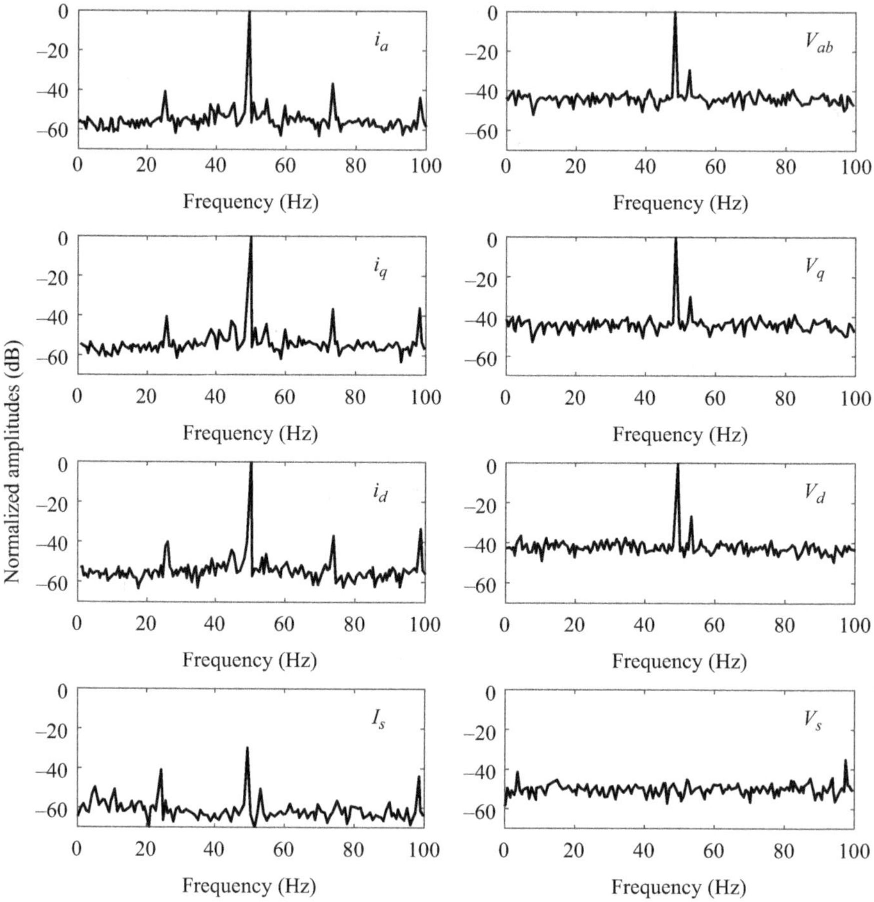


*Figure 9.16 Normalized amplitude of stator current ($i_a$), q-component of stator current ($i_q$), d-component of stator current ($i_d$), space vector of stator current ($I_s$), line voltage ($V_{ab}$), q-component of line voltage ($V_q$), d-component of line voltage ($V_d$) and space vector of line voltage ($V_s$) in DTC mode and low-frequency range (operating slip = 1.56%, static eccentricity = 46.2% and dynamic eccentricity 25.4%)*

where $I_{low}$, $I_{high}$ and $I_{no\ load}$ are the amplitudes of the left (low) sideband component, right (high) sideband component and no-load current, respectively. This index is more sensitive to the fault severity than the load level. So this is one of the advantages of the index. However, it is only useful for revealing the mixed eccentricity fault and is not able to separate the effects of static and dynamic faults. In fact, due to the presence of both static and dynamic inherent eccentricities, this index is always useful. On the basis of the comments in the section related to the sideband components, this index might not be useful for monitoring motors with

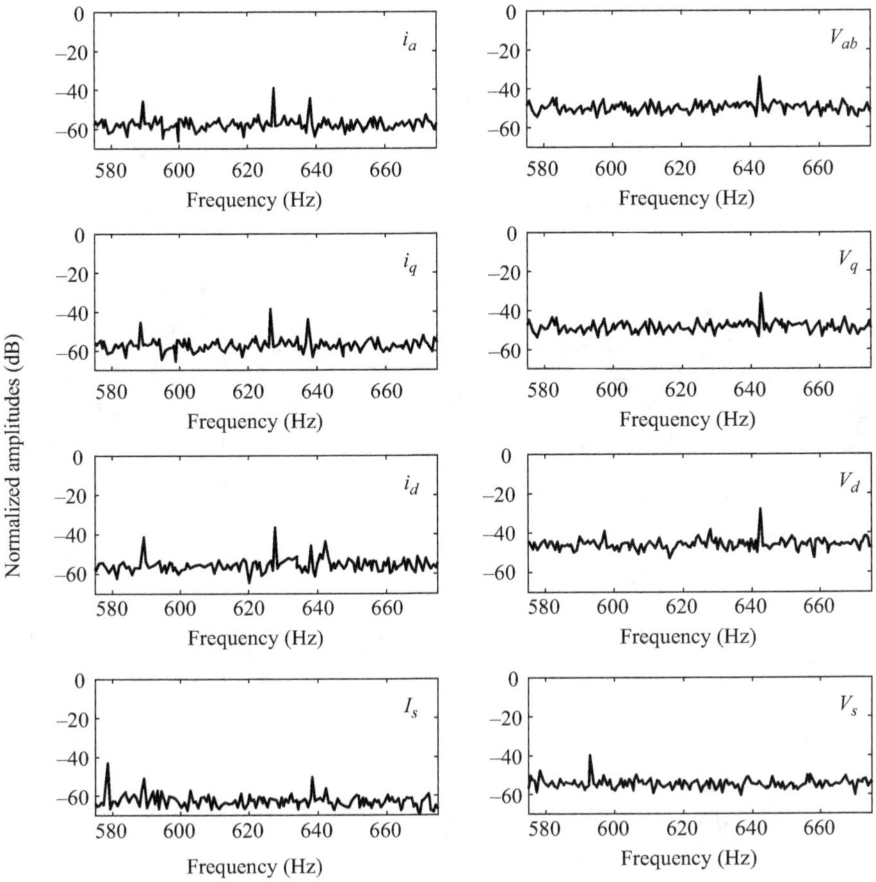


*Figure 9.17 Normalized amplitude of stator current ($i_a$), q-component of stator current ($i_q$), d-component of stator current ($i_d$), space vector of stator current ($I_s$), line voltage ($V_{ab}$), q-component of line voltage ($V_q$), d-component of line voltage ($V_d$) and space vector of line voltage ($V_s$) in DTC mode and high-frequency range (operating slip = 1.56%, static eccentricity = 46.2% and dynamic eccentricity 25.4%)*

low fault severities in an inverter-fed application since the sideband components might be masked completely by the noise. In the literature, it has been mentioned that the value of the index is not a function of the motor power rating and load which is a very significant aspect of this index. Figure 9.20 basically proves the claim. In this figure, the index is shown as a function of the motor slip. In addition, two mixed-eccentricity fault severities are investigated in order to explain the dependency of the index on the load and fault levels. The index shown in Figure 9.20(b) is shifted toward the larger values compared to the values reported

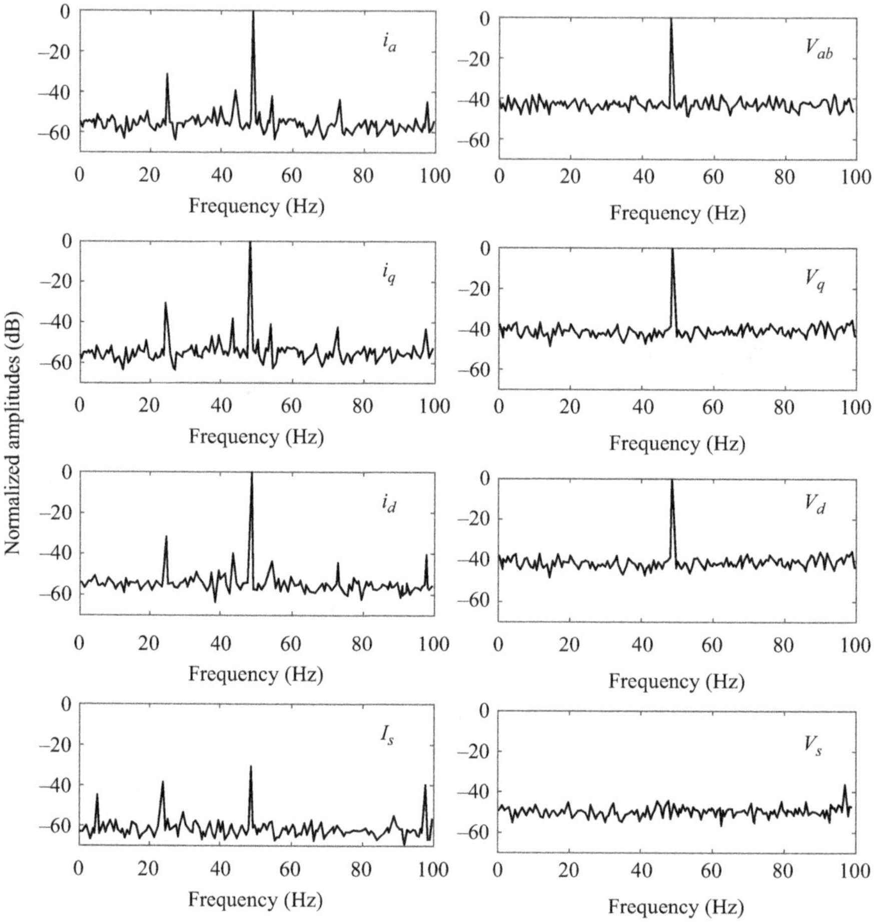


*Figure 9.18 Normalized amplitude of stator current ($i_a$), q-component of stator current ($i_q$), d-component of stator current ($i_d$), space vector of stator current ($I_s$), line voltage ($V_{ab}$), q-component of line voltage ($V_q$), d-component of line voltage ($V_d$) and space vector of line voltage ($V_s$) in CV/F mode and low-frequency range (operating slip = 1.56%, static eccentricity = 46.2% and dynamic eccentricity 25.4%)*

in Figure 9.20(a). This shows that increasing the fault level increases the index somehow independent of the load level. As seen, the indices corresponding to the line-start mode of operation exhibit rather flat variation trend as the slip increases. This is a very good demonstration of the independence of this index from the load variations in the line-start mode. However, the corresponding curves of the DTC mode show a decreasing trend. The CV/F mode has an increasing trend.

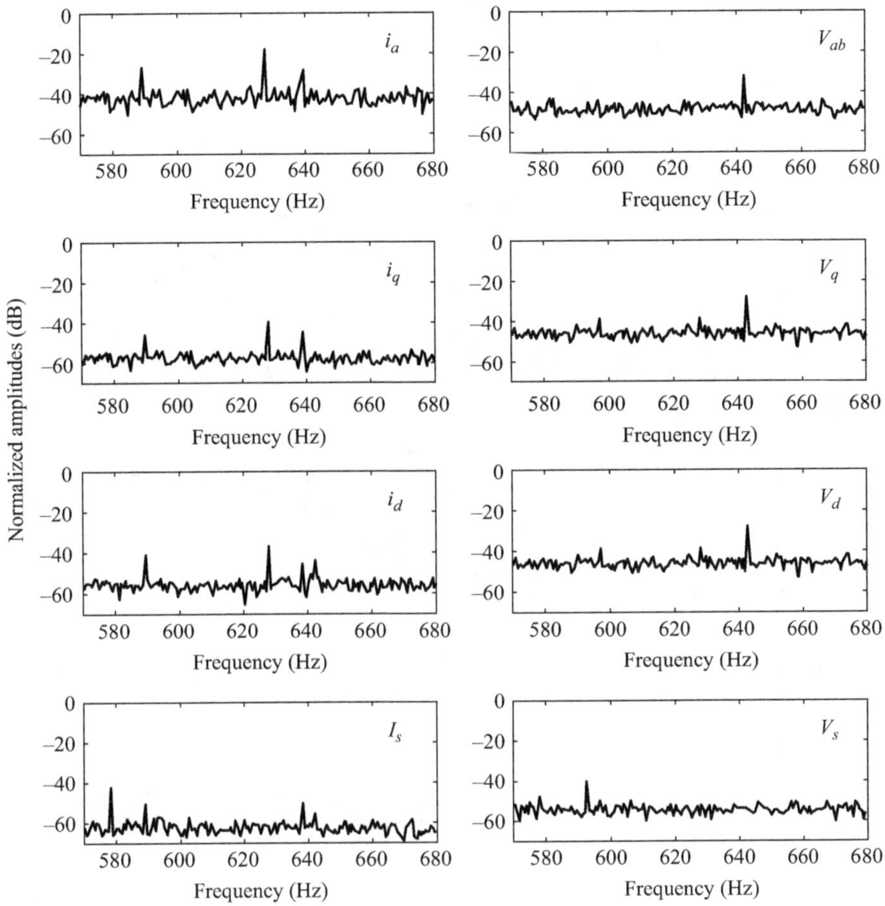


*Figure 9.19 Normalized amplitude of stator current ($i_a$), q-component of stator current ($i_q$), d-component of stator current ($i_d$), space vector of stator current ($I_s$), line voltage ($V_{ab}$), q-component of line voltage ($V_q$), d-component of line voltage ($V_d$) and space vector of line voltage ($V_s$) in CV/F mode and high-frequency range (operating slip = 1.56%, static eccentricity = 46.2% and dynamic eccentricity 25.4%)*

The effect of changing the motor speed on the index values is shown in Figures 9.21 and 9.22 for the DTC and CV/F modes, respectively. In these figures, the index is expressed as a function of the motor slip in two different speeds, namely the rated and almost half the rated speeds. In the DTC mode, the reduction in the motor speed changes the index to smaller amounts while the trend alters from increasing to decreasing as the motor slip increases from zero. Therefore, there is no fixed trend of variation in the CV/F mode.

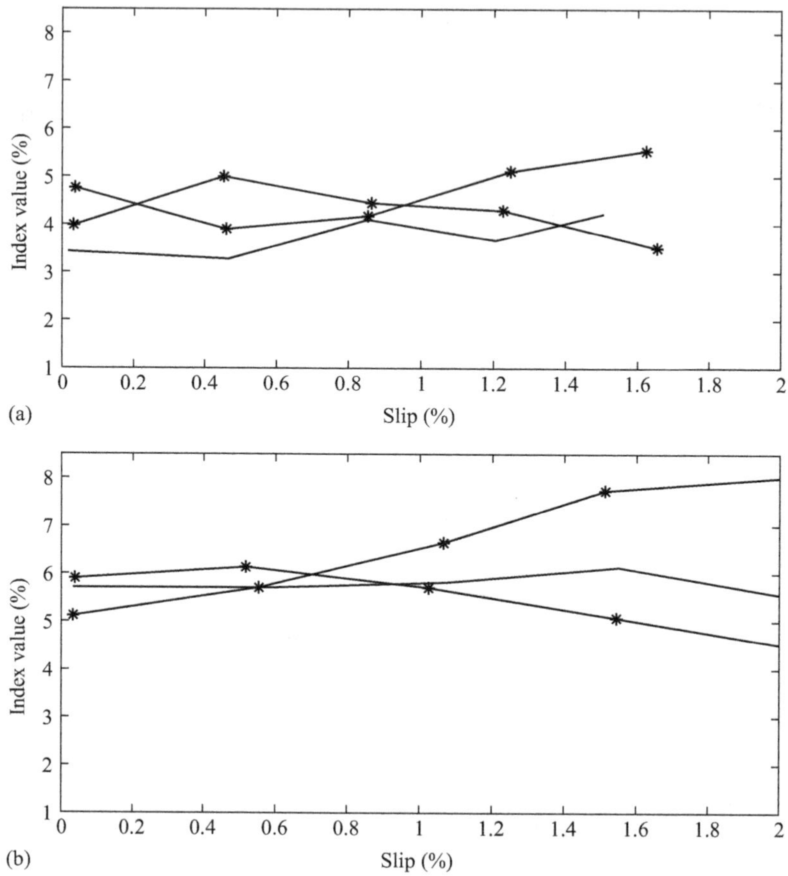


*Figure 9.20 Variation of index value versus slip level for line-start mode (–), DTC mode (–*) and CV/F mode (–o) (a) static eccentricity = 23% and dynamic eccentricity = 38% and (b) static eccentricity = 46% and dynamic eccentricity = 25%*

Figure 9.23 shows the variation of the index as the function of the motor speed in a fixed load level. Two operating modes, i.e., the DTC and CV/F modes, are investigated. The results indicate that the dependency of the index to the motor speed in the DTC mode is lower than that in the CV/F mode. It should be noted that the fault level (b) produces higher index values than the fault (a) revealing that the fault level (b) is severer than the fault level (a).

Considering the material discussed so far in terms of the last index, it is a promising mode-specific measure of the eccentricity fault. However, in inverter-fed modes, the load-dependency is rather high, so we should always take the index values as the functions of load and fault levels into account.

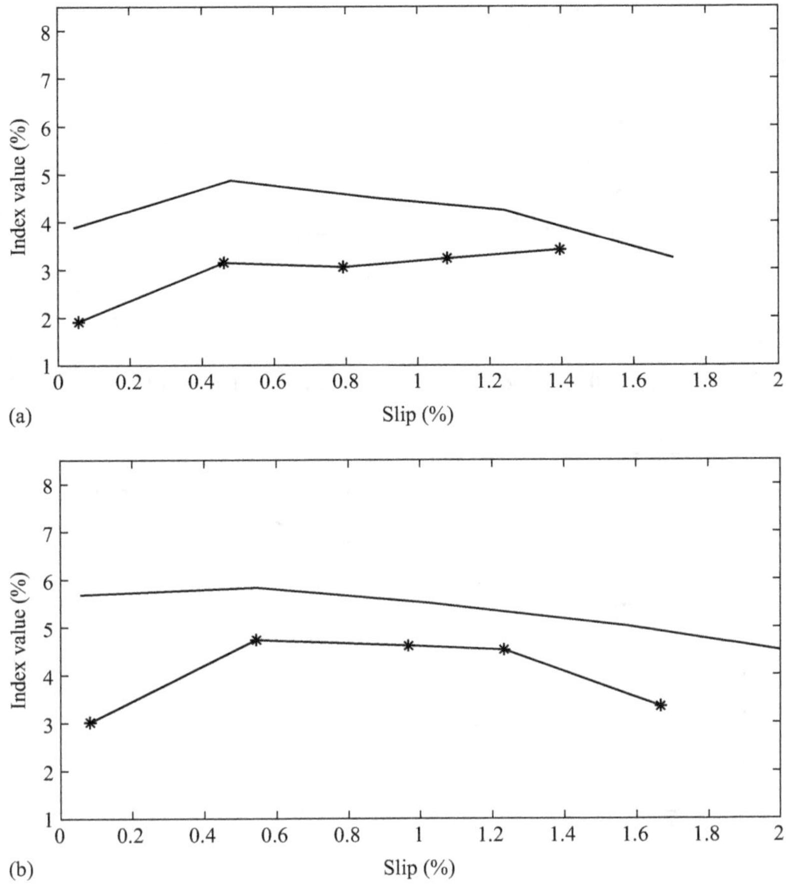


*Figure 9.21 Variation of index value versus slip level in rated speed (–) and 55% of rated speed (–*) in DTC mode (a) static eccentricity = 23% and dynamic eccentricity = 38% and (b) static eccentricity = 46% and dynamic eccentricity = 25%*

## 9.10 Negative sequence current

The experimental results of induction motors reveal the fact that the stator current consists of negative sequence components when the eccentricity fault takes place. Figure 9.24 demonstrates the variation trend of the negative sequence current (NSC) against the variation of the static eccentricity severity. The NSC is expressed in percentage of the fundamental component of the stator current. It is noticeable that the amplitude becomes larger when the fault severity increases. In addition, the index is less sensitive to the motor load because the motor speed changes in the line-start mode. An interesting representation of this phenomenon is seen in Figure 9.25. In this figure, the variation of the index is illustrated as a function of

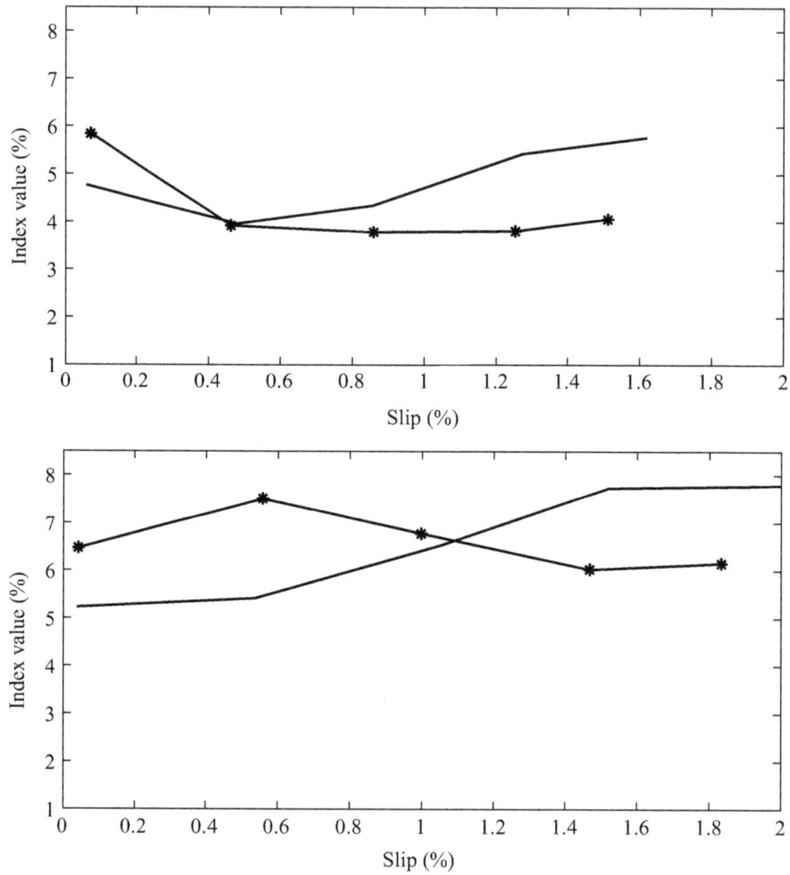


*Figure 9.22 Variation of the index value versus slip level in rated speed (–) and 55% of rated speed (–*) in CV/F mode (a) static eccentricity = 23% and dynamic eccentricity = 38% and (b) static eccentricity = 46% and dynamic eccentricity = 25%*

the motor speed and load. Increasing the motor load in the line-start mode causes the motor speed to decrease.

A very important point is that the NSC is one of the main indices to diagnose the short-circuit fault. However, there is a clear difference between the NSC in eccentricity and short-circuit faults. In fact, to discriminate between the types of faults, we should check the potential for the central node of the Y-connected stator windings. If the corresponding voltage is zero, the negative sequence is probably caused by the eccentricity fault. Otherwise, there is an unbalanced winding of the stator.

The NSC is based on the fact that the flux distribution of the stator windings is not symmetrical any more due to the presence of the eccentricity fault. This produces a NSC in the stator and so can be detected for the monitoring purposes. When

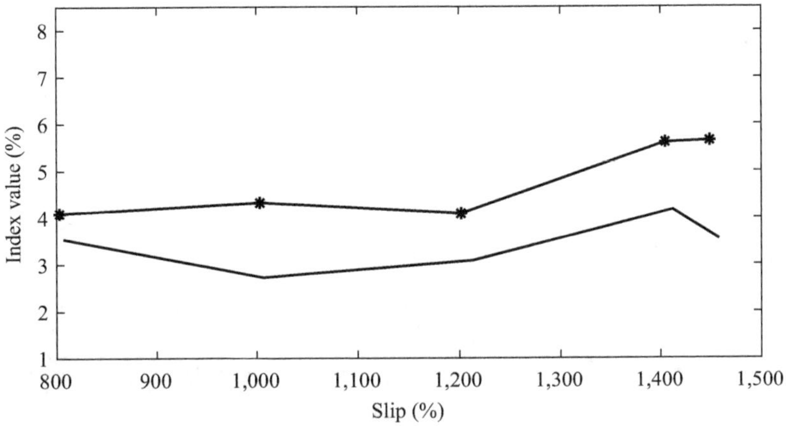


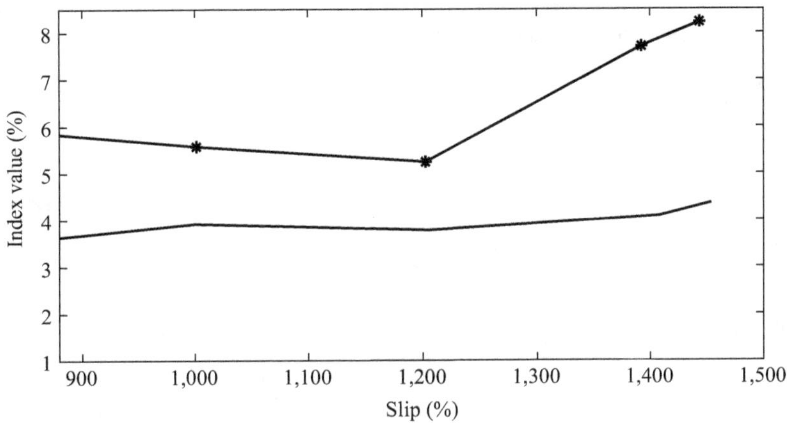


*Figure 9.23 Variation of the index value versus speed level in a fixed load level for DTC mode (–), CV/F mode (–*) and CV/F mode (–o) (a) static eccentricity = 23% and dynamic eccentricity = 38% and (b) static eccentricity = 46% and dynamic eccentricity = 25%*

the motor is supplied by a line-start mode, it is likely that the supply voltage contains some sort of low-level negative sequence components; hence, the discrimination between the fault and the unbalanced voltage is often impossible. Therefore, using the NSC in the line-start mode is not recommended. Instead, we switch our focus to the inverter-fed application in which the three-phase voltages are symmetrical and do not contain negative sequence components. On the other hand, in a closed-loop application, the operation of the current controller might produce NSCs with the same frequency as the fundamental one. Besides, the motor voltage is also affected in this way. Therefore, instead of using the NSC in its raw format, it is preferred to introduce a new index as the sum of the negative sequences of the current and voltage signals. Table 9.1 lists the index values in different operating points for a DTC controlled induction motor. In this table, $V_{nl}$, $V_{pl}$, $I_{nl}$ and

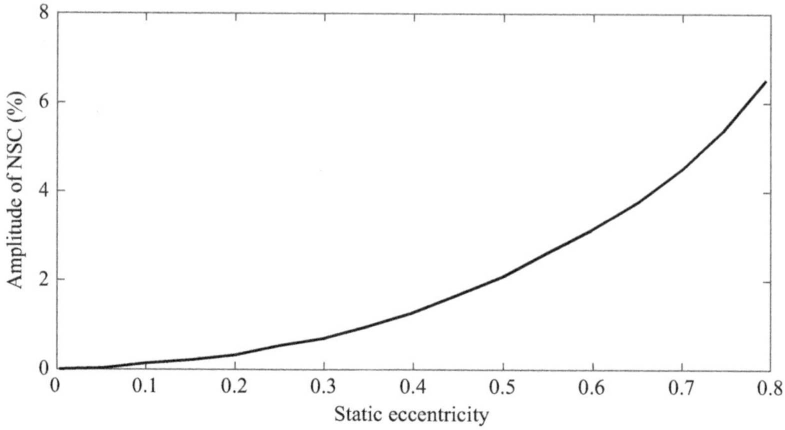


*Figure 9.24 Amplitude of negative sequence current versus static eccentricity level*

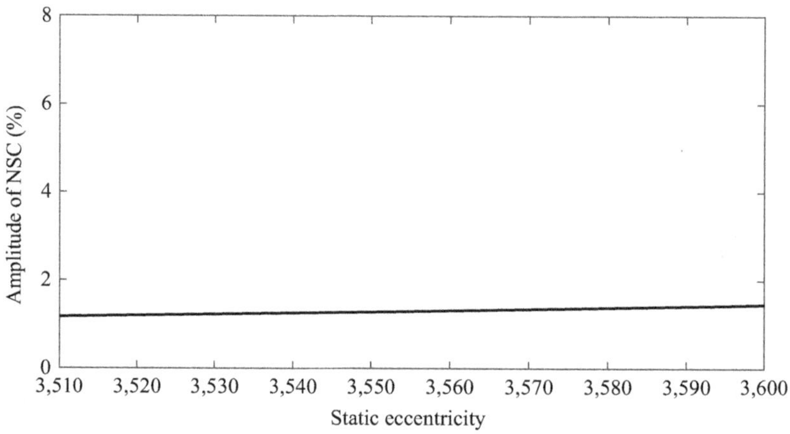


*Figure 9.25 Amplitude of negative sequence current versus motor speed which changes as a function of motor load level in line-start mode*

$I_{p1}$ are the negative sequence voltage (NSV), positive sequence voltage, NSC and positive sequence current, respectively.

Increasing the static eccentricity level up to 23% does not affect the NSV greatly. However, increasing the fault severity beyond 23% greatly changes the NSV so that its amplitude increases with the same rate as the fault severity increases. Increasing the load in a given fault level generally reduces the NSV. However, the NSC does not exhibit a specific variation trend as a function of the load or static eccentricity severity. The higher NSC for higher static eccentricity severities and its lower values for heavy loads are observed in Table 9.1.

Table 9.2 shows the corresponding variations in the CV/F mode.

*Table 9.1 Variation of sum of negative sequences of current and voltage signals versus static eccentricity severity in DTC mode*

| $\rho_s$ (%) | $\rho_d$ (%) | **Slip (%)** | $V_{n1}/V_{p1}$ (%) | $I_{n1}/I_{p1}$ (%) | **Index value** |
|---|---|---|---|---|---|
| Inherent | Inherent | 0.52 | 2.12 | 1.76 | 3.88 |
| 11.5 | | 0.46 | 2.19 | 1.06 | 3.25 |
| 23 | | 0.44 | 2.18 | 1.29 | 3.47 |
| 34.6 | | 0.47 | 4.11 | 0.71 | 4.82 |
| 46.2 | | 0.45 | 4.58 | 1.27 | 5.85 |
| Inherent | | 0.99 | 2.08 | 1.37 | 3.46 |
| 11.5 | | 0.86 | 2.13 | 1.01 | 3.15 |
| 23 | | 0.87 | 2.19 | 0.86 | 3.05 |
| 34.6 | | 0.90 | 3.73 | 1.02 | 4.75 |
| 46.2 | | 0.87 | 4.11 | 1.06 | 5.18 |
| Inherent | | 1.45 | 1.91 | 1.43 | 3.34 |
| 11.5 | | 1.28 | 2.11 | 1.50 | 3.60 |
| 23 | | 1.26 | 2.02 | 0.78 | 2.80 |
| 34.6 | | 1.30 | 3.50 | 1.07 | 4.56 |
| 46.2 | | 1.30 | 3.80 | 1.10 | 4.90 |
| Inherent | | 2.11 | 1.92 | 1.22 | 3.14 |
| 11.5 | | 1.69 | 2.00 | 1.50 | 3.50 |
| 23 | | 2.04 | 1.96 | 0.95 | 2.91 |
| 34.6 | | 1.64 | 3.26 | 1.24 | 4.51 |
| 46.2 | | 1.67 | 3.43 | 1.07 | 4.49 |

*Table 9.2 Variation of sum of negative sequences of current and voltage signals versus static eccentricity severity in CV/F mode*

| $\rho_s$ (%) | $\rho_d$ (%) | **Slip (%)** | $V_{n1}/V_{p1}$ (%) | $I_{n1}/I_{p1}$ (%) | **Index value** |
|---|---|---|---|---|---|
| Inherent | Inherent | 0.54 | 2.39 | 1.47 | 3.87 |
| 11.5 | | 0.47 | 2.45 | 0.89 | 3.34 |
| 23 | | 0.45 | 2.36 | 1.41 | 3.78 |
| 34.6 | | 0.49 | 4.13 | 0.99 | 5.13 |
| 46.2 | | 0.46 | 4.45 | 0.99 | 5.44 |
| Inherent | | 1.01 | 2.38 | 1.39 | 3.77 |
| 11.5 | | 0.87 | 2.48 | 1.27 | 3.75 |
| 23 | | 0.87 | 2.51 | 1.01 | 3.52 |
| 34.6 | | 0.91 | 4.11 | 0.81 | 4.92 |
| 46.2 | | 0.88 | 4.35 | 0.96 | 5.31 |
| Inherent | | 1.47 | 2.29 | 1.76 | 4.05 |
| 11.5 | | 1.30 | 2.44 | 1.30 | 3.73 |
| 23 | | 1.28 | 2.27 | 0.97 | 3.24 |
| 34.6 | | 1.32 | 3.84 | 0.77 | 4.61 |
| 46.2 | | 1.33 | 4.07 | 1.32 | 5.38 |
| Inherent | | 1.92 | 2.18 | 1.62 | 3.81 |
| 11.5 | | 1.72 | 2.32 | 1.12 | 3.44 |
| 23 | | 1.68 | 2.13 | 0.92 | 3.05 |
| 34.6 | | 1.70 | 3.50 | 0.88 | 4.38 |
| 46.2 | | 1.75 | 3.63 | 1.04 | 4.67 |

*Table 9.3 Variation of sum of negative sequences of current and voltage signals versus dynamic eccentricity severity in DTC mode*

| $\rho_s$ (%) | $\rho_d$ (%) | Slip (%) | $V_{n1}/V_{p1}$ (%) | $I_{n1}/I_{p1}$ (%) | Index value |
|---|---|---|---|---|---|
| Inherent | Inherent | 0.52 | 2.12 | 1.76 | 3.88 |
| | 11.5 | 0.44 | 4.15 | 1.09 | 5.24 |
| | 38 | 0.44 | 4.13 | 1.15 | 5.28 |
| | 49.6 | 0.46 | 4.16 | 0.93 | 5.09 |
| | Inherent | 0.99 | 2.08 | 1.37 | 3.45 |
| | 11.5 | 0.86 | 3.93 | 0.92 | 4.86 |
| | 38 | 0.84 | 4.03 | 0.71 | 4.74 |
| | 49.6 | 0.85 | 4.08 | 0.83 | 4.91 |
| | Inherent | 1.45 | 1.91 | 1.43 | 3.34 |
| | 11.5 | 1.24 | 3.65 | 0.82 | 4.47 |
| | 38 | 1.20 | 3.61 | 0.68 | 4.29 |
| | 49.6 | 1.24 | 3.80 | 0.88 | 4.68 |
| | Inherent | 2.11 | 1.92 | 1.22 | 3.14 |
| | 11.5 | 1.69 | 3.49 | 0.58 | 4.08 |
| | 38 | 1.64 | 3.42 | 0.69 | 4.12 |
| | 49.6 | 1.67 | 3.42 | 0.61 | 4.04 |

*Table 9.4 Variation of sum of negative sequences of current and voltage signals versus dynamic eccentricity severity in CV/F mode*

| $\rho_s$ (%) | $\rho_d$ (%) | Slip (%) | $V_{n1}/V_{p1}$ (%) | $I_{n1}/I_{p1}$ (%) | Index value |
|---|---|---|---|---|---|
| Inherent | Inherent | 0.54 | 2.39 | 1.47 | 3.87 |
| | 11.5 | 0.46 | 4.18 | 0.76 | 4.93 |
| | 38 | 0.45 | 4.16 | 1.03 | 5.19 |
| | 49.6 | 0.47 | 4.29 | 1.02 | 5.31 |
| | Inherent | 1.01 | 2.38 | 1.39 | 3.77 |
| | 11.5 | 0.84 | 4.17 | 1.16 | 5.33 |
| | 38 | 0.84 | 4.12 | 0.85 | 4.97 |
| | 49.6 | 0.85 | 4.37 | 0.94 | 5.31 |
| | Inherent | 1.47 | 2.29 | 1.76 | 4.05 |
| | 11.5 | 1.28 | 4.01 | 0.68 | 4.69 |
| | 38 | 1.24 | 3.95 | 0.91 | 4.86 |
| | 49.6 | 1.27 | 4.20 | 0.88 | 5.08 |
| | Inherent | 1.92 | 2.18 | 1.62 | 3.81 |
| | 11.5 | 1.65 | 3.71 | 0.88 | 4.58 |
| | 38 | 1.60 | 3.64 | 0.93 | 4.56 |
| | 49.6 | 1.65 | 3.91 | 1.00 | 4.91 |

Tables 9.3 and 9.4 list the corresponding indices for different dynamic eccentricity severities in various motor loads. Accordingly, in both inverter-fed modes, the small fault severity (11.5%) greatly affects the NSV while increasing the fault severity beyond 11.5% has no more effect on the index. Increasing the load generally reduces the NSV in both DTC and CV/F modes. Here again, the

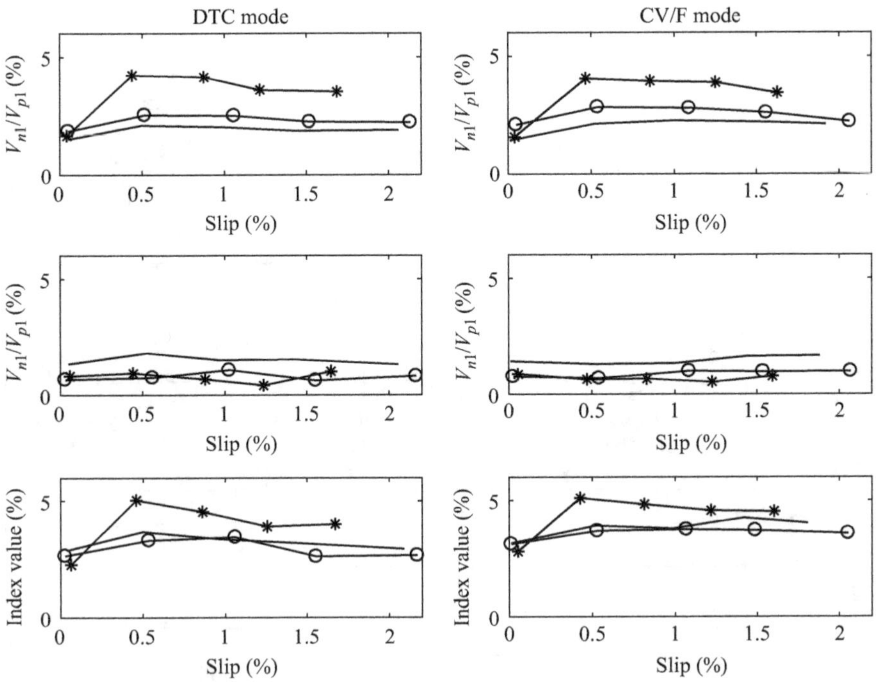


*Figure 9.26 Variation of NSV, NSC and (index: NSV + NSC) versus motor slip level in rated speed, healthy motor (–), static eccentricity = 23% and dynamic eccentricity = 38% (–*), static eccentricity = 46% and dynamic eccentricity = 25% (–o)*

NSC does not have a fixed increasing or decreasing trend as a function of the fault severity. The values reported in Tables 9.3 and 9.4 are of a very little difference for a given operating point which demonstrates the robustness of this technique against the supply mode.

Figure 9.26 shows the variation of the voltage and current sequences separately as well as the index which is equal to the sum of the amplitude of voltage and current sequences. Both the closed-loop DTC and open-loop CV/F modes are investigated in different load and fault severities. As seen, the eccentricity fault causes the NSV to increase considerably compared to the healthy motor. The second case of the eccentricity in which the static type is more powerful than the dynamic type, the index increases more than the first case. This shows the higher dependency of the NSV on the static eccentricity severity. The NSV is almost independent of the control mode. Both fault cases lead to a decrease in the value of NSC in both control modes, namely DTC and CV/F.

For all the loads, the index (NSC + NSV) increases if the first case of the fault, in which the dynamic part dominates the static part, takes place. This is true for both DTC and CV/F modes. For the second fault case, we observe a fixed

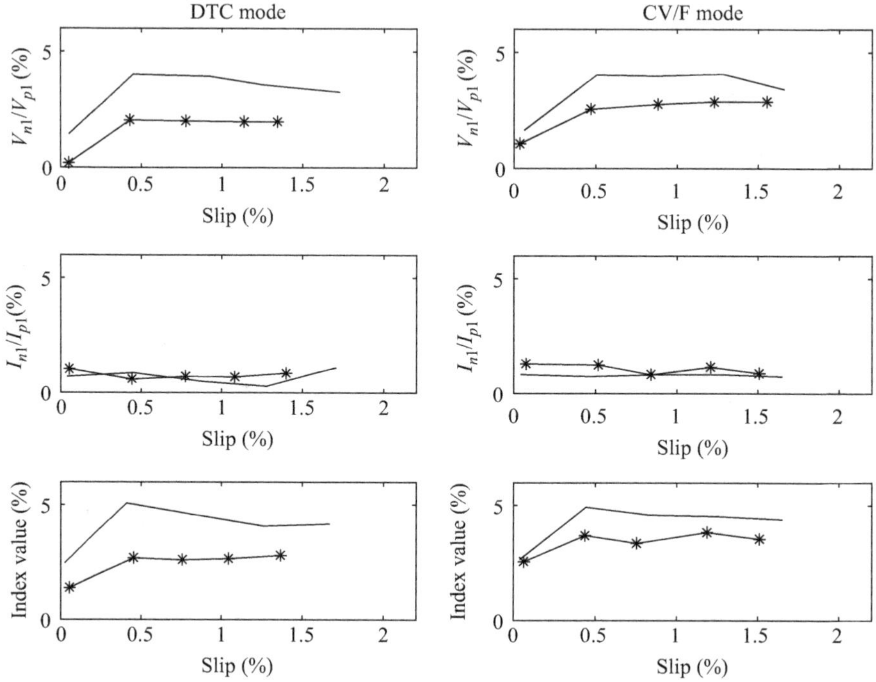


*Figure 9.27 Variation of NSV, NSC and (index: NSV + NSC) versus motor slip level in rated speed, static eccentricity = 23% and dynamic eccentricity = 38%, rated speed (–) and 55% of rated speed (–*)*

decreasing trend at some points of variation compared to the healthy motor in both DTC and CV/F modes.

Figure 9.27 illustrates the effect of reducing the motor speed to 55% of the rated one on the values of NSC, NSV and the introduced index (NSC+NSV) for a fixed eccentricity level. The results demonstrate that decreasing the speed leads to a decrease in the NSV, while it does not affect the NSC considerably. Therefore, it is expected to have a decreasing trend of variation also for the NSC+NSV. This is the case for both the closed-loop DTC and open-loop CV/F modes.

## 9.11 Harmonic components of instantaneous power

It is proved that a set of fault-related harmonic components is created in the motor power spectrum by the low-frequency sideband components of the stator current to the fundamental components of the motor voltage. The corresponding frequency patterns are $kf_r$ and $2f_s \pm kf_r$ where $f_s$ is the fundamental harmonic component of the motor supply, $f_r$ is the rotational frequency of the rotor and $k$ is an integer. These frequency patterns as indices are able to diagnose the mixed eccentricity fault.

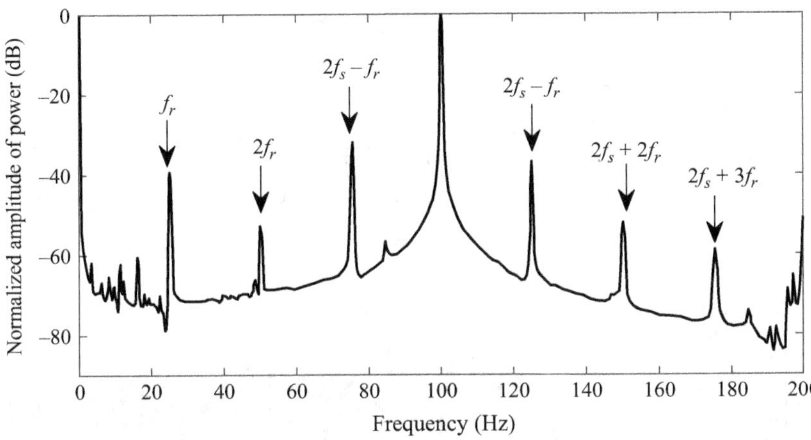


*Figure 9.28 Normalized amplitude of instantaneous power in line-start mode (slip = 0.6%, static eccentricity = 46.2% and dynamic eccentricity 25.4%)*

Figure 9.28 shows the normalized frequency spectrum of the instantaneous power for a fixed eccentricity severity.

As expected, $f_r$, $2f_s-f_r$ and $2f_s+f_r$ have the largest amplitudes among the patterns; hence, they are selected to be studied. A detailed illustration of variation of indexes is included in [35]. Here, we try to provide some of the guidelines based on the results shown in [35]. The static component of the eccentricity dominates the dynamic component. Except $f_r$, the other two frequency patterns reveal a decreasing trend as the motor load/slip increases. Reducing any of the eccentricity fault components, whether static or dynamic, greatly reduces the amplitude of all the frequency patterns in the figures. The rate of the change of $2f_s+f_r$ as a function of the fault is somehow independent of the load. The same conclusion is true for $2f_s-f_r$ in case if the static eccentricity severity is higher than the dynamic eccentricity severity. If the dynamic eccentricity dominates the static eccentricity, the rate of the change of index is much more than the previous case. Taking the mentioned discussions into account, it seems $2f_s+f_r$ is a better index for a mixed-eccentricity fault which is extracted from the instantaneous power of the motor.

While controlling the motor by a DTC mode, the improved noise level, along with the weakened harmonic components of the motor current, leads to the possibly complete elimination of the indices extracted from the motor power signal in higher loads. It is worth noting that unlike the broken bars fault in which increasing the load improves the corresponding index values, the eccentricity-related indices tend to diminish in the case of higher loads. Given that the noise level is around −40 dB in the DTC mode, the indices lower than −40 dB obtained for the DTC mode are not valid. Therefore, the corresponding curve of $f_r$ does not provide something special about the fault since the indices are below the noise level. Thus, working around $f_r$ component is not recommended in diagnosis procedures. In the case of

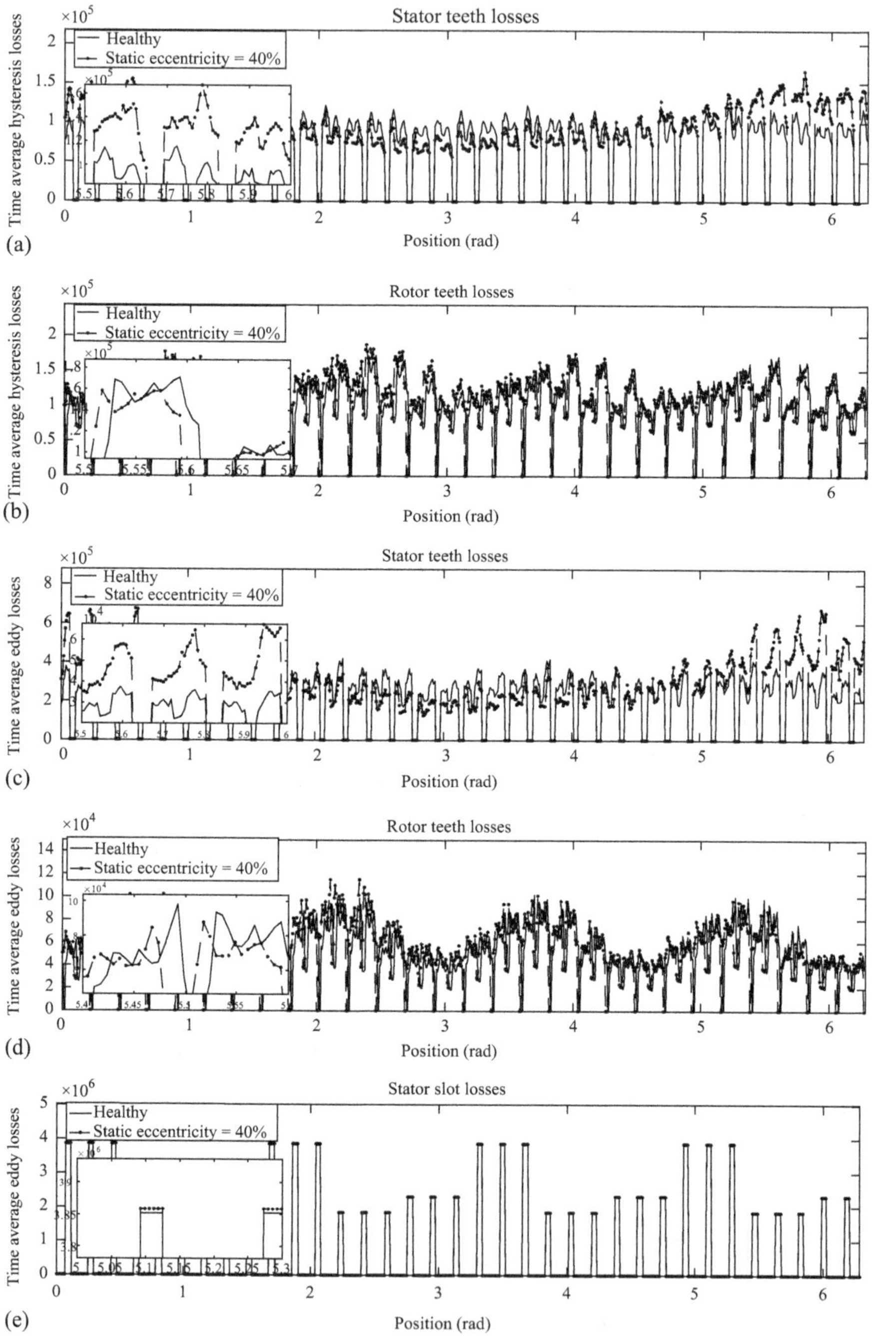


*Figure 9.29 Motor losses in different components (full-load, line-start, healthy and static eccentricity = 40%, simulation run time is equal to 500 ms) (a) stator teeth hysteresis losses, (b) rotor teeth hysteresis losses, (c) stator teeth eddy losses, (d) rotor teeth eddy losses, (e) stator slot ohmic losses and (f) rotor slot ohmic losses*

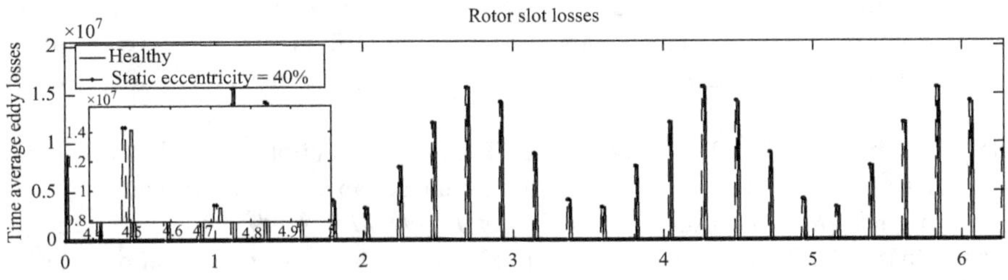


(f)

*Figure 9.29 (Continued)*

other two indices located at sidebands of $2f_s$ component ($2f_s + f_r$ and $2f_s - f_r$), the DTC mode does not have a great effect on the indices compared to the line-start mode. However, if the DTC-related values fall below the noise level at some load levels, those points must be ignored. The interesting point is that applying the DTC mode causes the amplitude of $2f_s + f_r$ to increase compared to the line-start mode. This is a promising behavior since it reduces the masking effect of noise which makes the pattern detection difficult and sometimes impossible.

The investigations prove that $2f_s + f_r$ components are eliminated in a CV/F control mode; hence, the only remaining patterns are $f_r$ and $2f_s - f_r$. The corresponding variation of these components in different operating modes for a CV/F drive is shown below.

## 9.12 Loss characterization of induction motors with eccentricity fault

Figure 9.29 illustrates the loss characterization curves of a 28-bar, 4-pole, an 11-kW induction motor in the healthy and static eccentricity operating conditions in the line-start mode.

## References

[1] Dorrell D.G., Thomson W.T. and Roach S. Analysis of airgap flux, current, and vibration signals as a function of the combination of static and dynamic airgap eccentricity in 3-phase induction motors, *IEEE Transactions on Industry Applications*, 1997; **IA-33**: 24–34.

[2] Jung J.H., Lee J.J. and Kwon B.H. Online diagnosis of induction motors using MCSA, *IEEE Transactions on Industrial Electronics*, 2006; **IE-53**: 1842–1852.

[3] Cameron J.R., Thomson W.T. and Dow A.B. Vibration and current monitoring for detecting air gap eccentricity in large induction motors, *IEE Proceedings B - Electric Power Applications*, 1986; **EPA-133**: 155–163.

[4] Nandi S., Ahmed S. and Toliyat H.A. Detection of rotor slot and other eccentricity related harmonics in a three phase induction motor with different rotor cages, *IEEE Transactions on Energy Conversion*, 2001; **EC-16**: 253–260.

[5] Benbouzid M.E.H., Vieira M. and Theys C. Induction motors' faults detection and localization using stator current advanced signal processing techniques, *IEEE Transactions on Power Electronics*, 1999; **PE-14**: 14–22.

[6] Zhang P., Du Y., Habetler T.G. and Lu B. A survey of condition monitoring and protection methods for medium-voltage induction motors, *IEEE Transactions on Industry Applications*, 2011; **IA-47**: 34–46.

[7] Nandi S., Bharadwaj R.M. and Toliyat H.A. Performance analysis of a three-phase induction motor under mixed eccentricity condition, *IEEE Transactions on Energy Conversion*, 2002; **EC-17**: 392–399.

[8] Liu Z., Yin X., Zhang Z., Chen D. and Chen W. Online rotor mixed fault diagnosis way based on spectrum analysis of instantaneous power in squirrel cage induction motors, *IEEE Transactions on Energy Conversion*, 2004; **EC-19**: 485–490.

[9] Thomson W.T., Rankin D. and Dorrell D.G. On-line current monitoring to diagnose air gap eccentricity in large three-phase induction motors-industrial case histories verify the predictions, *IEEE Transactions on Energy Conversion*, 1999; **EC-14**: 1372–1378.

[10] Akin B., Orguner U., Toliyat H.A. and Rayner M. Low order PWM inverter harmonics contributions to the inverter-fed induction machine fault diagnosis, *IEEE Transactions on Industrial Electronics*, 2008; **IA-55**: 610–619.

[11] Joksimovic G.M., Durovic M.D., Penman J. and Arthur N. Dynamic simulation of dynamic eccentricity in induction machines-winding function approach, *IEEE Transactions on Energy Conversion*, 2000; **EC-15**: 143–148.

[12] Faiz J. and Tabatabaei I. Extension of winding function theory for non-uniform air gap in electric machinery, *IEEE Transactions on Magnetics*, 2002; **MAG-38**: 3654–3657.

[13] Cardoso A.J.M. and Saraiva E.S. Computer-aided detection of air gap eccentricity in operating three-phase induction motors by Park's vector approach, *IEEE Transactions on Industry Applications*, 1993; **IA-29**: 897–901.

[14] Thomson W.T. and Barbour A. On-line current monitoring and application of a finite element method to predict the level of static airgap eccentricity in three-phase induction motors, *IEEE Transactions on Energy Conversion*, 1998; **EC-13**: 347–357.

[15] Faiz J., Ebrahimi B.M., Akin B. and Toliyat H.A. Finite-element transient analysis of induction motors under mixed eccentricity fault, *IEEE Transactions on Magnetics*, 2008; **MAG-44**: 66–74.

[16] DeBortoli M.J., Salon S.J., Burow D.W. and Slavik C.J. Effects of rotor eccentricity and parallel windings on induction machine behavior: a study

using finite element analysis, *IEEE Transactions on Magnetics*, 1993; **MAG-29**: 1676–1682.

[17] Kral C., Habetler T.G. and Harley R.G. Detection of mechanical imbalances of induction machines without spectral analysis of time-domain signals, *IEEE Transactions on Industry Applications*, 2004; **IA-40**: 1101–1106.

[18] Nandi S., Ilamparith T.C., Lee S.B. and Hyun D. Detection of eccentricity faults in induction machines based on nameplate parameters, *IEEE Transactions on Industrial Electronics*, 2011; **IE-58**: 1673–1683.

[19] Li X., Wu Q. and Nandi S. Performance analysis of a three-phase induction machine with inclined static eccentricity, *IEEE Transactions on Industry Applications*, 2007; **IA-43**: 531–541.

[20] Huang X., Habetler T.G. and Harley R.G. Detection of rotor eccentricity faults in a closed-loop drive-connected induction motor using an artificial neural network, *IEEE Transactions on Power Electronics*, 2007; **PE-22**: 1552–1559.

[21] Akin B., Choi S., Orguner U. and Toliyat H.A. A simple real-time fault signature monitoring tool for motor-drive-embedded fault diagnosis systems, *IEEE Transactions on Industrial Electronics*, 2011; **IE-58**: 1990–2001.

[22] Drif M. and Cardoso A.J.M. Discriminating the simultaneous occurrence of three-phase induction motor rotor faults and mechanical load oscillations by the instantaneous active and reactive rower media signature analyses, *IEEE Transactions on Industrial Electronics*, 2012; **IE-59**: 1630–1639.

[23] Drif M. and Cardoso A.J.M. Airgap eccentricity fault diagnosis, in three-phase induction motors, by the complex apparent power signature analysis, *IEEE Transactions on Industrial Electronics*, 2008; **IE-55**: 1404–1410.

[24] Faiz J., Ebrahimi B.M., Akin B. and Toliyat H.A. Comprehensive eccentricity fault diagnosis in induction motors using finite element method, *IEEE Transactions on Magnetics*, 2009; **MAG-45**: 1764–1767.

[25] Bangura J. F., Povinelli R.J., Demerdash N.A.O. and Brown R.H. Diagnostics of eccentricities and bar/end-ring connector breakages in polyphase induction motors through a combination of time-series data mining and time-stepping coupled FE-state-space techniques, *IEEE Transactions on Industry Applications*, 2003; **IA-39**: 1005–1013.

[26] Wu L., Huang X., Habetler T.G. and Harley R.G. Eliminating load oscillation effects for rotor eccentricity detection in closed-loop drive-connected induction motors, *IEEE Transactions on Power Electronics*, 2007; **PE-22**: 1543–1551.

[27] Morinigo-Sotelo D., Garcia-Escudero L.A., Duque-Perez O. and Perez-Alonso M. Practical aspects of mixed-eccentricity detection in PWM voltage-source-inverter-fed induction motors, *IEEE Transactions on Industrial Electronics*, 2010; **IE-57**: 252–262.

[28] Seera M., Lim C.P., Ishak D. and Singh H. Fault detection and diagnosis of induction motors using motor current signature analysis and a Hybrid FMM–CART model, *IEEE Transactions on Neural Networks and Learning Systems*, 2012; **NNLS-23**: 97–108.

[29] Dorrell D.G. Sources and characteristics of unbalanced magnetic pull in three-phase cage induction motors with axial-varying rotor eccentricity, *IEEE Transactions on Industry Applications*, 2011; **IA-47**: 12–24.

[30] Nandi S., Bharadwaj R.M. and Toliyat H.A. Mixed eccentricity in three phase induction machines: analysis, simulation and experiments, *Conference Record of the 2002 IEEE Industry Applications Conference. 37th IAS Annual Meeting (Cat. No.02CH37344)*, 2002; 1525–1532.

[31] Faiz J., Ebrahimi B.M. and Toliyat H.A. Effect of magnetic saturation on static and mixed eccentricity fault diagnosis in induction motor, *IEEE Transactions on Magnetics*, 2009; **MAG-45**: 3137–3144.

[32] Haji M. and Toliyat H.A. Pattern recognition-a technique for induction machines rotor fault detection eccentricity and broken bar fault, *Conference Record of the 2001 IEEE Industry Applications Conference. 36th IAS Annual Meeting (Cat. No.01CH37248)*, 2001; 1572–1578.

[33] Joksimovic G.M. Dynamic simulation of cage induction machine with air gap eccentricity, *IEE Proceedings – Electric Power Applications*, 2005; **EPA-152**: 803–811.

[34] Cruz S.M.A., Cardoso A.J.M. and Toliyat H.A. Diagnosis of stator, rotor and airgap eccentricity faults in three-phase induction motors based on the multiple reference frames theory, *38th IAS Annual Meeting on Conference Record of the Industry Applications Conference*, 2003; 1340–1346.

[35] Faiz J. and Ojaghi M. Instantaneous-power harmonics as Indexes for mixed eccentricity fault in mains-fed and open/closed-loop drive-connected squirrel-cage induction motors, *IEEE Transactions on Industrial Electronics*, 2009; **IE-56**: 4718–4726.

[36] Kim M.J., Kim B.K., Moon J.W., Cho Y.H., Hwang D.H. and Kang D.S. Analysis of inverter-fed squirrel-cage induction motor during eccentric rotor motion using FEM, *IEEE Transactions on Magnetics*, 2008; **MAG-44**:1538–1541.

[37] Dorrell D.G., Thomson W.T. and Roach S. Combined effects of static and dynamic eccentricity on airgap flux waves and the application of current monitoring to detect dynamic eccentricity in 3-phase induction motors, *Seventh International Conference on Electrical Machines and Drives (Conf. Publ. No. 412)*, 1995; 151–155.

[38] Riera-Guasp M., Pineda-Sanchez M., Perez-Cruz J., Puche-Panadero R., Roger-Folch J. and Antonino-Daviu J. A. Diagnosis of induction motor faults via gabor analysis of the current in transient regime, *IEEE Transactions on Instrumentation and Measurement*, 2012; **IM-61**: 1583–1596.

[39] Vitek O., Janda M., Hajek V. and Bauer P. Detection of eccentricity and bearings fault using stray flux monitoring, *8th IEEE Symposium on Diagnostics for Electrical Machines, Power Electronics & Drives*, 2011: 456–461.

[40] Faiz J. and Ojaghi M. Stator inductance fluctuation of induction motor as an eccentricity fault index, *IEEE Transactions on Magnetics*, 2011; **MAG-47**: 1775–1785.

[41] Nandi S. and Toliyat H.A. Detection of rotor slot and other eccentricity related harmonics in a three phase induction motor with different rotor cages, *International Conference on Power Electronic Drives and Energy Systems for Industrial Growth, 1998. Proceedings.*, 1998: 135–140.

[42] Hyun D., Hong J., Lee S.B., *et al.* Automated monitoring of airgap eccentricity for inverter-fed induction motors under standstill conditions, *IEEE Transactions on Industry Applications*, 2011; **IA-47**: 1257–1266.

[43] Puche-Panadero R., Pons-Llinares J., Roger-Folch J. and Pineda-Sanchez M. Diagnosis of eccentricity based on the Hilbert transform of the startup transient current, *IEEE International Symposium on Diagnostics for Electric Machines, Power Electronics and Drives*, 2009; 1–6.

[44] Cardoso A.J.M. and Saraiva E.S. Predicting the level of airgap eccentricity in operating three-phase induction motors, by Park's vector approach, *Conference Record of the 1992 IEEE Industry Applications Society Annual Meeting*, Houston, TX, 1992; 132–135.

[45] Ojaghi M. and Nasiri S. Modeling eccentric squirrel-cage induction motors with slotting effect and saturable teeth reluctances, *IEEE Transactions on Energy Conversion*, 2014; **EC-29**: 619–627.

[46] Faiz J. and Ojaghi M. Unified winding function approach for dynamic simulation of different kinds of eccentricity faults in cage induction machines, *IET Electric Power Applications*, 2009; **EPA-3**: 461–470.

[47] Tenhunen A., Benedetti T., Holopainen T.P. and Arkkio A. Electromagnetic forces in cage induction motors with rotor eccentricity, *IEEE International Electric Machines and Drives Conference, IEMDC'03*. 2003; 1616–1622.

[48] Wolbank T.M. and Macheiner P.E., Modulation of transient reactances of induction machines caused by different types of eccentricity, *IEEE International Symposium on Diagnostics for Electric Machines, Power Electronics and Drives*, 2007; 89–94.

[49] Kelk H.M., Eghbali A. and Toliyat H.A. Modeling and analysis of cage induction motors under rotor misalignment and air gap eccentricity, *Fourtieth IAS Annual Meeting. Conference Record of Industry Applications Conference*, 2005; pp. 1324–1328.

[50] Nandi S., Ahmed S. and Toliyat H.A. Detection of rotor slot and other eccentricity-related harmonics in a three-phase induction motor with different rotor cages, *IEEE Power Engineering Review*, 2001; **PER**: 62–62.

[51] Nandi S., Bharadwaj R.M., Toliyat H.A. and Parlos A.G. Performance analysis of a three phase induction motor under mixed eccentricity condition, *Proceedings of International Conference on Power Electronic Drives and Energy Systems for Industrial Growth*, 1998; 123–128.

[52] Faiz J., Ebrahimi B.M., Akin B. and Toliyat H.A. Dynamic analysis of mixed eccentricity signatures at various operating points and scrutiny of related indices for induction motors, *IET Electric Power Applications*, 2010; **EPA-4**: 1–16.

# Chapter 10
# Diagnosis of interturn short-circuit fault in induction motors

## 10.1 Motor current signature analysis

Motor current signature analysis (MCSA) is one of the most powerful methods of online motor diagnosis for detecting motor faults. MCSA technique in case of cage rotor induction motors means the analysis of stator-line current spectrum obtained by applying Fourier transform to the stator current waveform in steady-state conditions. The stator-line current frequency spectrum represents, a sort of speaking, an electrocardiogram of an electrical machine [1]. This technique is particularly useful because it is noninvasive, and the search coil is the stator winding of the machine, itself. Using MCSA has advantages such as no estimation of motor parameters and the simplicity of current sensors and their installation [2].

Although MCSA is one of the most powerful methods for diagnosing motor faults, it has some shortcomings:

- MCSA requires high precision of slip-frequency information;
- Stator current data should be sampled after motor speed arrives at the steady-state;
- Unspecified harmonic numbers induce ambiguous results from MCSA-based diagnosis methods [2].

In order to make the correct interpretation of changes in the spectrum as result of different faulty regimes, one should be familiar with the current spectrum of the healthy machine [1], which is not an easy task because of the complexity of electromagnetic processes, variable in space and time, which occur in the machine. In other words, it is necessary to have a powerful mathematical model that enables numerical modeling of the machine, taking the exact geometrical representation of the machine into account. Such a model that is widely recognized as a very powerful tool for modeling and analyzing different faults in induction machines is based on winding function approach [3]. In order to gain a deeper insight into the origin of different current harmonics, an appropriate analytical model should be used in addition to the numerical one. The most common approach for these purposes is the MMF—permeance wave approach. Therefore, the magnetic flux density waves in the air gap are obtained from multiplication of MMF and permeance waves [4]. The application of both models, numerical and analytical, which are verified by measurements, allows drawing conclusions on the order and magnitude of fault-induced current harmonics.

## 10.2 Healthy cage rotor induction motor

### 10.2.1 MMF space harmonics

Assuming uniform air gap length $g_0$ in induction motor, i.e., assuming constant value of air gap permeance as well as infinitely permeable iron, the flux density waves will have the same form as MMF waves. It should be mentioned that magnetic permeance $P$ is inversely proportional to the magnetic reluctance $R_m$. Assuming infinitely permeable iron, magnetic reluctance of air gap under one pole of induction motor is:

$$R_m = \frac{g_0}{\mu_0 S} = \frac{g_0}{\mu_0 \tau l} = \frac{g_0}{\mu_0 (\pi D/2p) l} = \frac{2pg_0}{\mu_0 \pi D l} \tag{10.1}$$

where $\tau$ is pole pitch, $D$ is mean air gap diameter and $l$ is axial length of machine. Therefore, magnetic permeance is

$$P = \frac{1}{R_m} = \frac{\mu_0 \pi D l}{2pg_0} \tag{10.2}$$

and it obviously has a constant value for constant uniform air gap length. In case of infinitely high permeability of iron, Ampere's circuital law could be applied only along the air gap length under one machine pole

$$\oint_C \mathbf{H} \cdot d\mathbf{l} = \int_S \mathbf{J} \cdot d\mathbf{S} \Rightarrow H \cdot g_0 = F \Rightarrow H = \frac{F}{g_0} \Rightarrow H \propto F \tag{10.3}$$

where $F$ is MMF value under one-machine pole. Magnetic flux density in air gap is also directly proportional to MMF:

$$B = \mu_0 H = \frac{\mu_0}{g_0} F \Rightarrow B \propto F \tag{10.4}$$

Therefore, magnetic flux density waves in machine air gap, in stator frame of reference are (see (4.105)–(4.107) in the fourth chapter):

$$B_1 = B_{1m} \cos(\omega_1 t - \nu p \theta_s) \tag{10.5}$$

$$B_2 = B_{2m} \cos\left(\left(1 - \frac{\lambda R}{p}(1-s)\right)\omega_1 t + \left(\frac{\lambda R}{p} - \nu\right) p\theta_s\right) \tag{10.6}$$

$$B_3 = B_{3m} \cos\left(\left(1 + \frac{\lambda R}{p}(1-s)\right)\omega_1 t - \left(\frac{\lambda R}{p} + \nu\right) p\theta_s\right) \tag{10.7}$$

These flux density waves could eventually induce electromotive forces (EMFs) and lead to currents in the stator windings at the same frequencies. Hence, all the flux density space harmonics from the stator, rotor reflects at the fundamental frequency and a series of two additional slip-dependent frequencies, located rather high in the stator-current spectrum – rotor slot harmonics (RSHs), lower

$$f_{L\_RSH} = \left(1 - \lambda\frac{R}{p}(1-s)\right)f_1 \tag{10.8}$$

and upper,

$$f_{U\_RSH} = \left(1 + \lambda\frac{R}{p}(1-s)\right)f_1 \tag{10.9}$$

where $\lambda = 1, 2, 3, \ldots$.

**For $\lambda = 1$, one has the first-order rotor slot harmonics (RSHs) or principal slot harmonics (PSHs).**

Obviously, frequencies given by (10.8) and (10.9) depend on machine load, i.e., slip. In other words, position of PSHs in stator current spectrum depends on rotor speed. This is what makes them interesting and is the reason why these harmonics have found wide application in the sensorless speed estimation of induction motor drives using numerous different techniques for digital signal processing [5]. However, the existence of stator current components at frequencies given by (10.8) and (10.9) depends on the number of pole pairs in the flux density waves in (10.6) and (10.7). In order that spectral component at the lower PSH frequency (10.8) exists in the stator current spectrum, it is required that the number of pole pairs in the flux density wave (10.6) be equal to the number of pole pairs produced by the stator winding itself. In other words, $R/p-\nu$ for some adequate value of $\nu$ must belong to the group $\mathbf{H} = (6k+1)$. Further, it means that:

$$R_{L\_RSH} = p(6(g+k)+2), \quad g = 0, \pm 1, \pm 2, \ldots \quad k = 0, \pm 1, \pm 2, \ldots \tag{10.10}$$

It can be easily concluded from (10.10) that for a rotor with $R$ bars, in a $p$-pole-pair motor, the lower PSH in the stator current spectrum will exist if the following condition is satisfied, [1]:

$$R_{L\_RSH} = (6n+2)p, \quad n = 0, 1, 2, 3, \ldots \tag{10.11}$$

For example, in case of four-pole machine, $p = 2$, $R_{L\_PSH} = 4, 16, 28, 40, \ldots$. On the other hand, the condition for existence of the upper PSH is satisfied for rotors with the following number of bars:

$$R_{U\_RSH} = -p(6(g+k)+2), \quad g = 0, \pm 1, \pm 2, \ldots \quad k = 0, \pm 1, \pm 2, \ldots \tag{10.12}$$

which is equivalent to the following expression:

$$R_{U\_RSH} = (6n-2)p, \quad n = 1, 2, 3, \ldots \tag{10.13}$$

This leads to $R_{U\_PSH} = 8, 20, 32, 44, \ldots$ for a four-pole machine.

In order that both PSHs exist in the stator current spectrum, the number of rotor bars in $p$-pole-pair motor must be equal to the mean value of (10.11) and (10.13):

$$R_{\text{BOTH\_RSH}} = 6np, \quad n = 1, 2, 3, \ldots \tag{10.14}$$

This leads to the well-known "forbidden" combinations of stator and rotor slot numbers [6,7], ($R_{\text{BOTH\_PSH}} = 12, 24, 36, 48, \ldots$ for a four-pole motor). Rotors with $R$ bars in $p$-pole-pair symmetrical induction motor, which do not satisfy any of conditions (10.11), (10.13) and (10.14), will not have any PSHs in stator current spectrum, under symmetrical voltage supply conditions.

As an example of stator current spectrum for an induction motor that satisfies condition (10.11), Figure 10.1 gives stator current spectrum in steady-state condition for $p = 4$ induction motor with $S = 48$ stator slots and $R = 32$ rotor bars. According to (10.11), only the lower RSH exists. For slip of $s = 2.73\%$, it means stator current component at 339 Hz.

As an example of stator current spectrum for induction motor that satisfies condition (10.13), Figure 10.2 gives stator current spectrum in steady-state regime for $p = 2$ induction motor with $S = 36$ stator slots and $R = 32$ rotor bars. According to (10.13), only the upper RSH exists. For slip value of $s = 2.6\%$, stator current component exists at 829 Hz.

As an example of stator current spectrum for induction motor that satisfies condition (10.14), Figure 10.3 gives stator current spectrum in steady-state regime for $p = 2$ induction motor with $S = 36$ stator slots and $R = 24$ rotor bars. According to (10.14), both RSHs exist at 534 and 634 Hz for slip value $s = 2.6\%$.

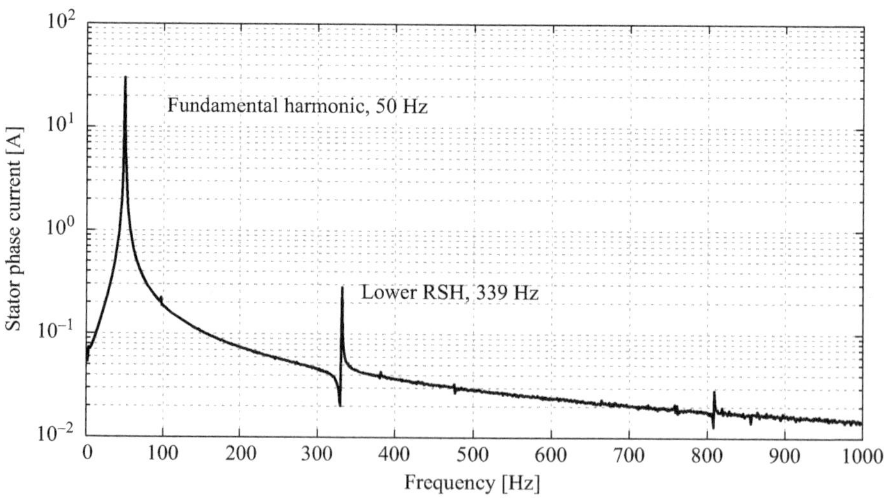


*Figure 10.1 Stator current spectrum for p = 4 induction motor with S = 48 stator slots and R = 32 rotor bars (according to (10.11), only the lower RSH exists in stator current spectrum)*

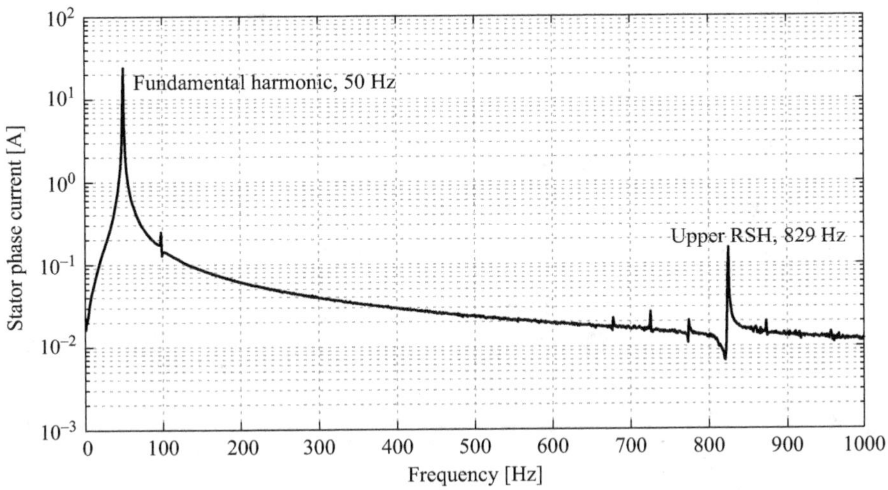


*Figure 10.2 Stator current spectrum for p = 2 induction motor with S = 36 stator slots and R = 32 rotor bars (only the upper RSH exists in stator current spectrum)*

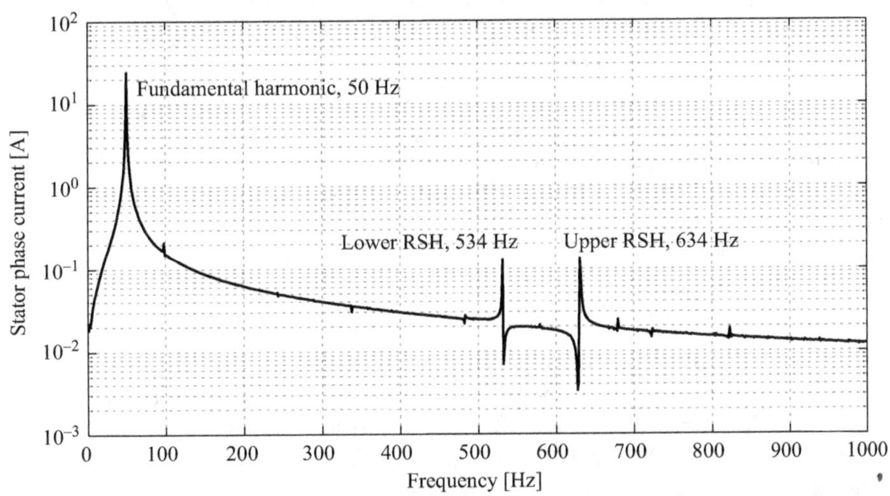


*Figure 10.3 Stator current spectrum for p = 2 induction motor with S = 36 stator slots and R = 24 rotor bars (both RSHs are clearly visible in stator current spectrum)*

Finally, as an example of stator current spectrum for induction motor that does not satisfy any of the previously defined conditions, Figure 10.4 gives stator current spectrum in steady-state regime for $p = 2$ induction motor with $S = 36$ stator slots and $R = 30$ rotor bars. It is now obvious that in the stator phase current spectrum, none of the principal slot harmonics (PSHs) exists.

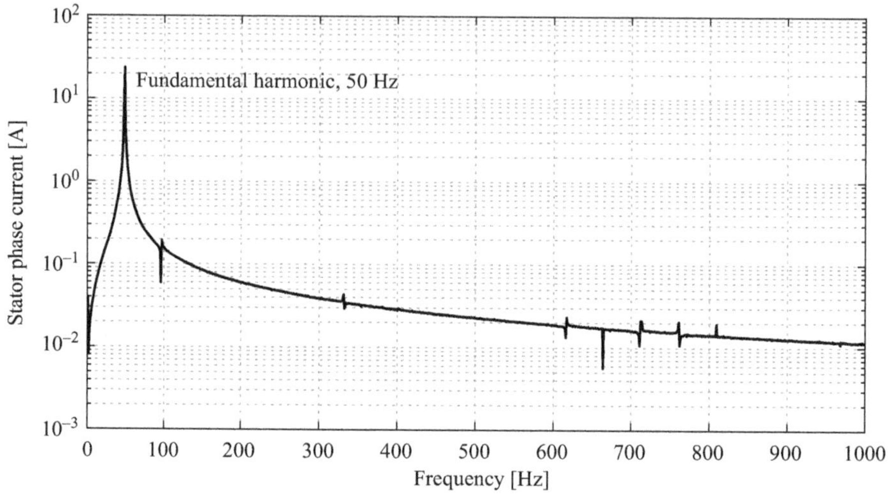


*Figure 10.4 Stator current spectrum for $p = 2$ induction motor with $S = 36$ stator slots and $R = 30$ rotor bars (none of the PSHs exists in spectrum)*

Here, it should be stressed that some of the PSHs that ordinarily could not exist in the spectrum of a healthy machine may be visible as a consequence of asymmetrical voltage supply and/or asymmetrical stator phase windings. In such cases, some amount of inverse-rotating magnetic flux waves in air gap exist, $\nu = -1$. Such a case can be seen, for example, in a motor with $S = 36$ stator slots, $R = 32$ rotor bars, and $p = 2$. Such a motor could develop only the upper PSH. However, for $\nu = -1$, such a motor could also develop, to some extent, the lower PSH as one can easily conclude from (10.6) where for $\nu = -1$, the number of pole pairs in magnetic flux density wave corresponds to the 17th space harmonic.

## 10.2.2 Slot permeance harmonics

Aside from the described stator or rotor MMF space harmonics, another effect exists in a real induction motor which generates EMFs and currents in stator windings at the same frequencies given by (10.8) and (10.9), as explained in the following. This is the stator and rotor slotting effect, i.e., the air gap length that varies with both space and time as a function of rotation of the rotor. As stated previously, both MMF space harmonics and slot permeance harmonics induce EMFs and lead to the stator currents at the same frequencies. Hence, the distinction between the contributions of these components is blurred. The main differentiation between these two effects is that rotor MMF space harmonics exist only in case where current flows through the cage winding, even in the no-load case. This is a consequence of high-frequency rotor currents induced by rotating flux density wave harmonics from the stator side. For the existence of rotor permeance slot harmonics, this precondition is not needed. Even a slotted rotor without any winding, rotating in the motor by some external means, e.g. an auxiliary motor, will lead to stator winding current components at frequencies given by (10.8) and (10.9) if the stator winding is connected to the voltage supply. In this case, the slip in (10.8) and (10.9) is simply the difference between the synchronous speed and the speed of the auxiliary motor. The reason for the existence of these stator current components is that the self-inductance of the stator windings becomes a function of time as a consequence of variable permeance due to the changing rotor angle, [1].

The actual induction motor air gap permeance is variable in both space and time as a consequence of stator and rotor slotting and their mutual position as a function of rotor angle. Taking only the fundamental harmonic of air gap permeance series into account, the air gap permeance function is commonly described by the following series [1,8]:

$$P(\theta_s, \theta_r) = \frac{1}{g_0} + P_s \cos(S\theta_s) + P_r \cos(R\theta_r) + P_{sr} \cos(S\theta_s - R\theta_r) + \cdots \quad (10.15)$$

where coefficients $P_s$, $P_r$ and $P_{sr}$ depend on the geometrical dimensions of the air gap and stator and rotor slot opening. The situation where stator MMF waves given by

$$F_{sv}(t, \theta_s) = F_{sv\max} \cos(\omega t - vp\theta_s) \quad (10.16)$$

act on the third term of (10.15) is considered next. As a result, two flux density waves are obtained, expressed in the stator frame of reference as follows:

$$B_{sv1}(t, \theta_s) = B_{sv1\max} \cos\left(\left(1 - \frac{R}{p}(1-s)\right)\omega t + \left(\frac{R}{p} - v\right)p\theta_s\right) \quad (10.17)$$

$$B_{sv2}(t, \theta_s) = B_{sv2\max} \cos\left(\left(1 + \frac{R}{p}(1-s)\right)\omega t - \left(\frac{R}{p} + v\right)p\theta_s\right) \quad (10.18)$$

that are identical with waves (10.6) and (10.7). On the other hand, stator MMF waves (10.16) acting on the fourth term in the permeance expression (10.15) lead to flux density waves of the same frequencies, but with different number of pole pairs given by:

$$B_{sv3}(t, \theta_s) = B_{sv3\max} \cos\left(\left(1 - \frac{R}{p}(1-s)\right)\omega t - \left(\frac{S-R}{p} + v\right)p\theta_s\right) \quad (10.19)$$

$$B_{sv4}(t, \theta_s) = B_{sv4\max} \cos\left(\left(1 + \frac{R}{p}(1-s)\right)\omega t + \left(\frac{S-R}{p} - v\right)p\theta_s\right) \quad (10.20)$$

As it will be shown, this difference in pole pair number (at certain combination of stator, rotor slot and pole pair number) can have a significant effect on the amplitudes of stator current components at these frequencies. These permeance harmonics could significantly amplify the already existed RSHs in stator current spectrum.

Let us consider, for example, the case of $p=4$ induction motor with $S=48$ stator slots and $R=40$ rotor bars. Figure 10.5 gives stator current spectrum in steady-state regime for this motor. It is obvious that the upper RSH is the most prominent higher harmonic in the spectrum. As motor is fully loaded, $s=2.6\%$,

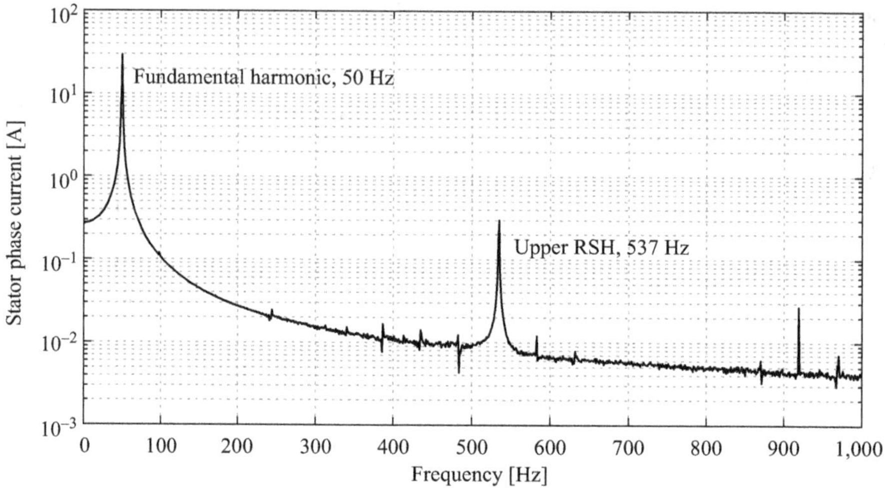


*Figure 10.5 Stator current spectrum for fully loaded $p = 4$ induction motor with $S = 48$ stator slots and $R = 40$ rotor bars (the upper RSH is the most prominent higher harmonic in the spectrum)*

with high intensity of rotor currents, the conclusion is that this RSH is emphasized due to the higher space harmonics of rotor MMF.

However, when the motor is almost unloaded (rotor currents are very small), this harmonic is also the most prominent higher harmonic in the spectrum, and this could be explained only by permeance harmonic effect (10.20). Normally, for $S = 48$, $R = 40$ and $p = 4$, the magnetic flux density wave (10.20) is a wave with fundamental number of pole pairs, and it could induce significant value of EMF in stator phase windings. Slip value for this operating regime is $s = 0.2\%$. It could be easily proven that in the first case, the amplitude of RSH is 1% of the amplitude of fundamental harmonic, while in the second case, this ratio is slightly smaller than 0.5%, Figure 10.6.

### 10.2.3 Saturation permeance harmonics

The saturation of the main magnetic flux path in a machine can be modeled by making the air gap length a function of the saturation level and spatial position, Figure 10.7 [9,10]. In this approach, the maximum value of the virtual air gap length corresponds to the maxima/minima of the main magnetic flux density wave while the minimum value of the virtual air gap length (equal to the effective value of the air gap length) corresponds to the zero value of the main magnetic flux density wave.

Accordingly, the air gap permeance function, neglecting the stator and rotor slotting, is described with the following expression:

$$P(\theta_s, \beta) = \frac{1}{g_0} - k_m - k_m \cos(2p\theta_s - 2\beta) \tag{10.21}$$

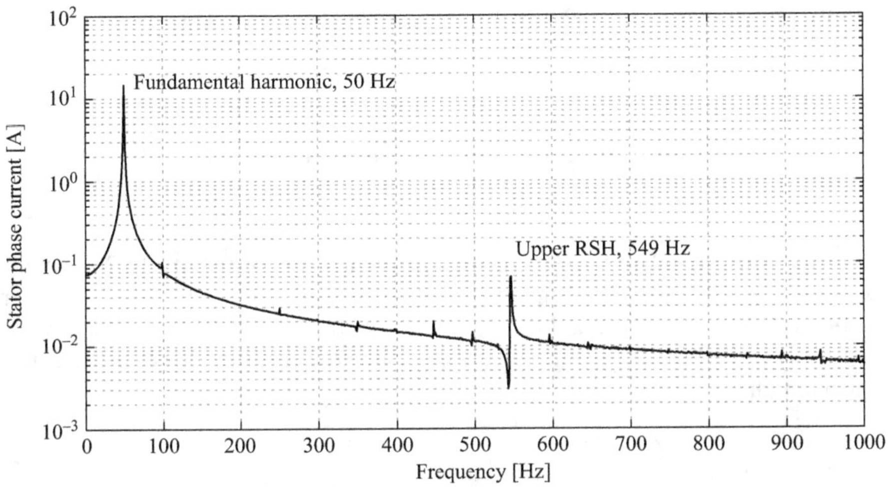


*Figure 10.6 Stator current spectrum for almost unloaded p = 4 induction motor with S = 48 stator slots and R = 40 rotor bars (the upper RSH is again the most prominent higher harmonic in the spectrum, s = 0.2%)*

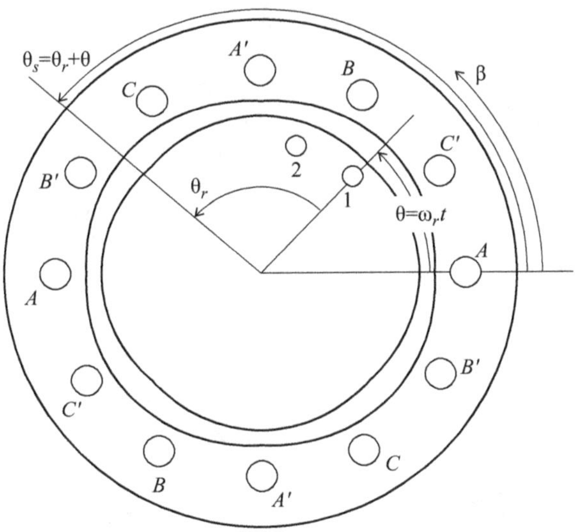


*Figure 10.7 Virtual air gap length variation along the machine circumference due to the main magnetic flux path saturation, p = 2, β = π/2 rad*

where $g_0$ is the effective air gap length and $p$ is the number of pole pairs of the machine. Angle $\theta_s$ (in mechanical radians) defines position along the stator circumference while $\beta$ (in electrical radians) describes the angular position of the air gap flux density maximum in the stator frame of reference. Factor $k_m$ correlates with the saturation factor in the following manner [10]. The product of the permeance function and the fundamental MMF wave $F_{1\max}\cos(\omega t - p\theta_s)$ gives two flux density waves:

$$B(t,\theta_s) = F_{1\max}\left(\frac{1}{g_0} - \frac{3}{2}k_m\right)\cos(\omega t - p\theta_s) - F_{1\max}\frac{k_m}{2}\cos(3\omega t - 3p\theta_s) \tag{10.22}$$

The maximum value of the fundamental flux density wave in (10.22) should correspond to the saturated condition of the machine ($B_{sat} \sim E_{sat} \sim U_{sat}$). In an unsaturated machine, the maximal value of the fundamental flux density wave ($B_{non\_sat} \sim E_{non\_sat} \sim U_{non\_sat}$) should correspond to the inverse value of the effective air gap length, i.e.

$$\frac{1}{g_0} : U_{non\_sat} = \left(\frac{1}{g_0} - \frac{3}{2}k_m\right) : U_{sat} \tag{10.23}$$

As the saturation factor is the ratio of the fundamental components of the air gap voltage in unsaturated and saturated conditions,

$$k_{sat} = \frac{U_{non\_sat}}{U_{sat}} \tag{10.24}$$

The previous relationship results in:

$$k_m = \frac{2}{3}\frac{1}{g_0}\frac{k_{sat} - 1}{k_{sat}} \tag{10.25}$$

The air gap permeance function (10.21) now attains a more compact form

$$P = P_1 + P_2 = P_1 + P_{2m}\cos(2p\theta_s - 2\beta) \tag{10.26}$$

where

$$P_1 = \frac{1}{g_0}\left(1 - \frac{2}{3}\frac{k_{sat} - 1}{k_{sat}}\right) \tag{10.27}$$

$$P_{2m} = -\frac{2}{3}\frac{1}{g_0}\frac{k_{sat} - 1}{k_{sat}} \tag{10.28}$$

The symmetrical three-phase stator winding produces a series of MMF waves, described by the well-known relation:

$$F_s = F_{sm\nu}\cos(\omega t - \nu p\theta_s) \tag{10.29}$$

where $\nu = 6g + 1$, $g = 0, \pm 1, \pm 2 \ldots$ These waves, in interaction with the air gap permeance function (10.26), produce numerous magnetic flux density waves.

It should be noted that in a steady-state condition, $\beta = \omega t$. In a saturated machine, two different effects can be noticed. First, rotor MMF waves

$$F_1 = F_{1m} \cos(s_\nu \omega t - \nu p \theta_r) \tag{10.30}$$

$$F_2 = F_{2m} \cos\left( s_\nu \omega t + \left( \frac{\lambda R}{p} - \nu \right) p \theta_r \right) \tag{10.31}$$

$$F_3 = F_{3m} \cos\left( s_\nu \omega t - \left( \frac{\lambda R}{p} + \nu \right) p \theta_r \right) \tag{10.32}$$

where

$$s_\nu = 1 - \nu(1 - s) \tag{10.33}$$

experience a variable air gap permeance. In interaction with $P_2$ (10.26), the rotor MMF waves produce the following flux density waves in stator frame of reference:

$$B_{11sat} = B_{11msat} \cos(\omega t + (\nu - 2) p \theta_s) \tag{10.34}$$

$$B_{12sat} = B_{12msat} \cos(3\omega t - (\nu + 2) p \theta_s) \tag{10.35}$$

$$B_{21sat} = B_{21msat} \cos\left( \left( 1 + \lambda \frac{R}{p} (1 - s) \right) \omega t - \left( \frac{\lambda R}{p} - (\nu - 2) \right) p \theta_s \right) \tag{10.36}$$

$$B_{22sat} = B_{22msat} \cos\left( \left( 3 - \lambda \frac{R}{p} (1 - s) \right) \omega t + \left( \frac{\lambda R}{p} - (\nu + 2) \right) p \theta_s \right) \tag{10.37}$$

$$B_{31sat} = B_{31msat} \cos\left( \left( 1 - \lambda \frac{R}{p} (1 - s) \right) \omega t + \left( \frac{\lambda R}{p} + (\nu - 2) \right) p \theta_s \right) \tag{10.38}$$

$$B_{32sat} = B_{32msat} \cos\left( \left( 3 + \lambda \frac{R}{p} (1 - s) \right) \omega t - \left( \frac{\lambda R}{p} + (\nu + 2) \right) p \theta_s \right) \tag{10.39}$$

As the number of pole pair in (10.35) is a multiple of three for any value of $\nu$, the third harmonic component in stator current spectrum cannot arise in a symmetrical machine connected to balanced three-phase voltages. Second, the stator MMF waves in a saturated induction machine, through variable air gap permeance $P_2$, induce flux density waves that do not exist in an unsaturated machine. These waves are

$$B_4 = B_{4m} \cos((1 - (2 - \nu)(1 - s))\omega t + (\nu - 2) p \theta_r) \tag{10.40}$$

$$B_5 = B_{5m} \cos((3 - (2 + \nu)(1 - s))\omega t - (\nu + 2) p \theta_r) \tag{10.41}$$

Due to these waves, new MMF waves in the cage rotor can appear. However, these new MMF waves, in the previously described manner, can induce the EMFs and lead to currents in the stator winding only at the same frequencies as in (10.34)–(10.39). Hence, in a saturated induction machine, new stator current components can be expected only at the following frequencies: the lower saturation-related harmonic ($L\_SH$)

$$f_{L\_SH} = \left(3 - \lambda\frac{R}{p}(1-s)\right)f_1 \tag{10.42}$$

and the upper saturation-related harmonic ($U\_SH$)

$$f_{U\_SH} = \left(3 + \lambda\frac{R}{p}(1-s)\right)f_1 \tag{10.43}$$

Similarly, as for the PSHs, in order that stator current components arise at frequencies given by (10.42), the number of pole pairs of the magnetic flux density waves (10.37) must be the same as the number of pole pairs that the stator winding itself produces. This means that, in order that stator current component at the lower saturation-related frequency exists, $R/p - (\nu + 2)$ must belong to group $\mathbf{H} = (6k+1)$, where $k = 0, \pm 1, \pm 2\ \ldots$. The same condition holds for the upper saturation-related PSH.

Figure 10.8 shows experimentally obtained stator current spectrum for an induction motor with $S = 36$ stator slots, $R = 28$ rotor bars and $p = 2$ pairs of poles. According to the predictions, only the lower PSH exists in the spectrum at 616 Hz, for $s = 4.86\%$. In addition, both saturation-related current components appear in the

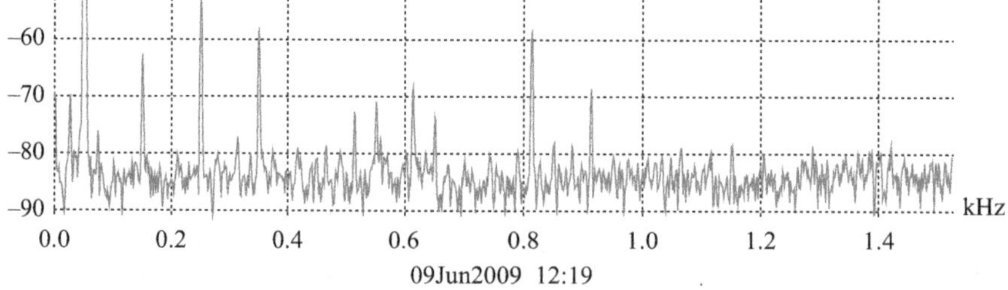


*Figure 10.8 An experimentally recorded stator current spectrum for cage rotor induction motor with S = 36 slots, R = 28 bars, p = 2 pole pairs @ s = 4.86%*

spectrum at predicted frequencies in (10.42) and (10.43), i.e., at 516 and 816 Hz for $s = 4.86\%$. Moreover, saturation-related component at 816 Hz is one of the most prominent harmonics in the spectrum. This phenomenon can be easily explained. Magnetic flux density wave (10.39) has fundamental number of pole pairs $p = 2$ for $\nu = -17$:

$$\frac{\lambda R}{p} + (\nu + 2) \xrightarrow{\lambda=1, R=28, p=2, \nu=-17} \frac{28}{2} + (-17 + 2) = 14 - 15 = -1$$

On the other hand, stator MMF space harmonic for $\nu = -17$ is rather high because this harmonic is the stator slot harmonic ($S = 36$ and $p = 2$). Thus, these two facts simultaneously lead to a rather high current component at the upper saturation-related frequency. The other saturation-related current component at 516 Hz is not so prominent because the magnetic flux density wave (10.37) could have fundamental number of pole pairs only for $\nu = 13$, i.e., for stator MMF space harmonic of rather small magnitude.

Figure 10.9 shows the stator current spectrum for loaded four-pole cage rotor induction motor with $S = 36$ stator slots and $R = 32$ rotor bars at $s = 6.4\%$.

Only the upper PSH exists in the spectrum according to predictions. Moreover, the upper PSH at 798 Hz is one of the most prominent higher harmonics in the whole stator current spectrum. Although the fundamental rotor current of the loaded motor has a significant effect on the PSH intensity, such a high PSH component can be explained only by the significant additional effect of slot permeance harmonics. Indeed, this motor has a certain number of stator and rotor slots so that

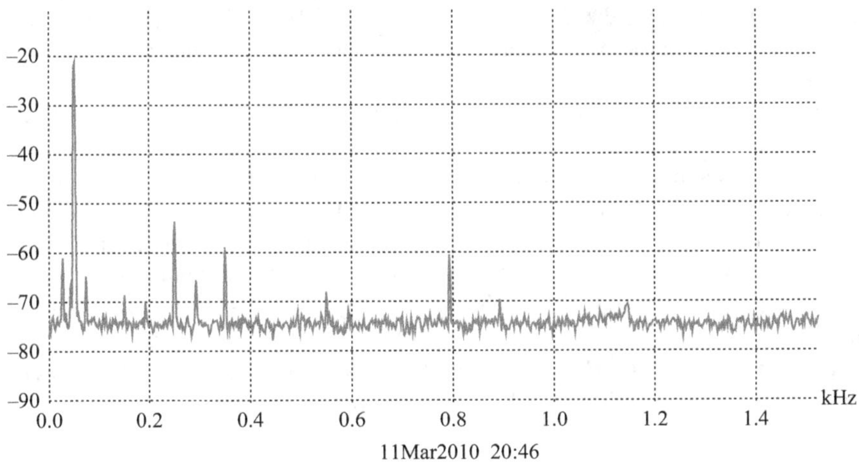


*Figure 10.9 An experimentally recorded stator-line current spectrum for loaded cage rotor induction motor with S = 36 slots, R = 32 bars and p = 2 pole pairs @ s = 6.4%*

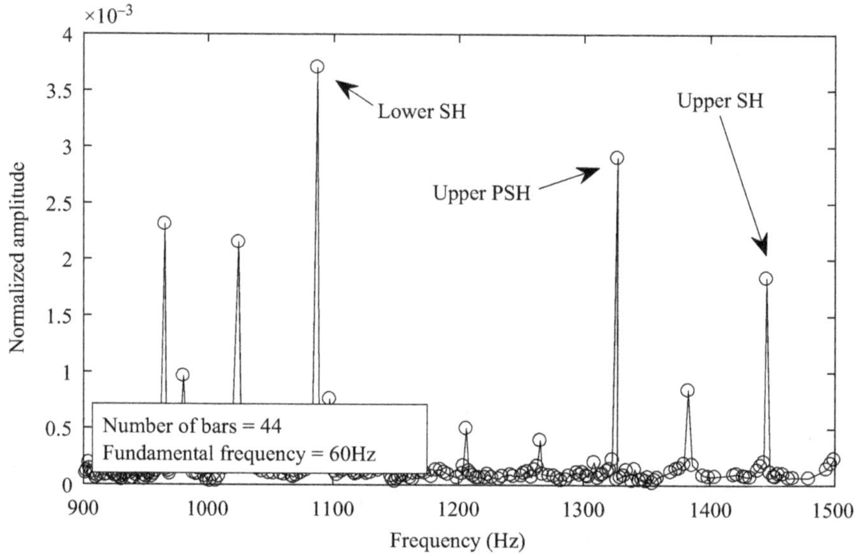


*Figure 10.10 Stator-line current spectrum for loaded 60-Hz cage rotor induction motor with S = 36 slots, R = 44 bars, and p = 2 pole pairs @ s = 4.16%*

depending on the logical relationship between the two clauses the slot permeance harmonics will also have a very significant effect on the same PSH amplitude. From (10.20), it follows that $(S - R)/p - \nu = (36 - 32)/2 - 1 = 1$ which means that the fundamental stator MMF wave through the slot permeance produces the upper PSH.

Figure 10.10 depicts stator current spectrum for a loaded 60 Hz four-pole cage rotor induction motor with $S = 36$ stator slots and $R = 44$ rotor bars [11]. Only the upper PSH exists in the spectrum at 1,325 Hz according to previously derived rules. The lower PSH at 1,205 Hz is almost invisible. The lower saturation-related harmonic at 1,085 Hz is the most prominent harmonic in this part of spectra because the pole pair number of flux density wave given by (10.37) is equal to one of the stator slot harmonics for $\nu = 1$, i.e., $(R/p - (\nu + 2)) = (44/2 - (1 + 2)) = 19$. On the other hand, for $\nu = 19$, $(R/p - (\nu + 2)) = (44/2 - 21) = 1$. The upper saturation-related harmonic at 1,445 Hz is not so pronounced because the pole pair numbers in flux density wave given by (10.39) cannot attain any significant value.

## 10.3 MCSA of an induction motor with interturn fault

Interturn short circuits in stator windings of cage induction motors are one of the most severe faults due to their consequences for motor life and, at the same time, the most difficult faults to detect. These faults have been analyzed from several points of views and tackled with different techniques for analysis, diagnosis and

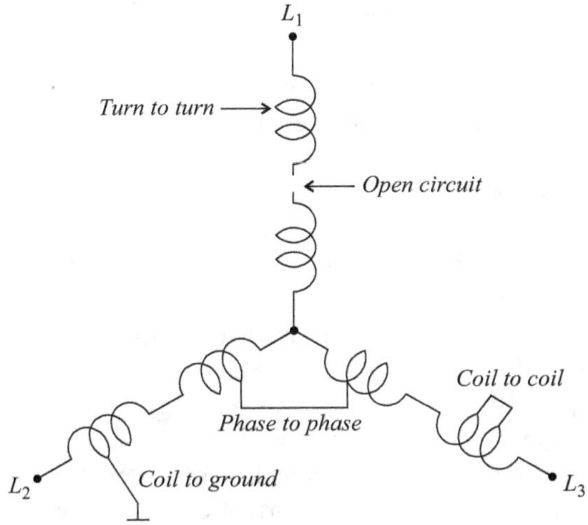


*Figure 10.11 Wye-connected stator showing possible failure modes*

detection, but it could be certainly said that detection of this type of fault has not been solved yet.

One of the first papers that has dealt with stator faults is the paper by Bonnet in 1978, [12]. Bonnet stated that the majority of winding failures are caused by various stresses which act on the stator winding and could be broken down into four groups: thermal, electrical, mechanical and environmental. In addition, he defined five key areas, which must be considered and related to one another to accurately diagnose the cause of winding failure, as follows: failure mode, failure pattern, appearance, application and maintenance history. He showed that the actual mode of failure could be broken down into the five groups illustrated in Figure 10.11.

Bonnet stated that the most probably initial fault is turn-to-turn fault—"A random wound motor is frequently started and due to excessive coil movement sustains a minor turn-to-turn short within one coil. As this condition progresses, excessive heating is generated within the shorted coil, resulting in insulation deterioration and eventually in a partial ground through the slot liner. Depending upon the type of motor protection, the motor may continue to run; more and more heat is generated in the damaged area until the phase or ground insulation is destroyed. At this point a direct phase-to-phase fault or ground fault occurs, and the motor is immediately dropped off the line.

Although inspection could reveal all five modes of failure, the turn-to-turn condition was the initial problem, and the others were the result of the problem. A turn-to-turn failure is usually very difficult to recognize due to the destructive nature of the final fault conditions."

One of the first attempts to detect and diagnose the interturn short-circuit fault, using MCSA, is presented by Joksimović and Penman [13], in 2000. A winding-function-based method for modeling a cage rotor induction motor with interturn short circuit in the stator winding is applied in that paper. This model is already presented in Chapter 5 of this monograph. Some observations with regard to the MCSA will be underlined here.

As it was stated before, as a consequence of the nature of cage rotor winding, all flux density waves from the stator side will be reflected at rotor slot frequencies in the stator current spectra. This is the case in a healthy machine as well as a machine with interturn short circuit. Assuming that interturn short circuit occurs between two turns in the same coil, a new series of MMF waves will appear under the interturn short circuit. That series consists of MMF and flux density waves with all number of pole pairs and in both directions of rotation. One of these waves is the wave with the same number of pole pairs as the basic flux density wave in machine, but with an opposite direction of rotation. This wave has no effect on the stator current spectra explicitly because it induces only base-frequency current component. This current component in stator winding is of inverse order.

However, this wave with fundamental number of pole pairs and inverse direction of rotation could produce new current component in stator current spectrum at 150 Hz (triplen time harmonic) through the following two effects that appear in real induction motors:

- first, this wave with $\nu = -1$ order produces third harmonic component at stator current spectrum through the saturation of magnetic circuit effect, (10.35);
- second, inverse-order currents in stator windings produce pulsating electromagnetic torque with $2f_1$ frequency. Through the speed ripple effect, as it is explained in [14], this pulsating electromagnetic torque can produce new current component in stator current spectrum at 150 Hz.

It should be stressed that the origin of third time harmonic (150 Hz) in a stator current spectrum could also be consequence of asymmetrical voltage supply, i.e., of some amount of inverse-rotating magnetic flux wave in the air gap, $\nu = -1$, that produces EMF and current at this frequency through the described saturation effect in the machine, (10.35). For example, none of the three-phase voltage supplies is ideally symmetrical as it is also the case with stator winding of induction machine.

On the other hand, as a consequence of nonideal voltage supply and nonideal stator winding distribution in real circumstances, small amounts of inverse-rotating magnetic flux waves produce pulsating electromagnetic torque component at frequency $2f_1$ and consequently, speed ripple effect [14]. It further means that this pulsating electromagnetic torque component produces magnetic flux pulsations that will induce additional EMF and currents in stator winding at $3f_1$ frequency.

As discussed previously, all other waves that are due to interturn short-circuit fault could only eventually induce EMFs and generate currents at RSH frequencies. One of the interturn short-circuit consequences could be the appearance of one of the RSHs in stator current spectrum that cannot appear in a healthy motor.

In order to underpin the previous discussion, some of the results from numerical model as well as experiment are provided in the following section.

- *Experimental and numerical results* ($S = 36$, $R = 32$, $p = 3$, delta connection)

The following three figures, Figure 10.12–10.14, give experimentally obtained stator-line current spectrum in case of healthy motor and interturn short circuit. Different phases are designated by three different labels: A, B and C. Internal stator phase winding connection is delta, and interturn short circuit is made on already wounded stator winding, so it was not known exactly in which phase interturn short circuit occurs.

All figures show the most significant changes in RSHs (468 and 568 Hz for slip $s = 2.875\%$) as well as changes in intensity of stator current component at 150 Hz. In a healthy motor, there should not exist any of the RSHs according to the previously defined rules, i.e., (10.11), (10.13) and (10.14). However, in stator current spectrum of a healthy machine, the small value of the upper RSH is visible at 568 Hz. In case of interturn short circuit, an additional lower slot harmonic in a small amount appears in all phases at 468 Hz. An additional prominent rise of current component at 150 Hz is visible in one of the line currents (the A one), but it is not so big in other two-line currents.

Similar results are obtained through the numerical model of the same machine. So, Figure 10.15 shows the stator-line current spectrum for healthy motor. It is clear that there are almost no higher current harmonic components as it is predicted by expressions (10.11), (10.13) and (10.14). In case of interturn short circuit, i.e., Figure 10.16, both of RSHs, as well as a very small amount of current at 150 Hz, appear in spectra.

- *Experimental results* ($S = 36$, $R = 32$, $p = 2$, delta connection)

As in the previous case, the stator-line current spectrum is recorded for a healthy machine and in cases where interturn short circuit exists in stator phase winding. For $p = 2$ and $R = 32$, it follows from (10.13) that only the higher RSH could exist in stator current spectrum in a healthy machine. This is evident from Figures 10.17–10.19 where the upper RSH is one of the most prominent harmonics in stator-line current spectrum at 806 Hz for $s = 5.5\%$. The only visible changes are significant rises of the upper RSH in faulty conditions and appearance of the lower RSH in one of the phases.

All the above experimental results show the existence of the fifth and seventh current time harmonics, at 250 and 350 Hz. This fact can be explained only as a consequence of the voltage time harmonics of the same order in the supply voltages. From previously given results and the analytical considerations, the following conclusions regarding the possibility of using MCSA for interturn short-circuit faults could be drawn:

- The nature of the cage rotor is such that as shown in Chapter 4, all MMF waves and therefore the magnetic flux density waves are mapped onto only two of the frequencies (and their multiples) in the spectrum of stator currents. As the result of the interturn short circuit, a larger number of magnetic flux density waves rotating in both directions in the air gap exists, and all of them will be mapped onto only two of the frequencies—at the rotor slot frequencies. So, the only thing that can be expected is the growth in these current components, and

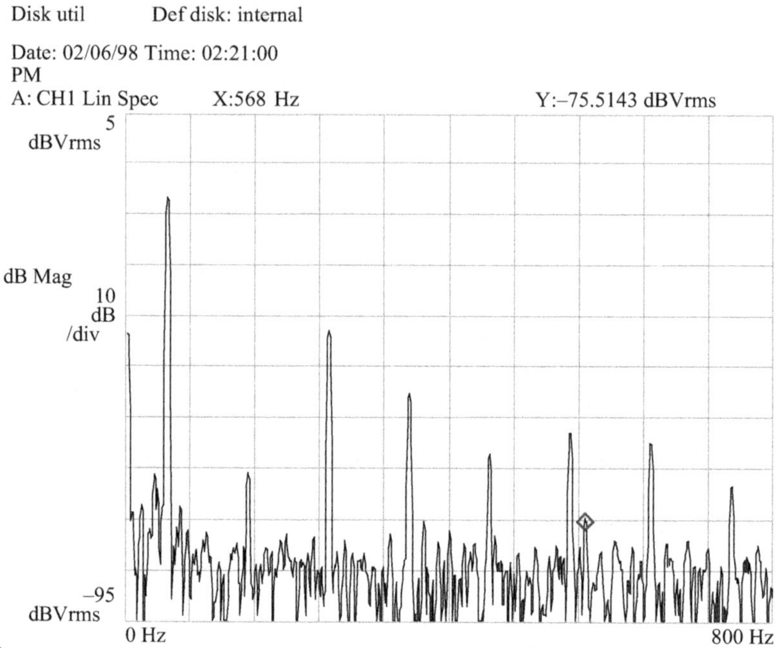


(a)

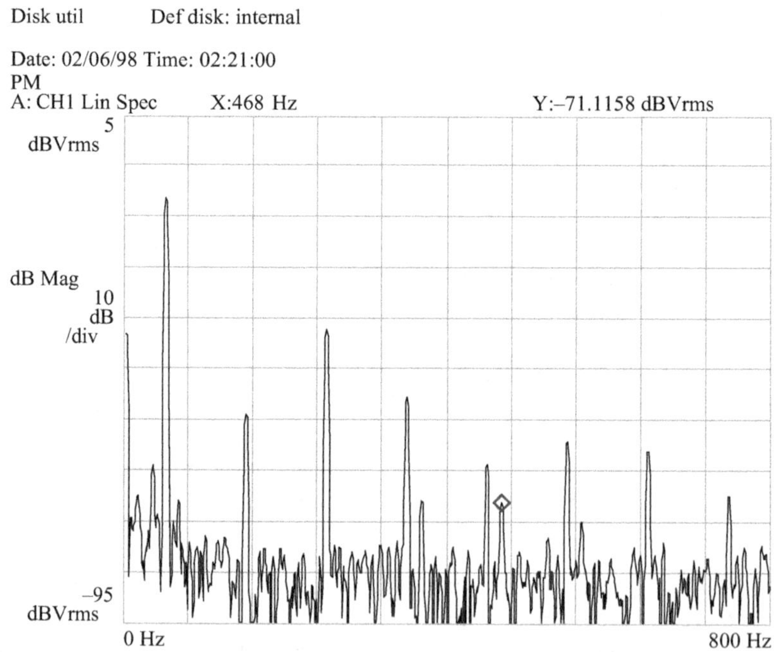


(b)

*Figure 10.12 Stator-line current spectrum (A phase) for loaded 50-Hz cage rotor induction motor with $S = 36$ slots, $R = 32$ bars, $p = 3$ pole pairs @ $s = 2.875\%$: (a) A phase, healthy motor and (b) A phase, interturn short circuit*

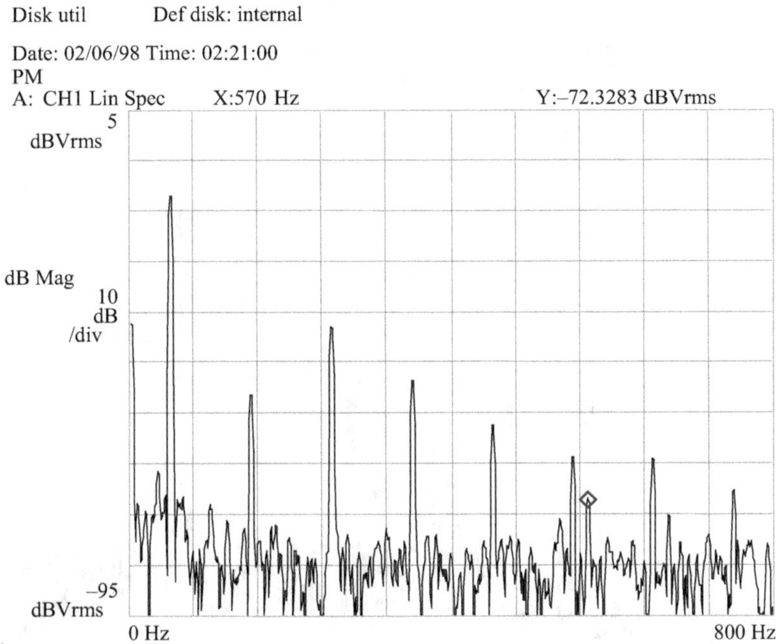


(a)

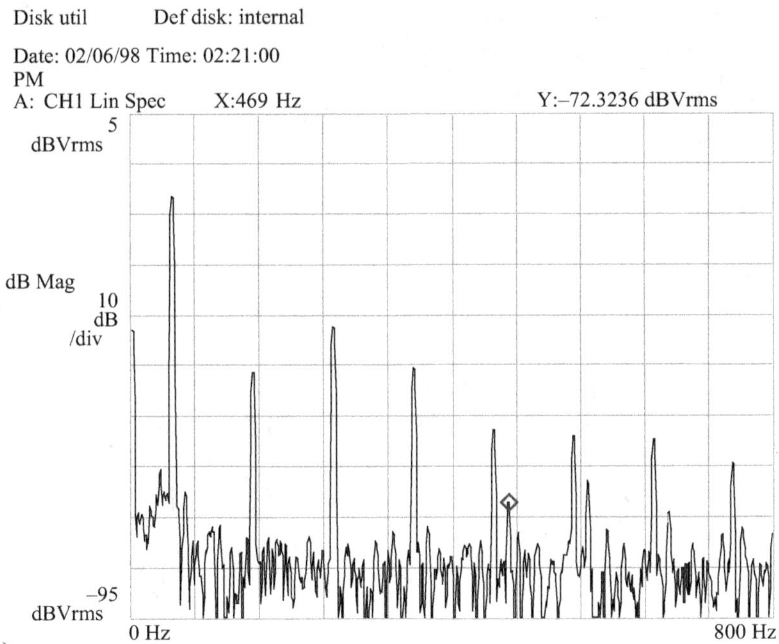


(b)

*Figure 10.13 Stator-line current spectrum (B phase) for loaded 50-Hz cage rotor induction motor with $S = 36$ slots, $R = 32$ bars, $p = 3$ pole pairs @ $s = 2.875\%$: (a) B phase, healthy motor and (b) B phase, interturn short circuit*

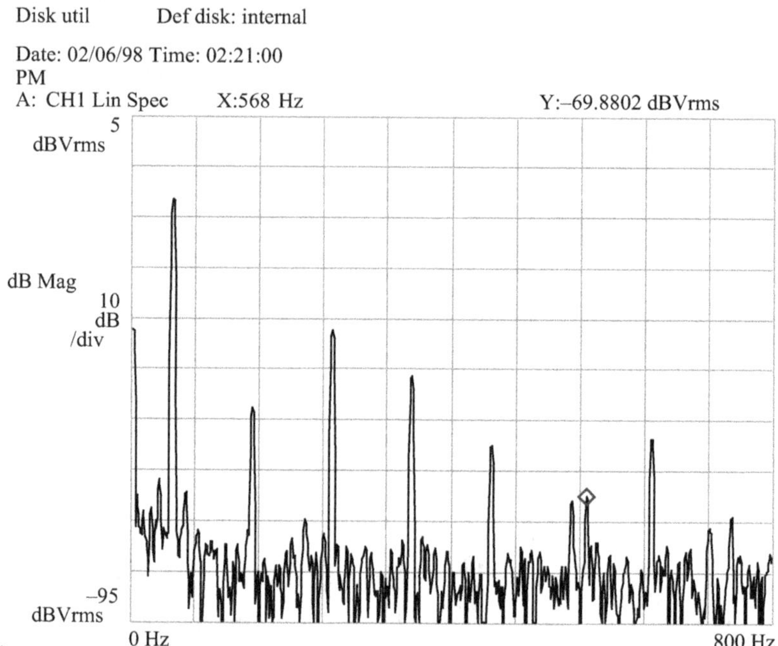

(a)

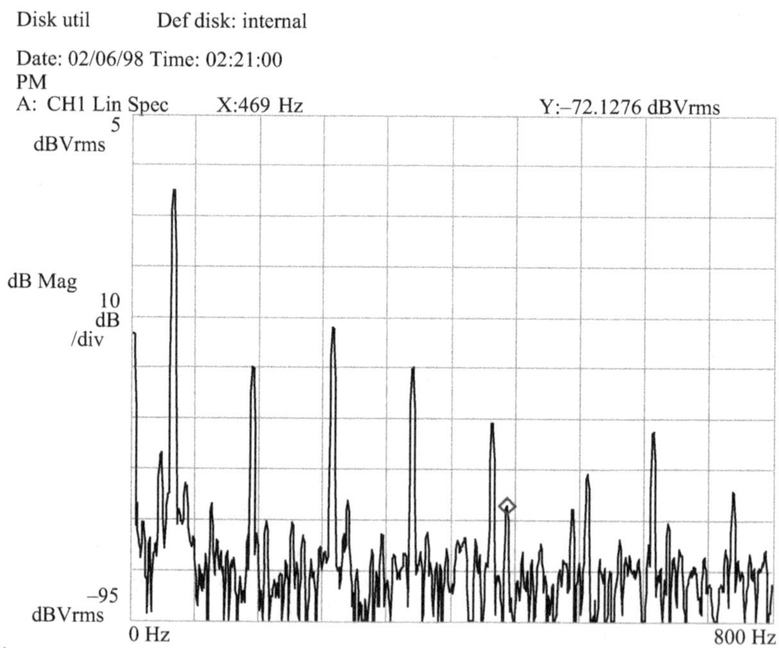

(b)

*Figure 10.14 Stator-line current spectrum (C phase) for loaded 50-Hz cage rotor induction motor with $S = 36$ slots, $R = 32$ bars, $p = 3$ pole pairs @ $s = 2.875\%$: (a) C phase, healthy motor and (b) C phase, interturn short circuit*

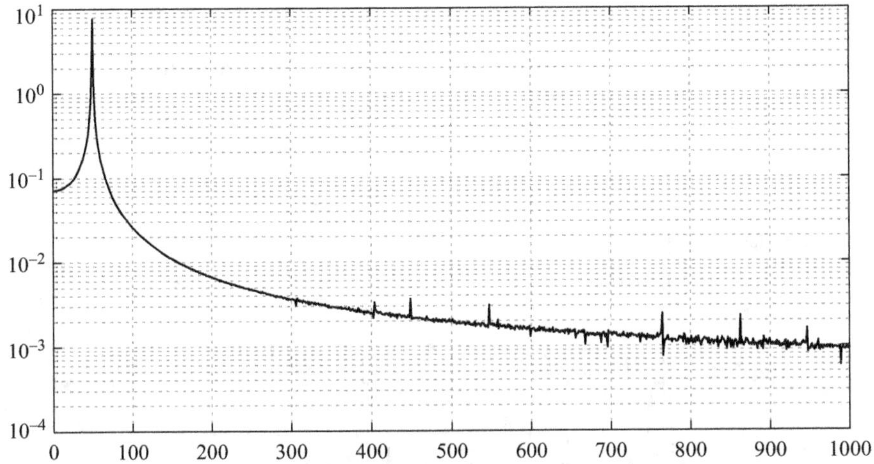


*Figure 10.15 Stator-line current spectrum for healthy, loaded 50-Hz cage rotor induction motor with $S = 36$ slots, $R = 32$ bars, $p = 3$ pole pairs @ $s = 6.25\%$. The results are obtained from numerical model*

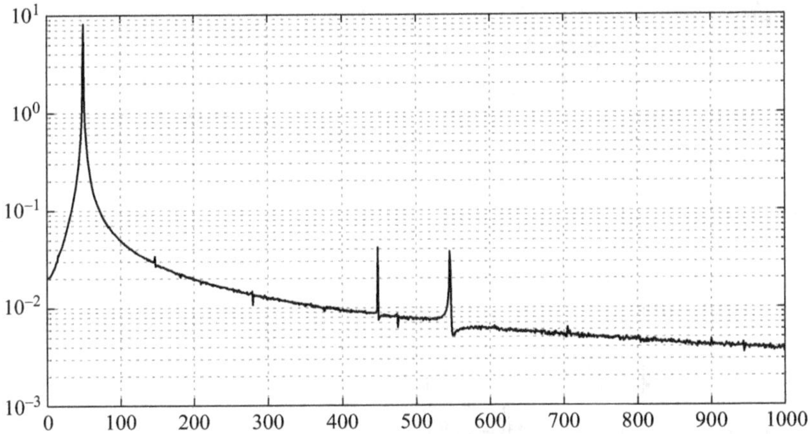


*Figure 10.16 Stator-line current spectrum for faulty, loaded 50-Hz cage rotor induction motor with $S = 36$ slots, $R = 32$ bars, $p = 3$ pole pairs @ $s = 6.25\%$. The results are obtained from numerical model*

there is two reasons for this: one is quantitative, so there is a larger number of waves that all have a smaller or greater effect on these frequencies and the other one is qualitative: some of these flux density waves have the same number of poles as fundamental waves; therefore, this effect becomes significant. The fact that as a result of interturn short circuit, any new frequency

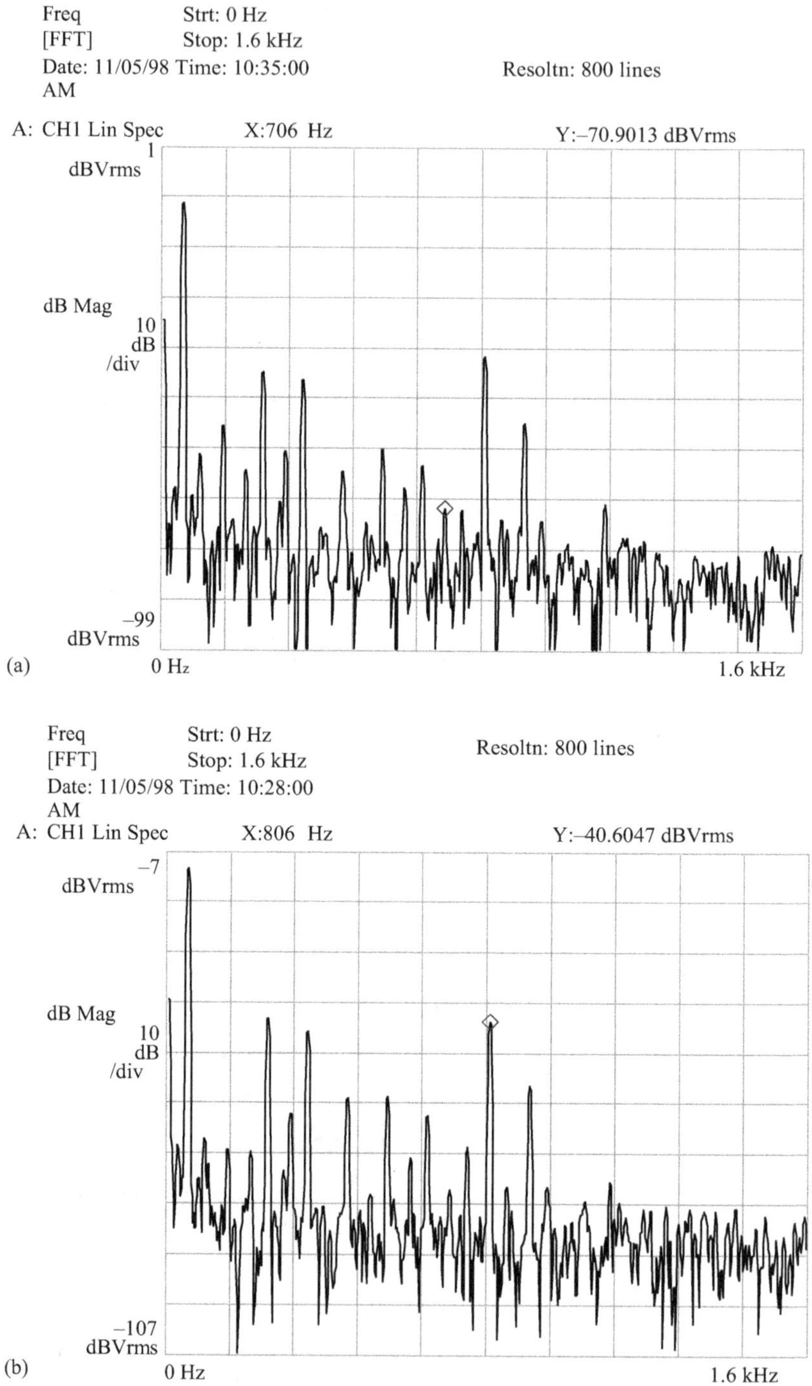


*Figure 10.17 Stator-line current spectrum (A phase) for loaded 50-Hz cage rotor induction motor with $S = 36$ slots, $R = 32$ bars, $p = 2$ pole pairs @ $s = 5.5\%$: (a) A phase, healthy motor and (b) A phase, interturn short circuit*

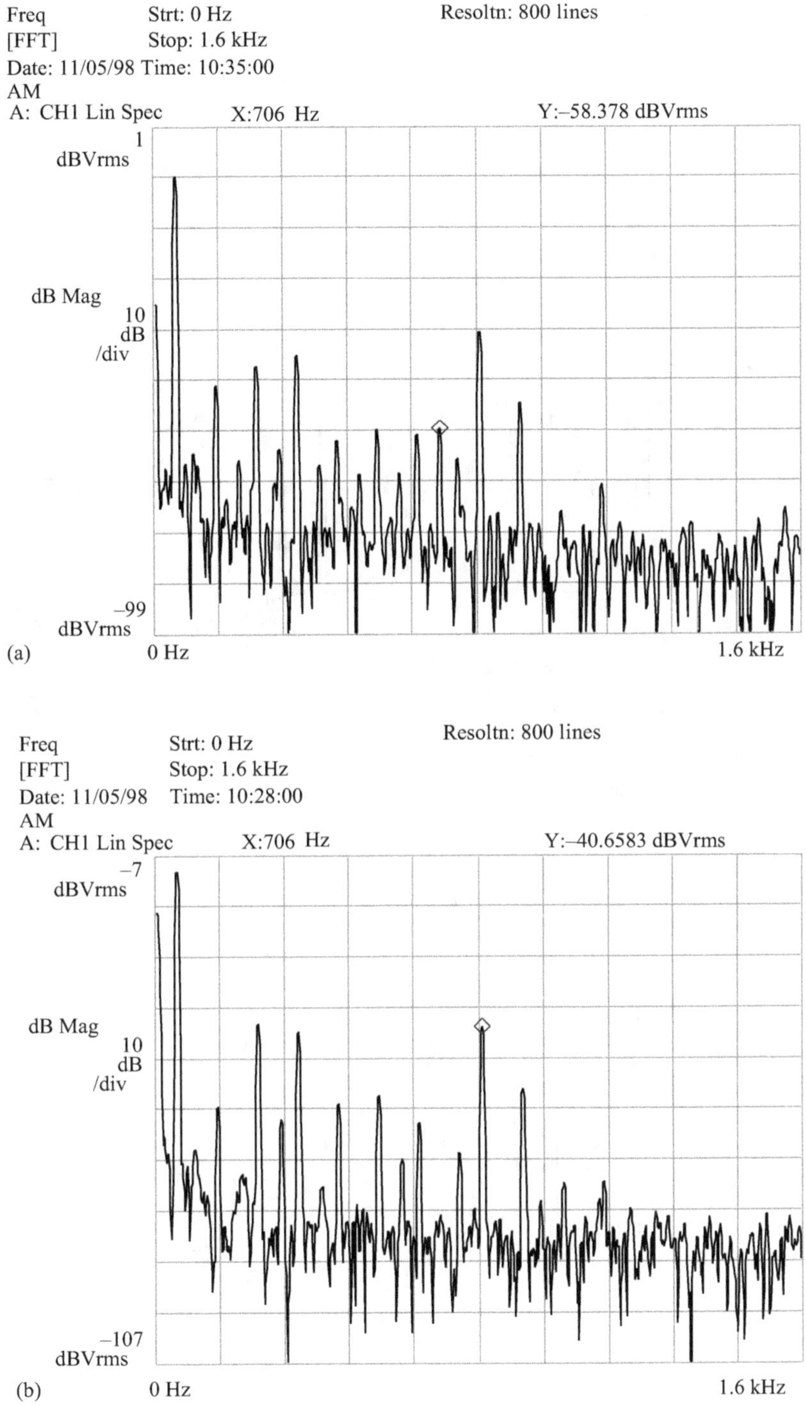


*Figure 10.18 Stator-line current spectrum (B phase) for loaded 50-Hz cage rotor induction motor with $S = 36$ slots, $R = 32$ bars, $p = 2$ pole pairs @ $s = 5.5\%$: (a) B phase, healthy motor and (b) B phase, interturn short circuit*

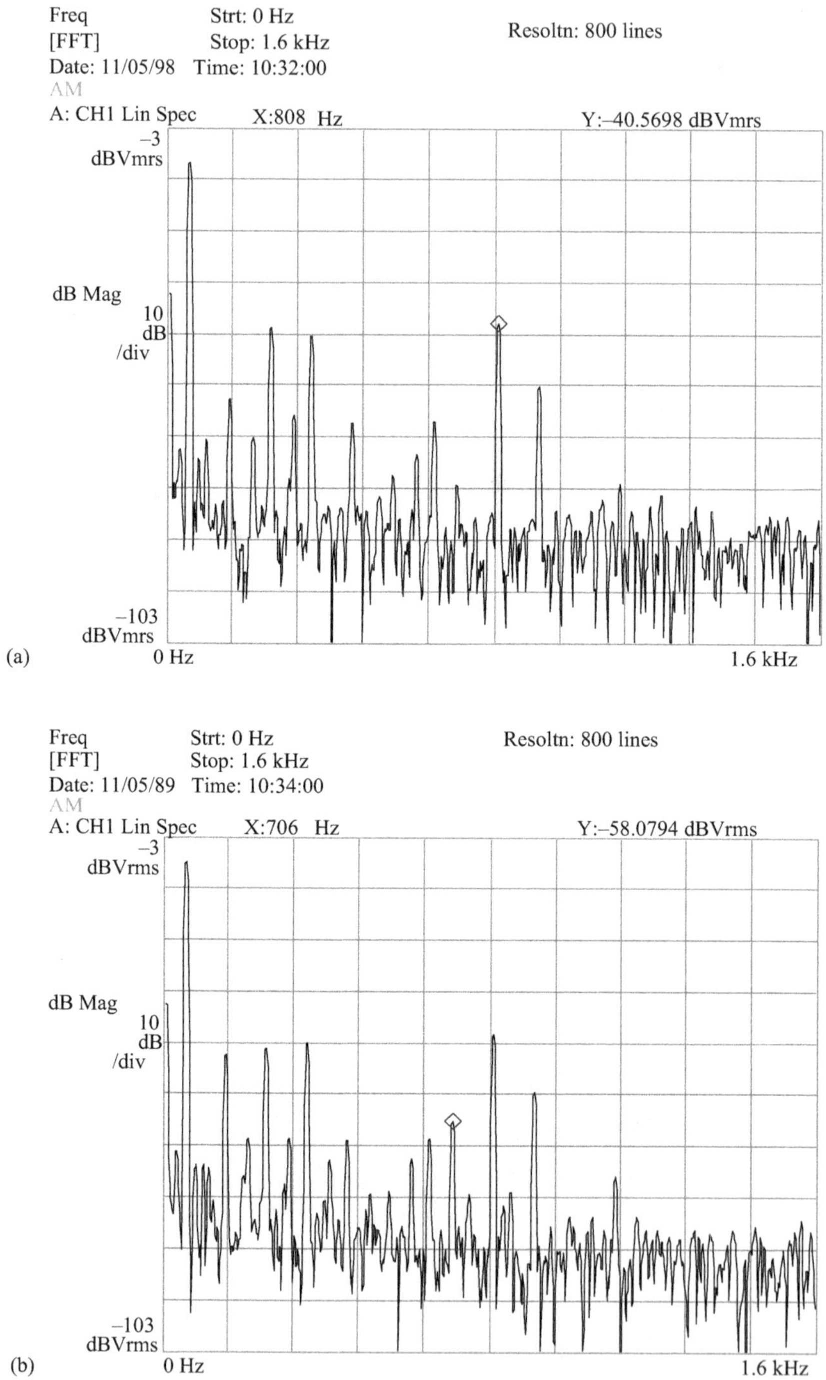


*Figure 10.19 Stator-line current spectrum (C phase) for loaded 50-Hz cage rotor induction motor with $S = 36$ slots, $R = 32$ bars, $p = 2$ pole pairs @ $s = 5.5\%$: (a) C phase, healthy motor and (b) C phase, interturn short circuit*

component does not appear, itself represents a significant limitation on detection of this fault condition, using MCSA.

- Second, the RSHs appear at frequencies that are the function of slip (speed, load) which is also undesirable for the purposes of detection and diagnosis.
- Third, the above components in the spectrum of stator currents are usually very small in value, and it is very difficult to measure them. For instance, it is necessary to exert greater number of consecutive measurements and looks for the mean value as a representative size.
- Fourth, in addition to these components in the stator current spectrum, as a result of interturn short circuit, a rise may occur in current component at 150 Hz, but unfortunately, this is still a component of stator current spectrum, as mentioned before, and is primarily a function of magnetic circuit saturation. Therefore, there cannot be a characteristic sign of the presence of interturn short-circuit in the stator windings.
- Fifth, the question of differentiation between this kind of failure and some other sorts of failures that also has an impact on these frequencies could be set forth. The static eccentricity condition is a good example of this fault, and only the changes in RSHs could be expected.

From all the above-mentioned points, it can be concluded that detection of interturn short-circuits in the stator windings of the induction motor with cage rotor through the stator current spectrum analysis does not give satisfactory results.

## 10.4 A review of interturn short-circuit fault detection techniques for induction motors

As one of the earliest attempts that provided a mathematical model of a cage rotor induction motor with stator winding faults are presented in [15]. The presented mathematical model was based on generalized version of harmonic analysis. The phase-to-neutral fault that is analyzed in this paper was fault in which one stator coil was short-circuited. In case of phase-to-phase fault, the short-circuited winding consisted of 18 coils. The main observation from this paper is that the negative sequence current in case of phase-to-neutral fault, as well as phase-to-phase fault as well as in case of phase-to-phase fault is almost constant valued: its value does not depend on the motor loading.

The effective negative sequence impedance detector as an online method to detect incipient failure of turn insulation in random wound motors is another diagnosis technique [16]. This index is computed from the voltage and current phasors at the motor terminals and does not interfere with the normal motor operation. Authors underline the fact that neither motor modeling nor knowledge of motor parameters is needed for implementation of the proposed detector. It is worth noting that both approaches, monitoring of negative sequence current or monitoring of negative sequence impedance, should be carefully implemented as unbalance in supply voltages, inherent motor windings asymmetry etc., also produce negative sequent voltages.

By inverting the well-known fact that the stator winding itself could serve as a search coil for problems associated with rotor, we can provide the concept of axial flux leakage monitoring—an idea that rotor itself could serve as search coil for problems from the stator side [17]. More precisely, search coil is placed concentric with the drive shaft of the motor or a series of coils axisymmetric to the shaft. Authors also proposed the method to locate the position of the shorted turn: four search coils should be located asymmetrically to the drive shaft in the end plane of the machine. Neither the detection nor the location techniques require the motor to be loaded.

Park's vector locus retrace is often used in order to conclude about the occurrence of interturn fault in stator winding [18]. This is a noninvasive approach based on computer-aided monitoring of the stator current Park's vector. The mathematical model presented in the paper is very simple as the proposed method needs only $d$ and $q$ component of Park's vector obtained from stator phase currents. The corresponding representation of Park's vector in case of a healthy motor is circular locus centered at the origin of the coordinates. In case of interturn short-circuits, this locus became more or less elliptical. However, the locus will be elliptical in any other case of asymmetry, i.e., in case of asymmetrical supply voltages. Moreover, it is difficult to establish some quantitative measure to correspond with fault level.

A novel frequency-domain-based technique for detection of incipient stator interturn faults in induction machines has been proposed in [19]. The presence of certain rotor-slot-related harmonics at the terminal voltages of an induction machine immediately after switch-off is the basis for diagnosis procedure. In the absence of supply voltage, issues such as voltage unbalance and time harmonics do not affect the measurements except at initial conditions. This is a very desirable feature when the machine is fed from an adjustable speed drive.

Using an external stray flux sensor for detection of stator winding faults can be a good alternative, [20]. The stray flux sensor is built around an air coil of several hundred turns which is placed near the machine body. The diameter of the air coil depends on the size of the machine and it should be much less than the total height of the machine body in order to cover the measurement of flux in an area covering the width of several stator slots. Then, the dimension of this coil is somewhat related to the induction machine size. It is mandatory to associate dedicated analog signal processing with this simple sensor. The method has been used at a standstill and when the machine is rotating without load torque and at the rated torque. The proposed method has proven to be efficient for both sinusoidal power supply and voltage source inverter (VSI).

As an important factor, drives have a great effect on the diagnosis process. For example, direct torque control (DTC) drives can directly affect the short-circuit fault-related components in the motor quantities [21]. Through a qualitative analysis of the phenomena involving the behavior of this type of drive after the occurrence of a stator fault in the motor, it was demonstrated that the flux and torque hysteresis controllers tend to introduce a significant third harmonic component in the motor supply currents. The presence of a strong third harmonic component in the motor supply currents is thus an indicator of the presence of a stator fault. It is not recommended to use this technique in case of high degree of intrinsic asymmetry of the motor such as the case of high-power motors because in that case, the third harmonic component will be present in the spectrum of the motor supply currents even in the absence of any stator fault.

Another technique has been introduced based on the concept of magnetic field pendulous oscillation [22]. The pendulous oscillation in a polar coordinate plot of $r$ and $\delta$ is formulated as follows:

$$r(t) = abs\left(\vec{i}_s(t)\right) = abs\left(I_{Lm}e^{j(\omega t-\psi)}\right) \tag{10.44}$$

$$\delta(t) = -\psi \tag{10.45}$$

and where $\vec{i}_s(t)$ is the space vector of motor terminal currents. Space vector of motor terminal voltages is

$$\vec{u}_s(t) = U_{Lm}e^{j\omega t} \tag{10.46}$$

Swing angle index as the interturn fault signature is introduced as the amplitude of the maximum thickness of the petal shape caused by the swing angle,

$$\Delta\delta = \max\left(\Delta\delta(r) = \delta^{\max} - \delta^{\min}\right) \tag{10.47}$$

This approach enables one to distinguish and classify an unbalanced voltage power supply and machine structure imperfections from an interturn fault. The authors concluded from experimental results that if the circulating current level in the shorted loop increases beyond the phase current level, an interturn fault can be easily detected using the proposed approach even in the presence of the motor manufacturing imperfection effects.

Analysis of three-phase stator current envelopes of induction motor, using reconstructed phase space transforms is a rather new method for the fault diagnosis of broken rotor bar and interturn short circuits. The fault classification method is based on machine learning techniques. The envelope signal is extracted from the experimentally acquired stator current signals and is used in conjunction with AI techniques based on Gaussian mixture models and reconstructed phase space to identify motor faults. This method creates signatures for each type of fault based on the three-phase stator current envelope. A signature for each newly acquired input set of three-phase stator currents, which are called "unseen signals," must be generated and compared with all the signatures that represent each type of fault learned from the previously acquired database [23].

Another strategy for stator winding fault detection in variable-speed induction motors with star connection using signal injection technique was proposed in [24]. This strategy is based on the effects of such faults when the motor is fed by an additional excitation to the fundamental one. The additional excitation consists of a predefined sequence of pulses applied by the inverter, and it was previously used for position estimation and for fault diagnosis such as broken bars and rings and air gap eccentricity. The excitation signal injection allows decoupling the diagnosis signal from the fundamental excitation aimed to generate the motor torque. Variations on the induction motor inductance values, under fault conditions, produce significant changes in the signals obtained from zero-sequence voltage, allowing the incipient fault diagnosis.

Instantaneous active $p(t)$ and reactive $q(t)$, power signature analysis is sometimes used for stator fault diagnosis in operating squirrel-cage three-phase induction motors either directly connected to the mains or fed from inverters is proposed in [25]. As the presence of the stator faults affects the electric, magnetic and mechanical quantities such as the motor torque (where it gives rise to a ripple at double supply frequency), the motor instantaneous active and reactive powers, where the latter quantities become oscillating at the same frequency. Based on this, the instantaneous active power signature analysis (IAPSA) and the instantaneous reactive power signature analysis (IRPSA) can be used for the diagnosis of such fault in induction motors.

The results show that stator fault conditions (both short- and open-circuit turns) can be effectively diagnosed by this new approach, whose operating philosophy relies on the analysis of the fault characteristic component at the disturbance frequency $2f$ in both spectra of the two considered quantities. Even for the case of a reduced number of stator shorted turns at no-load condition, this characteristic component is still visible, which reflects the capability and the sensitivity of the proposed detection technique. A normalized fault-severity factor defined as the ratio of the amplitude of the characteristic component $2f$ and the dc one, corresponding to the instantaneous reactive power, proved to be a good indicator of the condition of the machine independently of the load level when supplied by a sinusoidal voltage system or fed from inverters.

The stator fault conditions can be easily separated from other abnormalities by the combination of the active and reactive power media and the stator voltage modulus, which illustrates the effectiveness of the proposed approach, [25].

A new fault indicator, based on tracking of PSH under varying speed operation was proposed in [26]. A detailed study of the energy content of a PSH in an induction motor operating at variable slip is carried out. The machine is tested under different faulty conditions, namely broken rotor bar, mixed eccentricity and interturn short circuit. The use of a combined time–frequency analysis and particle filtering feature extraction procedure allows tracking the evolution of a PSH under different load profiles and integrating its energy. The proposed fault indicator provides values that are equivalent to the traditional one for stationary operation even in the case of pulsating load. The results show that the energy in the PSH reflects the overall state of the machine under these conditions. Furthermore, procedures to discern the different faults in real applications are proposed.

An inverse approach for interturn fault detection in asynchronous machines using magnetic pendulous oscillation technique has been proposed in [27]. Dynamic forward model is presented to analyze the performance of induction machines in healthy and interturn faulty conditions. The presented model calculates the speed, torque and stator currents in both transient and steady-state regimes.

As an interturn fault signature, the magnetic pendulous oscillation technique is implemented that gives specific petals shape to the relationship between the current and voltage space vectors. The area of that petal shape was used as the fault criterion that recovered by means of a coupled experimental–mathematical electromagnetic inverse problem. Numerical results were first obtained where the effect of noise was studied. Moreover, an experimental setup was built to validate the inverse approach, where acceptable fault detection results were successfully obtained. As a new fault criterion, which uniquely depends on the fault, the area of the petal shape is proposed:

$$\text{Area} = \frac{1}{2}\int_{\delta_{\min}}^{\delta_{\max}} (r(\delta))^2 d\delta \tag{10.48}$$

An improved complex-vector dynamic model for a three-phase induction machine with stator interturn short-circuit fault is presented recently in [28]. The inter-turn short-circuit in the stator winding of a three-phase induction machine was fully and theoretically analyzed, studied, modeled, simulated, experimentally tested and compared. The authors propose an alternative faulty leakage inductance approach with additional physical meaning compared with the ones presented in the literature. This leakage inductance model has more accurate behavior compared to the linear one, keeping simplicity of the modeling. The faulty model is completely deduced and verified by comparing simulation with experimental tests under different load conditions, regimes and variations of the fault severity factor and faulty resistance. These two variables define the magnitude of the faulty leakage current and thus damaging the induction machine. From the results, it can be concluded that the faulty model is accurate and behaves similar to the real asymmetric induction machine. Also, the article collects a series of equivalent circuits that is useful for future fault detection and identification techniques; however, they are more utilizable for improving two model-referenced approach techniques, such as off-line impedance signature and parameter estimation.

The sideband response of inverter switching harmonics in common-mode voltage (CMV) is recently employed to detect the stator interturn fault in inverter-fed induction machine in closed-loop control [29]. Thus, no additional injected signal is necessary. However, the pulse width modulation (PWM) nature harmonics introduce undesired CMV signals, depending on the inherent imbalance or resistance asymmetry. For this reason, the CMV amplitude is not suitable to indicate the fault directly. In this study, the behavior of switching harmonic is analyzed first; then a simple but efficient stator fault model is demonstrated. To suppress the adverse interference from inverter, an inter-turn fault indicator is proposed by utilizing both of the switching sidebands. This method can obtain fault location and severity with low sensitivity to steady operating condition and stator imbalance. The experimental results presented in paper validate the effectiveness of the proposed method.

Within the framework of the diagnosis by parameters estimation of the stator windings faults, in recently published paper [30], two approaches are proposed for off-line and on-line diagnosis of stator winding inter-turn short-circuit faults for squirrel-cage IM. Both on-line and off-line parameters estimation are used to perform faults detection and localization.

The first approach combines trust-region and Broyden–Fletcher–Goldfard–Shano methods to exploit both of their advantages. This algorithm is used for off-line minimization of the objective function represented by a quadratic criterion.

The second algorithm based on the moving horizon estimation, combines off-line measurement and on-line parameter estimation with high sampling time to monitor in real time the stator inter-turn faults in IMs. Since it requires a mathematical model suited for fault modeling, a faulty IM model is presented. Due to the parameters values of this model, the normalization of these parameters is proposed to obtain the sensitivity functions with the same magnitude order. The estimation results, which used simulated as well as experimental data, are presented to show the effectiveness and the advantage of the proposed algorithms for stator inter-turn short-circuit fault detection in IMs.

## 10.5 MCSA of wound-rotor induction motor

Wound-rotor induction machines are not as common as cage rotor machines in everyday use. However, they are irreplaceable as motors in some high-power applications. Furthermore, they found prominent use in generator mode especially as double-fed induction generators in high-power applications. Here its use in wind turbines is in the first place. For the sake of completeness of this chapter, the stator current spectrum content for wound rotor motor is also analyzed and presented here.

As mentioned before, the three-phase rotor winding in wound rotor machines is of the same nature as the winding from stator side. Along with the fundamental, higher-order MMF space harmonics of order $\nu = 6g + 1$, $g = 0, \pm 1, \pm 2, \ldots$ are present. In addition, rotor winding is always wounded for the same number of pole pairs as stator winding. Taking only the fundamental current in rotor windings, i.e., slip frequency current into account, the rotor MMF waves are

$$F_{r\nu}(t, \theta_r) = F_{r\nu\max} \cos(s\omega_1 t - \nu p\theta_r) \tag{10.49}$$

Assuming uniform air gap, the flux density waves that are the result of these MMF waves, referring to the stator side are

$$B_{r\nu}(t, \theta_s) = B_{r\nu\max} \cos((s + \nu(1 - s))\omega_1 t - \nu p\theta_s) \tag{10.50}$$

As $\nu$ is of the same order as $\nu$ that produces stator winding, it means that in stator current spectrum, all harmonics will exist at frequencies given by:

$$f_\nu = (s + \nu(1 - s))f_1 = \begin{cases} f_1 & \text{for } \nu = 1 \\ (5 - 6s)f_1 & \text{for } \nu = -5 \\ (7 - 6s)f_1 & \text{for } \nu = 7 \\ \vdots \end{cases} \tag{10.51}$$

These findings are illustrated by Figure 10.20 where stator current spectrum of healthy three-phase wound-rotor induction motor is provided. If any asymmetry in

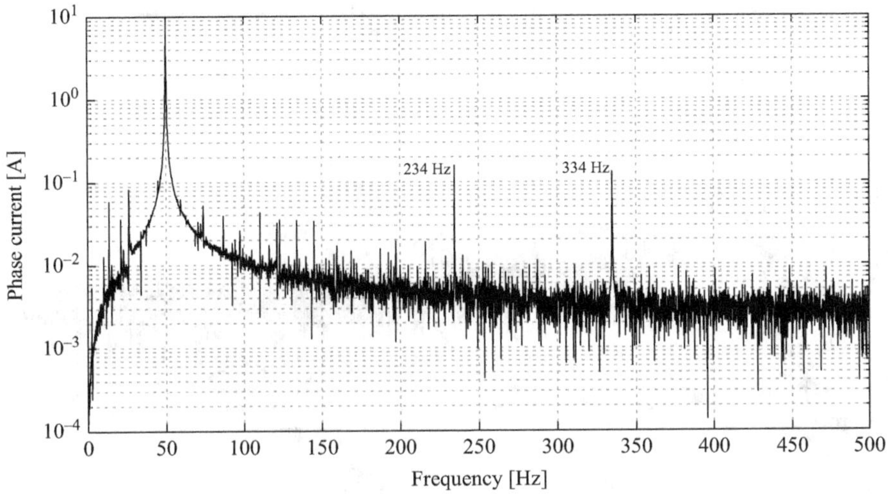


*Figure 10.20 Stator-line current spectrum for loaded 50 Hz wound-rotor induction motor with $S = 48$ slots, $R = 36$ slot and $p = 2$ pole pairs @ $s = 5.22\%$*

rotor winding exists, inverse-rotating MMF waves result in additional components in stator current spectrum at the following frequencies:

$$f_\nu = (s + \nu(1 - s))f_1 = \begin{cases} (1 - 2s)f_1 & \text{for } \nu = -1 \\ (5 - 4s)f_1 & \text{for } \nu = 5 \\ (7 - 8s)f_1 & \text{for } \nu = -7 \\ \vdots \end{cases} \tag{10.52}$$

One could find all these frequencies in stator current spectrum given in Figure 10.21 for the same wound-rotor induction motor. The asymmetry in the rotor is introduced by modeling the resistance of one of the phase windings which is 50% greater than that of the other two phases. One additional component that is not predicted by (10.52) is the current component at 56 Hz. As it is well known from the literature, this component is due to the pulsating electromagnetic torque that is, from the other side, the result of the stator phase current at frequency $(1-2s)f_1$.

In case of asymmetrical line-to-line voltages, stator current spectrum becomes richer for two more components, Figure 10.22. The asymmetry is modeled in such a way that one line-to-line voltage is 10% smaller than other two. One new component is the current component at 150 Hz, and the other one is somewhat lower, at 144 Hz. It can be shown that for different load levels, the current component at 150 Hz does not change its position while it is not the case with the other one. For example, for slip of $s = 2.53\%$, this slip-dependent component is at 147.7 Hz, i.e., this component approaches 150 Hz with lighter load level.

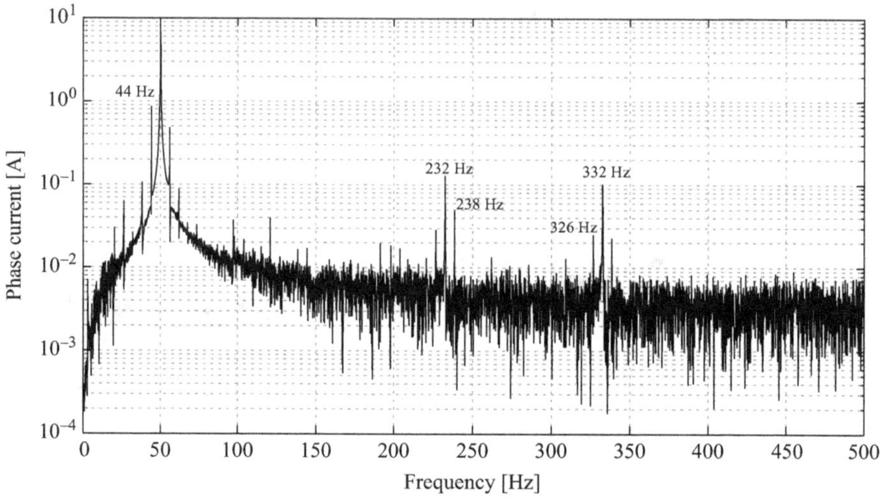


*Figure 10.21 Stator-line current spectrum for loaded 50 Hz wound-rotor induction motor with $S = 48$ slots, $R = 36$ slot and $p = 2$ pole pairs @ $s = 5.96\%$ (resistance of one rotor phase winding is 50% higher than that of the other two phases)*

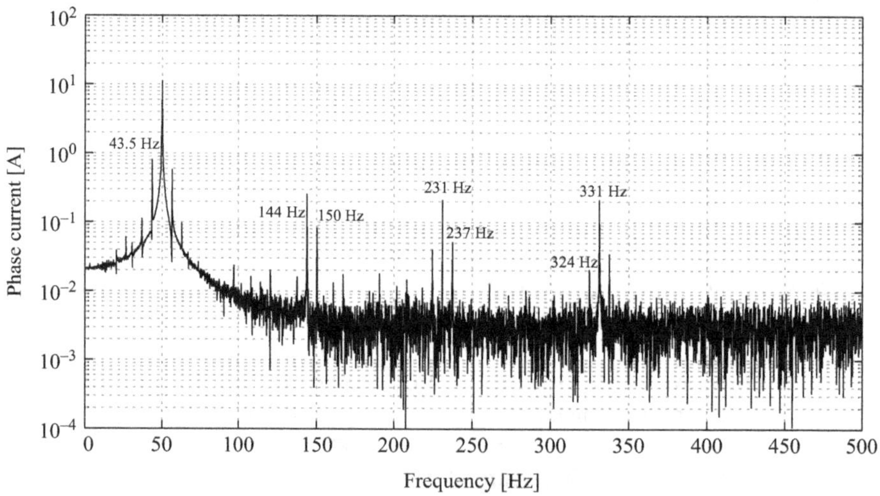


*Figure 10.22 Stator-line current spectrum for loaded 50 Hz wound-rotor induction motor with $S = 48$ slots, $R = 36$ slot and $p = 2$ pole pairs @ $s = 6.43\%$ (one line-to-line voltage is 10% lower than the other two voltages)*

## References


[1] Joksimović G.M., Riger J., Wolbank T.M., Perić N. and Vašak M. Stator-current spectrum signature of healthy cage rotor induction machines, *IEEE Transactions on Industrial Electronics*, 2013; **IE-60**: 4025–4033.

[2] Jung J.H. , Lee J.J., and Kwon B.H. Online diagnosis of induction motors using MCSA, *IEEE Transactions on Industrial Electronics*, 2006; **IE-53**: 1842–1852.

[3] Luo X., Liao Y., Toliyat H.A., El-Antalby A., Lipo T.A. Multiple coupled circuit modeling of induction machines, *IEEE Transactions on Industry Applications*, 1995; **IA-31**: 311–317.

[4] Blodt M., Bonacci D., Regnier J., Chabert M. and Faucher J. On-line monitoring of mechanical faults in variable-speed induction motor drives using the Wigner distribution, *IEEE Transactions on Industrial Electronics*, 2008; **IE-55**: 522–533.

[5] Keysan O. and Ertan H.B. Real-time speed and position estimation using rotor slot harmonics, *IEEE Transactions on Industrial Informatics*, 2013; **II-9**: 899–908.

[6] Alger P.L. Induction machines, New York, NY; Gordon and Breach Science Publishers, 1970.

[7] Boldea I. and Nasar S.A. The induction machine handbook, New York, NY; CRC Press, 2002.

[8] Nandi S. Modeling of induction machines including stator and rotor slot effects, *IEEE Transactions on Industry Applications*, 2004; **IA-40**: 1058–1065.

[9] Moreira J.C. and Lipo T.A. Modeling of saturated ac machines including air gap flux harmonic components, *IEEE Transactions on Industry Applications*, 1992; **IA-28**: 343–349.

[10] Joksimović G. Line current spectrum analysis in saturated three-phase cage induction machine, *Electrical Engineering (Archiv für Elektrotechnik)*, Springer, April 2010; **vol. 91**, no. 8: 425–437.

[11] Kim K., Parlos A. and Bharadwaj R. Sensorless fault diagnosis of induction motors, *IEEE Transactions on Industrial Electronics*, 2003; **IE-50**: 1038–1051.

[12] Bonnet A.H. Analysis of winding failures in three-phase squirrel cage induction motors, *IEEE Transactions on Industry Applications*, 1978; **IA-14**: 223–226.

[13] Joksimović G.M. and Penman J. The detection of inter-turn short circuits in the stator windings of operating motors, *IEEE Transactions on Industrial Electronics*, October 2000; **vol. 47**, no. 5: 1078–1084.

[14] Filippetti F., Franceschini Tassoni G.C. and Vas P. AI techniques in induction machines diagnosis including the speed ripple effect, *IEEE Transactions on Industry Applications*, 1998; **IA-34**: pp. 98–108.

[15] Williamson S. and Mirzoian K. Analysis of cage induction motors with stator winding faults, *IEEE Transactions on Power Apparatus and Systems*, 1985; **PAS-104**: 1838–1842.

[16] Sottile J. and Kohler J.L. An on-line method to detect incipient failure of turn insulation in random-wound motors, *IEEE Transactions Energy Conversion*, 1993; **EC-8**: 762–768.

[17] Penman J., Sedding H.G., Lloyd B.A. and Fink W.T. Detection and location of interturn short circuits in the stator winding of operating motors, *IEEE Transactions Energy Conversion*, 1994; **EC-9**: 652–658.

[18] Marques Cardoso A.J., Cruz S.M.A. and D.S.B. Fonseca, Inter-turn stator winding fault diagnosis in three-phase induction motors, by Park's vector approach, *IEEE Transactions Energy Conversion*, 1999; **EC-14**: 595–598.

[19] Nandi S. and Toliyat H.A. Novel frequency domain based technique to detect incipient stator inter-turn faults in induction machines, *Conf. Rec. IEEE-IAS Annu. Meeting*, 2000; Rome, Italy, 367–374.

[20] Henao H., Demian C. and Capolino, G.A. A frequency-domain detection of stator winding faults in induction machines using an external flux sensor, *IEEE Transactions on Industry Applications*, 2003; **IA-39**: 1272–1279.

[21] Cruz S.M.A., Marques Cardoso A.J. Diagnosis of stator inter-turn short circuits in DTC induction motor drives, *IEEE Transactions on Industry Applications*, 2004: **IA-40**: 1349–1360.

[22] Mirafzal B., Povinelli R.J. and Demerdash N.A.O. Interturn fault diagnosis in induction motors using the pendulous oscillation phenomenon, *IEEE Transactions on Energy Conversion*, 2006; **EC-21**: 871–882.

[23] da Silva A.M., Povinelli R.J. and Demerdash N.A.O. Induction machine broken bar and stator short-circuit fault diagnostics based on three-phase stator current envelopes, *IEEE Transactions on Industrial Electronics*, 2008; **IE-55**: 1310–1318.

[24] Bossio G.R., De Angelo C.H., de la Barrera P.M., Solsona J.A., Garcia G.O. and Valla M.I. Stator winding fault detection in induction motor drives using signal injection, *Conf. Rec.* SDEMPED 2011, pp. 92–97, September 2011.

[25] Drif M. and Marques Cardoso A.J. Stator fault diagnostics in squirrel cage three-phase induction motor drives using the instantaneous active and reactive power signature analysis, *IEEE Transactions on Industrial Informatics*, 2014; **II-10**: 1348–1360.

[26] Climente-Alarcon V., Antonino-Daviu J.A., Haavisto A. and Arkkio A. Diagnosis of induction motors under varying speed operation by principal slot harmonic tracking, *IEEE Transactions on Industry Applications*, 2015; **IA-51**: 3591–3599.

[27] Mahmoud H., A-E Abdallh A., Bianchi N., El-Hakim S.M., Shaltout A. and Dupre L. An inverse approach for inter-turn fault detection in asynchronous machines using magnetic pendulous oscillation technique, *IEEE Transactions on Industry Applications*, 2016; **IA-52**: 226–233.

[28] Berzoy A., Mohamed A.A.S. and Mohammed O. Complex-vector model of inter-turn failure in induction machines for fault detection and identification, *IEEE Transactions on Industry Applications*, 2017; **IA-52**: 226–233.

[29] Liu H., Huang J., Hou Z., Yang J. and Ye M. Stator inter-turn fault detection in closed-loop controlled drive based on switching sideband harmonics in CMV, *IET Electric Power Applications*, 2016; **EPA-11**: 178–186.

[30] Abdallah H. and Benatman K. Stator winding inter-turn short-circuit detection in induction motors by parameter identification, *IET Electric Power Applications*, 2016; **EPA-11**: 272–288.

# Index

www.ingramcontent.com/pod-product-compliance
Lightning Source LLC
LaVergne TN
LVHW011857200926
848484LV00002B/472